PLANT PATHOLOGY

Concepts and Laboratory Exercises

PLANT PATHOLOGY
Concepts and Laboratory Exercises

Edited by

Robert N. Trigiano
Mark T. Windham
Alan S. Windham

CRC PRESS

Boca Raton   London   New York   Washington, D.C.

Library of Congress Cataloging-in-Publication Data

Plant pathology : concepts and laboratory exercises / edited by Robert N. Trigiano, Mark
   T. Windham, Alan S. Windham.
       p. cm.
     Includes bibliographical references and index.
     ISBN 0-8493-1037-7 (alk. paper)
      1. Plant diseases—Laboratory manuals. 2. Trigiano, R. N. (Robert Nicholas), 1953–
   II. Windham, Mark Townsend, 1955– III. Windham, Alan S.
SB732.56.P63 2003
632'.3'078—dc21
                                        2003046138
                                              CIP

Visit the CRC Press Web site at www.crcpress.com

© 2004 by CRC Press LLC

No claim to original U.S. Government works
International Standard Book Number 0-8493-1037-7
Library of Congress Card Number 2003046138
Printed in the United States of America 1 2 3 4 5 6 7 8 9 0
Printed on acid-free paper

Preface

Plant Pathology: Concepts and Laboratory Exercises is intended to serve as a primary text for introductory courses and furnishes instructors and students alike with a broad consideration of this important and growing field. It also presents many useful protocols and procedures and should thus serve as a valuable reference to researchers and students in plant pathology and allied biological sciences. The book is intentionally written to be rather informal — it provides the reader with a minimum number of references, but does not sacrifice essential information or accuracy. Broad topic chapters are authored by specialists with considerable experience in the field and are supported by one or more laboratory exercises illustrating central concepts of the topic. Each topic begins with chapter concepts, highlighting some of the more important ideas contained within the chapter and signals students to read carefully for these primary topics. There is an extensive glossary for the bolded words and terms found in each of the concept chapters and some of the laboratory chapters. Collectively, the laboratory exercises are exceptionally diverse in nature, providing something for everyone — from beginning to advanced students. Importantly, the authors have successfully completed the exercises many times, often with either plant pathology or biology classes or in their own research laboratories. All the laboratory protocols are written as side-aside procedures that provide step-by-step, easy-to-follow instructions. A unique feature of this text is that the authors have provided in general terms what results should be expected from each of the experiments. At the end of each exercise, there is a series of questions designed to provoke individual thought and critical examination of the experiment and results. Our intentions are that instructors will not attempt to do all the experiments, but rather select one or two for each concept that serve the needs and interests of their particular class. For an advanced class, different experiments may be assigned among resourceful students. More advanced experiments following the general or beginning class exercises are embedded within some of the laboratory chapters. We caution instructors and students to obtain the proper documents for transport and use of plant pathogenic organisms and to properly dispose of cultures and plant materials at the conclusion of the laboratory exercises.

The textbook is divided into the following six primary parts: Introductory Concepts, Groups of Plant Pathogens, Molecular Tools for Studying Plant Pathogens, Plant–Pathogen Interactions, Epidemiology and Disease Control, and Special Topic. Each section combines related facets of plant pathology and includes one to several concept chapters, usually with accompanying laboratory exercises.

Part I introduces students to the basic concepts of plant pathology, including historical perspectives; fundamental ideas of what is disease; how disease relates to environment, the host, and time; and provides a very broad overview of organisms that cause disease. Part II includes chapters that detail the various disease-causing organisms. This section begins with a consideration of viruses, prokaryotic organisms, and nematodes. The next 10 chapters are devoted to the various phyla of fungi (classification primarily follows Alexopoulus, Mims, and Blackwell, *Introductory Mycology*, 4th ed.), followed by chapters that focus on the fungi-like Oomycota, parasitic seed plants and other biotic agents, and abiotic diseases. Part III introduces students to the basics of molecular tools and is illustrated with laboratory exercises that are adaptable for beginning as well as advanced students. Part IV explores plant–pathogen interactions, including treatments of attack strategies, extracellular enzymes, host defenses, and disruption of plant function. Part V is anchored with an extensive chapter outlining the basic ideas of epidemiology, which is followed in turn by several chapters detailing various strategies for disease control. This section also includes chapters on plant disease diagnosis. Part VI is devoted to an often-neglected topic in plant pathology texts — in vitro pathology. We have found that students are often fascinated with this topic as it combines several facets of investigation from biological and chemical disciplines.

The appendices illustrate career opportunities in plant pathology and closely related fields and a list of websites containing a wealth of supplementary information for the various topics covered in the book. The website URLs were checked and were accurate and functional as of February 1, 2003. Websites in Appendix II will be routinely updated and may be accessed at http://dogwood.ag.utk.edu/ and looking under "Fantastic Plant Pathology Websites."

It is our hope that students and instructors find the format of the book and level and amount of information contained in the book to be appropriate for an introductory course. The presentation style has been used very successfully in *Plant Tissue Culture and Laboratory Exercises* and with the addition of the glossary and concept statements, students should find the format stimulating and conducive to learning. We invite and welcome your comments and suggestions for improvements.

R.N. Trigiano, M.T. Windham, A.S. Windham
The University of Tennessee

Acknowledgments

We wish to acknowledge the efforts of all the contributing authors — their creativity, support, and patience throughout the conception and development of this project was nothing less than phenomenal; the Institute of Agriculture at the University of Tennessee and especially Dr. Carl Jones, head of Entomology and Plant Pathology, for providing the time and financial support necessary to complete the book; Dr. Gary Windham for his work behind the scenes and photographs; and Malissa H. Ament, whose computer skills made many of the illustrations in the book possible. We also thank our families for their patience and understanding throughout the project; a very special thanks to Kay Trigiano for her excellent technical editorial skills and applying them to each and every chapter; and thanks to John Sulzycki, Pat Roberson, and Joette Lynch at CRC Press, whose constant encouragement was essential for the completion of this text.

Editors

Robert N. Trigiano, Ph.D., is professor of ornamental plant biotechnology and plant pathology in the Department of Entomology and Plant Pathology at the University of Tennessee Agricultural Experiment Station at Knoxville.

Dr. Trigiano received his B.S. degree with an emphasis in biology and chemistry from Juniata College, Huntingdon, Pennsylvania, in 1975 and an M.S. in biology (mycology) from The Pennsylvania State University in 1977. He was an associate research agronomist in mushroom culture and plant pathology for Green Giant Co., Le Sueur, Minnesota, until 1979 and then a mushroom grower for Rol-Land Farms, Ltd., Blenheim, Ontario, Canada, in 1979 and 1980. He completed a Ph.D. in botany and plant pathology (comajors) at North Carolina State University at Raleigh in 1983. After concluding postdoctoral work in the Plant and Soil Science Department at the University of Tennessee, he was an assistant professor in the Department of Ornamental Horticulture and Landscape Design at the same university in 1987, promoted to associate professor in 1991, and to professor in 1997. He joined the Department of Entomology and Plant Pathology in 2002.

Dr. Trigiano is a member of the American Society for Horticultural Science (ASHS) and the Mycological Society of America, and the honorary societies of Gamma Sigma Delta, Sigma Xi, and Phi Kappa Phi. He has been an associate editor for the journals of the American Society of Horticultural Science and *Plant Cell, Tissue and Organ Culture* and editor for *Plant Cell Reports.* Dr. Trigiano is coeditor of *Critical Reviews in Plant Science* and the popular textbook *Plant Tissue Culture and Laboratory Exercises*, 2nd ed., both published by CRC Press. He has received the T.J. Whatley Distinguished Young Scientist Award (The University of Tennessee, Institute of Agriculture, 1991) and the Gamma Sigma Delta Research Award of Merit (The University of Tennessee, 1991). In 1998, he received the ASHS publication award for the most outstanding educational paper and the Southern region ASHS L. M. Ware distinguished research award.

Dr. Trigiano has been the recipient of several research grants from the United States Department of Agriculture (USDA), Horticultural Research Institute, and from private industries and foundations. He has published more than 140 research papers, book chapters, and popular press articles. He teaches undergraduate/graduate courses in plant tissue culture, plant disease fungi, DNA analysis, protein gel electrophoresis, and plant microtechnique. His current research interests include somatic embryogenesis and micropropagation of ornamental species, fungal physiology, population analysis, DNA profiling of fungi and plants, and gene discovery.

Mark T. Windham, Ph.D., is a professor of plant pathology and holds the distinguished chair in Ornamental Plant Diseases at the Institute of Agriculture, Department of Entomology and Plant Pathology with the University of Tennessee, Knoxville. He received his B.S. degree and M.S. degree in plant pathology and weed science from Mississippi State University. In 1983, Dr. Windham completed his Ph.D. in plant pathology with a minor in plant breeding from North Carolina State University. After graduation, he accepted a position as a visiting assistant professor at Colorado State University. In 1985, Dr. Windham accepted a position as an assistant professor at the University of Tennessee, Knoxville, and was promoted to professor in 1999.

Dr. Windham has taught introductory plant pathology since 1995. He also team-teaches two other courses: Diseases and Insects of Ornamental Plants and Plant Disease Fungi. Dr. Windham's research interests include diseases of ornamental plants, especially flowering dogwood. Dr. Windham has teamed with other scientists to release the first flowering dogwood cultivar resistant to dogwood anthracnose and to patent and release the first white blooming flowering dogwoods resistant to powdery mildew. Dr. Windham has published more than 100 research papers, book chapters, and popular press articles. He has also served as editor of the plant pathology section of the Southern Nursery Association Research Conference.

Dr. Windham's research has led to him to receiving the Porter Henegar Memorial Award from the Southern Nursery Association and the Research and Team Awards of Merit from Gamma Sigma Delta. He has coauthored *Dogwoods for American Gardens*, which was awarded the American Society for Horticultural Science Extension Publication Award.

Alan S. Windham, Ph.D., is professor of plant pathology in the Institute of Agriculture, Department of Entomology and Plant Pathology, with the University of Tennessee, Knoxville. Dr. Windham is stationed at the Plant and Pest Diagnostic Center at the Ellington Agricultural Center in Nashville, Tennessee. Dr. Windham received his B.S. degree with an emphasis in plant pathology in 1979 and an M.S. in plant pathology with a minor in botany in 1981 from Mississippi State University, Starkville. He completed his Ph.D. in plant pathology with a minor in soil science at North Carolina State University at Raleigh in

1985. After completing his graduate work, he accepted the position of assistant professor with the University of Tennessee in 1985, and was promoted to associate professor in 1989 and professor in 1995. Dr. Windham is a member of the American Phytopathological Society and was inducted into the honorary societies Gamma Sigma Delta and Sigma Xi. In 2000, he received the Gamma Sigma Delta Award of Merit for his work with the Dogwood Working Group (University of Tennessee). In 2002, he was awarded the American Society for Horticultural Science Extension Publication Award for *Dogwoods for American Gardens*. He has also served as editor for the plant pathology section of the Southern Nursery Association Research Conference. Dr. Windham has conducted educational programs and consulted with nursery, greenhouse, and turf industries nationally and internationally. He has published over 75 research papers, book chapters, and popular press articles. He has lectured in many undergraduate/graduate courses in plant pathology, mycology, floriculture, nursery, and turfgrass management. His current research interests include etiology and management of emerging plant diseases of ornamental plants and the development of disease-resistant ornamental plants.

Contributors

Emnet Abesha
Department of Biology
University of Oslo
Oslo, Norway

Malissa H. Ament
Department of Entomology and Plant Pathology
The University of Tennessee
Knoxville, Tennessee

Richard E. Baird
Department of Entomology and Plant Pathology
Mississippi State University
Mississippi State, Mississippi

Kira L. Bowen
Department of Entomology and Plant Pathology
Auburn University
Auburn, Alabama

Gustavo Caetano-Anollés
Department of Crop Sciences
University of Illinois
Urbana, Illinois

Martin L. Carson
USDA/ARS, Cereal Disease Laboratory
University of Minnesota
St. Paul, Minnesota

Marc A. Cubeta
Department of Plant Pathology
North Carolina State University
Raleigh, North Carolina

Kenneth J. Curry
Department of Biological Sciences
University of Southern Mississippi
Hattiesburg, Mississippi

Margery L. Daughtrey
Department of Plant Pathology
Cornell University
Ithaca, New York

James F. Green
Department of Chemistry
The University of Tennessee
Knoxville, Tennessee

Sharon E. Greene
Department of Entomology and Plant Pathology
The University of Tennessee
Knoxville, Tennessee

Ann B. Gould
Department of Plant Pathology
Rutgers University
New Brunswick, New Jersey

Dennis J. Gray
Mid-Florida Research and Education Center
University of Florida
Apopka, Florida

Kimberly D. Gwinn
Department of Entomology and Plant Pathology
The University of Tennessee
Knoxville, Tennessee

Ladare F. Habera
Department of Entomology and Plant Pathology
The University of Tennessee
Knoxville, Tennessee

Kathie T. Hodge
Department of Plant Pathology
Cornell University
Ithaca, New York

Clayton A. Hollier
Louisiana State University Agricultural Center
Louisiana State University
Baton Rouge, Louisiana

Subramanian Jayasankar
Department of Plant Agriculture – Vineland Campus
University of Guelph
Vineland Station, Ontario, Canada

Steven N. Jeffers
Department of Plant Pathology and Physiology
Clemson University
Clemson, South Carolina

George H. Lacy
Department of Plant Pathology, Physiology and Weed
   Science
Virginia Polytechnic Institute and State University
Blacksburg, Virginia

Marie A. C. Langham
Plant Science Department
South Dakota State University
Brookings, South Dakota

Larry J. Littlefield
Department of Entomology and Plant Pathology
Oklahoma State University
Stillwater, Oklahoma

Felix L. Lukezic
Department of Plant Pathology
The Pennsylvania State University
University Park, Pennsylvania

Gary Moorman
Department of Plant Pathology
The Pennsylvania State University
University Park, Pennsylvania

Sharon E. Mozley
Department of Botany
University of Georgia
Athens, Georgia

Jackie M. Mullen
Department of Entomology and Plant Pathology
Auburn University
Auburn, Alabama

James P. Noe
Department of Plant Pathology
University of Georgia
Athens, Georgia

Bonnie H. Ownley
Department of Entomology and Plant Pathology
The University of Tennessee
Knoxville, Tennessee

Jerald K. Pataky
Department of Crop Sciences
University of Illinois
Urbana, Illinois

David Porter
Department of Botany
University of Georgia
Athens, Georgia

Melissa B. Riley
Department of Plant Pathology and Physiology
Clemson University
Clemson, South Carolina

Nina Shishkoff
Department of Plant Pathology
Cornell University
Ithaca, New York

David J. Trently
Department of Entomology and Plant Pathology
The University of Tennessee
Knoxville, Tennessee

Larry E. Trevathan
Department of Entomology and Plant Pathology
Mississippi State University
Mississippi State, Mississippi

Robert N. Trigiano
Department of Entomology and Plant Pathology
The University of Tennessee
Knoxville, Tennessee

Alan S. Windham
Department of Entomology and Plant Pathology
The University of Tennessee
Knoxville, Tennessee

Mark T. Windham
Department of Entomology and Plant Pathology
The University of Tennessee
Knoxville, Tennessee

Table of Contents

Part I
Introductory Concepts

Chapter 1 Plant Pathology and Historical Perspectives.. 3
Mark T. Windham and Alan S. Windham

Chapter 2 What Is a Disease?..7
Mark T. Windham and Alan S. Windham

Chapter 3 Introduction to Plant Pathogens 11
Mark T. Windham

Part II
Groups of Plant Pathogens

Chapter 4 Plant Pathogenic Viruses 19
Marie A. C. Langham

Chapter 5 Mechanical Inoculation of Plant Viruses 33
Marie A. C. Langham

Chapter 6 Pathogenic Prokaryotes 41
George H. Lacy and Felix L. Lukezic

Chapter 7 Laboratory Exercises for Plant Pathogenic Bacteria 53
George H. Lacy and Felix L. Lukezic

Chapter 8 Plant-Parasitic Nematodes 61
James P. Noe

Chapter 9 Pathogenicity and Isolation of Plant Parasitic Nematodes 69
James P. Noe

Chapter 10 Plant Pathogenic Fungi... 75
Ann B. Gould

Chapter 11 Slime Molds and Zoosporic Fungi.. 91
Sharon E. Mozley, David Porter, and Marc A. Cubeta

Chapter 12 Laboratory Exercises with Zoosporic Plant Pathogens 99

Mark A. Cubeta, David Porter, and Sharon E. Mozley

Chapter 13 Archiascomycete and Hemiascomycete Pathogens 111

Margery L. Daughtrey, Kathie T. Hodge, and Nina Shishkoff

Chapter 14 The Powdery Mildews 117

Margery L. Daughtrey, Kathie T. Hodge, and Nina Shishkoff

Chapter 15 Ascomycota: The Filamentous Fungi Forming Perithecia, Apothecia, and Ascostromata 127

Kenneth J. Curry and Richard E. Baird

Chapter 16 Deuteromycota: The Imperfect Fungi 133

Richard E. Baird

Chapter 17 Laboratory Exercises Illustrating Some Fungi in the Deuteromycota 141

Richard E. Baird

Chapter 18 Smut and Rust Diseases 151

Larry J. Littlefield

Chapter 19 Fleshy and Other Basidiomycetes 165

Richard E. Baird

Chapter 20 Oomycota: The Fungi-Like Organisms 173

Malissa H. Ament and Robert N. Trigiano

Chapter 21 Laboratory Exercises with the Oomycetes 183

Robert N. Trigiano, Richard E. Baird, and Steven N. Jeffers

Chapter 22 Parasitic Seed Plants, Protozoa, Algae, and Mosses 193

Mark T. Windham and Alan S. Windham

Chapter 23 Abiotic Diseases 201

Alan S. Windham and Mark T. Windham

Part III
Molecular Tools for Studying Plant Pathogens

Chapter 24 Studying the Ecology, Systematics, and Evolution of Plant Pathogens at the Molecular Level 209

Emnet Abesha and Gustavo Caetano-Anollés

Chapter 25 Molecular Techniques Used to Study Systematics, Ecology, and Evolution of Plant Pathogens 217

Robert N. Trigiano, Malissa H. Ament, Ladare F. Habera, and Gustavo Caetano-Anollés

Part IV
Plant–Pathogen Interactions

Chapter 26 Pathogen Attack Strategies 235

Larry E. Trevathan

Chapter 27 Detecting and Measuring Extracellular Enzymes of Fungi and Bacteria 247

Robert N. Trigiano and Malissa H. Ament

Chapter 28 Host Defenses 261

Kimberly D. Gwinn, Sharon E. Greene, James F. Green, and David J. Trently

Chapter 29 Disruption of Plant Function 269

Melissa B. Riley

Part V
Epidemiology and Disease Control

Chapter 30 Plant Disease Epidemiology 281

Kira L. Bowen

Chapter 31 Host Resistance 295

Jerold K. Pataky and Martin L. Carson

Chapter 32 Cultural Control of Plant Diseases 313

Gary Moorman

Chapter 33 Chemical Control of Plant Diseases 319

Alan S. Windham and Mark T. Windham

Chapter 34 Biological Control of Plant Pathogens 327

Bonnie H. Ownley and Mark T. Windham

Chapter 35 Integrated Pest Management 337

Clayton A. Hollier

Chapter 36 Plant Disease Diagnosis 345

Jackie M. Mullen

Chapter 37 Diagnostic Techniques and Media Preparation 359

Jackie M. Mullen

Part VI
Special Topic

Chapter 38 In Vitro Plant Pathology 367

Subramanian Jayasankar and Dennis J. Gray

Glossary 379

Appendix I Careers in Plant Pathology 393

Alan S. Windham and Mark T. Windham

Appendix II Fantastic Plant Pathology Websites 395

Index 397

Part I

Introductory Concepts

1 Plant Pathology and Historical Perspectives

Mark T. Windham and Alan S. Windham

CHAPTER 1 CONCEPTS

- Plant pathology is composed of many other disciplines such as botany, microbiology, nematology, virology, bacteriology, mycology, meteorology, biochemistry, genetics, soil science, horticulture, agronomy, and forestry.

- Plant pathology is the study of what causes plant diseases, why they occur, and how to control them.

- Plant pathologists are usually interested in populations of diseased plants and not in individual diseased plants.

- Plant diseases have had a major impact on mankind. Diseases such as ergotism and late blight of potato have led to the deaths of thousands of people.

- Diseases such as coffee rust have changed the way people behave and/or their customs.

- Diseases such as southern corn leaf spot, chestnut blight, and dogwood anthracnose have appeared suddenly and caused millions of dollars to be lost in damage as the pathogens of the diseases spread through the ranges of the hosts.

BACKGROUND

Plants are the foundation of agriculture and life on this planet. Without plants, there would be nothing to feed livestock or humans. Plants are a primary component in building shelter and making clothing. Like humans and animals, plants are plagued with diseases, and these diseases may have devastating consequences on plant populations. Plant pathology is not a pure discipline such as chemistry, mathematics, or physics, but it embodies other disciplines such as botany, microbiology, nematology, virology, bacteriology, mycology, meteorology, biochemistry, genetics, soil science, horticulture, agronomy, and forestry. Plant pathology encompasses the study of what causes a plant disease; how a pathogen attacks the plant at the molecular, cellular, tissue, and whole plant levels of organization; how the host responds to attack; how pathogens are disseminated; how the environment influences the disease process; and how to manage plant pathogens and thereby reduce the effects of the disease on plant populations. Unlike physicians or veterinarians, who emphasize treatment of individuals, plant pathologists are usually interested in populations of plants and not individuals. An individual wheat plant has little worth to a farmer. If it dies from a disease, the plants on either side of it will grow into its space and their increased yield will compensate for the loss of the diseased plant. However, if entire fields become diseased or fields in a region are devastated by disease, economic losses can be staggering. The exception to emphasizing populations of plants to individual plants is specimen plants that include large shade trees or trees planted by a historical figure, such as an oak planted by George Washington at Mt. Vernon, or a southern magnolia planted on the White House lawn by Andrew Jackson. Extraordinary measures may be taken to protect or treat plants of high value or historical significance.

Because of the diversity of questions that plant pathologists are called on to answer, plant pathologists are a heterogeneous group of scientists. Some plant pathologists spend most of their time in the field studying how pathogens move over a large area and what environmental factors play a role in development of epidemics or determining which management tactics are most effective in controlling or reducing the impact of a disease. Other plant pathologists are interested in the processes by which a pathogen induces a disease or may look for genes that confer resistance in a plant and complete most of their professional activities in a laboratory. Some plant pathologists work in outreach programs such as the extension service or in private practice and diagnosis disease problems for producers and home gardeners and make recommendations as to how plant diseases may be managed. Still other plant pathologists work for private companies and are responsible for development of new products (biological control agents, chemicals, and new plant varieties) that reduce the impact of plant diseases on producers and consumers (Appendix I).

IMPACT OF PLANT DISEASES ON MANKIND

Plant diseases have impacted man's ability to grow plants for food, shelter, and clothing since he began to cultivate plants. Drawings and carvings of early civilizations in Central America depict corn plants with drooping ears and poor root systems. Crop failures for ancient man and through the middle ages were common, and plant diseases were often attributed to displeasure of various deities. The Roman god Robigus was thought to be responsible for a good wheat harvest and Romans prayed to him to prevent their wheat crop from being blasted with fire (rust). In more modern times (since 1800), plant diseases have destroyed military plans of monarchs; changed cultures; caused mass migrations of people to avoid starvation; resulted in loss of major components of forest communities; and bankrupted thousands of planters, companies, and banks. Some examples of the effects of some plant diseases on the history of mankind and the environment are illustrated in the following paragraphs.

ERGOTISM

Ergotism is caused by eating rye bread contaminated with sclerotia (hard survival structures shaped like the spur of a rooster) of *Claviceps purpurea*. Sclerotia are formed in the maturing heads of rye and may contain alkaloids, including LSD, a strong hallucinogenic compound. Symptoms in humans eating contaminated bread include tingling of extremities, including fingers, hands, and feet; a high fever; hallucinations; mental derangement; abortion; and loss of hands, feet, and legs due to restricted blood flow and subsequent gangrene. Death often follows consumption of large quantities of contaminated grain. In livestock fed contaminated grain, heifers may abort fetuses and livestock lose weight, quit giving milk, and lose hooves, tails, ears, etc., following gangrene. As in humans, death is likely when exposed to high doses of ergot. In the middle ages, thousands of people died from this disease in Europe and the disease was referred to as the Holy Fire, due to the high fever and the burning and tingling sensations in hands and feet of victims. An outbreak in France lead to the name St. Anthony's Fire, presumably because monks of the Order of St. Anthony successfully treated inflicted people by feeding them uncontaminated rye bread. The disease continued in Europe for centuries. A number of authors have concluded that the Salem Witch Trials were due to an outbreak of ergotism in the early American colonies because rye was the primary grain grown in the New England region. The behavior of the accused "witches" was similar to behavior associated with an outbreak of ergotism in human and livestock populations. In the 1950s, an outbreak of ergotism occurred in several small villages in France and demonstrated that even when it is known how ergotism is caused and how the sclerotia are introduced into grain, epidemics of ergotism are still possible.

IRISH POTATO FAMINE

Potatoes were one of the treasures taken from the New World back to Europe and were readily adapted to European farming practices. By 1840, potatoes had become the staple food crop in Ireland, and the average Irishman ate approximately seven pounds of potatoes daily. Because so many potatoes could be grown on a relatively small plot of land, the population of Ireland increased dramatically during the first four decades of the 19th century. In the early 1840s, an epidemic of a new potato disease was documented in the U.S., but little attention was paid to it in Europe. In 1845, an epidemic of potato cholera, later named late blight of potato and attributed to the pathogen *Phytophthora infestans*, swept across Europe. Although starvation was common at this time in Europe, it was spared the devastation that was found in Ireland because most of Europe had more diversity in its agricultural production and did not depend on one crop for survival as the Irish did. In Ireland, more than a million people starved to death because of an almost total destruction of the potato crop. Another million people migrated to the U.S., taking whatever jobs they could find in the New World. In cities such as Boston and New York, many of the jobs they took were low-paying, dangerous jobs, such as policemen and firemen. To this day, generations of descendants of these immigrants continue in the same line of work.

COFFEE RUST

In the 1700s and early 1800s, coffee was an expensive drink because of the monopoly Arab traders had on the coffee trade and the careful attention they paid to ensure that viable coffee beans (seeds) did not leave their domain. In the mid-1800s, some coffee beans were smuggled to Ceylon (present-day Sri Lanka) and the British began growing coffee. Coffee became the preferred drink of British citizens, and coffee houses became as common as pubs. By 1870, more than 400 plantations of coffee, comprising at least 200,000 ha, were found in Ceylon. In the 1870s, a new disease, coffee rust, caused by the fungal pathogen *Hemileia vastatrix*, struck Ceylon, resulting in terrible consequences and the destruction of Ceylon's coffee trade. Planters, banks, and shipping companies went bankrupt and caused widespread panic in British financial markets. By 1880, tea plants replaced 140,000 ha of the destroyed coffee trees. Great Britain became a country of tea drinkers, and this custom remains to date.

CHESTNUT BLIGHT

When the first colonists arrived in the New World, they found forests of eastern North America populated with American chestnut. Chestnut wood was resistant to decay, and the bark contained tannins that made the production of leather from animal hides feasible. In many areas, one of every four trees in the forest was an American chestnut. The nut crop from these trees was so prolific that the ground could be covered by nearly a foot of nuts. Nuts not only served as a food source for the colonists but were also a major mast crop for wildlife that the colonists depended on for meat. Many of the ships of the American shipping industry in the 19th century were made of rot-resistant chestnut timber. In the early 1900s, a new disease of chestnut, now known to be caused by Cryphonectria *parasitica*, was discovered in the northern Atlantic states and named chestnut blight. The disease spread rapidly south and westward, destroying chestnut stands as it went. The disease finally reached the southern and western extent of the chestnuts' range in the 1950s. By this time, millions of trees had been destroyed, which represented billions of dollars in lost timber. The effects of the disease on wildlife populations were also dramatic, as wildlife had to adapt to less reliable and nutritional mast crops such as acorns. There have been intensive breeding efforts to incorporate resistance to chestnut blight from Chinese chestnut into American chestnut. Resistant hybrids that have been backcrossed for some generations with American chestnut have resulted in a tree that is resistant to chestnut blight and that strongly resembles the American chestnut. Unfortunately, it will take more than a century before we see forests with the stately giants that Americans marveled at before the onset of chestnut blight.

SOUTHERN CORN BLIGHT

After the advent of hybrid seed corn, corn yields began to skyrocket to unheard of yields and came to dominate the corn seed market. To reduce labor costs in producing hybrid seed corn, seed companies began using breeding lines containing a sterility gene that was inherited through the cytoplasm of the female parent. The trait or gene was named the Texas cytoplasmic male sterility (cms) gene. Using this gene in the female parent meant substantial cost savings for the seed companies because they did not have to remove the tassels by hand when producing hybrid seed corn. This system worked for several years until an outbreak in 1970 of a new race of the fungus that is currently named *Cochliobolus heterostropus* (see Plate 7C following page 80). This new race caused a disease on

corn carrying the male sterility gene (practically all hybrid seed corn at that time), resulting in tan lesions that covered the leaves. Stalks, ear husks, ears, and cobs were also attacked and destroyed by the pathogen. The disease first appeared in Florida, spread northward, and destroyed approximately 15% of the U.S. corn crop. Losses were estimated to be in excess of $1 billion. Experts warned the country that nearly the entire U.S. corn crop would be lost in 1971 if substantial changes were not quickly made in the way hybrid seed corn was produced. Commercial seed companies leased almost all the available space in South America in the winter of 1970 and were able to produce enough hybrid seed corn that did not contain the Texas cms gene to save the corn crop of 1971.

DOGWOOD ANTHRACNOSE

Flowering dogwood, *Cornus florida*, is a popular tree in landscapes throughout much of the U.S. and is worth more than $100 million in wholesale sales to the U.S. nursery industry. It is also an important natural resource, and its foliage, high in calcium, is the preferred browse of lactating deer in early spring in the eastern U.S. Its bright red berries are high in fat and are an important mast crop to wildlife, including black bears, squirrels, turkeys, and more than 40 species of neotropical song birds. In 1977, a new fungal disease was reported in Seattle, WA, on flowering dogwood and Pacific dogwood, *C. nuttallii*. The following year, the disease was reported on flowering dogwood in the Brooklyn Botanical Garden in New York. The origin of this disease organism is unknown; however, genetic data suggest that the disease-causing fungus is exotic to the North American continent. Since these reports, dogwood anthracnose, caused by *Discula destructiva*, has destroyed millions of dogwoods on both coasts. In some areas of the Appalachians, flowering dogwood has nearly disappeared where it was once a common understory tree. The disease epidemic continues to spread, with new reports occurring in the Midwestern U.S. in 2001 through 2002.

CAUSES OF PLANT DISEASES

Plant diseases are caused by fungi, bacteria, mollicutes, nematodes, viruses, viroids, parasitic seed plants, algae, and protozoa. The largest group of plant pathogens is the fungi. This differs considerably from human pathogens, where the most common pathogen groups include bacteria and viruses. This is not to imply that other groups such as bacteria, mollicutes, nematodes, viruses, viroids, and parasitic seed plants do not cause important and destructive diseases — they do. For example, *Striga* species (witchweed) is the limiting factor in sorghum, sugarcane, and rice production in Africa, Asia, and Australia. Dwarf mistletoe, *Arceuthobium* species, severely limits conifer production in some areas of the western U.S. Millions of dollars are lost each year to nematode diseases such as root knot, bacterial diseases such as soft rot and crown gall, and viral diseases such as tobacco mosaic virus and impatient necrotic spot virus.

ABIOTIC STRESSES (ABIOTIC DISEASES)

Some abiotic stresses such as air pollution and nutrient deficiencies were earlier referred to as abiotic diseases. However, this terminology is no longer used in modern plant pathology. Plant stresses such as those previously listed above and others such as extremes in temperature, moisture, pH, and light levels, and exposure to herbicides are now referred to as abiotic stresses or environmental stresses that result in disease-like symptoms. Sometimes the symptoms that these stresses cause in plants, for example, chlorosis, wilting, necrosis, leaf spots, and blights, look like symptoms of diseases caused by plant pathogens.

WHERE TO GO FOR MORE INFORMATION ABOUT PLANT DISEASES

Most plant pathologists belong to professional societies such as the American Phytopathological Society, the Nematology Society, the Mycology Society of America, and the American Society of Horticultural Science. The most prominent society for plant pathology is the internationally recognized American Phytopathological Society. The society's web page (http://www.apsnet.org) is a clearing house of information on new and emerging disease problems, careers in plant pathology, a directory of plant pathology departments at universities in the U.S., and featured articles on plant diseases. The society also publishes several plant pathology journals such as *Phytopathology*, *Plant Disease*, *Molecular Plant-Microbe Interactions*, and *Plant Health Progress*, and other publications such as books and compendia on specific diseases or diseases affecting specific hosts. It also publishes the monthly newsletter *Phytopathology News*. Membership is open to professionals interested in plant pathology and to students at a very reduced rate.

SUGGESTED READINGS

Agrios, G.N. 1997. *Plant Pathology*, 4th ed. Academic Press, New York, 635 pp.

Campbell, C.L., P.D. Petersen and C.S. Griffith. 1999. *The Formative Years of Plant Pathology in the United States*. APS Press, St. Paul, MN, 427 pp.

Carefoot, G.L and E.R. Sprott. 1967. *Famine on the Wind*. Longmans, Ontario, 231 pp.

Horsfall, J.G. and E. Cowling (Eds.). 1977. *Plant Disease: An Advanced Treastise*, Vols.1–5. Academic Press, New York.

Large, E.C. 1940. *The Advance of the Fungi*. Henry Holt, New York, 488 pp.

Lucas, G.B., C.L. Campbell and L.T. Lucas. 1992. *Introduction to Plant Diseases: Indentification and Management*, 2nd ed. Van Nostrand Reinhold, New York, 364 pp.

Schumann, G.L. 1991. *Plant Diseases: Their Biology and Social Impact*. APS Press, St. Paul, MN, 397 pp.

2 What Is a Disease?

Mark T. Windham and Alan S. Windham

CHAPTER 2 CONCEPTS

- A disease is caused by the interactions of the pathogen, host, and environment.

- Diseases are dynamic (change over time). Injuries are discrete events.

- Plant stresses are usually caused by too much or too little of something.

- Host responses to disease are known as symptoms.

- Structures (e.g., mycelium, spores, nematode egg masses) of pathogens on a diseased host are known as signs.

- The interaction among the host, pathogen, and environment is known as the disease cycle.

- A disease cycle is made of a sequence of events including inoculation, penetration, infection, invasion, reproduction, and dissemination.

- Diseases with only a primary disease cycle are known as monocyclic diseases whereas diseases with secondary disease cycles are known as polycyclic diseases.

- Host plants can be infected by a pathogen, whereas soil or debris is infested by pathogens.

- Koch's postulates are used to prove that a pathogen causes a disease.

Before studying a plant disease, a framework or concept of what a plant disease is and, almost as importantly, what it is not, is essential. There are many definitions of plant disease. However, for this book, the following definition is used: A disease is the result of a dynamic, detrimental relationship between an organism that parasitizes or interferes with the normal processes of cells or tissues, or both, of the plant. The organism that incites or causes the disease process with the host is called a **pathogen**.

A pathogen may or may not be a **parasite**. A parasite is an organism that lives on or in another organism and obtains nutrients at the expense of the host. In contrast, pathogens may interfere with plant cell functions by producing toxins that disrupt or destroy cells; by producing growth plant regulators that interfere with the normal growth or multiplication of plant cells; by producing enzymes that interfere with normal cellular functions; or by absorbing water or nutrients that were intended for the cellular functions of the host. Pathogens may also incite disease by blocking the vascular system so that water and nutrients cannot be normally moved within the plant. Some pathogens disrupt normal functions of plants by

inserting portions of their DNA or RNA into host cells and interfering with replication of nucleic acids.

Pathogenicity is the ability of a pathogen to interfere with one or more functions within a plant. The rate at which or how well a pathogen is able to interfere with cell functions is referred to as **virulence**. A virulent pathogen can be very aggressive and incite disease over a wide range of environmental conditions. An avirulent pathogen is an organism that is rarely able to interfere with normal cellular functions of the host or does so under very specific environmental conditions. The ability of the pathogen to survive in the environment where the host is grown is a measure of **pathogen fitness**.

Plant stresses or injuries are not diseases because they are not dynamic; that is, they do not change over time. If lightning strikes a tree, the tree may be damaged or killed. However, the lightning does not get hotter or more dangerous to the tree over time. It happens in a discrete instant in time and is therefore an injury and not a disease. The same thing can be said for using a lawn mower on turf. Cutting off 30 to 50% of the leaf area may severely impede ability of the grass to grow, but the cut is done in a discrete instant of time. Therefore, it is not a disease but an injury.

Disease Triangle

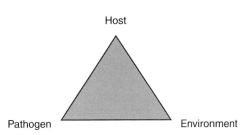

FIGURE 2.1 Disease is dependent on three components: host, pathogen, and environment. The area within the triangle represents the interaction of these components, referred to as a disease.

A plant stress is usually too much or to little of something. Giving the plant either too much (flooding) or too little (drought) water can cause water stress. Other examples of plant stresses can be extremes in temperature, improper pH, and nutrient deficiency or excess. Pollutants, pesticides, and road salt may also cause stresses to plants.

Plant pathogens that cause diseases include organisms such as fungi, prokaryotes (bacteria and mollicutes), viruses, viroids, nematodes, protozoa, algae, and parasitic seed plants. These organisms can detrimentally affect the host in diverse ways. Pathogens can be classified into several groups. **Biotrophs** are pathogens that require living host tissue to complete their life cycle. Examples of biotrophs include fungi, such as powdery mildews (Chapter 14) and rusts (Chapter 18) and some members of the Oomycota, such as downy mildews and white rusts (Chapter 20); prokaryotes, such as some species of *Xylella* and mollicutes (Chapter 6); viruses and viroids (Chapter 4); phytoparasitic nematodes (Chapter 8); and protozoa and dwarf mistletoe (Chapter 22). Many pathogens can be parasitic on a host under some conditions and at other times be saprophytic, living on organic matter. A pathogen that often behaves as a parasite but under certain conditions behaves as a saprophyte, is a **facultative saprophyte**. A pathogen that often behaves as a saprophyte but under some conditions becomes a parasite is known as a **facultative parasite**. Nonbiotrophic organisms kill before feeding on the cells or cellular contents. These organisms that live on dead tissues are known as **necrotrophs**.

Diseased plants are infected by a pathogen. However, in some cases, once disease is incited, symptoms may continue to develop even if the pathogen is no longer present (e.g., crown gall caused by *Agrobacterium tumefaciens*). Inanimate objects such as soil, pots, or debris are not infected by the pathogen, but infested by the pathogen.

The host plant, pathogen, and environment interact with each other over time and this interaction is referred to as disease (Figure 2.1). The sequence of events that takes place during the course of disease, sometimes at set

or discrete time intervals, is known as the **disease cycle**. The disease cycle is not to be confused with the pathogen's life cycle. Sometimes these two cycles follow similar paths, but the cycles are different.

The steps of the disease cycle are inoculation, penetration, infection, invasion, reproduction, and dissemination. **Inoculation** is the placement of the pathogen's infectious unit or propagule on or in close proximity to the host cell wall. The propagule then penetrates the cell wall of the host. In fungi, the propagule may germinate and the germ tube may penetrate the wall directly or indirectly through a wound or natural opening. Once the pathogen is through the cell wall, a food relationship with the host may develop and the cell is considered infected. After infection takes place, the pathogen may grow and invade other parts of the host or reproduce. The pathogen will continue to reproduce and the new propagules will be dispersed or disseminated by a variety of means such as wind, rain, within or on vectors, seed, or on contaminated debris or equipment.

Some diseases are monocyclic diseases (Chapter 30), meaning that there is only one disease cycle in a growing cycle. Inoculum that is produced during the disease cycle does not contribute or fuel the disease during the current growing season. The sequence of events of a monocyclic disease cycle is given in Figure 2.2. In many diseases, the inoculum produced during much of the disease cycle contributes to continuing the current disease cycle or epidemic. This inoculum actually fuels the epidemic, and the disease cycle expands to include many more host plants, which in turn contribute more and more inoculum to the disease cycle. In diseases where there are more than one

Primary Disease Cycle
Monocyclic Disease

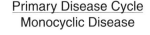
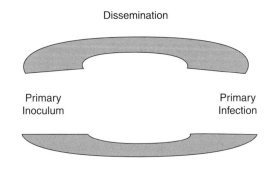

FIGURE 2.2 In a monocyclic disease, the primary disease cycle is composed of discrete events where inoculation and penetration lead to infection. Propagules produced during the disease cycle overwinter and become the primary inoculum (inoculum that begins a new disease cycle) for the next disease cycle. The inoculum is disseminated at the beginning of the next cycle.

Secondary Disease Cycle
Polycyclic Disease

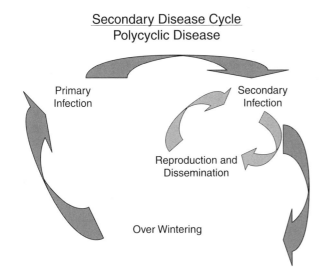

FIGURE 2.3 Polycyclic diseases have primary inoculum that penetrates and infects the plant. This is a part of the primary disease cycle. Inoculum produced after invasion is disseminated and causes more infections during the current growing season. The inoculum contributes to the secondary disease cycle, and the secondary cycle may be repeated many times.

disease cycles, the primary cycle often has a repeating phase known as the secondary disease cycle (Figure 2.3). Such diseases are known as polycyclic diseases.

Plant pathologists study disease cycles to determine where cultural or other types of disease control tactics can be applied to interfere with the disease, and thereby interrupt the processes. Elimination of infested or infected seed may reduce the primary inoculum used to start a primary disease cycle. Use of resistant cultivars (Chapter 31) that are able to wall off a plant infection and prevent invasion of the host can also stop a disease cycle or prevent the formation of secondary cycles. Elimination of plant debris may reduce the ability of a pathogen to overwinter. By understanding the disease cycle and the series of events that are parts of that cycle, plant pathologists may attack the disease processes and reduce the disease's affects on society.

Host responses to infection are known as **symptoms.** Symptoms include leaf spots, blights, blotches, twig blights, cankers, galls, seed, root and stem rots. Symptoms of a plant disease may occur on only a small portion of the plant and result in little disruption of the plant's functions or symptoms may cover the entire plant. Definitions of common plant symptoms are given in Table 2.1.

Structures of the pathogen are referred to as **signs**. Signs can include spores, mycelia, resting structures such as sclerotia, nematodes, and bacterial streaming into water. In some cases, symptoms and signs are present together. For example, a plant that is infected with *Sclerotium rolfsii* may exhibit symptoms of wilting and show signs (sclerotia) of the fungus.

TABLE 2.1
Common Symptoms Associated with Plant Host Response to Disease

Blight. Extensive area of diseased flowers or leaves.

Butt rot. Basal trunk rot.

Burl. Swelling of a tree trunk or limb differentiated into vascular tissue — contrast with gall.

Canker. A sunken area in a fruit, stem, or limb caused by disease.

Chlorosis. Yellow-green color of foliage due to destruction or lack of production of chlorophyll.

Dieback. Generalized shoot death.

Flagging. Scattered or isolated dead or dying limbs.

Gall. Swollen area of nondifferentiated tissues (tumor) caused by an infection. Galls can arise from hypertrophy (cell enlargement) or hyperplasia (increase in cell division) — contrast with burl.

Leaf spot. Lesion on leaf.

Lesion. A necrotic (dead) or chlorotic spot that occurs on all plant organs. Anthracnose lesion is a necrotic lesion with a reddish or purplish border. Local lesion is a necrotic or chlorotic lesion in which infection is limited to a small group of cells and the infection does not spread to other parts of the tissue.

Mosaic. Chlorotic pattern, ringspots, and mottles in leaves, petals, or fruit. Mosaics are usually associated with virus infections.

Mummy. Shriveled desiccated fruit.

Necrosis. Dead tissue.

Rot. Portion of plant destroyed by disease. Root rot is rotted roots.

Wilt. Loss of turgor in a plant or plant part.

Once we recognize that a disease is occurring, it is also important to be able to prove that a pathogen is causing a particular disease. To do this, we use a series of rigid rules or postulates known as Koch's postulates (Chapter 36). Koch's postulates are as follows:

1. The pathogen must be found associated with all symptomatic plants.
2. The pathogen must be isolated and grown in pure culture and its characteristics described. If the pathogen is a biotroph, it should be grown on another host plant and have the symptoms and signs described.
3. The pathogen from pure culture or from the test plant must be inoculated on the same species or variety as it was originally described on and

it must produce the same symptoms that were seen on the originally diseased plants.

4. The pathogen must be isolated in pure culture again and its characteristics described exactly like those observed in Step 2.

In conclusion, plant pathologists study what causes diseases; how plants are affected by diseases; how plants resist pathogens; and how the host, pathogen, and environment interact with each other. Through their investigations, the concepts of what is a plant disease; which organisms cause diseases; how those organisms are classified; and how a pathogen, a susceptible host, and the environment interact in a disease relationship is not static. For example, until a few years ago, diseases such as downy mildews and late blight of potato were thought to be caused by fungi. However, it has been found from molecular studies that the pathogens that cause these diseases are now classified in the kingdom Stramenopila (Chromista) instead of in the kingdom Fungi. Although the inter-actions of the disease triangle endpoints (host, environment, and pathogen) will continue to be reevaluated and redefined, plant pathologists still agree that without a conducive environment, a susceptible host, and a pathogen, no disease will take place.

SUGGESTED READINGS

Agrios, G.N. 1997. *Plant Pathology.* Academic Press, San Diego, CA, 635 pp.

Andrews, J.H. 1984. Life history strategies of plant parasites. *Adv. Plant Pathol.* 2:105–130.

Horsfall, J.G. and E.B. Cowling (Eds.). 1977. *Plant Disease,* Vols. 1–5. Academic Press, New York.

Vanderplank, J.E. 1963. *Plant Diseases: Epidemics and Control.* Academic Press, New York, 349 pp.

Zadoks, J.C. and R. Schein. 1979. *Epidemiology and Plant Disease Management.* Oxford University Press, New York, 427 pp.

3 Introduction to Plant Pathogens

Mark T. Windham

CHAPTER 3 CONCEPTS

- Fungi are achlorophyllous, filamentous, eukaryotic organisms that reproduce by spores and have walls that contain chitin.

- Anamorphic spore types include sporangiospores, conidia, and chlamydospores.

- Spores in a sporangium are formed by cleavage of the cytoplasm.

- Sexual spores of fungi are zygospores, ascospores, and basidiospores.

- Members of Oomycota have oospores and contain cellulose in their cell walls.

- Prokaryotic pathogens include bacteria, which have cell walls, and mollicutes, which have a cell membrane but no cell wall.

- Phytopathogenic nematodes have mouthparts called stylets.

- Viruses are nucleoproteins and are too small to be seen with light microscopy.

Plant pathogens belong to very diverse groups and are found in the kingdoms Animalia, Stramenopila (Chromista), Fungi, Prokaryotae, Plantae, and Protozoa. The largest group of plant pathogens is found in the Fungi. This chapter is intended to very superficially acquaint students with the various groups of pathogens and some of the specialized language and terms associated with each of them. A more complete description of the different groups follows in subsequent chapters.

THE FUNGI

Fungi are achlorophyllous and eukaryotic. They are generally filamentous, branched organisms that normally reproduce by spores and have walls made of chitin and other polymers. Most of their life cycles are spent in the haploid (N) or dikaryotic ($N + N$) state. The thread-like filaments of the fungus are known as **hyphae** (sing. **hypha**; Figure 3.1). Hyphae of one body or thallus are known as **mycelia** (sing. **mycelium**). Spores are the reproductive or propagative bodies of fungi. In some fungi, spores of the **anamorph** (asexual) stage are formed by cleavage of cytoplasm within a sac or **sporangium** (pl. **sporangia**; Figure 3.2). Spores produced in the sporangium are referred to as **sporangiospore**s, which may be motile (have one or more flagella) or nonmotile. Nonmotile sporangiospores are usually dis-

seminated by air currents. Motile spores within a sporangium are called **zoospores** and are dispersed in water. Asexual spores of other fungi are borne on the tips or sides of specialized hyphae called **conidiophores**. Spores borne in this fashion are referred to as **conidia** (sing. **conidium**; Figure 3.3). Conidia vary in shape, size, color, and number of cells. Some conidia are borne on naked conidiophores whereas others are borne on conidiophores that are contained in specialized structures. A **pycnidium** (pl. **pycnidia**) is an asexual fruiting body that is flask shaped and contains conidia and conidiophores. Pycnidia (Figure 3.4) usually have a hole (**ostiole**) from which conidia are pushed out of the structure. An **acervulus** (pl. **acervuli**) is an asexual fruiting body that is found under the cuticle or epidermis of the host. As conidiophores and conidia form, the epidermis or cuticle, or both, rupture and spores are released. Spores may be released in a gelatinous matrix referred to as a **cirrhus** (pl. **cirrhi**). Some acervuli have **setae** or sterile hairs (Figure 3.5).

Conidia may be borne singularly or in clusters on branched or unbranched conidiophores. A number of conidiophores may be fused at the base to form a structure known as a **synnema** (pl. **synnemata**; Figure 3.6A). In other fungi, short conidiophores may be borne on mats of hyphae known as **sporodochium** (pl. **sporodochia**; Figure 3.6B).

0-8493-1037-7/04/$0.00+$1.50

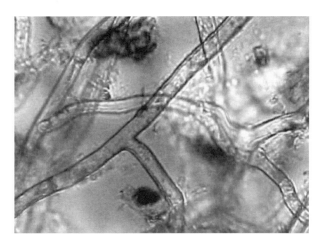

FIGURE 3.1 Hyphae of *Rhizoctonia solani* have cross walls know as septa.

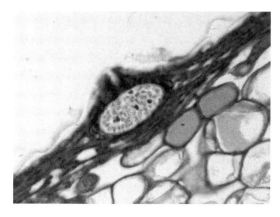

FIGURE 3.4 Pycnidium of *Phoma* species. The hole at the top of the structure is referred to as the ostiole.

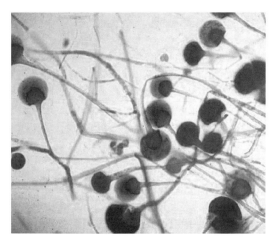

FIGURE 3.2 Sporangia of *Rhizopus niger.* The specialized hyphae or stalk that is attached to each sporangium is a sporangiophore.

FIGURE 3.5 An acervulus of a *Colletotrichum* species with many setae or sterile hairs.

Fungi also form a number of asexual survival structures. **Chlamydospores** are thick-wall resting spores; some may even have a double cell wall. **Sclerotia** (sing. **sclerotium**) are composed of hyphae that are so tightly packed that they have lost their individuality. They are very hard and extremely resistant to harsh environmental conditions.

Fungi are usually classified by their **telemorph** (sexual) stage. In some fungi, **gametes** (sex cells) unite to produce a zygote (Figure 3.7). This is usually how the more primitive fungi reproduce. The fusion of gametes that are of equal size and appearance results in a zygote, referred to as a **zygospore**, and are classified in the Zygomycota (Chapter 10).

In other fungi, there are no definite gametes and instead one mycelium may unite with another compatible mycelium. In the **Ascomycota**, **ascospores**, usually eight

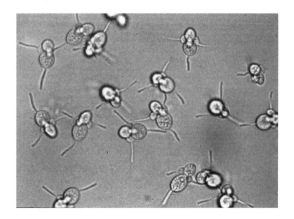

FIGURE 3.3 Conidia of *Entomosporium* species. Conidia can be made of a single cell or be multicellular as are these conidia.

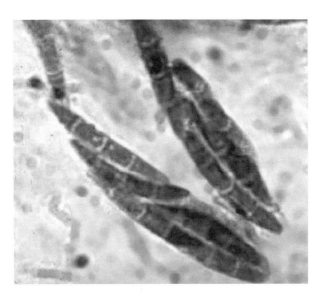

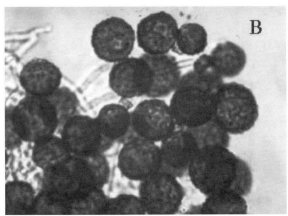

FIGURE 3.8 Asci containing spindle-shaped, multicellular ascospores of *Gibberella* species.

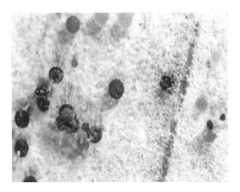

FIGURE 3.6 (A) Fused conidiophores comprise the synnema of *Graphium* species. (B) Sporodochium of *Epicoccum* species is a mat of densely packed, short conidiophores.

FIGURE 3.9 Cleistothecia (ascomata) of *Erysiphe* species.

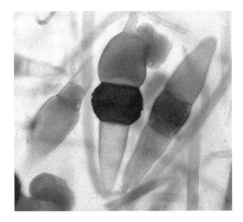

FIGURE 3.7 Zygosprorangium containing a single zygospore (zygote) of *Rhizopus niger*.

in number, are produced within a zygote cell, the **ascus** (pl. **asci**; Figure 3.8). Asci may be produced naked (not in any structure) or in specialized structures. **Cleistothecia** (sing. **cleistothecium**) are enclosed structures with asci located at various levels (Figure 3.9). Cleistothecia

typically not released until the cleistothecium ruptures or is eroded by the environment. Ascocarps of powdery mildew have been referred to traditionally as cleistothecia but are perithecia (Chapter 14). **Perithecia** (sing. **perithecium**) are usually flask-shaped structures with an opening (Figure 3.10). Asci and ascospores are formed in a single layer. Ascospores are pushed or forcibly ejected through the opening. They are dispersed via air currents, insects, and water. Some members of the phylum form asci in open, cup-shaped structures known as **apothecia** (sing. **apothecium**), and ascospores are disseminated by modes similar to those found in fungi that produce perithecia.

In other fungi, sexual spores are produced on the outside of the zygote cell or **basidium** (pl. **basidia**) and are called basidiospores (usually four in number). Fungi that reproduce in this manner are placed in the **Basidiomycota**. In this group, the basidia and basidiospores may be borne naked (rusts and smuts) or formed in structures such as mushrooms, puffballs, and conks. Mushrooms are

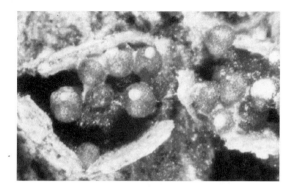

FIGURE 3.10 Perithecia of *Nectria coccinea* var. *faginata*.

fleshy, sometimes tough, umbrella-like structures, whereas puffballs are white to light tan (dark brown to black when mature) spongy, spherical bodies formed on the soil surface. Conks are shelf or very hard, bracket-like fruiting bodies and are usually found on stumps, fallen logs, or living trees.

PLASMODIOPHOROMYCOTA — PROTOZOA

Parasitic slime molds are placed in the kingdom Protozoa and the phylum **Plasmodiophoromycota**. They are unicellular and produce **plasmodia** in root cells (Figure 3.11). Plasmodia are amoeba-like cells without cell walls inhabiting the lumens of host cells. Parasitic slime molds produce zoospores that can function as gametes.

STRAMENOPILA — FUNGI-LIKE ORGANISMS

Other fungal-like organisms are found in kingdom Stramenopila (Chromista) and include those in phylum Oomycota. These organisms were traditionally characterized as fungi because of filamentous growth, lack of chlorophyll, and reproduction by spores. However, with the advent of modern molecular techniques, they are now classified in a different kingdom, which includes the brown algae. These pathogens cause some of the most destructive plant diseases and include the downy mildews and species in the genera *Phytophthora* and *Pythium*. They produce anamorphic spores in sporangia and the spores may be motile (zoospores). The gametes are of different size and shape (**antheridia** in male and **oogonia** in female) as illustrated in Figure 3.12. The sexual spore is an oospore and functions as a survival structure. Members of Oomycota have cellulose in their cell walls and the majority of their life cycle is diploid (2*N*).

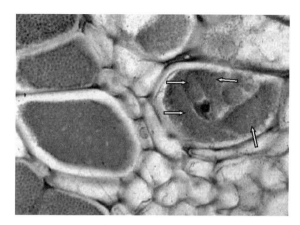

FIGURE 3.11 Plasmodia of *Plasmodiophora brassicae* in a cabbage cell. Arrows point to a single plasmodium.

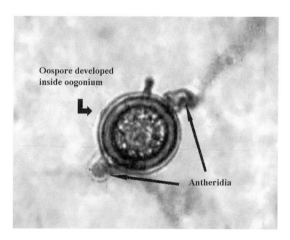

FIGURE. 3.12 Antheridia and oospore inside oogonium in a *Pythium* species.

BACTERIA AND MOLLICUTES

Prokaryotic organisms such as bacteria and mollicutes (Chapter 6) cause some of the most important plant diseases. Bacteria are prokaryotic (have no nucleus or double-membrane-bound organelles) and have a rigid cell wall that is enveloped in a slime layer. Most of the DNA in bacteria is present as a single circular chromosome. Additional DNA is found in many bacteria as independently reproducing plasmids composed of smaller amounts of DNA. Most plant pathogenic bacteria are Gram negative, with the exception of *Clavibacter* (*Corynebacterium*). Phytopathological bacteria are either rod or filamentous shaped, may or may not be flagellate, and reproduce by binary fission. Traditionally, bacteria were classified based on Gram stain, cell shape, cultural morphology, and substrate utilization. Today, bacteria are grouped by molecular analysis of genetic material.

Pathogenic bacteria are known as wound pathogens because they usually do not penetrate the host directly. They may also enter through natural plant openings such as nectaries, hydathodes, and stoma. They are disseminated by air currents, water, insects, plant materials, and contaminated equipment.

Mollicutes are smaller than bacteria, do not have a cell wall, and are delimited only by a plasma membrane. Most of the mollicutes are round or elongated and are referred to as phytoplasms. A few members of this group have a helical form and are termed spiroplasms. They are very difficult to culture. Some of the more important diseases caused by this group include aster yellows and x-disease of peach and apple. Mollicutes are typically disseminated by insects, budding, and grafting.

VIRUSES AND VIROIDS

Viruses and viroids are much smaller than bacteria, cannot be seen with light microscopy, and require the host plant's replication machinery for multiplication. Viruses are nucleoproteins — their nucleic acid (either DNA or RNA) is surrounded by a protein coat. Viruses may or may not be encapsulated with a lipid layer. In some viruses, the genome is spread between more than one particle. Viruses may be spherical or shaped like long or short rods that may be rigid or flexible. They may be very sensitive to environmental conditions such as heat and light or may be very stable under most environmental conditions. Viruses are disseminated by budding, grafting, insects, wounding, or infected plant materials. **Viroids** have many attributes of viruses, but differ in being naked strands of RNA that have no protein coat.

NEMATODES

Plant parasitic nematodes are small worm-like animals that have a cuticle made of chitin and a piercing mouthpart called a stylet, whereas free-living species, which are more common in soil samples, do not have stylets. Nematodes vary in shape: they may be very elongated, kidney-shaped, or globose. They reproduce sexually or parthenogenetically because many species do not have males. Plant pathogenic nematodes may be migratory and ectoparasitic (feed from outside of the root) or endoparasitic (feed inside the root). In some cases they are sedentary. Plant nematodes are disseminated in water, soil, plant materials, insects, and contaminated equipment. Although most nematodes are parasitic on roots, some nematodes are parasitic on aerial portions of plants.

OTHER PATHOGENS

Other types of pathogens include parasitic seed plants, flagellate protozoa, algae, and mosses. Some of these pathogens (parasitic seed plants such as dwarf mistletoe and flagellate protozoa) may cause severe crop losses. Diseases caused by algae and mosses seldom cause damage that lead to economic losses.

This brief chapter has only touched on the incredibly diverse nature of organisms that cause plant diseases. Plant pathologists, especially extension pathologists, must be well versed in many different types of organisms. The succeeding chapters explore these organisms and diseases in greater detail and hopefully whet the appetite for more advance study.

SUGGESTED READINGS

Agrios, G.N. 1997. *Plant Pathology*, 4th ed. Academic Press, San Diego, CA, 635 pp.

Alexopoulos, C.J., C.W. Mims and M. Blackwell. 1996. *Introductory Mycology*, 4th ed. John Wiley & Sons, New York, 868 pp.

Horsfall, J.G. and E.B. Cowling. (Eds.). 1977. *Plant Disease*, Vols. 1–5. Academic Press, New York.

Tainter, F.H. and F.A. Baker. 1996. *Principles of Forest Pathology*. John Wiley & Sons, New York.

Part II

Groups of Plant Pathogens

4 Plant Pathogenic Viruses

Marie A. C. Langham

CHAPTER 4 CONCEPTS

- Viruses are usually composed of RNA or DNA genomes surrounded by a protein coat (capsid).

- Viruses are unique, submicroscopic obligate pathogens.

- Plant viruses replicate through assembly of previously formed components, and replication is not separated from the cellular contents by a membrane.

- Virus species are named for the host from which they were originally associated and the major symptom that they cause. Virus species may be grouped into genera and families.

- Plant viruses are vectored by insects, mites, nematodes, parasitic seed plants, fungi, seed and pollen.

- Plant viruses can be detected and identified by biological, physical, protein, and nucleic acid properties.

WHEN POSSIBLE, PLANT VIRUS CONTROL FOCUSES ON HOST RESISTANCE

Plant virology is one of the most dynamic research areas in phytopathology. During the last quarter of the 20th century, our understanding of plant viruses and their pathogenic mechanisms has exceeded the imagination of early virologists. Today, new plant viruses are identified rapidly and our awareness of their pathological impact continues to increase. This impact is most clearly seen in yield and other economic losses. Plant viruses generate economic loss for farmers, producers, and consumers by adversely affecting plant growth and reproduction; causing death of host tissues and plants, sterility, reduction of yield or quality, crop failure, increased susceptibility to other stresses, loss of aesthetic value, quarantine and eradication of infected plants; and increasing the cost of control and detection programs (Waterworth and Hadidi, 1998). Viruses are also unique in the deceptive simplicity of their structure. However, this simplicity leads to a greater dependency on the host and a highly intricate relationship exists between the two. This complicates strategies for control of plant viruses and the losses caused by them. Control programs depend on our understanding of the virus–host relationship, and control remains one of the greatest challenges for the future of plant virology.

HOW ARE VIRUSES NAMED?

Plant viruses were typically named for the host that they were infecting when originally described and for the principal symptom that they cause in this host. The word virus follows these two terms. For example, a virus causing a mosaic in tobacco would be *Tobacco mosaic virus* (TMV). This is the species name for the virus. The use of the species concept in plant virology began in recent years following much debate regarding what constitutes a virus species (van Regenmortel et al., 2000). Following the first use of the species name, the virus is referred to by the abbreviation that is given in parentheses after the first use of the species name. Two levels of taxonomic structure for grouping species are the genus, which is a collection of viruses with similar properties, and the family, which is a collection of related virus genera. *Cowpea mosaic virus* (CPMV) is a member of the genus *Comovirus* and the family *Comoviridae* (van Regenmortel et al., 2000). Table 4.1 lists some of the virus genera and families. Virus species may also be subdivided into strains and **isolates**. **Strains** are named when a virus isolate proves to differ from the type isolate of the species in a definable character, but does not differ enough to be a new species (Matthews, 1991). For example, a virus strain may have altered reactions in an important host, such as producing a systemic reaction in a host that previously had a local lesion reaction, or the strain may have an important serological difference. Strains represent mutations or adaptations in the type virus. Isolates are any propagated culture of a virus with a unique origin or history. Typically, they do not differ sufficiently from the type isolate of a virus to be a strain.

0-8493-1037-7/04/$0.00+$1.50
© 2004 by CRC Press LLC

TABLE 4.1
Families and Genera of Some Plant-Infecting Viruses with Their Morphology and Basic Genome Characters

Family or Genus	Morphology	Nucleic Acid	Number of Strands	Vector
Single-Stranded RNA				
Allexivirus	Filamentous	+ Linear	1	Eriophyid mites
Benyvirus	Rigid rod	+ Linear	4 or 5	Fungi
Bromoviridae				
Alfamovirus	Bacilliform; Isometric	+ Linear	3	Aphids
Bromovirus	Isometric	+ Linear	3	Beetles
Cucumovirus	Isometric	+ Linear	3	Aphids
Ilarvirus	Quasi-isometric	+ Linear	3	Thrips, pollen
Oleavirus	Quasi-isometric to bacilliform	+ Linear	3	Mechanical
Bunyaviridae				
Tospovirus	Spherical[a]	+/−Circular	3	Thrips
Capillovirus	Filamentous	+ Linear	1	Seed and grafting
Carlavirus	Filamentous	+ Linear	1	Aphids; whiteflies and seed in some
Closteroviridae				
Closterovirus	Filamentous	+ Linear	1	Aphids, mealybugs, and whiteflies
Crinivirus	Filamentous	+ Linear	2	Whiteflies
Comoviridae				
Comovirus	Isometric	+ Linear	2	Beetles
Fabavirus	Isometric	+ Linear	2	Aphids
Nepovirus	Isometric	+ Linear	2	Nematodes
Foveavirus	Filamentous	+ Linear	1	Graft
Furovirus	Rigid rod	+ Linear	2	Fungi
Hordeivirus	Rigid rod	+ Linear	3	Seed and pollen
Idaeovirus	Isometric	+ Linear	3	Seed and pollen
Luteoviridae				
Enamovirus	Isometric	+ Linear	1	Aphids
Luteovirus	Isometric	+ Linear	1	Aphids
Polerovirus	Isometric	+ Linear	1	Aphids
Marafivirus	Isometric	+ Linear	1	Leafhoppers and vascular puncture
Ophiovirus	Filamentous	− Circular	3	Vegetative propagation
Ourmiavirus	Bacilliform	+ Linear	3	Mechanical and seed
Pecluvirus	Rigid rod	+ Linear	2	Fungi and seed
Pomovirus	Rigid rod	+ Linear	3	Fungi
Potexvirus	Filamentous	+ Linear	1	Mechanical
Potyviridae				
Bymovirus	Filamentous	+ Linear	2	Fungi
Ipomovirus	Filamentous	+ Linear	1	Whiteflies and grafting
Macluravirus	Filamentous	+ Linear	1	Aphids
Potyvirus	Filamentous	+ Linear	1	Aphids
Rymovirus	Filamentous	+ Linear	1	Eriophyid mites
Tritimovirus	Filamentous	+ Linear	1	Eriophyid mites
Rhabdoviridae				
Cytorhabdovirus	Bullet	− Linear	1	Leafhoppers, planthoppers, and aphids
Nucleorhabdovirus	Bullet	− Linear	1	Leafhoppers and aphids
Sequiviridae				
Sequivirus	Isometric	+ Linear	1	Aphids when helper virus is present
Waikavirus	Isometric	+ Linear	1	Leafhoppers and aphids
Sobemovirus	Isometric	+ Linear	1	Beetle
Tenuivirus	Filamentous	+/− Linear	4 or 5	Planthoppers
Tobamovirus	Rigid rod	+ Linear	1	Mechanical and seed

TABLE 4.1 (continued)
Families and Genera of Some Plant-Infecting Viruses with Their Morphology and Basic Genome Characters

Family or Genus	Morphology	Nucleic Acid	Number of Strands	Vector
Tobrabvirus	Rigid rod	+ Linear	2	Nematodes
Tombusviridae				
Aureuvirus	Isometric[a]	+ Linear	1	Mechanical
Avenavirus	Isometric	+ Linear	1	Mechanical; possibly fungi
Carmovirus	Isometric	+ Linear	1	Beetles and seed; fungi and cuttings in some
Dianthovirus	Isometric	+ Linear	2	Mechanical
Machlomovirus	Isometric	+ Linear	1	Beetles, seed, and thrips
Necrovirus	Isometric	+ Linear	1	Fungi
Panicovirus	Isometric	+ Linear	1	Mechanical, infected sod, and seed
Tombusvirus	Isometric	+ Linear	1	Mechanical and seed
Trichovirus	Filamentous	+ Linear	1	Grafting and seed
Tymovirus	Isometric	+ Linear	1	Beetles
Umbravirus	RNP complex[b]	+ Linear	1	Aphids when helper virus is present
Vitivirus	Filamentous	+ Linear	1	Grafting and mealybugs; scales and aphids in some

Double-Stranded RNA
Partiviridae

Family or Genus	Morphology	Nucleic Acid	Number of Strands	Vector
Alphacryptovirus	Isometric	Linear	2	Seed and pollen
Betacryptovirus	Isometric	Linear	2	Seed and pollen
Reoviridae				
Fijivirus	Isometric	Linear	10	Planthoppers
Phytoreovirus	Isometric	Linear	12	Leafhoppers
Oryzavirus	Isometric	Linear	10	Planthoppers
Variscosavirus	Rigid rod	Linear	2	Fungi

Double-Stranded DNA
Caulimoviridae

Family or Genus	Morphology	Nucleic Acid	Number of Strands	Vector
Badnaviruses	Bacilliform	Circular	1	Mealybugs
Caulimovirus	Isometric	Circular	1	Ahpids

Single-Stranded DNA
Geminiviridae

Family or Genus	Morphology	Nucleic Acid	Number of Strands	Vector
Begomovirus	Geminate	Circular	2	Whiteflies
Curtovirus	Geminate	Circular	1	Leafhoppers and treehoppers
Mastrevirus	Geminate	Circular	1	Leafhoppers and vascular puncture
Nanovirus	Isometric	Circular	6–9	Aphids; not mechanically transmissible

Note: Genera that have been grouped into families are listed under the family

[a]Spherical particles are more variable in size and morphology than isometric particles.
[b]RNP complexes are composed of nucleic acid inside a membrane. These particles lack the traditional protein coat unless they are in mixed infection with a helper virus and are able to use the helper's coat protein.

Sources: From Brunt, A.A., K. Crabtree, M.J. Dallwitz, A.J. Gibbs, L. Watson and E.J. Zurcher (Eds.). 1996. http://biology.anu.edu.au/Groups/MES/vide/; Hull, R. 2002. *Matthews' Plant Virology*, 4th ed. Academic Press, New York; and van Regenmortel, M.H.V., C.M. Fauquet, D.H.L. Bishop, E.B. Carstens, M.K. Estes, S.M. Lemon, J. Maniloff, M.A. Mayo, D.J. McGeoch, C.R. Pringle and R.B. Wickner. 2000. *Seventh Report of the International Committee on Taxonomy of Viruses*. Academic Press, San Diego, CA.

WHAT IS A PLANT VIRUS?

Plant viruses are a diverse group infecting hosts from unicellular plants to trees. Despite this diversity, plant viruses share a number of characteristics. A good definition of plant viruses focuses on characteristics that all plant viruses have in common.

SIZE

Viruses are ultramicroscopic. Their visualization requires an electron microscope, which uses a beam of electrons instead of visible light. The narrower wavelength of the electron beam allows the resolution of smaller objects such as cell organelles and viruses.

GENOMES

Viruses have nucleic acid **genomes**. This nucleic acid may be either ribonucleic acid (RNA) or deoxyribonucleic acid (DNA). In plant viruses, no virus has been discovered that includes both types of nucleic acids. However, there are many variations in the structures of viral genomes. The nucleic acid may be single stranded (ss) or double stranded (ds), and it may be either linear or circular. The genome may be on a single piece of nucleic acid or on multiple pieces. These pieces may be encapsulated in a single particle or in multiple particles.

CAPSIDS

Viruses have one or more protein coats or **capsids** surrounding their perimeter. These capsid layers are composed of **protein subunits**. The subunits may be composed of the same or different types of protein. For example, *Tobamoviruses* have one type of protein subunit in their capsids, whereas *Comoviruses* have two types of protein subunits and *Phytoreoviruses* have six to seven types of structural protein subunits in their capsids (van Regenmortel et al., 2000). Some viruses may also have a lipoprotein layer associated with them.

OBLIGATE PARASITES

Viruses are **obligate parasites**. The simplicity of the virus leads to its dependence on the host for many functions. Viruses have no systems for the accumulation of metabolic materials. They have no systems for energy generation (mitochondria), protein synthesis (ribosomes), or capture of light energy (chloroplasts). Thus, viruses are dependent on their host for all these functions plus the synthesis of nucleic acids and amino acids.

CELLULAR ASSOCIATION

Viruses and their hosts have a more basic relationship than do other pathogen systems. Viruses are not separated from their host by a membrane during replication (Matthews, 1991). Viruses infect the cellular structure of the host and control a part of the subcellular systems of the plant.

REPLICATION BY ASSEMBLY

Viral replication is dependent on assembly of new particles from pools of required components (Matthews, 1991). These components are synthesized as separate proteins or nucleic acids, using the host enzyme systems and the infecting viral genome. New particles are then assembled by using these materials. This type of replication contrasts strongly with binary fission or other methods of replication found in prokaryotic and eukaryotic organisms.

VECTOR

Plant viruses are not capable of causing an entry wound in order to infect a plant. Thus, they cannot disperse from plant to plant without the assistance of a vector. Plant viruses are dependent on vectors to breach the epidermal layer of the plant and to place them within a living host cell. Virus vectors include insects, mites, nematodes, fungi, seed, and dodder and also humans, animals, and other organisms that transfer viruses through mechanical transmission.

WHAT ARE THE IMPORTANT HISTORICAL DEVELOPMENTS IN PLANT VIROLOGY?

One of the first references to a disease caused by a plant virus occurred during Tulipomania in 15th century Holland (1600 to 1660; Matthews, 1991). This is an unusual case of a virus that increased the value of infected plants. It began with the importation of tulips into Holland from Persia. The beautiful flowers quickly became popular and were grown throughout Holland. People began noticing that some flowers were developing streaks and broken color patterns. These tulips were called bizarres and people quickly learned that they could produce more bizarres by planting a bizarre tulip in a bed or by rubbing the ground bulb onto plain tulip plants. They did not realize that they were transmitting a pathogen. Bizarres became so popular that a single bizarre bulb was worth large sums of money, thousands of pounds of cheese, or acres of land. One case is recorded where a man offered his daughter in marriage in exchange for a single bizarre tulip bulb.

In 1886, Adolf Mayer scientifically confirmed a primary principle of plant virology when he transmitted TMV to healthy tobacco plants by rubbing them with sap from infected plants. The newly rubbed plants displayed the same symptoms as the original infected plants (Scholthof et al., 1999). Mayer's research established the contagious nature of plant viruses and the first procedure for mechanical transmission of a virus. Today, mechanical transmission enables virologists to transmit viruses for experimental purposes and to evaluate plants for resistance and tolerance to viral diseases.

The independent experiments of the Russian scientist Dmitri Ivanowski in 1892 and the Dutch scientist M. W. Beijerinck in 1898 first indicated the unique nature of viral pathogens. Both scientists extracted plant sap from tobacco plants infected with TMV. The sap was then passed through a porcelain bacterial filter that retained bacteria and larger pathogens. If bacteria or other organisms were the cause of the disease, the filter would have retained the pathogen and the sap that had been passed through the filter would not transmit the disease. When Ivanowski or Beijerinck filtered sap from TMV-infected plants and inoculated the filtered sap onto healthy tobacco plants, the plants became diseased. Beijerinck recognized that this indicated the unique nature of this disease and declared that TMV was a new type of pathogen that he called a *contagium fluidium vivium* (a contagious living fluid; Scholthof et al., 1999). Later, scientists discovered that the ultramicroscopic particle nature of viruses was not truly fluid. However, this was the first indication that viral pathogens represented a new and unique type of pathogen.

Clues to this unique nature would wait until the research of W. M. Stanley in 1935. Working with TMV, Stanley extracted gallons of plant sap and used the newly developed technique of fractionation of proteins by precipitation with salts and other chemicals. Each fraction was tested by inoculation on susceptible host to determine where the infectivity remained. Gradually, Stanley isolated the infective fraction into a pure form in which he crystallized TMV. Stanley was the first person to purify a virus, and in 1946, he was awarded the Nobel Prize for this accomplishment (Scholthof, 2001).

Stanley's first analyses of purified TMV solution found protein only. Bawden and Pirie found the presence of phosphorous, which indicated that the solution contained nucleic acid (Bawden et al., 1936). Fraenkel-Conrat was able to isolate the RNA from TMV in 1956 and used it to infect healthy tobacco plants, proving that RNA was the source of infectivity and establishing the RNA contained genome of TMV (Frankel-Conrat, 1956; Creager et al., 1999).

This section has discussed only a few of the most important historical principles in plant virology. However, the brevity of this section should not be used to judge the importance of historical research. The theory and accomplishments of today's research in plant virology is built on the accomplishments of many researchers who preceded today's researchers and their contributions are the foundation for tomorrow's research.

WHAT ARE THE SYMPTOMS FOUND IN PLANTS INFECTED WITH VIRUS?

Symptoms, the host's response to infection, are what attracts the attention of the pathologist, farmer, producer, or homeowner. However, viruses are sometimes referred to as the great imposters because symptoms are often mistaken for other diseases or conditions. Viral diseases may be misidentified as nutritional deficiencies, genetic abnormalities, mineral toxicities, pesticide damage, environmental stresses, insect feeding, or infection by other classes of plant pathogens (Chapter 23). Symptomatology can provide strong first indications of a possible virus infection. However, using symptomatology as the sole basis for diagnosis should be avoided due to the confusion with other possible disease and nondisease conditions.

Classic symptoms of virus infections can be grouped by their similarities. These include the following categories.

SYMPTOMS CAUSED BY LOCALIZED INFECTION

Local Lesions

Local lesions occur when the virus infection fails to spread systemically due to host response. The virus may overcome this initial reaction to spread systemically or it may never spread beyond the initial infection site. Local lesions may be either chlorotic or necrotic. Local lesions can often be used to quantify the infectivity in a virus solution.

SYMPTOMS BASED ON CHANGES IN CHLOROPHYLL OR OTHER PIGMENTS

Mosaics and Mottles

Patterns of lighter and darker pigmentation are referred to as mosaics or mottles. Areas of lighter pigmentation may be pale green, yellow, or white and are caused by decrease in chlorophyll, decrease or destruction of the chloroplasts, or other damage to the chlorophyll system. Other theories suggest that mosaics and mottles may also be due to increased pigmentation or stimulation in the areas of darker green. TMV is a classic example of a virus that can produce mosaic in systemically infected hosts.

Stripes and Streaks

When mosaics occur on monocots, the changes in pigmentation are restricted in spread by the parallel venation of the leaves. Thus, the mosaic appears as a stripe or streak either in short segments (broken) or may extend the entire length of the leaf. Many cereal viruses, such as *Wheat streak mosaic virus* (WSMV) or *Maize dwarf mosaic virus* (MDMV), produce streaks or stripes (see Plate 3A and Plate 3B following page 80).

Ringspots and Line Patterns

Ringspots form as concentric circles of chlorotic or necrotic tissue. *Tomato spotted wilt virus* (TSWV) produces vivid ringspots in many hosts (see Plate 3D following page 80). Line patterns are extensions of this response. They occur as symptomatic patterns near the edge of the leaf and often follow the outline of the leaf. For example, *Rose mosaic virus* produces a striking yellow line pattern on rose.

Vein Banding

Vein banding occurs when areas of intense pigmentation form bordering the veins of the leaves. These typically are seen as dark-green bands along major leaf veins.

Vein Clearing

Vein clearing is a loss of pigmentation (clearing or translucence of tissue) in the veins and can be best observed by allowing light to shine through the leaf. This clearing is caused by enlargement of the cells near the vein in some viruses (Hull, 2002). Vein clearing sometimes precedes the formation of a mosaic.

SYMPTOMS CAUSED BY GROWTH ABNORMALITIES

Stunting and Dwarfing

Reduction in the size of the infected host plant (stunting or dwarfing) can be one of the most prominent symptoms resulting from a viral infection. Some viruses stunt the plant only slightly whereas others affect the host dramatically. Stunting may include changes in sizes of all plant parts such as leaves in addition to height. Stunting that includes shortening of the internodes is often referred to as a bushy stunt.

Tumors and Galls

Hyperplasia (enlargement in cell size) and hypertrophy (increase in cell number) of infected cells can result in large overgrowths or tumors. *Wound tumor virus* (WTV) is an example of a virus that produces tumors in its host.

Distortion

Changes in the lamina of the host leaves result in areas that are twisted, deformed, or distorted. The deformations may be described as blisters, bubbles, rumpling, rugosity, or twisting. *Bean pod mottle virus* (BPMV) produces distortion in infected soybean leaves (see Plate 3C following page 80).

Enations

Enations are small overgrowths occurring on the leaf. *Pea enation mosaic virus* (PEMV) is named for the enations that it produces.

SYMPTOMS AFFECTING REPRODUCTION

Sterility

Host plants infected with plant viruses may lose their ability to produce viable seed. This may be linked with decreases in flowering, seed development, or seed set. The production of sterility is linked with changes in the metabolism of the plant and changes in its biochemical signaling.

Yield Loss

Yield loss can take many forms. It may be a reduction in the total reproduction of the plant, or seed or fruit produced may be shriveled, reduced in size, distorted, or inferior in quality. *Wheat streak mosaic virus* (WSMV) is a good example of a virus that causes yield loss due to undersized and shriveled grain in addition to reducing the total number of seeds set. *Plum pox virus* (PPV) is an example of a virus that not only disfigures the fruit but also reduces the carbohydrate level in the fruit.

WHAT DO VIRUSES LOOK LIKE?

The shape of virons is one of the most fundamental properties of the virus. Plant viruses are based on four types of architecture or morphology. First, icosahedral (isometric) viruses seem to be basically spherical in shape. However, on closer examination, they are not simply smooth but are faceted. Icosahedral viruses have 20 facets or faces. Second, rigid rod viruses are all based on protein coats surrounding a helical nucleic acid strand. Rigid rod viruses are typically shorter and have a greater diameter than flexuous rods. Also, the central canal, an open region in the center of the viral helix, is more apparent in rigid rod viruses. Third, flexuous rods are typically very flexible and may bend into many formations. They are narrow in diameter and are longer than rigid rods. Lastly, bacilliform viruses are short, thick particles (rods) that are rounded on both ends. When particles are found with only one rounded end, they are referred to as bullet shaped.

HOW ARE VIRUSES TRANSMITTED?

Plant viruses cannot penetrate the plant cuticle, epidermis, and cell wall. For experimental transmission, viruses are often transferred by mechanical methods. This is discussed in more detail in Table 4.2 and diagrammatically shown in Figure 4.1. In nature, viruses depend on vectors

TABLE 4.2
Summary of Mechanical Transmission of Plant Viruses

- Mechanical transmission allows the transmission of plant viruses without a vector.
- An abrasive is used to make wounds that penetrate the epidermal layers, cell walls, and membranes. Silica carbide is the most widely used abrasive, but sand, bentonite, and celite have also been used.
- These wounds must not cause the cell to die, because the viruses are obligate parasites and require a living cell. This type of wounding is described as nonlethal.
- Mechanical transmission is effective for viruses that infect epidermal cells. Other viruses, which are limited to other tissues such as phloem, are not mechanically transmissible.
- Some viruses that are highly stable, such as TMV, may be mechanically transmitted through accidental contact between healthy and infected plants wounding the plants and allowing infected plant sap to be transmitted to the healthy plants. Also, contacting healthy plants after handling plants infected with stable viruses may allow you to become the vector by transmitting infected plant sap to the healthy plants.
- Mechanical transmission allows scientists to study the effects of plant viruses without using the vector. It is also used to evaluate new plant cultivars for viral disease resistance.

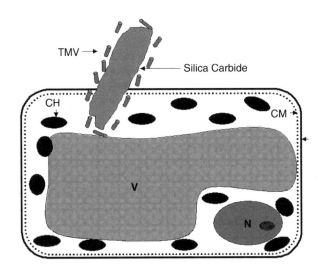

FIGURE 4.1 Mechanical inoculation of plant viruses requires wounding of the plant cell by an abrasive. Silica carbide particles rupture the cell wall and cell membrane (CM) and allow virus to enter the cell. Other structures shown in the cell include nucleus (N), chloroplasts (CH), and central vacuole (V).

to breach these defenses and to allow entry to living cells. Vectors may be insects, mites, nematodes, fungi, or parasitic seed plants. Each virus evolves a unique and specific relationship with its vector. Viruses are dependent on this complex interaction and develop many methods for capitalizing on the biology of their vectors. Thus, understanding the relationship between virus and vector is vital in developing control programs.

INSECTS AS VECTORS

Aphids

Aphids transmit more viruses than any other vector group. Viruses have developed four types of interactions with aphids. These interactions are dependent on the plant tissue infected, the virus association with the vector, and virus replication (or lack of replication) in the vector.

Nonpersistent Transmission

Viruses that are transmitted in a nonpersistent manner infect the epidermal cells of the host plant. This type of transmission is dependent on the sampling behavior of the aphid that quickly probes in and out of the cells in order to determine host suitability. The virus forms a brief association with two sites. The first site is at the tip of the **stylet** and the second is found just before the **cibareal pump** at the top of the stylet. This association lasts only

as long as the next probe of the aphid when it is flushed from the stylet during the process of **egestion**. Nonpersistent aphid transmission requires only seconds for acquisition and transmission and is increased by preacquisition starvation or by any other condition that increases sampling behavior of the aphid. Nonpersistent transmission in some viruses requires the presence of an additional protein called helper component that is theorized to assist in the binding of the virus to the aphid stylet. Potyviruses are among the most important viruses transmitted in a nonpersistent manner.

Semipersistent Transmission

Semipersistent viruses form an association with the lining of the foregut of the aphid. The acquisition of these viruses requires phloem feeding, leading to longer acquisition times than for nonpersistently transmitted viruses, transmission requires minutes, and retention of the virus typically lasts for hours. Virus retention does not persist through the aphid's developmental molts as the lining of the foregut is shed with the rest of the cuticular exoskeleton. Semipersistent transmission may require the presence of a helper component or a helper virus. *Cauliflower mosaic virus* (CaMV) is semipersistently transmitted by aphids and has been studied because of its helper component for transmission (Hull, 2002).

Persistent Circulative Nonpropagative Transmission

All viruses transmitted in this manner must be taken from the phloem of an infected plant and placed in the phloem of a healthy plant. Thus, they are dependent on the phloem-probing behavior of aphids. Approximately 20 min are required for aphids to establish phloem probes.

Thus, the minimum acquisition and transmission times for this type of transmission are 20 min. Viruses transmitted in a persistent circulative nonpropagative manner form a close association with their aphid vectors. The virus moves through the intestinal tract to the hindgut where it passes into the aphid's **hemolymph** and begins to circulate throughout the hemoceol. The virus moves from the hemoceol into the **salivary glands** by passing through the basal membrane of the salivary membrane. The virus is then injected into the healthy plant with the saliva during egestion. Viruses transmitted in this manner are retained for days to weeks. Retention time is correlated with the amount of virus in the hemoceol and the level of virus in the hemolymph is related to the acquisition time that the vector feeds on the infected host. *Barley yellow dwarf virus* (BYDV), one of the most widely distributed viruses in the world, is transmitted in this manner.

Persistent Circulative Propagative Transmission
Persistent circulative propagative transmission has many characteristics in common with persistent nonpropagative virus transmission. Viruses transmitted in this matter are phloem-limited viruses. Transmission is dependent on phloem probes and passage of the virus into the hemolymph. However, after entering the hemolymph, the virus infects and replicates in the aphid. Many tissues of the aphid may be infected. The virus can also pass through the ovaries into the offspring, which are **viruliferous** when they are born. This is termed transovarial passage of the virus. Some viruses in the *Rhabdoviridiae*, such as *Lettuce necrotic yellows virus* (LNYV), are examples of viruses that multiply in their aphid vector (Hull, 2002).

Leafhoppers

Leafhopper transmission of virus parallels aphid transmission with one exception. Nonpersistent transmission does not occur in leafhoppers. Leafhoppers transmit viruses by semipersistent transmission, persistent circulative nonpropagative transmission, or persistent circulative propagative transmission. Important examples of viruses transmitted by leafhoppers include *Beet curly top virus* (BCTV), *Potato yellow dwarf virus* (PYDW), and *Maize stripe virus* (MSV).

Beetles

Beetle transmission is unique among all the other types of virus transmission in that the specificity of virus transmission is not in the ability of the beetle to acquire the virus, but in the interaction of the virus and host after transmission. Beetles can acquire both transmissible and nontransmissible viruses. Both types of viruses may be found in the hemolymph of some beetles. Other viruses are found only in the gut lumen and mid-gut epithelial cells (Wang et al., 1994). However, the specificity of trans-

mission does not depend on these factors. Beetles spread a layer of predigestive material known as **regurgitant** on the leaves as they feed. This layer contains high concentrations of deoxyribonucleases, ribonucleases, and proteases. When viruliferous beetles spread this layer, they also deposit virus particles in the wound at the feeding site. Beetle-transmissible viruses are able to move through the vascular system to an area away from the wound site with its high level of ribonuclease in order to establish infection (Gergerich et al., 1984; Gergerich and Scott, 1988a,b). Nontransmissble viruses are retained at the wound site where the ribonuclease levels inhibit their ability to infect the plant (Field et al., 1994). *Comovirus* and *Sobemovirus* are among the most important viral genera transmitted by beetles.

Whiteflies

The mode of transmission of viruses in whiteflies varies with the genus of virus. *Begomovirus* is transmitted in a persistent circulative manner, resembling aphid transmission of *Luteovirus*. However, this relationship may be more complex than it appears due to the extended retention lengths and the transovarial passage of some species of this virus genus (Hull, 2002). In contrast, *Closteroviruses* and *Criniviruse*s are transmitted in a foregut-borne semipersistent manner (Hull, 2002). Regardless of the virus transmitted, whiteflies present constant challenges as virus vectors due to their dynamic population increases, resistance to control, and changes in their biotype.

Thrips

Virus transmission by thrips has been a dynamic area of research in recent years. The rapid increase in the importance of *Tospovirus* and the diseases that they cause in both greenhouse and field situations has stimulated much of this research. *Tospovirus* is persistently and propagatively transmitted by thrips. Thrips' **larvae** acquire the virus while feeding on virus-infected tissue, and the virus crosses through the midgut barrier and enters the salivary glands. The virus must be acquired by immature thrips (Moyer et al., 1999) and there is no evidence of virus in thrip hemolymph. Thus, viruses cannot travel the same path through the thrip that persistent circulative viruses use in aphids and leafhoppers (Hull, 2002). Thrips retain infectivity for their lifetime and the virus titer has been shown to increase as the virus replicates in the thrip. However, no evidence of virus passage through the egg (transovarial passage) has been found (Moyer et al., 1999). Thrips have also been shown to transmit three viral genera, *Ilarvirus, Sobemovirus*, and *Carmovirus*, by movement of virus-infected pollen. The virus from the infected pollen is then transmitted to the host plant through wounds caused by thrip feeding (Hull, 2002).

OTHER VECTORS

Mites

Eriophyid mites are tiny **arthropods** (0.2-mm length) known to transmit several plant viruses, including WSMV. Mites acquire the virus during their larval stages. As in thrips, adult mites cannot acquire the virus, but both the larvae and adults transmit the virus (Slykhuis, 1955). Mites have been shown to remain infective for over 2 months (Hull, 2002). WSMV particles have been found in the midgut, body cavity, and salivary glands of the mite (Paliwal, 1980). However, there has been no evidence to prove replication of virus in the mite.

Nematodes

Nematodes that transmit plant viruses are all **migratory ectoparasites** (Chapter 8). Three genera of nematodes, *Longidorus, Xiphinema,* and *Trichodorus,* are primarily associated with transmission of viruses. Nematodes feeding on virus-infected plants retain virus on the stylet, buccal cavity, or esophagus. When the nematodes are feeding on healthy host plants, the retained virus is released into the feeding site to infect the new host. Viruses in the *Tobravirus* and the *Nepovirus* genera are transmitted by nematodes.

Fungi and Fungi-Like Organisms

The Chytridiomycete *Olpidium* and the Plasmodiophoromycetes fungal-like *Polymyxa* and *Spongospora* transmit viruses as they infect the root systems of their hosts. Zoospores released from infected plants may carry virus either externally or internally. Viruses absorbed to the external surface of the zoospore, such as the *Tombusviridae,* are released to infect the new plant. Viruses absorbed to the zoospore flagellum can enter the zoospore when its flagellum is retracted to encyst (Hull, 2002). The process through which viruses are carried internally is undefined and remains a topic for future research. *Bymovirus* and *Furovirus* are examples of viral genera that are transmitted in this manner (Hull, 2002). Rhizomania of sugarbeets, caused by *Beet necrotic yellow vein virus* (BNYVV) transmitted by *Polymyxa,* is a good example of a viral disease transmitted in this manner and is a major economic problem.

Seed and Pollen

Viruses may be transmitted in seed by two methods. In the first method, the virus infects the embryo within the seed and the seedling is already infected when it emerges. This is often referred to as true seed transmission. The second transmission method is through contamination of the seed, especially the seed coat. As the germinating seedling emerges from the seed, the virus infects the plant through wounds or through microfissures caused through cell maturation. TMV is transmitted to tomato seedlings by contamination of the seed coat.

Pollen may also transmit viruses. In addition to infecting the ovule during pollination, infected pollen may be moved to uninfected plants and infected during pollination (Hull, 2002). It may also carry virus into wounds. *Barley stripe mosaic virus* (BSMV) is an example of a virus transmitted by pollen.

Dodder

The parasitic seed plant dodder (*Cuscuta* species) sinks haustoria into the phloem of the plants that it parasitizes. This connection allows the carbohydrates and other compounds to move into the dodder's phloem. When a dodder plant connects a healthy and virus-infected host plant, viruses can be transmitted from the infected plant through the dodder to the phloem of the noninfected plant.

Vegetative Propagation and Grafting

Viruses that systemically infect plants can be transmitted by vegetative propagation of a portion of the infected plant. This portion can range from leaves, stems, branches, and roots to bulbs, corms, and tubers. Grafting is a form of vegetative transmission, and transmission occurs through the newly established vascular system linking the graft and the scion (Hull, 2002).

HOW ARE PLANT VIRUSES DETECTED AND IDENTIFIED?

Detection and identification of plant viruses are two of the most important procedures in plant virology. Correct identification of viruses is critical to establish control tactics for the disease. Procedures for the identification of plant viruses can be divided into the following categories.

BIOLOGICAL ACTIVITY

Infectivity assays, indicator hosts, and host range studies are all types of bioassays that are based on defining the interaction of the viral pathogen and its hosts. Infectivity assays measure the number of host plants infected at different dilutions of the virus suspension. Indicator hosts are certain species of plants that have known reactions to a wide range of viruses. *Chenopodium quinoa* and different species and cultivars of *Nicotiana* are widely used as indicator hosts. Host range studies test the ability of the virus to infect different plant species. Some types of viruses have very narrow host range and infect only closely related plants. *Maize dwarf mosaic virus* (MDMV) infects only monocots. TMV infects a wide range of host plants, including plants from the Solanaceae, Chenopodiaceae, and Compositae. Although indicator plants and bio-assays were once used as the principle method of virus identification, they are currently employed only for the primary characterization of new viruses. Another type of biological activity that is an important identification characteristic is the mode of transmission. Identification of the vector association helps indicate the relationship of the unknown virus to characterized groups.

PHYSICAL PROPERTIES

The most important physical property used in detection and identification of viruses is morphology, which is usually determined by electron microscopy. Viruses can be visualized by negative stains of sap extracts or purified virus solutions, thin sections of infected tissue to localize the virus within the cellular structure, and immunospecific electron microscopy that combines electron microscopy and **serology** to capture virus particles on coated electron microscope grids. Visualization of the virus provides the viron shape and size. It can also determine the presence of features such as spikes on the capsid surface.

Other physical properties that have been classically used to identify viruses measure the stability of the particle outside the host. These properties include thermal inactivation point, the temperature at which a virus loses all infectivity; longevity in vivo, the length of time a virus can be held in sap before it loses its infectivity; and dilution endpoint, the greatest dilution of sap at which the titer of virus is capable of causing infection in a susceptible host. Many of these characters take several weeks to obtain and are no longer used as routine diagnostic techniques. However, they continue to be used as part of the official virus description.

PROTEIN

Serology

Serology is based on the ability of the capsid protein to elicit an antigenic response. Areas of the capsid with unique shape and amino acid composition are capable of inducing immune responses in avians and mammals. These uniquely shaped areas are referred to as epitopes and are only a few amino acids in length. Animals injected with plant viruses produce antibodies (**immunoglobulins**) that can recognize and attach to the epitopes. These antibodies can be isolated from the serum or used as the serum fraction (**antiserum**). These antibodies are termed polyclonal antibodies due to the presence of many **antibody** types in the serum. Antisera can be exchanged between researchers around the world in order to compare diseases from many countries. Antisera can also be frozen and used to compare viruses for many years. This allows

researchers to compare viruses or virus strains across time to follow the evolution and epidemiology of the disease.

Monoclonal antibodies are produced by the fusion of an isolated spleen cell from a mouse immunized to the plant virus and a murine myeloma cell. The resulting hybridoma cell line produces only one type of antibody. Advantages to monoclonal antibodies include a single antibody that can be well characterized, identification of the eliciting epitope, production of large amounts of antibody, and the ability of the producing cell line to be multiplied and frozen.

Antisera have allowed the development of many rapid and widely used detection assays. The most commonly used serological assay is enzyme-linked immunosorbant assay (ELISA). Many variations exist in the ELISA procedure, but the most widely adopted protocol is the double-antibody sandwich (Figure 4.2). This procedure starts by trapping a layer of antibodies on the well surfaces of a polystyrene microtiter plate. Attaching the antibodies to a solid surface is important because it allows all reactants that do not attach to the antibody and well to be washed away between steps. The attached antibodies are used to trap virus particles from sap solutions. A second layer of antibodies conjugated to an enzyme is then used to label the virus. Alkaline phosphatase is the most widely used enzyme, but other enzymes, such as horseradish peroxidase, may also be used. After the final washing, a substrate solution is added to the wells. Substrates are chosen because they change color after being acted on by the conjugated enzyme. The color change can be quantified by reading the absorbance of a known wavelength of light when passed through each well. The amount of virus in the test solution is proportional to the light absorbance. ELISA is the basis for the development of many of the rapid diagnostic tests used by growers, producers, and agricultural consultants.

Molecular Properties

The molecular weight and number of proteins forming the capsid layer are also important characteristics of plant viruses and can be determined by standard gel electrophoresis procedures. Determination of the amino acid content and sequence is also important in understanding particle structure and function.

NUCLEIC ACID

The ability to rapidly compare nucleic acid characteristics opened a new dimension in the identification of plant viruses. Identification based on the nucleic acid genome has enhanced the ability to identify strains and new viruses with similar characteristics. Comparisons of viral genomes have allowed the organization of plant viruses

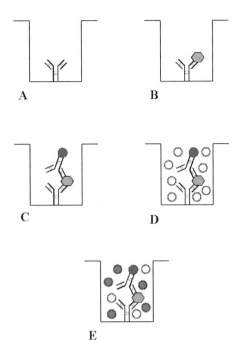

FIGURE 4.2 The principal steps in a double-antibody sandwich (DAS). (A) Antibodies (isolated immunoglobin-G) to a virus are attached to a polystyrene microtiter plate well by incubation with an alkaline pH carbonate buffer. Unattached antibodies are removed from the well by washing with phosphate buffered saline with Tween 20 added. (B) Sap extracted from plant samples is diluted and added to the well. Virus contained in infected samples binds to the matching antibody. Washing removes unbound materials from the well. (C) A second layer of antibodies that have been conjugated to an enzyme is added to the well and attach to the virus. If the virus has not been trapped by the primary layer of antibody, these detecting antibodies are removed by the wash and thus are not present to react in the remaining steps. (D) A substrate solution is added to the wells. (E) The enzyme attached to the detecting antibody causes a color change in the substrate. The intensity of the color is proportional to the amount of enzyme present and can be quantified by spectrophotometry.

into the current taxonomic system and have had important implications in concepts of viral evolution. Nucleic acid tests provide some of the best tools for viral identification. The most powerful of these new tools is the polymerase chain reaction (PCR).

Polymerase Chain Reaction

Polymerase chain reaction amplifies a portion of a nucleic acid genome. This area is defined by the use of short sequences of matching nucleotides called primers. Initially, the use of PCR was limited to DNA viruses; however, the use of reverse transcriptase to generate DNA strands (cDNA) from RNA viral genomes has expanded our ability to use this technology with RNA viruses.

HOW ARE PLANT VIRUSES CONTROLLED?

Control of plant viruses has a different primary focus than that for many other pathogens because there are no practical therapeutic or curative treatments for plant viruses. Thus, plant virus control focuses on preventative measures.

HOST PLANT RESISTANCE

Host plant resistance (Chapter 31) is the major approach to control of viral diseases, but has not been identified in many crop species. It is typically the most economical control measure because it requires low input from the producer. Also, resistant plants eliminate the need for controlling the vector and are selective against the primary pathogen (Khetarpal et al., 1998). Resistance to plant viruses can be due to the inability to establish infection; inhibited or delayed viral multiplication; blockage of movement; resistance to the vector and viral transmission from it (Jones, 1998); and resistance to symptom development, also known as tolerance (Khetarpal et al., 1998). Development of plant cultivars with viral resistance is the primary goal of plant virology and breeding projects.

In addition to traditional breeding, two other methods are available for increasing the resistance of the host plant. In **cross protection**, a viral strain that produces only mild or no symptoms is inoculated as a protecting strain. When a second more severe strain is used as a challenge inoculation, the presence or effects of the first virus block its infection. This effect was first observed by McKinney (1929) and it has been used successfully in the control of some viruses, such as *Citrus tristeza virus* (CTV) in Brazil and South Africa (Lecoq, 1998). Limitations to the usefulness of cross protection include the inability of viruses to cross protect, yield reductions caused by mild strains, and the possible mutation of the mild strain to a more virulent form.

Genetic engineering provides a second method for enhancing host plant resistance. It is particularly valuable in situations where no natural source of resistance has been identified. Genes coding for coat protein, replicase (Kaniewski and Lawson, 1998), **antisense** RNA, and **ribozymes** (Tabler et al., 1998) have been used to confer resistance to different viruses. This technology holds much promise for the future. However, problems with stability of the inserts, expression of the inserted genes, unanticipated effect on the host plant, and public acceptance of genetic engineering remain challenges to this technique.

VECTOR CONTROL

Controlling pests to minimize viral transmission differs from simply controlling pests. To control pests, the principal goal is to reduce the levels of pests below the levels that cause economic damage. However, low levels of pests may still be significant in the transmission of viral diseases (Satapathy, 1998). To determine the suitability of vector control, it is necessary to determine the vector and virus affecting the host (Matthews, 1991). This information establishes the type of relationship involved in transmission. For example, aphids transmitting MDMV require only seconds to accomplish this because the virus is transmitted in a nonpersistent manner. *Oat blue dwarf virus* (OBDV), in contrast, is transmitted by leafhoppers in a persistent circulative propagative manner that requires the leafhopper to feed on the plant for over 20 min. Thus, chemical treatments may be more effective on the leafhoppers transmitting OBDV than on the aphids transmitting MDMV. Also, effects of the controls on the feeding behaviors of the hosts should be considered. Vectors moving into the field from surrounding hosts or due to seasonal migration may affect control measures. Each of these affects the probability for successful control of the virus.

QUARANTINE AND ERADICATION

Control of viral movement between fields, states, countries, and continents is often difficult and time consuming. **Quarantine** acts as the first line of defense against the introduction of many foreign viruses. However, the ability to detect viruses and other submicroscopic pathogens for exclusion is limited in comparison to other pathogens (Foster and Hadidi, 1998). Viruses may enter into new areas through importation of seeds, nursery stock, viruliferous vectors, plants, or experimental material (Foster and Hadidi, 1998). When exclusion of these pathogens fails, the cost of control and **eradication** of the virus (if possible) is often overwhelming. A current example of this is the PPV outbreak in the Northeastern U.S. Costs for this outbreak have already included loss of hundreds of nursery trees, cost of the quarantine (detection, identification, and tree removal), and loss of income to the growers.

CULTURAL CONTROLS

Cultural controls for viruses can be referred to as "wise production practices." These include any production practice or method that eliminates or significantly reduces the threat of viruses. Virus control is based on breaking the bridge of living hosts that is necessary to complete the seasonal movement of a virus. Any cultural practice that modifies the time or spatial interaction of virus and host can be an effective control. One common cultural control is modifying the date of planting. In South Dakota, delaying the planting of winter wheat until the seasonal end of wheat curl mite movement is one of the most effective controls available. However, other concerns such as early planting of winter wheat to increase the amount of residue

cover for the winter conflicts with virus control. Thus, producers must decide the wise balance point between these goals for their farm. Other common cultural controls include production of virus-free seeds and plants through indexing and **seed certification** programs, elimination of volunteer plants, elimination of alternative host plants, elimination of overlap between the planting of two crops, and sanitation of equipment.

Plant virology is a diverse area of plant pathology. Research with plant viruses includes determining the basic molecular structure and mechanisms of viruses, studying the cellular interaction, understanding vector relationships, or working with farmers and producers in the field. However, the ultimate goal of assisting farmers and producers across the world in the control of plant viruses remains the same.

REFERENCES

Bawden, F.C., N.W. Pirie, J.D. Bernal and I. Fankuchen. 1936. Liquid crystalline substances from virus-infected plants. *Nature* 138: 1051–1055.

Brunt, A.A., K. Crabtree, M.J. Dallwitz, A.J. Gibbs, L. Watson and E.J. Zurcher (Eds.). 1996. Plant Viruses Online: Descriptions and Lists from the VIDE Database. Version: 20 August 1996. http://biology.anu.edu.au/Groups/MES/vide/.

Creager, A.N.H., K.B.G. Scholthof, V. Citovsky and H.B. Scholthof. 1999. Tobacco mosaic virus: Pioneering research for a century. *Plant Cell* 11: 301–308.

Field, T.K., C.A. Patterson, R.C. Gergerich and K.S. Kim. 1994. Fate of virus in bean leaves after deposition by *Epilachana varivestis*, a beetle vector of viruses. *Phytopathology* 84: 1346–1350.

Fraenkel-Conrat, H. 1956. The role of the nucleic acid in the reconstitution of active tobacco mosaic virus. *J. Am. Chem. Soc.* 78: 882–883.

Foster, J.A. and A. Hadidi. 1998. Exclusion of plant viruses, pp. 208–229 in *Plant Virus Disease Control*, Hadidi, A., R.K. Khetarpal and H. Koganezawa (Eds.), APS Press, St. Paul, MN.

Gergerich, R.C., H.A. Scott and J.P. Fulton. 1984. Evidence that ribonuclease in beetle regurgitant determines the transmission of plant viruses. *J. Gen. Virol.* 67: 367–370.

Gergerich, R.C. and H.A. Scott. 1988a. The enzymatic function of ribonuclease determines plant virus transmission by leaf-feeding beetles. *Phytopathology* 78: 270–272.

Gergerich, R.C. and H.A. Scott. 1988b. Evidence that virus translocation and virus infection of non-wounded cells are associated with transmissibility by leaf-feeding beetles. *J. Gen. Virol.* 69: 2935–2938.

Hull, R. 2002. *Matthews' Plant Virology*, 4th ed. Academic Press, New York.

Jones, A.T. 1998. Control of virus infection in crops through breeding plants for vector resistance, pp. 41–55 in *Plant Virus Disease Control*, Hadidi, A., R.K. Khetarpal and H. Koganezawa (Eds.), APS Press, St. Paul, MN.

Kaniewski, W. and C. Lawson. 1998. Coat protein and replicase-mediated resistance to plant viruses, pp. 65–78 in *Plant Virus Disease Control,* Hadidi, A., R. K. Khetarpal and H. Koganezawa (Eds.), APS Press, St. Paul, MN.

Khetarpal, R.K., B. Maisonneuve, Y. Maury, B. Chalhoub, S. Dinant, H. Lecoq and A. Varma. 1998. Breeding for resistance to plant viruses, pp.14–32 in *Plant Virus Disease Control,* Hadidi, A., R.K. Khetarpal and H. Koganezawa (Eds.), APS Press, St. Paul, MN.

Lecoq, H. 1998. Control of plant virus diseases by cross protection, pp.33–40 in *Plant Virus Disease Control,* Hadidi, A., R.K. Khetarpal and H. Koganezawa (Eds.), APS Press, St. Paul, MN.

Matthews, R.E.F. 1991. *Plant Virology*, 3rd ed. Academic Press, San Diego, CA.

McKinney, H.H. 1929. Mosaic diseases in the Canary Islands, West Africa and Gibraltar. *J. Agric. Res.* 39: 557–578.

Moyer, J.W., T. German, J.L. Sherwood, and D. Ullman. 1999. An Update on Tomato Spotted Wilt Virus and Related Tospoviruses. APSnet Feature (April 1999). http://www.apsnet.org/online/feature/tospovirus/Top.html.

Paliwal, Y.C. 1980. Relationship of wheat streak mosaic and barley stripe mosaic viruses to vector and nonvector eriophyid mites. *Arch. Virol.* 63: 123–132.

Satapathy, M.K. 1998. Chemical control of insect and nematode vectors of plant viruses, pp. 188–195 in *Plant Virus Disease Control,* Hadidi, A., R.K. Khetarpal and H. Koganezawa (Eds.), APS Press, St. Paul, MN.

Scholthof, K.B. 2001. 1898 — The Beginning of Virology ... Time Marches On. *The Plant Health Instructor*, DOI: 10.1094/PHI-I-2001–0129–01.

Scholthof, K.B., J.G. Shaw and M. Zaitlin (Eds.). 1999. *Tobacco Mosaic Virus — One Hundred Years of Contributions to Virology.* APS Press, St. Paul, MN.

Slykhuis, J.T. 1955. *Aceria tulipae* Keifer (Acarina: Eriophyidae) in relation to the spread of wheat streak mosaic virus. *Phytopathology* 45: 116–128.

Tabler, M., M. Tsagris and J. Hammond. 1998. Antisense RNA and ribozyme-mediated resistance to plant viruses, pp. 79–93 in *Plant Virus Disease Control,* Hadidi, A., R.K. Khetarpal and H. Koganezawa (Eds.), APS Press, St. Paul, MN.

van Regenmortel, M.H. V., C.M. Fauquet, D.H.L. Bishop, E.B. Carstens, M.K. Estes, S.M. Lemon, J. Maniloff, M.A. Mayo, D.J. McGeoch, C.R. Pringle and R.B. Wickner. 2000. Virus taxonomy — Classification and nomenclature of viruses. *Seventh Report of the International Committee on Taxonomy of Viruses.* Academic Press, San Diego, CA.

Wang, R.Y., R.C. Gergerich and K.S. Kim. 1994. The relationship between feeding and virus retention time in beetle transmission of plant viruses. *Phytopathology* 84: 995–998.

Waterworth, H.E. and A. Hadidi. 1998. Economic losses due to plant viruses, pp.1–13 in *Plant Virus Disease Control,* Hadidi, A., R.K. Khetarpal and H. Koganezawa (Eds.), APS Press, St. Paul, MN.

5 Mechanical Inoculation of Plant Viruses

Marie A. C. Langham

Mechanical transmission or inoculation is a widely used technique in experimentation with plant viruses. It can be used to propagate plant viruses, determine **infectivity**, test **host range**, and evaluate plants for host plant resistance. However, successful **mechanical inoculation** is dependent on the condition of the virus-infected tissue source, condition of the host plants for inoculation, inoculum preparation, and the use of an **abrasive**.

The ideal propagation host for the preparation of inoculum is a young systemically infected plant with a high concentration (titer) of virus in the tissue. The host should not develop necrotic symptoms. It should contain neither a high concentration of **inhibitors** that interfere with the establishment of infection nor **tannins** nor other phenolic compounds that can precipitate virus particles. Unfortunately, it is not always possible to use an ideal host. Propagation hosts with less desirable characteristics may require modification of inoculation protocols; for example, using a host with a low virus concentration or one that produces local lesions may require modification of buffer-to-tissue ratios. Additionally, poor growing conditions of an ideal propagation host may cause changes in the virus concentration or in compounds present in the tissue.

Release of virus from the host cellular structure requires **maceration** of infected tissue (Figure 5.1). Small amounts of tissue may be ground in a mortar and pestle. Large amounts are macerated in blenders. Disrupting plant cells not only releases the virus but also breaks the **central vacuole** and other smaller vacuoles, releasing their contents into the extracted sap. The first effect of the vacuole contents is to increase the acidity (lowering the pH) of the plant sap extract. As the pH of the sap decreases, proteins (including viruses with their protein capsid) precipitate out of solution. Thus, buffers are added to the tissue during grinding (Figure 5.2) to stabilize the extract at a pH favorable to virus infectivity, integrity, and suspension. Although the majority of viruses can be transmitted in the pH 7 (neutral) range, the desired pH varies with the inoculating virus. This is not the only role that buffers serve in preparing inoculum. The ionic composition of the buffer contributes to the success of the inoculation. Addition of phosphate to an inoculation buffer increases the infection rate during inoculation, and phosphate is one of the most commonly used buffers (Hull, 2002). Other ions may be added to the inoculation buffer if required for virus stability. For example, some viruses require the presence of calcium or magnesium to maintain their structure and these ions must be added to the inoculation buffer.

Reducing agents, **chelating agents**, and **competitors** are other categories of chemicals that may be added to inoculation buffers for viruses with specific requirements. Sodium sulfite, 2-mercaptoethanol, sodium thioglycollate, and cysteine hydrochloride are examples of frequently used reducing agents. Sodium sulfite is often used in inoculation buffers for tomato spotted wilt virus. Quinones, tannins, and other phenolic compounds in plant sap can cause the precipitation of plant viruses. This activity is countered by chelating agents and competitors. Chelating agents such as diethyldithiocarbamate are added to bind the copper ions found in polyphenoloxidase, the enzyme that oxidizes phenolic compounds in the plant. Competitors such as hide powder have been used to dilute the effects of phenolic compounds by serving as alternative sites for phenolic activity (Matthews, 1991).

Finally, mechanical inoculation is based on our ability to create nonlethal wounds in plant cells for virus entry (Figure 5.3 and Figure 5.4). Wounds in the cell wall and membrane are created by the use of an abrasive. The most commonly used abrasive is silica carbide; however, celite, bentonite, and acid-washed sand have also been used.

EXERCISES

EXPERIMENT 1. TRANSMISSION OF *TOBACCO MOSAIC VIRUS* (TMV)

The objective of this exercise is to demonstrate the mechanical inoculation of a plant virus.

Materials

Each team of students requires the following items:

0-8493-1037-7/04/$0.00+$1.50

FIGURE 5.1 Infected plant material is ground in a mortar and pestle to macerate the tissue as the beginning step in the preparation of sap extracts for inoculation.

FIGURE 5.2 Sap extract is prepared by adding buffer to macerated leaf tissue and grinding.

- TMV-infected host plants (*Nicotiana tabacum* is a good systemic host for providing inoculum)
- Healthy host plants (*N. tabacum*)
- Mortars and pestles
- Inoculation pads [sterile cheesecloth rectangles (approximately 2–3 cm × 10–15 cm)]
- Silica carbide (600 mesh)
- 0.02 *M* phosphate buffer, pH 6.8 to 7.2 (to make add 3.48 g of K_2HPO_4 and 2.72 g of KH_2PO_4 to 1 l of distilled water)
- *Nicotiana tabacum* plants with 3 to 4 well-expanded leaves (4 plants per student or group)

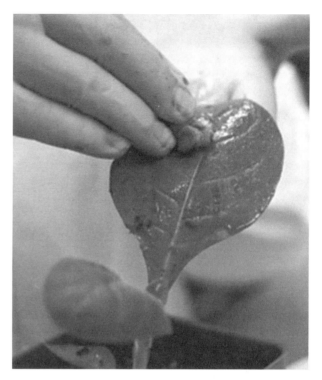

FIGURE 5.3 Mechanical inoculation of *Phaseolus vulgaris* cv. Provider is accomplished by gently rubbing the surface of plant leaves with a mixture of sap extract and silica carbide.

- Plants of *Chenopodium quinoa*, *N. glutinosa*, and *N. tabacum* cv. Samsun NN, and *Phaseolus vulgaris* cv. Pinto
- 10-ml pipettes
- Top-loading balance
- Paper tags and string

Follow the protocol outlined in Procedure 5.1 to complete this part of the experiment.

Procedure 5.1 Mechanical Transmission of Tobacco Mosaic Virus (TMV)

1. Control plants — Collect 0.1 g of healthy plant tissue. Grind the tissue thoroughly in a mortar and pestle. Add 9.9 ml of phosphate buffer. Grind the mixture again. Remove any plant material that has not been thoroughly macerated. Add 1% silica carbide to the sap extract (silica carbide can also be dusted on the leaves with a sprayer; however, adding it to the sap mixture decreases the amount that may be accidentally inhaled). Stir the mixture well and thoroughly wet the inoculation pad. Silica carbide settles to the bottom of the sap extract and must be stirred each time the inoculation pad is soaked. With the soaked inoculation pad, firmly but

gently rub the upper surface of leaves on two tobacco plants. Excessive pressure will result in lethal damage to the epidermal cells from the silica carbide; however, too little pressure will not wound the cells for viral entry.

2. Experimental plants — Repeat Step 1, using infected plant tissue.

3. Observe plants for the next 2 to 4 weeks for symptom development.

Anticipated Results

Plants should be observed closely for two to three weeks. Mechanical damage due to excessive pressure during inoculation appears first and can be observed as necrotic damage to the epidermis. It appears on both the plants inoculated with virus and those inoculated with extracted healthy tissue. Local lesion hosts begin to display lesions in 3 to 5 days. Development of the local lesions from a pinpoint size to a larger size can be observed. Systemic symptoms begin from 10 days to 2 weeks, depending on growth conditions. Symptoms first appear in the newly expanding leaves.

FIGURE 5.4 Mechanical inoculation of tobacco demonstrating the technique for supporting the leaf and coating it with inoculum during inoculation.

Dilution of Inoculum

Experiments to compare the effects of inoculum dilution can be accomplished by using the above procedure. Follow the protocols outlined in Procedure 5.2 to complete this part of the experiment.

Procedure 5.2 Dilution of Plant Extract and Calculation of Infectivity Curve

1. Grind the infected tissue and buffer at the lowest dilution ratio to be used for the experiment as previously described. The remaining dilutions for the series can be prepared by diluting the initial sap extract. For TMV, the dilution series can be 1:50, 1:100, 1:500, 1:1,000 and 1:10,000. An alternative for this experiment is to use *Bean pod mottle virus* (BPMV) or *Tobacco ringspot virus* (TRSV). Dilutions of 1:10, 1:20, 1:50, 1:100 and 1:500 would be good choices for these viruses.

2. With the materials from Procedure 5.1, prepare the initial sap extract by macerating tissue in buffer at a 1:10 ratio. To make a 1:50 ratio, add 1 ml of the sap extract to 4 ml of buffer. Make the remaining dilutions.

3. Hosts that produce local lesions should be chosen for this experiment. For TMV, *Chenopodium quinoa*, *Nicotiana glutinosa*, and *N. tabacum* cv. Samsun NN are good local lesion hosts. For BPMV, *Phaseolus vulgaris* cv. Pinto is a good choice for local lesion assays. If a local lesion-producing tobacco plant with at least four expanded leaves is used, half leaves can be used for comparison in a Latin square design. When using half leaves, a paper tag on a string loop can be used to designate the solution used on each side.

4. Inoculate plants as described in Procedure 5.1. An inoculation of a half leaf with buffer only should be included as a control.

5. Plants should be examined for developing local lesions in 3 to 4 days. Plot number of local lesions against the dilution of sap extract to obtain the **infectivity curve** of the virus.

Anticipated Results

This experiment demonstrates the effects of inoculum dilution on the infectivity of the virus. As the dilution factor of the inoculum increases, the number of local lesions produced by inoculation decreases. The infectivity curve from plotting these results demonstrates this relationship.

TABLE 5.1
Examples of Plant Species Used in Host Range Experiments

Plant species	Reaction[a]
TMV	
Chenopodium amaranticolor	Susceptible
C. quinoa	Susceptible
Cucumis sativus	Susceptible
Cucurbita pepo	Susceptible
Lactuca sativa	Susceptible
Nicotiana glutinosa	Susceptible
N. tabacum	Susceptible
Avena sativa	Not susceptible
Pisum sativum	Not susceptible
Spinacia oleracea	Not susceptible
Triticum aestivum	Not susceptible
Zea mays	Not susceptible
Zinnia elegans	Not susceptible
BPMV	
Glycine max	Susceptible
Lens culinaris	Susceptible
Phaseolus vulgaris cvs. *Black Valentine, Tendergreen,* or *Pinto*	Susceptible
P. sativum	Susceptible
Vigna unguiculata	Susceptible
C. sativus	Not susceptible
N. glutinosa	Not susceptible
N. tabacum	Not susceptible
Petunia × hybrida	Not susceptible
TRSV	
Beta vulgaris	Susceptible
C. amaranticolor	Susceptible
C. sativus	Susceptible
Cucurbita pepo	Susceptible
N. tabacum	Susceptible
Petunia × hybrida	Susceptible
P. vulgaris	Susceptible
Brassica campestris species *Napa*	Not susceptible
Helianthus annuus	Not susceptible
Secale cereale	Not susceptible
Trifolium hybridum	Not susceptible
T. aestivum	Not susceptible

[a]Variation in reaction is possible due to virus isolate and inoculation conditions.

Host Range

Experiments to study host range can be done by using the inoculation procedure in the transmission exercise to inoculate a number of different host plants, including **susceptible** and nonsusceptible ones. Table 5.1 provides some examples of plants for some sample viruses, and the VIDE provides a listing of useful host plants for many viruses (Brunt et al., 1996). Use five plants per species (for each student or group) for the experimental and five for the control group inoculated with extract from healthy sap or with buffer. Follow the protocols listed in Procedure 5.3 to complete this part of the experiment.

Procedure 5.3 Inoculation and Assessment of Host Range

1. Prepare the sap extract as described in Experiment 1 and inoculate in the same manner. Changing inoculation pads between species is a good procedure in case one species is contaminated with a virus. Plants with viscous sap, such as *Chenopodium*, should be inoculated last as the sap may contaminate the extract and inhibit the virus.
2. Observe the plants daily. Record the date of symptom appearance, symptom type, and changes in symptoms as plants mature.

Anticipated Results

The inoculated plants should be observed closely over the next two to three weeks, noting the type of symptom, the date of appearance, and which plant species display symptoms. Compare host ranges among students for variation in these factors. Virus or virus strain variation and variation in inoculation efficiency produce differences in host range. The list of infected plants should be correlated with the plant families represented to note other possible susceptible or nonsusceptible species.

Questions

- Why is nonlethal wounding required for plant virus infection?
- Why do local lesion infections differ from systemic infections?
- Can a single virus particle infect a plant?
- How can host ranges assist in predicting the susceptibility of an untested plant species?
- What is needed to prevent people from becoming plant virus vectors by mechanical in-field or greenhouse situations?
- How can mechanical inoculation be used to develop host plant resistance in crop plants?
- Why should plants inoculated with healthy plant sap be included in all mechanical inoculation studies?

EXPERIMENT 2. OUCHTERLONY DOUBLE DIFFUSION TECHNIQUE TO DETECT AND IDENTIFY VIRUSES

The objective of this exercise is to confirm the presence of stable icosahedral viruses, such as *Bean pod mottle virus* (BPMV) or *Tobacco ringspot virus* (TRSV). This protocol was originally developed from Ball (1990) and from personal communication with Drs. Howard Scott and Rose Gergerich, University of Arkansas.

Materials

Each team of students requires the following items:

- **Agarose** (low EEO)
- 1% solution of sodium azide. *Caution*: Sodium azide is a toxic chemical. Avoid contact with this solution or the agarose gel containing it.
- Two 100-mm × 15-mm petri dishes. (Allow at least two dishes per student team as students often take more than one attempt to make a double diffusion plate.)
- Antisera (diluted to experimental concentrations with phosphate buffered saline)

- Distilled water (or phosphate buffered saline)
- Well punch, cork bore, etc. for making wells
- Humidity box or ziplock bags with damp paper towels
- Pipettes (borosilicate glass) and bulbs
- Light source for observing plates. (Although light boxes or specially constructed double diffusion light boxes are excellent, a flashlight will work well and provides a low-cost alternative.)

Follow the protocol listed in Procedure 5.4 to prepare agarose Ouchterlony dishes.

Procedure 5.4 Preparation of Agarose Plates

1. Mix 1 g of agarose with 98 ml water in a flask (more than twice the volume of the liquid). Record the weight of the flask with solution in it. Heat the mixture slowly in a microwave with an additional beaker of water, stopping to swirl the mixture occasionally. Agarose crystals become clear before they dissolve. Thus, observe the solution carefully while swirling to determine whether the crystals have dissolved. The flask should be observed during the entire heating period to avoid boiling over.
2. When all crystals have dissolved, allow the flask to sit until the steaming has slowed. Weigh the flask a second time and use the heated water in the second beaker to return the agarose solution to its original weight. Water has been lost as steam and its replacement maintains the original concentration of agarose.
3. Add 2 ml of sodium azide to the mixture and swirl gently to avoid bubbles. Use a level surface to pour the plates. Measure 15 ml of the agarose solution into each petri dish. The agarose should solidify without being moved or replacing the lids to keep condensation from accumulating. When cool, cover and store the petri dishes in a humid container.

Preparation of Sap Extracts

Plant sap extracts can be obtained by grinding the tissue in a mortar and pestle in combination with a buffer. They may also be prepared by expressing the sap with a hand squeezer or with a sap extractor. Another alternative is to seal the tissue and buffer in a plastic bag and use a heavy smooth object to macerate the tissue by rubbing the surface. With this method, avoid sealing excessive amounts of air in the bag as this will make maceration more difficult.

Follow the protocols in Procedure 5.5 to complete the experiment.

Procedure 5.5 Preparation of Experimental Ouchterlony Plates

1. Using a pattern of six wells surrounding a central well, punch and carefully remove (by suction or other means) the wells in the agarose plates prepared in Procedure 5.4. Damage to the walls of the wells can result in leaking wells or confused reactions.

2. Mark the well where numbering originates with a small mark on the plate bottom. Mark the alignment of the top and bottom of the dish if using the top lid for marking the wells. In each pattern of six wells, leave one well empty and fill one well with healthy sap as checks for false and healthy plant reactions. Fill wells by using a small amount of fluid in the pipette until the surface of the fluid is level with the surface of the agarose. The level can be determined by watching the light reflectance off the fluid surface. When the fluid is level, the light reflectance will dull rather than shine as it does when the surface is convex or concave. Fill the outer wells first. Add antiserum to the central well last. Determine optimum concentrations for sap extracts and antisera prior to the exercise. After filling, do not move plates until the extracts and antisera have begun to be absorbed into the agarose. Gently place the plate in a humid container and allow it to incubate undisturbed.

3. Bands of precipitation begin to form within 24 hours and continue to grow stronger over the next week (Figure 5.2). The observation of bands is improved by shining a light through the bottom of the plate. Record the reactions and any spur formation that is observed.

Anticipated Results

After filling the plate, diffusion of the extracts and antiserum begins. When compatible antibodies and antigens come together, the antibody attaches to the antigen or antigens. The typical antibody has two attachment sites, making it possible for it to bind to more than one antigen. The virus particle repeats antigenic sites on each identical protein subunit that allows it to bind to multiple antibodies. As more and more antibody and antigen bind, the mass begins to precipitate. This precipitant forms the typical white band in positive reactions. An example of typical results in an Ouchterlony can be found in Figure 5.5. Bands begin to form within 24 h and grow more intense over time. An extension of the precipitin line beyond where it meets the line formed by a neighboring well is sometimes seen. This is caused when one of the test sam-

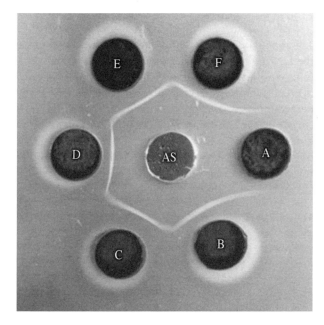

FIGURE 5.5 Double diffusion plate with six wells surrounding the center antiserum (AS) well demonstrates the white precipitin bands formed in the plate by a positive reaction. The antiserum used was made to detect *Bean pod mottle virus* (BPMV). Sample A is sap from a healthy bean plant. Sample B to Sample F are from beans infected with BPMV.

ples contains some antigens recognized by the antiserum that are not in the neighboring sample. This extension is called a spur and its formation can indicate a strain difference in the virus contained in the two samples.

Questions

- Why should a healthy sample be included in each test?
- What advantages does serological identification of the virus provide?
- Why would a precipitin line form closer to the antiserum well?
- Why would a precipitin line form closer to the antigen well?
- How can this technique be used in addition to an inoculation technique to provide more information about an unknown virus sample?
- What would happen if a test sample were infected with more than one virus?

Notes

These exercises were written as examples of exercises that can be used in a laboratory and are based on exercises that have been used in my classes. They can be modified to meet the requirements of your classes. One of the most important modifications could be the choice of virus. Examples of common viruses have been provided. However, these viruses may not be available to you. The VIDE provides information on host range that can be used in modifying the given exercises. If you do not have any plant viruses available, TMV can often be isolated by grinding the tobacco from a cigarette that has not been treated to add flavor or reduce the tar or nicotine and inoculating the extract on a susceptible tobacco. Dilution may also require adjustment depending on your infected plant tissue.

If you desire to also use a viral detection exercise, the choice of virus for use in this exercise should be based on the antisera and virus combination available to your class, because this is often a limiting factor in this exercise. Double diffusion is applicable for many icosahedral viruses. However, some viruses, such as the *Cucumovirus*, require additional ions to form precipitin bands. Flexuous and rigid rod viruses diffuse through the agarose gel very poorly and require the addition of sodium dodecyl sulfate (SDS) to remove the capsid from the particles. SDS also solubilizes the precipitin band that forms in plates after approximately 24 h and makes these plates less useful in classes with laboratories meeting once a week. For those who would like to demonstrate how Ouchterlony double diffusion works but do not have the antisera available, the following protocol, provided by Dr. Robert N. Trigiano, demonstrates double diffusion by forming a chemical precipitate in the agarose.

Simulation of the Ouchterlony Technique without Viruses and Antisera

Antisera to specific viruses can be expensive and cost prohibitive to use in introductory plant pathology with many students. Also, maintaining virus-infected plants may not be possible or desirable for any number of reasons. The simple technique described in this experiment demonstrates the basic principles of diffusion and precipitation found in the Ouchterlony test. It requires a minimum amount of materials that are inexpensive and readily available.

Materials

Each student or team of students requires the following materials to complete the experiment:

- 0.1 M silver nitrate solution (1.70 g in 100 ml of distilled water). *Caution*: Silver nitrate is a strong oxidizing agent — wear gloves and avoid contact with this solution.
- 0.1 M sodium chloride solution (0.58 g in 100 ml of distilled water)
- Number 3 cork borer or plastic drinking straws
- Two eyedroppers or bulbs with glass Pasteur pipettes

- Plastic 60-mm diameter petri dishes containing 0.6% water agar (6 g agar per liter of distilled water. Autoclave at 121°C for 15 min and dispense about 15 ml per dish.)
- Distilled water

Follow the protocol in Procedure 5.6 to complete this experiment.

Procedure 5.6 Demonstration of the Ouchterlony Double Diffusion Technique with Silver Nitrate and Sodium Chloride

1. Using a Number 3 cork borer or plastic straw, remove agar plugs from two petri dishes as depicted in Figure 5.2.
2. For petri dish 1, fill the center well with silver nitrate solution. Next, fill three of the outside reservoirs with sodium chloride solution and the remaining three outside wells with distilled water. Use different pipettes for each of the fluids.
3. For petri dish 2, fill the center well with sodium chloride solution. Next fill three of the outside reservoirs with silver nitrate solution and the remaining three wells with distilled water. Use different pipettes for each of the fluids.
4. Label drawings of each of the dishes and record the results.

Anticipated Results

White precipitate should form between the center well and the surrounding wells within 15 to 30 min, depending on the temperature of the room. The precipitate is analogous to the interaction of antisera with the antigen on the coat protein of the virus.

ELISA

A basic **ELISA** protocol is not included in these exercises (see Chapter 36). For laboratories that have the antisera and supplies, ELISA protocols are available from many sources, such as Converse and Martin (1990). Any ELISA protocol requires modification based on the activity of available antisera. Another option for laboratories with antisera is dot **immunobinding** assays (Hammond and Jordan, 1990). These assays require shorter incubation periods and are often more adaptable to laboratory schedules. For laboratories that are not doing ELISA routinely, the equipment, antisera, supplies, and time investment for a basic protocol is considerable. To minimize the investment needed with ELISA for laboratories that must buy all components or for ELISA with a virus for which the antiserum is not available, AGDIA produces a large variety of kits utilizing ELISA with several enzyme detection systems or immunostrips that adapt easily to classrooms. AGDIA can be contacted at the following address: AGDIA, Inc., 30380 County Road, 6 Elkhart, IN 46514. Tel.: 1–574–264–2014 or 1–800–62-AGDIA; fax: 1–574–264–2153; e-mail: info@agdia.com.

REFERENCES

Ball, E.M. 1990. Agar double diffusion plates (Ouchterlony): Viruses, pp. 111–120 in *Serological Methods for Detection and Identification of Viral and Bacterial Plant Pathogens,* Hampton, R., E. Ball and S. DeBoer (Eds.), APS Press, St. Paul, MN.

Brunt, A.A., K. Crabtree, M.J. Dallwitz, A.J. Gibbs, L. Watson and E.J. Zurcher (Eds.). 1996. Plant Viruses Online: Descriptions and Lists from the VIDE Database. Version: 20 August 1996. http://biology.anu.edu.au/Groups/MES/vide/.

Converse, R.H. and R.R. Martin. 1990. ELISA methods for plant viruses, pp. 179–196 in *Serological Methods for Detection and Identification of Viral and Bacterial Plant Pathogens,* Hampton, R., E. Ball and S. DeBoer (Eds.), APS Press, St. Paul, MN.

Hammond, J. and R.L. Jordan. 1990. Dot blots (viruses) and colony screening, pp. 237–248 in *Serological Methods for Detection and Identification of Viral and Bacterial Plant Pathogens,* Hampton, R., E. Ball and S. DeBoer (Eds.), APS Press, St. Paul, MN.

Hull, R. 2002. *Matthews' Plant Virology,* 4th ed. Academic Press, New York.

Matthews, R.E.F. 1991. *Plant Virology,* 3rd ed. Academic Press, San Diego, CA.

6 Pathogenic Prokaryotes

George H. Lacy and Felix L. Lukezic

CHAPTER 6 CONCEPTS

- Phytopathogenic prokaryotes are a diverse group including over 30 genera.

- Fastidious phytopathogenic prokaryotes are recognized as more economically important as methods improve for their detection and identification.

- Epiphytic, nonpathogenic resident phase of plant colonization is important for pathogen survival and disease control.

- Biofilms composed of bacteria and extracellular polysaccharides, such as zoogloea, are important in pathogen colonization, survival, and dissemination.

- Chemical control of diseases caused by phytopathogenic prokaryotes is made difficult because of short generation times.

- Cultural control of diseases caused by phytopathogenic prokaryotes is critical because chemical controls may not be adequate.

CHARACTERISTICS OF PHYTOPATHOGENIC PROKARYOTES

All **prokaryotes** have cell membranes, cytoplasmic 70S ribosomes, and a nonmembrane-limited nuclear region (Figure 6.1). Phytopathogenic prokaryotes have a unique character — the ability to colonize plant tissues. Pathogens increase from a low to a high number in a remarkably short time within a diseased host plant. The essence of pathogenicity is this ability to live and multiply within plant tissues.

TAXONOMY

Taxonomy is a classification devised by humans of living organisms according to their current understanding of phylogenetic groupings. Current understanding of phylogenetics has been determined by a variety of methods, each with its own strengths and weaknesses. Therefore, taxonomies have strengths and weaknesses and change in response to development of the new technologies used to explore natural relationships among prokaryotes. Listed next are some groups of phytopathological prokaryotes, a brief description of the groups, and some of the diseases they cause.

GENERA OF PLANT PATHOGENIC PROKARYOTES

Gram-Negative Aerobic/Microaerophilic Rods and Cocci

- *Acetobacter* — Ellipsoid- to rod-shaped organisms motile by peritrichous or lateral flagella that oxidize ethanol to acetic acid. *Acetobacter aceti* and *A. diazotrophicus* cause pink disease of pineapple.
- *Acidovorax* — Rods motile by a polar flagellum. *Acidovorax avenae* subspecies are pathogenic on oats (subsp. *avenae*), watermelon (subsp. *citrulli*), and orchids (subsp. *cattleyae*).
- *Agrobacterium* — Rods motile by 1 to 6 peritrichous flagella. *Agrobacterium* species cause proliferations on many plants: smooth and rough galls (*A. tumefaciens*), root proliferations (*A. rhizogenes*), and galls on berry roots (*A. rubi*). Tumorigenic *Agrobacterium* species carry parasitic plasmids vectoring tumor DNA (T-DNA).
- *Bradyrhizobium* — Rods motile by one polar or subpolar flagellum. *Bradyrhizobium japonicum* is a phytosymbiotic, nitrogen-fixing bacte-

0-8493-1037-7/04/$0.00+$1.50
© 2004 by CRC Press LLC

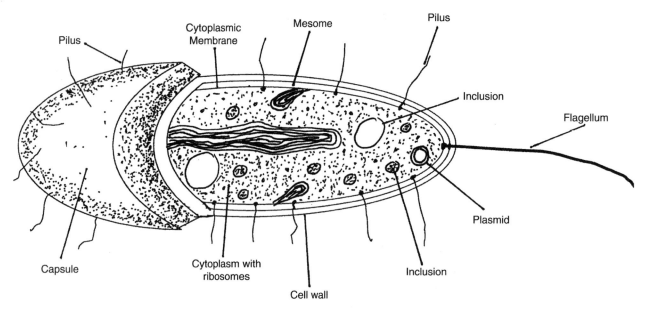

FIGURE 6.1 Three-dimensional view of a bacterium.

rium on soybean. Under high soil nitrogen conditions, it produces phytotoxic rhizobitoxin and damages its host.

- *Burkholderia* — Straight rods motile by polar flagella. *Burkholderia cepacia* is an onion pathogen and *B. gladioli* is pathogenic on gladiolus and iris. *Burkholderia cepacia* may cause disease in persons with cystic fibrosis. It is not known whether strains causing plant disease also cause human disease.
- *Gluconobacter* — Ellipsoid- to rod-shaped, occurring singly and in pairs, either nonmotile or motile by several polar flagella, and oxidizing ethanol to acetic acid. *Gluconobacter* species causes discolorations in apples, pears, and pineapple fruit.
- *Herbaspirillum* — Usually vibroid, occasionally helical, motile by 1 to 3 flagella at one or both poles. Strict aerobes, associated with roots. *Herbaspirillum rubrisubalbicans* is a pathogen of cereal roots, and was previously known as *Pseudomonas rubrisubalbicans*.
- *Pseudomonas* — Rods motile by one or more polar flagella. Many plant pathogens produce water-soluble pigments that fluoresce light blue to greenish-yellow with ultraviolet light. *Pseudomonas marginalis* is a soft-rotting pathogen of many plants. *Pseudomonas syringae*, subdivided into many pathovars (pvs.), includes the leaf spot pathogens on soybean (pv. *glycinea*) and field beans (pv. *phaseoli-*

cola). *Pseudomonas syringae* pv. *syringae* causes warm temperature frost damage (WTFD).

- *Ralstonia* — Rods motile by one or more polar flagella. *Ralstonia solanacearum* causes wilts on solanaceous crops (potato, tobacco, and tomato), peanuts, and banana.
- *Rhizobacter* — Rods nonmotile or motile by polar or lateral flagella. *Rhizobacter daucus* causes bacterial gall of carrot.
- *Rhizomonas* — Rods motile by one lateral, subpolar, or polar flagellum and accumulates polyhydroxybutyrate granules. *Rhizomonas suberifaciens* causes corky root of lettuce.
- *Xanthomonas* — Rods motile by one polar flagellum. Colonies are yellow due to xanthomonadin pigments. *Xanthomonas campestris* has over 140 pathovars, including important pathogens of rice (pv. *oryzae*), field beans (pv. *phaseolus*), and cole crops such as cabbage (pv. *campestris*).
- *Xylella* — Long flexuous rods, nonmotile, nutritionally fastidious inhabitants of insects and plants. *Xylella fastidiosa* is a leafhopper-vectored, xylem-limited bacterium causing scorches of almond, elm, and oak as well as citrus leaf blight (see Plate 2C following page 80).
- *Xylophilus* — Rods motile by one polar flagellum. *Xylophilus ampelinus* causes bacterial necrosis and canker of grapes.

Facultative Anaerobic Gram-Negative Rods

- *Brennaria* — Rods occur singly, in pairs, and occasionally in short chains. Various species cause diseases of alder, walnut, banana, oak, and cricket-bat willow.
- *Enterobacter* — Rods motile by peritrichous flagella. *Enterobacter cloacae* causes brown discoloration of papaya fruits. *Enterobacter cloacae* is also an opportunistic pathogen of humans, causing problems during burns and wounds and in urinary tracts. It is not known whether strains causing plant disease also cause human disease.
- *Erwinia* — Rods motile by peritrichous flagella. This group is divided into the "true erwinias," including nonpectolytic species such as *E. amylovora*, which causes fire blight of pear and apple, (see Plate 2B following page 80) and the soft-rotting "pectobacteria" (described later).
- *Klebsiella* — Nonmotile, straight rods. These pectolytic plant-associated bacteria (*K. planticola* and *K. pneumoniae*) are rarely associated with soft rots. *Klebsiella pneumoniae* might also cause human disease, but it is unknown whether strains that cause plant disease also cause human disease.
- *Pantoea* — Rods motile by peritrichous flagella. *Pantoea stewartii* causes Stewart's wilt of sweet corn. Many strains of epiphytic *P. agglomerans* (e.g., *E. herbicola*) cause warm temperature frost damage to many plants.
- *Pectobacterium* — Rods motile by peritrichous flagella. This group is separated from the "true" erwinias. *Pectobacterium* species are strongly pectolytic. *Pectobacterium* (*Erwinia*) *carotovora* and *P. chrysanthemi* cause soft rot in numerous plant hosts.
- *Serratia* — Straight rods usually motile by peritrichous flagella. *Serratia marcesens* causes yellow vine of cucurbits. *Serratia marcesens* is also a pathogen of insect larvae and an opportunistic pathogen of humans, causing septicemia and urinary tract problems. It is not known whether strains causing plant disease also cause human disease.

Irregular, Nonsporing Gram-Positive Rods

- *Arthrobacter* — Young cultures contain irregular rods often V-shaped with clubbed ends. In older cultures, rods segment into small cocci. Cells are nonmotile, Gram-positive, but easily destained. *Arthrobacter ilicis* causes bacterial blight of American holly.
- *Clavibacter* — Irregular, wedge-, or club-shaped, nonmotile rods arranged in pairs or in V shapes or palisades. *Clavibacter michiganensis* and subspecies (*michiganensis*) cause tomato canker [ring rot of potato (*C. sepedonicus*)].
- *Curtobacterium* — Young cultures have small, short, irregular, nonmotile rods in pairs arranged in V shapes. In older cultures, rods segment into cocci are Gram-positive, but easily destain. *Curtobacterium flaccumfaciens* pv. *poinsettiae* causes poinsettia canker.

Nocardioform Actinomyces (Gram-Positive, Acid Fast, and Nonmotile)

- *Nocardia* — Partially acid-fast, branching, mycelium-like hyphae with coccoid spore-like units. Colonies have sparse aerial mycelia. *Nocardia vaccinii* cause galls on blueberry.
- *Rhodococcus* — Rods to extensively branched substrate mycelium. Weakly acid-fast. *Rhodococcus fascians* is the causes fasciation of sweet pea, geranium, and other hosts.

Streptomyces (Gram-Positive, Fungus-Like Bacteria)

- *Streptomyces* — Vegetative, extensively branched hyphae. Aerial mycelium matures to form three or more spores in chains. *Streptomyces scabies* and *S. acidiscabies* cause potato scab.

Endospore-Forming, Motile Gram-Positive Rods

- *Bacillus* — Large rods motile by peritrichous flagella with oval central endospores. Strongly Gram-positive, aerobic to facultative anaerobes. *Bacillus* species may cause rots of tobacco leaves, tomato seedlings, and soybean and white stripe of wheat.
- *Paenibacillus* — Resemble *Bacillus* and cause rots of terminal buds in date palms.
- *Clostridium* — **Obligate anaerobes**, rods motile by peritrichous flagella with subpolar to polar endospores. Plant pathogenic, *Clostridium* species are usually Gram-negative despite typical Gram-positive cell walls. Pectolytic *Clostridium* species are associated with wetwood of poplar and elm and soft rots of potato and carrot.

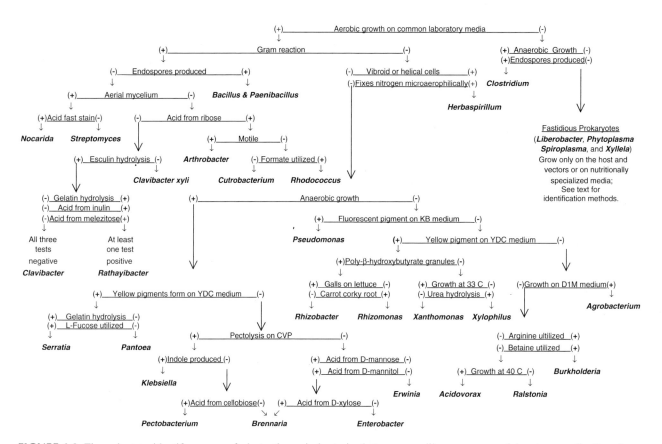

FIGURE 6.2 Flow chart to identify genera of phytopthogenic bacteria that grow readily on common laboratory media [based on Schaad et al. (2001) and Holt et al. (1994)]. Fastidious bacteria are not dealt with here. The genera *Acetobacter* and *Gluconobacter* contain opportunistic bacteria that affect fruit only and are not treated here. This scheme assumes that the bacteria tested are pathogens. In other words, Koch's postulates have been satisfied or approximated.

Mollicutes

Mollicutes lack cell walls, phloem-limited, fastidious mollicutes vectored usually by leafhoppers.

- Phytoplasma — Polymorphic. The aster yellows phytoplasma (see Plate 2D following page 80) causes proliferation or yellows in over 120 host plants. These organisms are identified using PCR-amplified elements of ribosomal RNA (rRNA).
- Sprioplasma — Spiral shape supported around a central protein "rod" embedded in the membrane. Plant pathogenic spiroplasmas cause stubborn disease of *Citrus* species (*S. citri*) and corn stunt (*S. kunkelii*).

Organisms of Uncertain Affiliation

- Phloem-Limited Bacteria — A group of insect-transmitted, phloem-limited bacteria have been

poorly studied because they cannot be cultured in the laboratory. Apparently, they represent a wide selection of bacteria. *Liberobacter* species include the citrus-greening pathogens.

RAPID IDENTIFICATION OF PHYTOPATHOGENIC PROKARYOTES

Rapid identification of prokaryotes for disease control differs from taxonomic description. Speed rather than precision is the goal. The clinician identifies pathogens only to the level required to understand what is needed for disease control. Dr. Norman W. Schaad has compiled the *Laboratory Guide for Identification of Plant Pathogenic Bacteria*. This authoritative reference presents a combination of selective media, diagnostic tests, and serological and PCR methods to guide the clinician to successful identification of prokaryotes from diseased plants. Figure 6.2 shows an abbreviated flow chart for identifying plant pathogenic bacteria.

ANATOMY AND CYTOCHEMISTRY OF PROKARYOTES

FLAGELLA (SING. FLAGELLUM)

Flagella are narrow, sinuous structures 12 to 30 nm wide and 3.7 mm long composed of flagellin (protein) subunits arranged helically. Flagella propel bacteria through water at speeds approaching 50 μm/sec. This is accomplished by the counterclockwise motion of flagella, which pulls (rather than pushes) the bacterium. A flagellar "motor" called the basal body is located in the cell wall and requires biological energy equivalent to 64 NADH oxidations or the formation of 128 ATP molecules per flagellar rotation. A system that requires such high energy must be significant in the disease process. Flagella of a *Pseudomonas fluorescens* stimulating plant growth enhanced colonization of potato roots. Motile strains of ice-nucleation-active (INA) *P. syringae* pv. *syringae* more efficiently colonized expanding bean leaves. Motile strains of *Erwinia amlyovora* caused a greater incidence of disease in apple.

PILI (SING. PILUS)

Pili are rod-like protein structures (length 0.2 to 2 mm and diameter 2 to 5 nm) composed of helically arranged pilin units. Pili are often correlated with parasitic plasmids in bacteria, and the genes controlling pili production are often located on these plasmids. For the plasmid, pili function in the process of bacterium-to-bacterium transfer of plasmid DNA (type IV secretion of proteins and nucleic acids). The transfer mechanism for T-DNA from tumorigenic *Agrobacterium* species and strains to plants is also accomplished by type IV secretion. Other pili are encoded within the hypersensitive response/pathogenicity (hrp/pth) region of the bacterial genome and involved with secretion of proteins into or close to plant cells.

EXTRACELLULAR POLYSACCHARIDES (EPSS)

EPSs are high-molecular-weight carbohydrates either loosely attached (slime polysaccharides are easily removed by washing) or more tightly attached (capsular polysaccharides are less easily removed). *Erwinia amlyovora* EPS is made up of 98% galactose. Bacterial cells embedded in EPS are extruded from lenticels and dried into very fine filaments, which are dispersed by air currents over long distances. *Pseudomonas syringae* pv. *glycinea* produces acetylated alginate EPSs in diseased leaves. The best-studied EPS is xanthan gum (MW 10^6 Da) produced by *Xanthomonas campestris* pv. *campestris* and composed of D-glucose, D-mannose, and D-glucuronic acid with some pyruvate. Xanthan gum, important for black rot pathogen of crucifers, is also used commer-

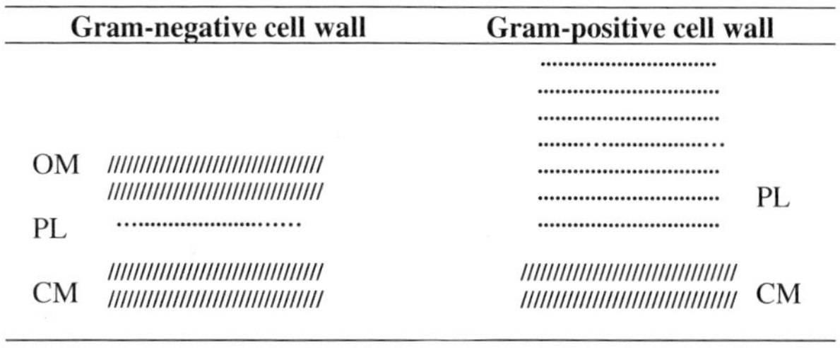

FIGURE 6.3 Schematic comparison of Gram-negative and Gram-positive cell walls. Gram-negative cell walls are 15 to 20 nm thick and composed of a 1.4-nm-thick peptidoglycan layer sandwiched in the periplasm, a space between the 7.5 nm-thick cellular lipid bi-layer (////) outer (OM) and inner cellular membranes (CM). Gram-positive cell walls are thicker (22 to 25 nm) with a dense peptidoglycan layer (15 nm) enclosing a single 7.5–nm-thick cellular membrane.

cially as an emulsifer in foods and is found in ice cream, salad dressings, and sauces. EPS is a primary pathogenic determinant in wilting plants. *Clavibacter michiganense* subsp. *insidiosum* (alfalfa wilt), *Ralstonia solanacearum* (Granville wilt of banana, peanut, and tobacco), *Erwinia amylovora* (fire blight pathogen of pear and apple), and *Pantoea stewartii* (Stewart's wilt of maize) block xylem vessels in wilting plants with high-molecular-weight EPSs.

CELL WALL

Cell walls are Gram-negative, Gram-positive, or absent in prokaryotes. Cell walls protect bacteria against osmotic extremes. Low osmotic events occur during water dispersal or colonization of xylem sap (<1 mg/ml dissolve nutrients). Phytoplasmas and spiroplasmas lack cell walls but live in protected high osmotic phloem sap (30 mg/ml dissolved nutrients) and insect hemolymph. Most plant pathogenic bacteria are Gram-negative. The cell wall of Gram-negative bacteria consists of an outer lipid membrane sandwiching a peptidoglycan layer over an inner lipid membrane. The peptidoglycans provide cell rigidity. Gram-positive bacteria have much thicker peptidoglycan layers and lack the outer membrane. Figure 6.3 schematically compares Gram-negative and Gram-positive cell walls.

PROTEINS

Proteins play many important roles in plant–pathogen interactions. For instance, acetosyringone from wounded plants diffuses to and interacts with receptor proteins to trigger a cascade of T-DNA-interacting proteins that configure *Agrobacterium tumefaciens* to transfer T-DNA into host plants. Porins and permeases allow bacteria to move nutrients and wastes across cell membranes. Binding proteins concentrate substrates in proximity to permeases. Enzymes associated with the cell wall, such as ATPase, provide energy for secretion. Pectate lyases and proteases from soft rot pathogens [*Pectobacterium* (*Erwinia*) *carotovora*] digest pectic materials and plant proteins, respectively, releasing nutritional components to the pathogen. Other proteins are structural in nature. Lipoproteins bind cell membranes to the peptidoglycan layer, and actins in *Spiroplasma citri* form the central spindle around which the spiral-shaped cell is supported. Finally, in *Pseudomonas syringae* pv. *syringae* and *Pantoea agglomerans*, causal agents of WTFD, ice-nucleation proteins and cell-surface proteins catalyze ice formation. Plant cells broken by ice crystals leak nutrients to the pathogens.

LIPOPOLYSACCHARIDES (LPSs)

LPSs are anchored into the outer membrane of bacteria. Specificity is imparted depending on which sugars are incorporated and linkage of the sugars one to another. LPS bind bacteria to their plant hosts if the charges and shape of bacterial LPSs complement sugar moieties on plant glycoproteins. The interaction between plant glycoproteins (also known as lectins) and bacterial lipopolysaccharides gave rise to the lectin hypothesis for binding bacteria to host plants. For instance, soybean binds most *Bradyrhizobium* strains at high frequencies. However, soybean lacking lectin is also nodulated at low frequencies by the bacterium, suggesting that LPSs are not wholly responsible for bacterial binding.

MEMBRANE-BOUND PIGMENTS

Bacteria contain lipid-soluble carotenoid pigments. In xanthomonads, pantoeas, and curtobacteria, carotenoid pigments have been shown to protect cells from light damage. The major protection mechanisms are quenching of triplet sensitizer, quenching of singlet oxygen, and inhibition of free radical reactions. Other roles suggested for pigments include modifying membrane permeability, altering antibiotic sensitivity, electron transport, and enhancing enzyme activity.

CYTOPLASM

Bacterial cytoplasm seems homogeneous; however, on careful examination, several structures are apparent. Mesosomes, invaginations of the cytoplasmic membrane, occur with septa formation; bacterial nuclear region, skeins of double-stranded DNA (dsDNA) fibrils are visible by transmission electron microscopy; storage granules may be observed and include glycogen, poly-hydroxbutyric acid (HBA, or lipid) and metachromatic granules (polymerized inorganic metaphosphate); and endospores are present. Endospores are special survival structures composed of a thin exosporium; layers of protein; a cortex

of dipicolinic acid, peptidoglycan, and calcium; a cytoplasmic membrane; and desiccated cytoplasm containing ribosomes, mRNA, and DNA. Endospores of *Bacillus*, *Paenibacillus*, and *Clostridium* survive at 80°C for 15 min.

HOW PROKARYOTES DAMAGE PLANTS

Plant pathogenic prokaryotes cause soft rots with enzymes; tissue proliferations with phytohormones; wilts with extracellular polysaccharides; and leaf spots, blights, and necroses with toxins and enzymes.

ENZYMES

Enzymes are catalytic proteins. Cell degrading enzymes (CDEs) produced by plant pathogenic bacteria reduce plant components to compounds useful for pathogen nutrition (Table 6.1; see also Chapter 27). Enzymes may be located extracellularly and secreted into the environment, intracellularly in the cytoplasm, inserted into membranes or periplasmically (between the cytoplasmic and the outer membranes of Gram-negative bacteria or within the cell wall of Gram-positive bacteria). Coordinated batteries of enzymes are required for plant pathogenesis.

Pectic Enzymes

Pectate lyase is the major pectolytic enzyme involved in bacterial soft rot pathogenesis by species of *Pectobacterium* (*Erwinia*), *Pseudomonas*, and *Xanthomonas*. *Pectobacteria* have a battery of several pectate lyases that have different locations in the cell (extracellular or periplasmic), pH optima, endo- or exoabilities, and digestion products (pentamers, dimers, or monomers of galacturonic acid). Pectate lyases are major determinants of disease. X-ray diffraction studies suggest that efficient folding makes pectate lyase molecules small enough to diffuse into spaces (about 4 nm) between cellulose fibrils in plant cell walls. Pectin-degrading enzymes also have a role in pathogenesis. Pectate lyases cannot degrade pectin (methylated pectic acid). Pectin methylesterase demethylates pectin to pectic acid. Some bacteria, including *Ralstonia solanacearum*, have polygalacturonases that degrade pectin directly.

Cellulases

Degradation of crystalline cellulose requires C_1 cellulase to cleave cross-linkages among cellulose fibrils, C_2 endocellulase to break primary cellulose polymers, C_x cellulase [endo-(β1,4 glucanase] to cleave soluble cellulose into cellibiose, and cellobiase [β-glucosidase (β-glucanase)] to degrades cellobiose to D-glucose. The Granville wilt pathogen, *Ralstonia solanacearum*, has several cellulases (endoglucanases).

TABLE 6.1
Plant Cell Structures, Putative Nutrients or Macromolecular Building Blocks Provided to the Pathogen, and Corresponding Cell-Degrading Enzyme Found in the Pathogen

Affected Plant Cell Structure	Nutrient for Pathogen	Degradative Enzyme
Cuticle		
Cutin	Fatty acid peroxides	Cutinase
Suberin		
Suberin	Fatty acid polyesters	Suberin esterase
Cell wall		
Pectic substances	Galacturonans	
	Nonmethylated	Pectate lyase
		Oligogalacturonase
	Methylated	Pectin methylesterase
		Pectin lyase
		Polygalacturonase
Cellulose	Glucose monomer	Cellulases
Native cellulose cross links		C_1 cellulase
Native cellulose main strand		C_2 cellulase
Soluble cellulose to cellibiose		C_x cellulase
Cellibiose to glucose		β-glucanase
Hemicellulose	β-1.4-linked xylans	Xylanases
Proteins		
Cytoplasmic membrane (and other organelle membranes)	Complex structure	
Proteins	Polypeptides	Proteases, proteinases
Phospholipids	Phospholipids	Phospholipase
Phosphatidyl compounds	Phospholipids	Phosphatidase
DNA	3-deoxy polynucleotides	Deoxyribonucleases (DNases)
RNA	Ribopolynucleotides	Ribonucleases (RNases)

Proteases (Proteinases)

These enzymes cleave proteins into peptones and peptides. Protein degradation products such as di- and tripeptides are assimilated by bacteria. Proteins in plants occur in cell walls, cell membranes, and cytoplasm. Proteases apparently coordinate with pectate lyases to produce a synergistic increase in soft rot damage by digesting extensins, hydroxyproline-rich cell wall structural proteins, and loosening plant cell wall so that endopectate lyases penetrate more effectively into its cellulose–pectin matrix.

Cutinases and Suberin Esterases

Cutinases and suberin esterases digest cuticular waxes and suberin, respectively. These enzymes are produced by bacteria as diverse as *Pseudomonas syringae* pv. *tomato* and *Streptomyces scabies* and aid in bacterial penetration of the host and nutrition.

Toxins

Bacteria produce many metabolites that may harm plants. However, toxins are those metabolites present during pathogenesis in concentrations great enough to cause plant damage. Although toxins extend the amount of damage a pathogenic bacterium may cause, the presence of a toxin alone is not enough to make a bacterium a pathogen. Toxins enter plants by several methods. Syringotoxin, produced by *Pseudomonas syringae* pv. *syringae*, is a cyclic peptide with a hydrophobic lipid tail. The lipid tail dissolves into the plant cell membrane and the cyclic peptide causes an aberration in the membrane, causing it to leak. Phaseolotoxin, produced by *P. syringae* pv. *phaselolicola*, and tabtoxin, produced by *P. syringae* pv. *tabici*, resemble di- and tripeptides and are taken up by the plant as nutrients. Tagetitoxin, produced by *P. syringae* pv. *tagetis*, affects chloroplast thylakoid membranes and inhibits chloroplast RNA polymerase. Rhizobitoxin, produced by *Bradyrhizobium japonicum*, inhibits production of homocysteine and blocks ethylene production. The toxic degradative product of phaseolotoxin, octicidin or psorn, inhibits production of citrulline, causing ornithine to accumulate. Ornithine blocks arginine and citrulline biosynthesis, resulting in disruption of chloroplast membranes. Tabtoxinine, the toxic degradative product of tabtoxin, inhibits glutamine synthetase, causing ammonia accumulation and chlorosis.

WILTS

Xylem sap, with a constant flow of dilute nutrients (<1 mg/ml aqueous sugars, organic acids and amino acids), is an excellent habitat for a plant pathogenic bacterium. Bacteria that colonize xylem vessels and cause wilts are a heterogeneous group. These organisms include *Pantoea stewartii*, causing Stewart's wilt of maize; *Clavibacter michiganense* subsp. *michiganense*, causing tomato canker; *C. michiganense* subsp. *insidiosum*, causing bacterial wilt of alfalfa; *Ralstonia solanacearum*, causing Granville wilt of tobacco, other solanaceous plants, bananas, and peanut; and *Xanthomonas campestris* pv. *campestris*, causing blackrot of crucifers such as cabbage. Wilts are caused by high-molecular-weight polysaccharides restricting the flow of xylem sap by collecting on endplates and lateral pits of xylem vessel cells.

TUMORIGENESIS

This group of pathogens includes those that produce tumors (galls) as well as overproduce (proliferating) roots or shoots. Common to production of galls and proliferations of roots or shoots are plant growth regulators, especially auxins and cytokinins. Tumor-inducing bacteria include *Agrobacterium tumefaciens,* causing crown galls on many plants (see Plate 2A following page 80); *Rhizobium* and *Bradyrhizobium* species, causing nitrogen-fixing nodules on legume roots; *Pseudomonas syringae* pv. *savastanoi*, causing olive and oleander knot; and *Nocardia vaccini*, causing galls on blueberries. Bacteria that cause proliferations of plant organs include *A. rhizogenes* (root proliferation on many plants), *Rhodococcus fascians* (proliferation of lateral shoots on many plants including pea), and, most likely, phytoplasmas that cause "witches' broom."

ECOLOGY OF PLANT PATHOGENS

EPIPHYTES

Epiphytic bacteria obtain their nutrition on the surfaces of plants. Because most plant pathogenic bacteria lack specialized resistant structures to survive environmental conditions not conducive to pathogenesis, plant pathogens must coexist as epiphytes either on their host or on other plants without causing disease. Thus, **endophytic** growth probably represents a survival stage (Figure 6.4). Examples of bacteria with endophytic residence phases include *Pectobacterium* (*Erwinia*) *carotovora*, which colonizes potato lenticels and causes bacterial soft rot, and *Pseudomonas syringae* and *Pantoea agglomerans*, agents of WTFD.

ENDOPHYTES

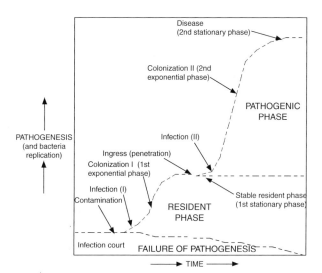

FIGURE 6.4 Pathogenesis and the resident phase of bacteria. This figure indicates steps in pathogenesis based on the growth curves of bacteria (dashed lines and consisting of lag, exponential and stationary phases). One sequence, including contamination, infection (I), and colonization (I; first exponential phase), leads to establishment of a stable resident phase population on or in plants (first stationary phase). This resident phase may extend for days or months before the pathogenic phase is indicated. A second sequence, ingress, infection (II) colonization (II; second exponential phase) and disease (indicated by the second stationary phase), occurs when conditions are correct for pathogenesis. In this figure, infection and the establishment of a nutritional association with the host is indicated by the end of the lag phases of growth and the points where bacterial growth curves begin to increase.

Endophytic plant pathogens colonize epidermal cells (e.g., *Streptomyces scabies*, causal agent of scab of potato), apoplast or the space outside living protoplasts, including cell walls and free space (e.g., *Pseudomonas syringae* pv. *phaseolicola*, causal agent of haloblight of bean), xylem vessels (e.g., *Ralstonia solanacearum*, causal agent of Granville wilt), and phloem (e.g., *Spiroplasma citri*, causing citrus stubborn disease). The pathogens causing WTFD (*Pseudomonas syringae* and *Pantoea agglomerans*) do not fit the four ecological niches indicated. Instead, in WTFD, pathogenesis may be a joint effort of epiphytic and apoplastic bacteria. Ice nucleation most likely occurs on the phylloplane, but the cells of the pathogen that benefit from nutrients leaking from ice-damaged cells are those colonizing the apoplast.

Nutritional Classes of Endophytes

Phytopathogenic prokaryotes are nutritionally nonfastidious (may be cultured on general laboratory media), nutritionally fastidious (require special, chemically complex ingredients) or obligate endophytes (cannot be cultured on media at this time). Nonfastidious endophytic prokaryotes may be cultured on simple agar media containing chemically defined inorganic or simple organic compounds. They have some portion of their disease cycle separate from the host plant and insect vector — as plant epiphytes or soil saprophytes. They include toxin-producing pathogens (e.g., *Pseudomonas syringae* pv. *phaseolicola*), enzyme-producing pathogens [e.g., *Pectobacterium* (*Erwinia*) *carotovora*], tumor-inducing pathogens (e.g., *P. syringae* pv. *savastanoi*), and wilt-inducing pathogens (e.g., *Clavibacter michiganense* subsp. *michiganense*). Fastidious phloem- or xylem-limited prokaryotes vectored by insects do not have any part of their life cycle outside of their plant and insect hosts. These fastidious organisms include *Xyllela fastidiosa* (leaf scorch of many hosts) and *Spiroplasma citri* (citrus stubborn disease) as well as obligate endophyte *Liberobacter asiaticum* (citrus greening) and *Phytoplasma* species (yellows and proliferations).

PHYLLOSPHERE

The **phyllosphere** is the environment created by leaf surfaces. For this discussion, the phyllosphere encompasses all open plant surfaces, including stems and flowers, in contact with the atmosphere.

NATURE OF PROKARYOTES IN THE PHYLLOSPHERES

Bacteria found in the phyllosphere originate from soil, water, other plants, seeds, pollen, or dust, or they may be vectored by insects. Population numbers are reduced compared to the soil and vary from 10^{2-6}/cm^2. Casual inhabitants are bacteria derived from airbiota deposited from the soil and water. Populations of these organisms, although occasionally very high, do not establish nutritional relationships with plants and decrease with time. Provided a nutritional relationship is established and populations are maintained over a lengthy period of time, organisms are classified as resident. Because populations of residents may fluctuate greatly over time, this relationship may be difficult to recognize.

ENVIRONMENT

The phyllosphere is a harsh environment. Temperature fluctuates up to 35°C because the atmosphere does not provide as much insulation as soil. Ultraviolet and infrared radiation may be lethal to bacteria. On the upper (adaxial) leaf surfaces, infrared radiation causes heating, and ultraviolet radiation, at 256 nm, is bactericidal and mutagenic. Bacteria escape some radiation by persisting on the lower (abaxial) leaf surfaces. Bacteria require free water for replication, but on the leaf surface, the relative humidity (RH) also varies greatly each day. Environmental rigors

cause wide fluctuations in bacterial populations on plant surfaces. Surprisingly, the epiphytes *Pseudomonas syringae* pv. *syringae* and *Pantoea agglomerans* multiply even during times of low humidity. Hydrophilic and absorptive bacterial EPSs may be important in maintaining holding water. The nutrient supply is also limited. Water of guttation from hydathodes supplies less than 1.0 mg/ml combined amino acids, carbohydrates, and organic acids. Additional nutritional input from the plant is derived from leakage via ectodesmata directly through the waxy cuticle and deterioration of the cuticle itself.

Leaf attachment is important for survival and successful pathogenesis. For example, *Pseudomonas syringae* pv. *lachrymans*, causal agent of angular leafspot of cucumbers (*Cucumus sativus*), adheres more tightly to its host than to nonhost plants. On host plant leaves inoculated by dipping into a suspension of a pathogen, up to 90% of the colony forming units (cfu) are retained after gently washing. However, 60 to 89% cfu are removed from leaves of nonhost cultivars.

Biofilms are important to attachment as well as survival. In the phyllosphere, biofilms are limited to aggregations of cells embedded in EPS. EPSs are important for attachment to phylloplane surfaces because they retain absorbed water when plant surfaces are dry. As bacteria require free water for growth and replication, EPSs are an extremely important part of survival. For dissemination, bacterial ooze (EPSs mixed with bacterial cells) is exuded from stoma, lenticels, nectaries, and hydathodes, where it is available for water splash dispersal.

RHIZOSPHERE AND RHIZOPLANE

The **rhizosphere** is the volume of soil affected biologically, chemically, or physically by the presence of roots. The **rhizoplane** is the two-dimensional surface of the root. Soil is a rich environment. Agricultural soils contain more than 10^9 bacteria/g. Bacteria colonize anticlinal intercellular depressions on the surfaces of elongating roots. Only about 10^2 bacteria/cm^2 root are on the rhizoplane because of the geometry of the surface and bacterial microcolonies. Rhizosphere soil is enriched for Gram-negative bacteria and contains fewer actinomyces than soil away from roots. Competition is intense, with predatory bacteria, protozoa, nematodes, and mites harvesting bacteria. Plant products affect the metabolism of root-associated phytopathogenic bacteria. Exudates, including most commonly low-molecular-weight metabolites and inorganic ions, leak passively from living cells. Pectins and hemicelluloses are secreted into the root cap and the cells in the elongation zone. Lysates are compounds released by the autolysis of the older epidermal cells and include sugars, amino acids, nucleotides, enzymes, growth factors, and terpines. Diffusion of substances from roots stimulates bacterial motility by nutrient concentration gradients. Motility is restricted to water films clinging to soil particles and is limited especially at the low water potentials encountered near the root.

PENETRATION OF PLANT SURFACES

PASSIVE PENETRATION

Bacteria enter plants mostly through breaches in physical barriers. Natural openings such as stoma, hydathodes, and lenticels also provide avenues of entrance as well as areas of higher moisture for bacterial colonization. During plant development, wounds develop by the growth of lateral or adventitious roots, abscission of leaves, or the breaking of root hairs during root elongation among abrasive mineral particles and provide sites for bacterial entry. Wounding, whether by developmental processes or other processes, bypasses the integrity of the host surfaces.

ACTIVE PENETRATION

Zoogloea are masses of phytopathogenic bacteria embedded in EPSs living in the apoplastic spaces in the roots, stems, leaves, and vascular systems of plants. These zoogloea represent biofilms trapped among parenchymous cells. Zoogloea expand with bacterial growth and replication and EPS production. Therefore, zoogloea exert hydrostatic pressure that forces bacteria into adjacent spaces among cells. Very likely, enzymes that degrade the cell wall aid zoogloeal separation of cells.

Other enzymes may aid direct penetration. Cutinases degrade cutin protecting epidermal plant cells. The potato scab pathogen *Streptomyces scabies* produces cutinase that allows direct penetration of tuber epidermal cells.

DISEASE CONTROL

HOST RESISTANCE

Plants are not universally compatible with disease development. A pathogen on its susceptible host cultivar comprises a compatible interaction and disease will develop provided environmental conditions permit. Conversely, a pathogen on a resistant or nonsusceptible host cultivar yields an incompatible interaction. Breeding or engineering host plants for disease resistance is the most important method to control plant diseases.

PREFORMED RESISTANCE

Plants remain attractive ecological niches and are underexploited nutrient sources practically free of competition

from microorganisms. The barriers to basic compatibility are strong because only about 200 of 2400 named species of bacteria cause plant disease. Preformed resistance, or passive plant defense to entry and colonization, resembles ancient defenses of plants to prokaryotes. Physical barriers include the cuticle, epidermis, plant cell walls, turgor and hydrostatic pressure within vascular elements, special anatomic features of natural plant openings, and inhibitory chemicals. These defense mechanisms have broad activity and protect plants against entry and colonization by most prokaryotes. Plants have developed methods to prevent damage and possible entry of plant pathogens by making plant surfaces less appealing for colonization. Modified trichomes (leaf hairs) form gland cells containing antimicrobial terpenoids. The root tip is protected by rapidly dividing cells in the root cap. Their lubricating ability proceeds from the ease with which they are sloughed off and the softness of the pectic gel containing them. These cells are sacrificed to protect the actual root tip from abrasion damage by mineral particles or attachment or penetration by pathogens. Plants may also develop protective symbioses. Epiphytic microbial populations in the rhizosphere and phyllosphere often have protective properties, which resist through microbial competition or antagonism colonization by the resident phases of phytopathogenic bacteria or fungi. Ecto- and endomycorrhizal associations provide additional protection from pathogens.

Preformed chemical barriers protecting plants from pathogens must occur in the plant in a form not damaging to the plant itself, occur at a concentration effective against the target pathogen, and be either in its active form or be able to be readily released in an active antimicrobial form. In resistant pea cultivars, antimicrobial homeoserine occurs in concentrations inhibitory to *Pseudomonas syringae* pv. *pisi*. Glycosylation, the addition of sugar moieties to toxic compounds, renders these compounds nontoxic within the cytoplasm of the plant. Glucosidases, stored in plant vacuoles, deglycosylate or remove the sugars. Thus, in pathogen-damaged cells, the contents of vacuoles mix with the cytoplasm resulting in the formation of the aglycone (deglycosylated) toxin. The aglycone suppresses pathogen growth.

INDUCED NONSPECIFIC RESISTANCE SYSTEMS

In the coevolution of plants with their pathogens, some pathogens have developed the ability to evade passive or preformed defenses. In response, plants had to develop resistance in a different way. In induced resistance, challenge of the plant by living or dead pathogens or certain chemicals induces a general resistant response against several types of pathogens. It is now clear that at least three types of nonspecific induced resistance occur: systemic acquired resistance (SAR) signaled by salicylic acid, induced systemic resistance (ISR) signaled by jasmonic acid, and riboflavin-induced systemic resistance (RIR). These systems often induce pathogenesis-related proteins, including chitinases and endoglucanases in plants (Chapter 28).

Specific Induced Resistance

Specific induced resistance, or the hypersensitive reaction (HR), to pathogens requires that the plant chemically recognize the pathogen to trigger its defenses (see also Chapter 28). Systems in which HR occur typically involve specialization of both pathogens and hosts, including the development of races of the pathogen and resistant plant lines or cultivars within the susceptible plant species. Incompatible pathogens are recognized by the plant and induced HR prevents disease development. Compatible pathogens are not recognized by the plant and may cause disease because no HR is triggered. HR is apoptosis (programmed plant cell death) triggered by the interaction between a plant and a pathogen incompatible with disease development in that plant. Plant cells in the immediate vicinity of the site of bacterial ingress die rapidly, preventing further spread of the pathogen into the plant.

Recognition of an incompatible pathogen by a plant results from the identification of a chemical signal (effector) from the pathogen by the host. Basic to the concept is the existence of specific receptor proteins on the plant surface that detect the effector and trigger a multiplying cascade of signals inducing many genes in the resistant response. Genes for phytoalexins (antibiotic compounds) as well as pathogenesis proteins are induced. Bacterial effectors are encoded by and secreted through a type III secretion system of the hypersensitive response and pathogenicity (hrp/path) region of xanthomonads, pseudomonads, erwinias, and pectobacteria. Effectors are secreted through pili either very close to or within plant cells. Effectors such as harpin, encoded by hrpN in *Erwinia amylovora*, may have separate roles in pathogenesis.

Engineering resistance into plants

The importance of receptor proteins (derived from R genes; see Chapter 31) is clear. The Pto locus (an R gene) was engineered into a cultivar susceptible to tomato spot. Transfer of Pto caused the susceptible line to become resistant to *Pseudomonas syringae* pv. *tomato* strains containing the effector protein locus avrPto. Similarly, a rice R gene, Xa21, was introduced to rice cultivars grown on over 9 million hectares (22 million acres) in Asia and Africa and provides resistance to rice blight caused by *Xanthomonas oryzae* pv. *oryzae*.

Cultural Control

Diseases caused by bacteria occur very rapidly due to short generation times of the pathogen and tend to follow wet weather closely, when fields and orchards are difficult or impossible to work with heavy machinery and are usually sequestered in tissue and protected from treatment with bactericides. Consequently, they are difficult to manage by chemicals alone. Fastidious prokaryotes are sequestered in phloem and xylem and are often difficult to eradicate as chemical delivery to these locations is irregular, resulting in pockets of surviving pathogens. Therefore, prophylactic management by cultural methods is very important for bacterial diseases of plants. Cultural controls include using pathogen-free propagating materials, controlling weed hosts, removing diseased plants (roguing or pruning), doing crop rotation, avoiding overhead irrigation, and planting wind breaks.

Chemical Control

A few chemical controls, including copper- and zinc-containing fungicides, have utility against bacteria. Antibiotics are low-molecular-weight compounds usually produced by microbes as antagonists of other microbes. Streptomycin, which interferes with the translation of bacterial proteins, has been useful as a control for fireblight, caused by *Erwinia amylovora* and pepper spot, caused by *Xanthomonas campestris* pv. *vesicatoria*. High-level resistance arises in bacteria in a single mutation of ribosomal subunit protein gene. In many apple- and pear-growing regions, streptomycin is now useless for fire blight control. Oxytetracycline (OTC) is less effective than streptomycin in fire blight control. OTC also inhibits protein synthesis, but low-level resistance develops slowly by stepwise chromosomal mutation. Currently, gentamycin has the best potential for controlling fire blight; however, environmental and health reviews and registration are incomplete.

SUGGESTED READINGS

Andrews, J.H. and R.F. Harris. 2000. The ecology and biogeography of microorganisms on plant surfaces. *Annu. Rev. Phytopathol.* 38: 145–180.

Barras, F., G. van Gijsegem and A.K. Chaterjee. 1994. Extracellular enzymes and pathogenesis of soft-rot erwinia. *Annu. Rev. Phytopathol.* 32: 201–234.

Beattie, G.A. and S.E. Lindow. 1995. The secret life of foliar bacterial pathogens on leaves. *Annu. Rev. Phytopathol.* 33: 145–172.

Burr, T.J. and L. Otten. 1999. Crown gall of grape: Biology and disease management. *Annu. Rev. Phytopathol.* 37: 53–80.

Denny, T.P. 1995. Involvement of bacterial polysaccharides in plant pathogenesis. *Annu. Rev. Phytopathol.* 33: 173–197.

Dow, M., M.-A. Newman and E. von Roepenak. 2000. The induction and modulation of plant defense responses by bacterial lipopolysaccharides. *Annu. Rev. Phytopathol.* 38: 241–261.

Hahn, M.G. 1996. Microbial elicitors and their receptors in plants. *Annu. Rev. Phytopathol.* 34: 387–412.

Hammerschmidt, R. 1999. Phytoalexins: What have we learned after 60 years? *Annu. Rev. Phytopathol.* 37: 285–306.

Holt, J.G., N.R. Krieg, P.H.A. Sneath, J.T. Staley and S.T. Williams. 1994. *Bergey's Manual of Determinative Bacteriology,* 9th ed. Williams & Wilkins, Baltimore.

Kinkel, L.L. 1997. Microbial population dynamics on leaves. *Annu. Rev. Phytopathol.* 35: 327–347.

Madigan, M.T., J.M. Martinko and J. Parker. 2000. *Brock: Biology of Microorganisms,* 9th ed. Prentice Hall, Upper Saddle River, NJ, p. 991.

Mount, M.S. and G.H. Lacy (Eds.) 1982. *Phytopathogenic Prokaryotes,* Vols. 1–2, Academic Press, New York.

Romantschuk, M. 1992. Attachment of plant pathogenic bacteria to plant surfaces. *Annu. Rev. Phytopathol.* 30: 225–243.

Salmond, G.P.C. 1994. Secretion of extracellular virulence factors by plant pathogenic bacteria. *Annu. Rev. Phytopathol.* 32: 181–200.

Schaad, N.W., J.B. Jones and W. Chun. 2001. *Laboratory Guide for Identification of Plant Pathogenic Bacteria,* 3rd ed. American Phytopathological Society Press, St. Paul, MN.

Young, J.M., Y. Takikawa, L. Gardan and D.E. Stead. 1992. Changing concepts in the taxonomy of plant pathogenic bacteria. *Intl. J. System Bact.* 30: 67–105.

7 Laboratory Exercises for Plant Pathogenic Bacteria

George H. Lacy and Felix L. Lukezic

Phytopathogenic prokaryotes are very diverse. Further, several different phytopathogenic prokaryotes and fungi cause similar-appearing diseases that confuse identification of the specific pathogens. Finally, phytopathogenic prokaryotes may easily be overgrown by saprophytic bacteria and fungi. Therefore, they are surprisingly difficult to isolate from plant materials.

Identification of the actual pathogen is essential for selecting the best method(s) for disease control. Methods for isolation, purification, and characterization, including pathogenicity assays, of phytopathogenic prokaryotes are very important. This chapter illustrates some of these basic methods. See Schaad et al. (2001) for more details.

EXERCISES

EXPERIMENT 1. OBSERVATIONS OF BACTERIA IN PLANT TISSUE

Direct microscopic observation for the pathogen is important to determine whether bacteria are involved in plant disease. Selection of the diseased tissues is important because pathogenic bacteria may occupy different locations in the plant. For instance, leaf spotting bacteria are found in the apoplastic spaces in leaves, wilting pathogens are found in vascular tissues, and soft rotting pathogens are found among macerated and dead cells. Some prokaryotes, especially the mollicutes (including phytoplasmas and spiroplasmas), which lack a cell wall, and phloem-limited bacteria are very small and difficult to observe microscopically unless their DNA is stained with 4', 6'-diamidino-3-phenylindole (DAPI; Lee et al., 2001). Mature phloem sieve tubes, the plant host habitat for these pathogens, lack nuclei, allowing detection of the pathogen DNAs. DAPI staining is beyond the scope of this exercise. Simple methods for visualizing bacteria from plant tissues are the subject of this protocol.

Materials

Each student or team of students will require the following materials:

- Compound light microscope with 40× or 60× objective lens
- Microscope slides and coverslips
- Razor blade
- Eyedropper bottle with distilled water
- Infected plant material — We suggest the following bacteria and diseases for this exercise, although other materials will also be appropriate: *Ralstonia solanacearum* (Granville wilt of tobacco), *Pseudomonas syringae* pv. *angulatum* (angular leaf spot of tobacco) and/or *Pectobacterium* (*Erwinia*) *carotovora* (hand rot of tobacco/soft rot of potato).

Follow the protocols outlined in Procedure 7.1 to complete the experiment.

Procedure 7.1 Observing Bacteria in Diseased Tissue

1. Tissue selection — Lesions are often colonized by secondary organisms that may be either saprophytes living on dead material or secondary pathogens. To detect the primary pathogen, select tissue that has recently been colonized. Select tissue at or near the active margin of lesions to avoid observing saprophytic bacteria.
2. Cut tissues — With a single blade razor, slice strips of tissue thin enough to observe microscopically. (*Hint*: Make oblique sections of stems, fruits, tubers, and flowers.) Place the strips on a microscope slide in a drop of water.
3. Microscopic evaluation — Observe the tissue at 400× to 600× magnifications. Locate and focus sharply on the interface of the cut edge of the tissue and the water. Bacteria are not resolved at this magnification; they are only visible by refracted light and the Brownian movement of cells. Reduce the amount of light by either closing the iris beneath the stage or lowering the condenser. Bacteria streaming out of the plant tissue will be visible as bright points of light

by refraction. Use apparently healthy tissue (lacking bacteria) as a control.

Questions

- What is Brownian motion and how does it differ from cell motility?
- Why would microscopic evaluation of tissue for bacterial streaming not be used for crown gall disease caused by *Agrobacterium tumefaciens*?

EXPERIMENT 2. ISOLATION OF BACTERIA FROM PLANT TISSUE

A critical step for working with plant pathogenic bacteria is obtaining a pure culture of the organism. Pure cultures are absolutely essential for pathogenicity assays and characterizing the pathogen for identification. General methods for isolation and purification are the subject of this protocol.

Materials

Each student or team of students will require the following items:

- Infected plant material — We suggest the following bacteria and diseases for this exercise, although other materials will also be appropriate: *Ralstonia solanacearum* (Granville wilt of tobacco), *Pseudomonas syringae* pv. *angulatum* (angular leaf spot of tobacco) and/or *Pectobacterium* (*Erwinia*) *carotovora* (hand rot of tobacco/soft rot of potato).
- 10% commercial bleach solution (90 ml of water and 10 ml of bleach; equal to 0.5 to 0.6% sodium hypochlorite)
- Distilled or sterile distilled water
- Several 12- × 75-mm snap cap sterile test tubes
- Glass rod
- Several sterile transfer pipettes
- 1.5-ml eppendorf tubes and micro centrifuge
- Nutrient agar [8 g powdered medium (Difco, Detroit, MI), 20 g agar and 1 l of water]
- Bacterial loop
- Incubator set for 25°C (optional)

Follow the protocol outlined in Procedure 7.2 to complete this experiment.

Procedure 7.2 Isolation of Pathogenic Bacteria

1. Select tissue to be sampled. Tissue selected for sampling can be any plant part with symptoms. Remove any unnecessary material. For example, do not use the whole leaf, but a portion of the leaf with the suspect lesion surrounded by some apparently healthy tissue. Select the portion to fit the container used in Step 3.
2. Surface-disinfest plant organs with 0.5 to 0.6% sodium hypochlorite for 30 to 60 sec, and then rinse three times with freshly deionized, distilled, or autoclaved water.
3. Immerse the tissue in 1 to 2 ml of distilled water in a clean or sterile 12- × 75-mm snap cap tube. Crush the tissue with a clean glass rod or other convenient implement. Let the mixture stand for 5 to 10 min to allow any bacteria to diffuse out of the tissue and into the water.
4. Agitate the mixture briefly to suspend any bacteria evenly in the water. Allow the mixture to stand until the larger portions of remaining tissue settle to the bottom of the tube. Using a transfer pipette, remove as much of the supernatant as possible to one or more sterile, 1.5-ml eppendorf tubes.
5. Briefly centrifuge at 14 to 17 K for 30 to 60 sec in a microcentrifuge to pellet bacteria. Gently pour or pipette the supernatant off allowing only the pellet and about 50 to 100 µl to remain.
6. Streak directly from the remaining volume onto either a general bacteriological medium such as nutrient agar. Consult Schaad et al. (2001) for the medium and temperature of incubation recommended for the pathogen you expect to isolate. Incubate the medium for 72 to 96 h at 25°C.
7. Select colonies that are well isolated and represent the most numerous colony morphology (Figure 7.1). Generally, pathogens are in high populations in lesions.
8. Restreak the selected colony at least once onto fresh isolation medium. Each time, select isolated colonies representing the most numerous colony morphology. Bacteria isolated from nature may be contaminated with saprophytic species; restreaking for isolation ensures a pure culture.
9. Bacterial cells from isolated colonies may be suspended in 15% sterile aqueous glycerol (w/v) and stored at −20 to −80°C in snap-cap, 1.5-ml conical tubes. It is wise to make several glycerol tubes so that you have a fresh tube for each experiment. The bacterial strain isolated from plant tissue is ready for pathological studies and identification.

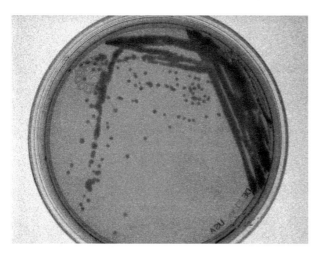

FIGURE 7.1 Isolation (streaking) of bacteria. Note the individual colonies used to make final isolations and cultures.

Anticipated Results

Growth of bacteria on the initial dish of nutrient agar may be heavy and individual colonies may be difficult to select. Restreaking a small portion of an individual colony onto the second dish should result in growth of a number of individual colonies (Figure 7.1). Pure cultures should be obtained with as few as one or two restreakings.

Questions

- Ideally during isolation, how many bacteria form an individual colony on a nutrient agar dish?
- Why is it important to have pure cultures of plant pathogenic bacteria?

EXPERIMENT 3. PATHOGENICITY ASSAYS

Bacteria require free water for growth and they are surprisingly fragile when subjected to extremes of temperature, drying, or light intensity. Therefore, pathogenicity assays are most successful when these conditions for growing host plants are stabilized. Temperature is usually maintained between 16°C and 25°C (61°F and 75°F) and lighting is subdued. In glasshouses, shade inoculated plants with one to two layers of cheesecloth and inoculated plant organs, such as rhizomes, tubers and roots, should be maintained in the dark. Moisture should be maintained such that the surfaces of the inoculated plants do not dry out for 48 to 96 h after inoculation. A mist chamber may be inexpensively constructed of plastic sheeting (painting drop cloths) stretched over a frame constructed of snap-fit polyvinylchloride (PVC) pipes and moisturized by a bedside humidifier.

Susceptible Plant

Generally, pathogenesis is encouraged by inoculating pathogenic bacteria onto succulent, young tissue of compatible plants. New, fully expanded leaves and new stems are usually ideal. Matching plant species and cultivar with the diseased host from which the pathogen was isolated avoids genetic resistance. The same organ that was affected in the host of isolation should be inoculated (e.g., inoculate flowers if flowers were blighted; inoculate leaves if leaf spots developed; inoculate tubers if tubers were rotted; or inoculate stems if the plant wilted).

Materials

Each student or team of students will require the following items:

- Pure culture of pathogenic bacteria isolated in Procedure 7.2
- Nutrient agar or nutrient broth (8 g powder medium and 1 l of water) or both
- Rotary shaker
- Spectrophotometer set at 550 nm and disposable cuvettes
- Graduated cylinder and 1- and 5-ml pipettes
- Host plants — See introduction for host plants and diseases
- Microliter pipette and tips
- Hypodermic syringe (1-ml tuberculin) and 23- to 26-gauge needle
- Sewing needles (#7 sharp)
- Potting soil

Follow the protocol listed in Procedure 7.3 to complete the experiment.

Procedure 7.3 Pathogenicity Assays

1. Inoculum preparation — From a freshly restreaked, isolated colony from Procedure 7.2, inoculate either an agar dish or broth. Incubate the dishes statically or the broth cultures with shaking (150 to 200 rpm/min) at a temperature similar to the temperature at which the inoculated plants will be incubated. Grow the bacteria until colonies form on the agar (48 to 96 h) or the broth culture reaches the stationary phase of growth (turbidity reaches an optical density (OD) of 0.7 to 1.0 at 550 to 600 nm as read on a colorimeter with an optical path of 1.0 cm). Suspend bacteria from colonies in water such that OD at 550 to 600 nm = 0.7 to 1.0. Dilute broth-grown or agar-grown bacterial suspensions 100-fold with water. The final concentration of bac-

terial cells will be approximately 10^6 to 10^7 colony forming units (CFU)/ml. Use this suspension for inoculation immediately.

2. Inoculation — Inoculation methods vary with the organs affected by the pathogen.

- Leaves — Using a sprayer (window cleaner sprayer from grocery store disinfested with 70% isopropanol and rinsed carefully with freshly distilled water), adjust the spray to a fine mist and spray the underside (abaxial) surface of the leaf until water-soaked areas are obvious. Water-soaked areas are where the bacterial suspension has been forced through stomata into the leaf, which appear darker and more translucent to light than surrounding leaf tissue.

- Flowers — With a microliter pipettor, deliver 10:1 of bacterial suspension to the area of the nectartodes.

- Stems — Mildly stress the plants for water before inoculation. Stress plants until leaves are slightly flaccid without being so wilted that they do not revive with watering. With a hypodermic syringe, slide the needle (23 to 26 gauge) about 2 to 5 mm into an axil where a leaf petiole joins the stem. With the needle in the plant, apply gentle pressure to the plunger until a drop of bacterial suspension forms in the axil. Remove the needle and water the plant. The bacterial suspension will be drawn into the plant through the needle wound by transpiration.

- Tubers, rhizomes, and fruits — Fill the eye of a sewing needle (#7 sharp) with either bacteria from a colony on an agar dish or a turbid broth culture (OD at 550 to 600 nm = 0.7 to 1.0) and press the eye of the needle 0.5 to 1.0 cm into the organ, rotate the needle and remove.

- Roots — Uproot young, mildly water-stressed seedlings and gently rinse their roots free of potting medium. Immerse their roots in a bacterial suspension (10^6 to 10^7 CFU/ml) for 1 to 5 min. Replant in fresh potting medium and water.

3. Symptoms — Incubate inoculated plants in a moist chamber for 48 to 72 h. Remove plants to a shaded area in the glasshouse and observe for symptom development.

Anticipated Results

Generally, blighting and rotting occur in 2 to 6 days, leaf spots appear in 5 to 14 days, wilting occurs in 10 to 21 days, and proliferation of plant tissue happens in 14 to 28 days. Well-watered plants often do not wilt. Mildly water-

stress the plants by increasing the rate of transpiration with a small electric fan.

Questions

- Why is it necessary to reinoculate a host plant with the isolated bacterium?
- What would be the next step in this experiment to complete Koch's postulates?

EXPERIMENT 4. BACTERIAL CELL WALL

Two methods are available for determining bacterial cell wall structure: the KOH lysis technique and Gram's stain. Both techniques are easy to perform and take little time. Students may need some instruction on how to use oil immersion. Cellular morphology (rods, helical, vibroid, mycelial, etc.) is important to identify prokaryotic plant pathogens (see Chapter 6). The Gram stain is usually the standard for visualizing cellular morphology. Although the KOH method damages or lyses cells and is not suitable for microscopic examination, it is a rapid and reliable method for determining cell wall structure. In Gram-positive reactions, the higher amounts of peptidoglycans in the cell wall may retain the crystal violet–iodine complex more completely than in Gram-negative cells. In Gram-negative cells, the crystal violet–iodine complex is more easily removed by alcohol treatment. Excessive destaining of Gram-positive cells can cause them to appear to be Gram-negative. Likewise, inadequate destaining can make Gram-negative cells appear to be Gram-positive. Use of controls (mixed Gram-negative and Gram-positive bacteria) aid pathologists in determining the optimum amount of destaining.

The malachite green method for staining endospores also depends on differential retention of dye. Steaming fixes the malachite green dye in the tough endospores of clostridia and bacilli. Washing removes the malachite green dye from the surrounding cytoplasm. Counterstaining with safranin red allows the pathologist to determine whether the endospores are central, subterminal, or terminal in the cells — these are important criteria for bacterial identification.

Materials

Each student or team of students will require the following items:

- KOH (3 g in 100 ml water)
- Microscope slides
- Alcohol lamp and bacterial loop

- Cultures of *Pectobacterium* (*Erwinia*) *carotovora*, *Bacillus subtilis*, and *Escherichia coli*
- Crystal violet solution (Fisher)
- Gram's iodine solution (Fisher)
- Decolorizer solution (Fisher)
- Safranin solution (Fisher)
- Compound microscope with oil immersion lens and lens paper
- Paper towels
- Malachite green (5 g malachite green dye dissolved in 100 ml water)
- Safranin (0.25 g safranin dissolved in 100 ml of 10% ethanol)

Follow the protocols in Procedure 7.4 to complete this exercise. Anticipated results are provided as comments in the protocols.

Procedure 7.4 Determination of Bacterial Cell Wall Structure

1. KOH procedure — In this test, mix a loopful of bacteria with 3% (w/v) potassium hydroxide. Gram-negative bacteria lyse, releasing DNA, and causing the mixture to become very viscous. Strands of viscous DNA are visualized by lifting the inoculating loop out of the mixture. Although the Gram-positive cells also lyse, their thick peptidoglycan cell walls trap DNA and strands do not form.
2. Gram's stain procedure — On a clean slide, dry a very thinly spread bacterial film in air without heat. (*Hint*: Use controls. On the same microscope slide with the unknown bacterium, prepare flanking smears of Gram-positive *Bacillus subtilis* and Gram-negative *Escherichia coli*.) Then lightly flame the underside of the slide twice to fix the bacteria to the slide. Flood the smear with crystal violet solution for 1 min. Wash in tap water a few seconds. Drain off excess water and lightly blot dry on a paper towel. Flood the smear with iodine solution for 1 min. Wash in tap water a few seconds blot dry. Decolorize with decolorizer solution until the solvent flows colorlessly from the slide (10–15 sec). Blot dry. (If decolorizer is used longer the Gram-positive bacteria may lose color.) Rinse in tap water for about 2 sec. Counterstain for 30 sec with safranin solution. Wash briefly in tap water. Blot dry and examine with an oil immersion lens. Gram-positive bacteria retain color after decolorizing whereas Gram-negative bacteria rapidly decol-

orize and are counterstained with the red dye safranin.

3. Endospores — Endospores of clostridia and bacilli may be detected by steaming slides flooded with malachite green. Wash the smear under running water until the green color no longer runs off the slide. Counterstain 30 to 60 sec with safranin. Examine preparation by using oil immersion. Endospores appear green and the cells surrounding them appear red. Use as controls endospore-forming *Bacillus subtilis* and non-spore-forming *Escherichia coli*.

Question

- What property of the Gram-positive bacterial cell wall allows the stain to be retained?

EXPERIMENT 5. SOME PHYSIOLOGICAL PROPERTIES OF PLANT PATHOGENIC BACTERIA

Physical and morphological characteristics that can be used to distinguish bacterial genera are limited. Therefore, phytobacteriologists have relied on some simple physiological tests to characterize different groups. Although with molecular techniques, including the PCR methods outlined in Schaad et al. (2001), reflecting phylogenetic groupings of bacteria are being developed, currently relatively few phytopathogens can be identified by these procedures. No doubt this will change in the near future. Therefore, the information gained from physiological tests is still critical for pathogen identification. This protocol demonstrates several basic methods for physiological testing of bacteria that are also important for taxonomical identification.

Materials

- King's medium B (Difco, Detroit, MI)
- Small disposable test tubes and bacterial loop
- Cultures of *Pseudomonas fluorescens*, *Escherichia coli*, *Pectobacterium carotovora*, *Pseudomonas syringae*, *Burkholderia cepacia*, *Ralstonia solanacearum*, and bacterium isolated in Procedure 7.2
- YS broth (800 ml water; 0.5 g HNH_4PO_4, 0.5 g K_2HPO_4, 0.2 g $MgSO_4 \cdot 2H_2O$, 5 g NaCl, 1 g yeast extract, and 16 mg cresol red) to which add two parts urea stock solution (20 g urea in 200 ml water and filter sterilized)

- CVP medium — Using a blender at low speed, rapidly mix 500 ml of boiling water, 4.5 ml of 1 N NaOH, 3 ml of 10% aqueous $CaCl_2 \cdot 2H_2O$, 1 g $NaNO_3$, 2 g agar, and 1.0 ml 0.075% aqueous crystal violet (w/v). While blending at high speed, add 10 g sodium polypectate. Pour into a 2-l flask containing 0.5 ml 10% aqueous sodium dodecyl sulfate (SDS), mix, and autoclave. Pour into 10-cm-diameter petri dishes.
- DM1 medium — 5 g cellibiose, 1 g NH_4Cl, 1 g NaH_2PO_4, 1 g K_2HPO_4, 3 g $MgSO_4 \cdot 7H_2O$, 10 mg malachite green, 15 g agar, and 1 l water
- Carbohydrate testing medium — Add 0.1% (w/v) filter-sterilized carbohydrate (e.g., galactose, glucose, or lactose) to a minimal medium consisting of 0.5 g HNH_4PO_4, 0.5 g K_2HPO_4, 0.2 g $MgCl_2 \cdot 7H_2O$, 5 g NaCl, and 1 l water. Adjust to pH 7.2.
- Dye's medium C — Add 0.5% (w/v) filter-sterilized carbohydrate (e.g., glucose, galactose, or lactose) to basal Dye's medium C consisting of 0.5 g HNH_4PO_4, 0.5 g K_2HPO_4, 0.2 g $MgCl_2 \cdot 7H_2O$, 5 g NaCl, 1 g yeast extract, and 0.7 ml 1.5% bromocresol purple (w/v) in 95% ethanol. Adjust pH to 6.8.
- Kovac's reagent — Add 75 ml amyl or isoamyl alcohol, 5 g p-dimethylaminobenzaldehyde, and 25 ml concentrated hydrochloric acid — make under fumehood.
- PHB — 0.2 g $(NH_4)_2SO_4$, 0.2 g KCl, 0.2 g $MgCl_2 \cdot 7H_2O$, and 5 g DL-hydroxybutyrate per liter of water.
- Sudan black B (0.3% dye in 70% ethanol)
- 0.5% aqueous safranin

Follow the protocols in Procedure 7.5 to complete this experiment. Anticipated results are provided as comments in the protocols.

Procedure 7.5 Physiological Characteristics of Bacteria

1. Fluorescent pigments — Fluorescent pyoverdine siderophores may be detected by growing bacteria in test tubes on King's medium B, which is high in glycine-rich peptones that chelate iron. Creating an iron-depleted environment induces siderophore synthesis in the bacteria. Under long-wave ultraviolet light, water-soluble pyoverdine pigments fluoresce green to blue. (*Hint:* Use fluorescent *Pseudomonas fluorescens* and nonfluorescent *Escherichia coli* as controls.)

2. Urease hydrolysis — Urease activity may be detected by growing bacteria on modified YS broth. Use a tube without urease as a color control. An increase to about pH 9.0 accompanied by the development of a magenta color indicates urease activity.

3. Pectinase — Streak bacteria onto CVP medium and incubate at 25°C for 1 week. Use the following controls: *Pectobacterium carotovora* and *Pseudomonas fluorescens* create conical and flat pits, respectively, and nonpectolytic *Escherichia coli* does not pit CVP medium.

4. Selective growth medium — Streak bacteria onto D1M medium and incubate at 25°C. Use the following controls: *Agrobacterium* species grow on D1M whereas *Pseudomonas, Xanthomonas,* and *Erwinia* species do not.

5. Utilization of carbon sources — Dispense 5 ml of sterile carbohydrate testing medium into test tubes, inoculate with a loop of bacteria, and incubate at 25°C. Make three serial transfers on this medium. Turbidity will develop if the substrate is utilized and growth occurs. Use a control that has no added carbohydrate.

6. Acid production from carbohydrates — Dispense 5 ml of Dye's medium into test tubes. Inoculate the medium with a loop of bacteria and incubate at 25°C. The medium turns yellow if acid is produced from the carbohydrate.

7. Indole production — Culture organism(s) in broth (5 g yeast extract, 10 g tryptone, and 1 l water) at 25°C. Add 1 ml of turbid broth to 1 ml Kovac's reagent. Indole is produced from tryptophan and results in a cherry-red reaction with Kovac's reagent

8. Poly β-hydroxybutyrate (PHB) granules — Culture bacteria in 10 ml of PHB broth. Smear, air dry, and fix the cells to a microscope slide. Flood with Sudan black B for 10 to 15 min. Drain and blot dry and decolorize with water. Counter stain with safranin. Examine cells with oil immersion. Cells appear red or pink and PHB inclusions are black. *Burkholderia cepacia* or *Ralstonia solanacearum* produce PHB granules and *Pseudomonas syringae* does not.

Questions

- What product is important in the identification of phytopathogenic bacteria produced on King's medium B?
- D1M semiselective medium is important for the identification of what important group of phytopathogenic prokaryotes?
- Why do the carbohydrate sources in Dye's medium C vary?
- What bacterial staining agent collects in granules of polyβ–hydroxybutyrate?

Experiment 6. The Hypersensitive Test for Pathogens

Hypersensitive response (HR) in tobacco is useful to determine whether a bacterium may be a plant pathogen. Briefly, pathogens of plants (other than tobacco) cause the HR in tobacco whereas pathogens of tobacco cause symptoms of disease. Saprophytes or nonpathogens of plants cause no symptoms of disease and do not elicit a response. The HR test takes advantage of the specific induced resistance mechanism determined genetically by plants. The HR test is especially useful to quickly screen possible pathogens before the more time- and labor-intensive pathogenicity assays are performed. Although HR tests are extremely useful for a large number of leaf spotting and blighting pathogens (e.g., erwinias, pseudomonads, ralstonias, and xanthomonads) other pathogens may not be identified readily using this test, such as soft-rotting bacteria (e.g., pectobacteria) and agrobacteria. For those pathogens, pathogenicity assays are necessary.

Materials

Each student or team of students will require the following items:

- Nutrient agar and nutrient broth (Difco, Detroit, MI)
- Rotary shaker
- Spectrophotometer and cuvette
- Centrifuge and tubes
- Tobacco plants
- Syringe with 23- to 26-gauge needle
- Spray bottle
- Cultures of *Ralstonia solanacearum, Pectobacterium (Erwinia) carotovora, Pseudomonas pisi, P. syringae* pvs. *angulatum, syringage,* or *tabaci,* and *Escherichia coli*

Follow the protocols in Procedure 7.6 to complete the experiment. Anticipated results are provided as comments in the protocols.

Procedure 7.6 Hypersensitive Test for Bacterial Pathogens

1. Grow bacterial cells in NA broth overnight or on NA agar for 48 h. Suspend cells grown on agar in sterile distilled water. Transfer centrifuge tubes and centrifuge at 14,000 rpm. Pour off water and resuspend the cells in sterile water to a density of 10^6 cfu/ml. (See Procedure 7.3 for determining density by spectrophotometry.)

2. Grow tobacco plants in styrofoam drinking cups with a hole in the bottom for drainage. Tobacco seedlings (one per cup) may be grown in peat potting media under fluorescent lights or in window boxes. Plants may be reused, using a different leaf for each test.

3. Soak a leaf panel (the space between two major veins perpendicular to the mid-vein) with a bacterial suspension. Water-soaked areas (with water forced through stomata into the intercellular region of the leaf) appear dark and are more translucent than normal leaf tissue. If the leaf is large enough, use four leaf panels on the same leaf to accommodate and control organisms. Inoculate three panels, one each with (a) an incompatible (hypersensitive) pathogen such as *Pseudomonas pisi*, a pathogen of pea and a nonpathogen of tobacco; (b) a compatible (disease) pathogen such as *Pseudomonas syringae* pvs. *angulatum* or *tabaci*, pathogens of tobacco; and (c) a saprophytic (nonplant pathogen) control, *Escherichia coli*. Inoculate the fourth panel with an unknown bacterium.

4. Inoculation methods — Three methods are available: (1) Gently press the flat surface of a 25-gauge needle against the abaxial (bottom) side of a leaf stretched over your finger. With much practice a whole panel may be flooded without tearing the leaf or injuring your finger. (2) With a 1-ml tuberculin syringe with the needle removed, gently press the tip against the abaxial (bottom) side of a leaf stretched over your finger. With some practice a panel may be flooded without crushing the leaf. (3) By a hand-pumped sprayer (window cleaner pump), adjust the nozzle for a fine mist. With practice, an area of leaf panel may be water soaked without tearing the leaf.

5. Incubate plants in the window box or under fluorescent lights for 24 to 96 h. Complete collapse of leaf tissue with drying to a paper-like consistency is typical of an incompatible or hypersensitive reaction. In 72 to 120 h, lesions with water-soaked edges with or without chlorotic halos indicate a compatible or disease interaction. After incubation, a normal appearing leaf without water soaking or dry, dead tissue indicates no reaction with a saprophytic bacterium.

Question

- In the hypersensitivity test, what reactions would you expect tobacco leaves to exhibit when treated separately with each of the following organisms: a pathogen of tobacco, *Escherichia coli*, and a pathogen of lilac?

REFERENCES

Lee, I.-M., R.E. Davis and J. Fletcher. 2001. Cell-wall free bacteria, pp. 283–320 in *Laboratory Guide for Identification of Plant Pathogenic Bacteria*, 3rd ed. Schaad, N.W., J.B. Jones and W. Chun (Eds), American Phytopathological Society Press, St. Paul, MN.

Schaad, N.W., J.B. Jones and W. Chun. 2001. *Laboratory Guide for Identification of Plant Pathogenic Bacteria*, 3rd ed. American Phytopathological Society Press, St. Paul, MN.

8 Plant-Parasitic Nematodes

James P. Noe

CHAPTER 8 CONCEPTS

- Plant-parasitic nematodes are important pests in most plant production systems. There are many different species of plant-parasitic nematodes. Almost every plant is attacked by some species of nematode and many species have very wide host ranges.

- Root-knot nematodes (*Meloidogyne* species) are the most common and widespread plant-parasitic nematodes. They cause diagnostic galls or knots on roots.

- Plant damage resulting from nematodes is dependent on the number of nematodes or eggs present at planting. Soil and root assays can provide information on nematode populations prior to planting that may be used to predict crop loss and formulate management strategies.

Plant-parasitic nematodes are members of a primitive group of animals called nonsegmented roundworms and are all obligate parasites, meaning that they can only feed on living plants. They are usually found in the soil and in plant roots, but a few species may attack above-ground parts of the plant. Most species are microscopic, with lengths ranging from 300 to 4000 μm and diameters of 15 to 35 μm, which are within the range of large fungal hyphae. Nematodes undergo four molts during their life cycles. The adults are larger than juveniles and may be shaped differently. Most nematodes may be observed easily under a dissecting microscope at a magnification of 40× to 60×. The amount time taken for plant-parasitic nematodes to complete their life cycles ranges from a few weeks to more than a year depending on the nematode species, plant host status, and ambient temperatures.

All plant-parasitic nematodes have a hardened spear-like feeding structure, called a stylet, in the anterior portion of their head region (Figure 8.1). This stylet is used to penetrate plant tissues directly and to feed on plant cells. The stylet secretes compounds produced by the nematodes into the plant cells and removes nutrients from the cells. Plant cells fed on by nematodes usually are not killed. Nematodes may feed on roots from outside, referred to as **ectoparasites**, or from inside the root, referred to as **endoparasites**. Within both feeding types, some species are migratory and move around throughout their life cycles, whereas others are sedentary and settle in one location to establish a permanent feeding site.

Free-living nematodes (nonplant-parasitic nematodes) are often found in the soil samples in large numbers along with plant-parasitic nematodes (Figure 8.2). These nematodes can be distinguished from plant-parasitic nematodes by their lack of stylets or by the absence of knobs or flanges at the base of their stylets. Free-living nematodes feed on dead organic matter or microorganisms associated with the organic matter. These nematodes recycle nutrients in soil and are an important component of the ecosystem. Types and numbers of free-living nematodes in soil samples are often used by ecologists as an indicator of the health of an ecosystem.

Soil texture and structure are of primary importance in determining the number and types of plant-parasitic nematodes found on a suitable host plant. On the same host, for example, cotton, root-knot nematodes (*Meloidogyne incognita*) are usually a problem in light, sandy soils, whereas reniform nematodes (*Rotylenchulus reniformis*) are found more often in clay soils with high bulk densities. Within the soil, plant-parasitic nematodes can move only a short distance, whereas long-distance dissemination is usually by movement of soil, water, or plant-propagative parts.

Plant-parasitic nematodes attack most economically important plants in agriculture, horticulture, ornamentals, and turf. Nematode problems are usually more prevalent in warmer, humid climates, but some species attack plants even in the coldest and most arid climates. Most plant-parasitic nematodes do not produce any obvious or easily distinguishable symptoms on their host plants. Because most nematodes are microscopic and feed below ground, many nematode problems may remain undiagnosed. An assay of soil samples completed by technicians trained in **nematology** is usually the only way to confirm a nematode diagnosis. Plant-parasitic nematodes are typically found in the root zones of suitable host plants with most of the population occupying the top 15 cm of soil.

Above-ground symptoms attributed to plant-parasitic nematodes include stunting, yellowing, and wilting. Root

0-8493-1037-7/04/$0.00+$1.50
© 2004 by CRC Press LLC

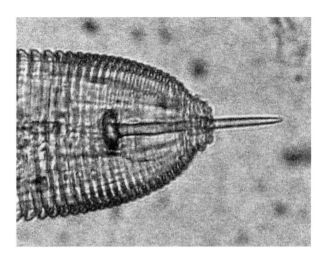

FIGURE 8.1 Head of a ring nematode. Note the stylet (or spear) used to penetrate plant tissues when feeding.

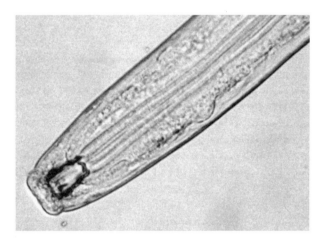

FIGURE 8.2 Free-living nematode. These nematodes are commonly found in soil and play an important role in nutrient recycling.

FIGURE 8.3 White clover root systems infected with the southern root-knot nematode. The root system on the left is severely galled and stunted. (Courtesy of Dr. G. Windham, USDA, Mississippi State, MS.)

symptoms may include galls, lesions, stunting, stubby appearance, and excessive branching. Crop losses from nematodes range from 10 to 60% of yield, depending on species, host status, and environmental conditions. Many nematodes enhance crop losses by interacting with other soilborne pathogens such as fungi and bacteria.

ROOT-KNOT NEMATODES

Root-knot nematodes (*Meloidogyne* species) are a widespread and diverse group of plant-parasitic nematodes. Although more than 100 species have been described, four root-knot species (*M. arenaria, M. hapla, M. incognita,* and *M. javanica*) cause most of the damage reported on agricultural crops. Of these four, *M. incognita* is the most common and has a very wide host range. Root-knot nematodes are found worldwide, but are more common in warmer climates and in light, sandy soils. Plants infected

with these nematodes may be stunted, chlorotic, and wilted during the daylight hours. Galls or knots form on host roots due to nematode infection and are often used to diagnose the disease (Figure 8.3 and see Plate 4C following page 80).

Root-knot nematode females lay eggs in a gelatinous matrix and first-stage juveniles undergo one molt while still in the egg. Worm-shaped, or vermiform (diameter 15 μm; length 400 μm), second-stage juveniles hatch from the eggs, using their stylets to break through the tough egg casing. Second-stage juveniles are the only infective stage of root-knot nematodes. After penetrating suitable hosts at the root tips, the juveniles migrate to the developing vascular cylinder and begin feeding on several cells near the endodermis. These cells enlarge, become multinucleate, and serve as feeding cells for the rest of the

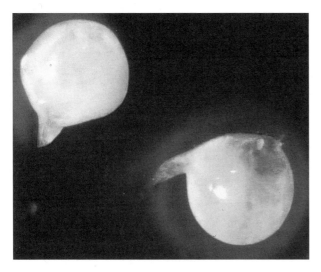

FIGURE 8.4 Mature root-knot nematode females that have been excised from galled roots. (Courtesy of Dr. G. Windham, USDA, Mississippi State, MS.)

nematode life cycle. A gall rapidly begins to develop around the feeding juvenile as a result of cell division and enlargement. Once juveniles begin feeding, they undergo a series of three additional molts. The third- and fourth-stage juveniles are short-lived stages and are slightly swollen to sausage shaped in appearance. Root-knot nematodes exhibit **sexual dimorphism**. Females (Figure 8.4) become enlarged and spherical (diameter 400 μm; length 700 μm), whereas males molt to a relatively large vermiform shape (diameter 30 μm; length 1400 μm) and migrate from the root. The posterior end of the adult female usually protrudes from the surface of the root gall, where an egg mass containing 300 to 500 eggs is produced (Figure 8.5). The life cycle typically requires 21 to 50 days, depending on root-knot species, plant host, and environment. Root-knot nematodes survive intercrop periods primarily as eggs in the soil. Survival rates are enhanced by protection from the egg-mass matrix and host plant debris.

CYST NEMATODES

Cyst nematodes (*Heterodera*, *Globodera*, and *Punctodera* species) are sedentary endoparasites and infect vegetables, small grains, soybean, corn, and legumes. *Heterodera glycines*, the soybean cyst nematode is one of the most studied plant-parasitic nematodes (see Plate 4D following page 80). This nematode is found in most soybean production areas and is the most important disease of soybean. Soybean plants infected with *H. glycines* may be stunted, exhibit foliar chlorosis, and have necrotic roots and suppressed bacterial root nodulation. Dead seedlings or plants may be found in fields with high numbers of the soybean cyst nematode. Other important cyst nematodes are the potato cyst nematodes (*G. rostochiensis* and *G. pallida*) and the sugar-beet nematode (*H. schachtii*).

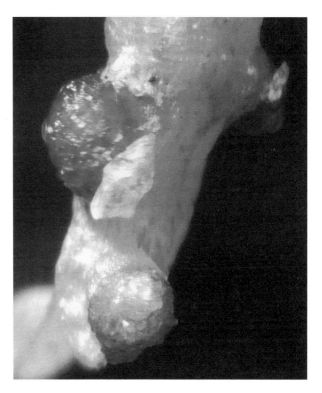

FIGURE 8.5 Root-knot nematode egg masses on root galls. Egg masses may contain as many as 500 eggs. (Courtesy of Dr. G. Windham, USDA, Mississippi State, MS.)

The life cycle of soybean cyst nematodes is similar to that of root-knot nematodes in several ways. These nematodes are another example of sexual dimorphism, juveniles undergo four molts before becoming adults, and eggs are deposited in a gelatinous matrix. Second-stage juveniles penetrate roots behind the root tip, migrate to vascular tissues, begin feeding, and become sedentary. Nematode feeding results in the production of a specialized food cell called a **syncytium** (Figure 8.6). The bodies of adult females eventually erupt through the root surface and can be easily seen with the naked eye or with the aid of a hand lens. Adult males are necessary for reproduction. Many nematodes such as root-knot nematodes do not require the presence of males for reproduction. The color of the adult female body or cyst changes from white to yellow, becoming brown at death and is filled with 200 to 600 viable eggs (Figure 8.7). Cysts protect the eggs from unfavorable environmental conditions and are readily spread by contaminated soil, wind, and water.

LESION NEMATODES

Lesion nematodes (*Pratylenchus* species) attack many crops, including vegetables, row crops, legumes, grasses, and ornamentals (Figure 8.8). Important species include *P. brachyurus*, *P. penetrans*, *P. scribneri,* and *P. vulnus*. These nematodes can be found in host roots in large num-

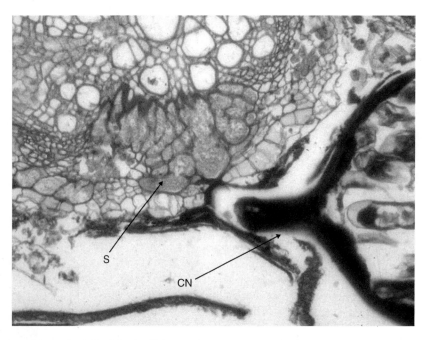

FIGURE 8.6 A syncytium (S), or specialized food cell, in a soybean root system initiated by feeding of soybean cyst nematode (CN). (Courtesy of Dr. R. Hussey, University of Georgia.)

FIGURE 8.7 The body of a soybean cyst nematode female becomes brown at death and is filled with 200 to 600 viable eggs. (Courtesy of Dr. G. Windham, USDA, Mississippi State, MS.)

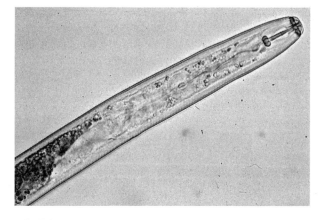

FIGURE 8.8 Anterior end of a lesion nematode. Note the well-developed stylet of this migratory endoparasite. (Courtey of Dr. G. Windham, USDA, Mississippi State, MS.)

bers and cause necrotic lesions on fibrous or coarse roots. Lesions may cover the entire root system, and root pruning may occur in heavily infested fields. Stunted plants may be localized in fields or a general suppression of growth may be evident throughout an entire field. Lesion nematodes have been associated with **disease complexes** involving other plant pathogens. *Pratylenchus penetrans* and the fungus *Verticillium dahliae* cause potato early death that can cause a severe reduction in potato yields.

The life cycle of lesion nematodes is typical for many plant-parasitic nematodes. Adult females lay eggs singly in root tissue or in the soil. The first-stage juvenile molts to the second stage within the egg. The second-stage juvenile hatches from the egg and then molts three more times to become an adult. The presence of host roots has been shown to stimulate egg hatch. All juvenile stages outside the egg and adults can infect host roots. Lesion nematodes are classified as migratory endoparasites because they enter and exit roots numerous times to feed during a growing season. Invasion of roots may occur at the root tip, the root hair region, and occasionally in young lateral root junctions.

STING NEMATODES

The sting nematode (*Belonolaimus* species) can have a devastating effect on the growth and yields of vegetables, row crops, ornamentals, and turf (see Plate 4B following page 80). The most important species of sting nematode is *B. longicaudatus*. Sting nematodes are nearly always found in very sandy soils that have a sand content of greater than 80%. Plants parasitized by these nematodes may be severely stunted and appear to be nutrient deficient. Seedlings in heavily infested areas may be severely stunted and devoid of a root system, which can lead to plant death. Below-ground symptoms may include roots with coarse, stubby branches and necrotic lesions may develop on the root surface. These nematodes are such severe pathogens that the presence of one sting nematode in a soil assay sample may necessitate the use of **nematicides**.

Sting nematodes are large, slender worms that reach a length of nearly 2.5 mm and have very long stylets, which may be 60 to 150 μm (Figure 8.9). The life cycle of this nematode is similar to other plant-parasitic nematodes. Reproduction is bisexual, with males comprising about 40% of the population. Adult females deposit eggs in soil and the first-stage juvenile molts once before emerging from the egg. Sting nematodes are considered ectoparasites because they feed from the root surface and rarely enter the root tissue. These nematodes feed on root tips and along the sides of succulent roots by inserting their long, slender stylets into epidermal cells.

DAGGER NEMATODES

Dagger nematodes (*Xiphinema* species) are ectoparasites and are commonly found associated with fruit or nut trees and in vineyards. Agriculturally important dagger species include *X. americanum*, *X. californicum*, *X. index*, and *X. rivesi*. Root systems of plants parasitized by dagger nematodes may be stunted, discolored, and have a limited number of feeder roots. In 1958, *X. index* was documented as the first soil nematode to **vector** a plant virus (grapevine fanleaf virus). Species found in the *X. americanum* group transmit the following North American *Nepoviruses*: *Cherry rasp leaf virus, Tobacco ringspot virus, Tomato ringspot virus,* and *Peach rosette mosaic virus.*

Dagger nematodes are slender, up to 2-mm long, have a dorylamoid esophagus, and have a flanged **odontostyle** that may be 130 μm long (Figure 8.10). Females lay eggs singly in the soil near host plants and the nematodes hatch as first-stage juveniles. These nematodes may have three or four juvenile stages that can be recognized by the

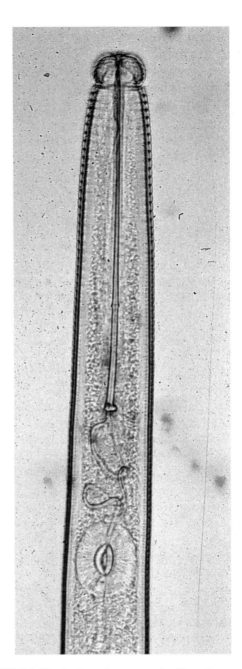

FIGURE 8.9 Head of a sting nematode. Note the very long stylet that the nematode uses to feed on root tips. (Courtesy of Dr. G. Windham, USDA, Mississippi State, MS.)

lengths of their developing stylets. Male dagger nematodes are very rare. Dagger nematodes can live up to 3 years under favorable environmental conditions.

STUBBY-ROOT NEMATODES

Stubby-root nematodes, *Paratrichodorus* and *Trichodorus* species, are important parasites of corn, vegetable crops, and turf in the coastal plain region of the southeastern U.S. The most important species is *P. minor*. These nem-

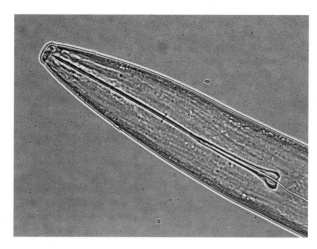

FIGURE 8.10 Head of a dagger nematode. Dagger nematode stylets have flanges and are called odontostylets. (Courtesy of Dr. G. Windham, USDA, Mississippi State, MS.)

atodes are most widely distributed in sandy soils, but have also been found in organic-based soils. The most characteristic symptom of damage by stubby-root nematodes is a stunted, stubby root system. Above-ground symptoms are similar to plants deprived of a root system. Plants are stunted and chlorotic, wilt easily, and have very little ability to withstand drought. *Paratrichodorus* species are vectors of the following tobraviruses: *Tobacco rattle virus*, *Pea early browning virus*, and *Pepper ringspot virus*.

Stubby-root nematodes are ectoparasites and feed almost exclusively at the root tips. When feeding, their head very rarely becomes embedded in the root. Feeding by stubby-root nematodes is very different from that by other plant-parasitic nematodes in that food is not ingested through the stylet. These nematodes have a dorsally curved stylet referred to as an **onchiostyle**. The life cycle of this nematode consists of an egg stage, four juvenile stages, and the adult stage and can be completed in 16 to 17 days. Soil populations of stubby-root nematodes can increase rapidly and then decline with equal abruptness. These nematodes have commonly been found at deep soil levels, which may create sampling problems and limit detection of these parasites.

ADDITIONAL PLANT-PARASITIC NEMATODES

The awl nematode, *Dolichodorus heterocephalus*, is an ectoparasite of vegetables and is similar to the sting nematode in general appearance. This nematode is generally found in fields with high soil moisture and causes symptoms on roots similar to those by the sting nematode. The lance nematodes (*Hoplolaimus* species) are serious pests of cotton and soybean in the coastal plain regions of Geor-

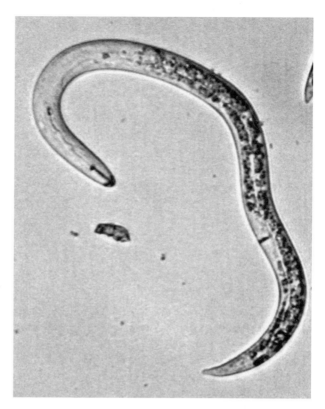

FIGURE 8.11 A preadult reniform nematode. After the nematode enters the root and begins feeding, the posterior portion of the body becomes kidney shaped. (Courtesy of T. Stebbins, University of Tennessee.)

gia, South Carolina, and North Carolina. Lance nematodes have heavily sclerotized stylets with large, tulip-shaped knobs. These nematodes are classified as semiendoparasites or as migratory endoparasites. The reniform nematode, *Rotylenchulus reniformis,* has become the major nematode pest of cotton in Southern U.S. (Figure 8.11). The anterior portion of the adult female remains embedded in the root and the posterior, kidney-shaped portion is found on the external root surface. The ring nematodes, *Criconemella* or *Mesocriconema* species, are ectoparasites and attack turf, cotton, peanut, and peach trees (Figure 8.12). The cuticle of these nematodes is coarsely striated, giving the nematodes a segmented or ring-like appearance.

A number of plant-parasitic nematodes feed within above-ground plant tissues. The seed gall nematode, *Anguina tritici*, invades florets of rye and wheat and can survive for long periods in seed-like galls. *Aphelenchoides* species, the bud and leaf nematodes, attack the foliar parts of ornamentals, potato, strawberry, and onions (see Plate 4A following page 80). The stem and bulb nematode, *Ditylenchus dipsaci*, is an important pest of alfalfa, onion, sweet potato, strawberry, and nursery crops.

NEMATODE MANAGEMENT

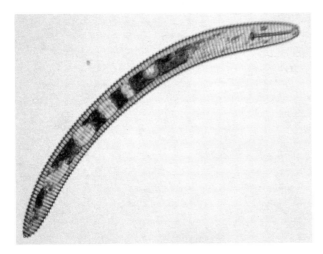

FIGURE 8.12 A ring nematode. The cuticles of these nematodes give them a segmented or ring-like appearance. (Courtesy of D. Cook, University of Tennessee.)

Control methods should be used to keep nematode numbers below damaging levels instead of trying to **eradicate** these organisms. Host plants are often able to compensate for damage caused by moderately high numbers of nematodes, especially where the plants are grown under optimum environmental conditions. Although nematodes such as root-knot nematodes may establish hundreds of galls on a root system, very large numbers of these nematodes are usually required to cause significant crop losses. This density-dependent damage relationship for plant-parasitic nematodes is the foundation of nematode advisory programs. In these programs, data from soil assays are used to predict the nematode hazard to anticipated crops and to recommend control practices if necessary. To accurately advise farmers of nematode hazards, a relationship between nematode numbers in the soil and crop performance must first be established. This information is best obtained from infested fields, but controlled studies in greenhouse pots are also useful.

In agricultural fields, nematode management usually relies on the application of chemical pesticides to the soil. These pesticides, termed **nematicides**, are expensive, exceedingly dangerous to apply, and are cause for severe environmental concerns. Nematicides do not eliminate all the nematodes from a field. Fields treated with nematicides may have higher numbers of nematodes at the end of the growing season compared to fields not treated with chemicals. In greenhouses and limited acreage high-value crops, soil sterilization with heat or application of a broad-spectrum biocidal fumigant is commonly practiced.

For most field crops, resistance is the only practical means for controlling plant-parasitic nematodes. Resistant crops are economical in that little or no additional costs are assessed to the grower and no special equipment is needed to use these resistant plants. Also, from the standpoint of impact on the environment, resistant cultivars are considered the best choice of control practices. Cultivars resistant to endoparasitic nematodes such as root-knot, lesion, and cyst nematodes are available. However, use of resistant plants is complicated by the existence of races among root-knot nematodes and the soybean cyst nematode. Very little progress has been made in identifying plants resistant to ectoparasites.

Other methods used to control plant-parasitic nematodes include **sanitation**, **cultural practices** (Chapter 32), **biological control** (Chapter 34), and the use of organic soil amendments. Sanitation methods prevent the spread of nematodes from infested fields or plant materials to uninfested fields. Cultural control practices include crop rotation with nonhosts and use of tillage practices to destroy plant roots at the end of the growing season to eliminate nematode overwintering sites. Biological control methods, which use bacterial or fungal pathogens to control nematodes, are still under development, but may be used in combination with other control methods in the future to enhance nematode management. Incorporation of organic amendments into nematode-infested soil stimulates population increases of soil organisms that attack plant-parasitic nematodes. Additional options for the control of nematodes are badly needed. Nematode problems continue to worsen globally as more agricultural production is concentrated on smaller areas of arable land.

ACKNOWLEDGMENT

The author and editors gratefully acknowledge the contributions of Dr. Gary Windham to this chapter.

SUGGESTED READINGS

Barker, K.R, C.C. Carter and J.N. Sasser (Eds.). 1985. *An Advanced Treatise on Meloidogyne. Volume 2: Methodology.* North Carolina State Graphics, Raleigh, NC, 223 pp.

Barker, K.R., G.A. Pederson and G.L. Windham (Eds.). 1998. *Plant and Nematode Interactions.* Monograph 36. American Society of Agronomy, Madison, WI, 771 pp.

Dropkin, V.H. 1989. *Introduction to Plant Nematology.* John Wiley & Sons, New York, 304 pp.

Mai, W.F. and H.H. Lyon. 1975. *Pictorial Key to Genera of Plant-Parasitic Nematodes.* Cornell University Press, Ithaca, NY, 219 pp.

Southey, J.F. (Ed.). 1986. *Laboratory Methods for Work with Plant and Soil Nematodes.* Her Majesty's Stationery Office, London, 202 pp.

Zuckerman, B.M., W.F. Mai and M.B. Harrison. 1985. *Plant Nematology Laboratory Manual.* University of Massachusetts Agricultural Experimental Station, Amherst MA, 212 pp.

9 Pathogenicity and Isolation of Plant Parasitic Nematodes

James P. Noe

Plant-parasitic nematodes are found in the root zone of most plants grown in urban and rural areas. A diverse mixture of plant-parasitic and free-living nematodes (non-parasitic) is present in most soil samples. These nematodes can be readily isolated from both soil and root samples for viewing and identification. Damage to plants caused by plant-parasitic nematodes in the field can be easily reproduced in the greenhouse. The amount of damage caused by nematodes can be related to the number of nematodes (or eggs) present at planting or at the start of an experiment. The laboratory exercises in Experiment 1 will allow students to observe symptoms caused by root-knot nematodes and to observe various life stages. In Experiment 2, students will isolate nematodes from soil samples and observe the diverse nematode populations present in several settings.

EXERCISES

EXPERIMENT 1. THE EFFECTS OF ROOT-KNOT NEMATODES ON PLANT GROWTH

This experiment will require two laboratory periods about 6 to 8 weeks apart. In the first part of the exercise, tomato plants will be inoculated with eggs of root-knot nematodes and in the second part the effects of the nematodes on the plants will be observed. This experiment will familiarize students with the root-knot nematode and demonstrate the effect of various concentrations of inoculum (eggs) on plant growth and nematode reproduction. Tomato plants inoculated in the first laboratory will have sufficient time to show the effects of the nematode for the second lab exercise in 6 to 8 weeks. Additional objectives are to observe the various life-stages of this plant-parasitic nematode and to learn basic methods for inoculation and evaluation of host plants.

Materials

Each student or group of students will need the following materials:

- One each of root-knot nematode resistant and susceptible tomato plants. Check with your local cooperative extension office for variety recommendations. Tomato cultivar names change frequently and this step will ensure that you have cultivars that are locally available and appropriately classified. Start plants in seedling trays 6 weeks before the laboratory exercise. Transplant to pots of 15-cm diameter in nematode soil mix 3 to 4 weeks after germination.
- Two greenhouse pots of 15-cm diameter
- Sterile greenhouse soil mix suitable for nematode experiments (>85% sand, low organic content) in sufficient amount to fill required number of pots of 15-cm diameter. Avoid common potting mixtures, because the nematodes will not do well and it is very difficult to clean the roots for observation.
- Metal sieves for preparation of nematode inoculum. Two sieves are required, with a pore opening of 75 µm (200-mesh) and 26 µm (500-mesh), each with a frame of 15-cm diameter.
- Several cultures of root-knot nematodes (*M. incognita*). These nematodes must be maintained on living plants, preferably on a nematode-susceptible tomato cultivar, such as Marglobe or Rutgers. Start at least three cultures for the laboratory about 60 days before they are needed in Experiment 1. At least one million eggs can be expected from each tomato plant if they are originally inoculated with 10,000 or more eggs. Do not keep cultures too long (>75 days) as they may deteriorate rapidly, causing recovered egg numbers to drop dramatically.
- Household bleach and a 2- to 3-l flask or large plastic bottle for collecting nematode eggs
- Greenhouse bench space adequate for the number of pots anticipated. Usually, 150 ft^2 is sufficient for 30 to 35 students.
- Phloxine B stain solution (150 mg/l tap water)

- Stereo dissecting microscope and light
- Large beaker and plastic wash bottles
- Rubber aprons
- Cork borer
- Permanent marker and pot labels
- 10-ml pipettes and small beakers

General Considerations — Inoculum Preparation

To prepare nematode inoculum, cut the shoots off from stock tomato plants at the soil line and gently remove the roots from the pot. Wash the roots well in water and cut galled sections into 5- to 10-cm lengths. Immerse and shake sections in 0.5% NaOCl solution (100 ml household bleach in 900 ml tap water) in a 2- to 3-l flask or plastic bottle. (*Caution:* Wear rubber apron — even diluted bleach will damage clothing.) This solution will dissolve the egg masses and free the eggs. Remove the eggs from the solution immediately because leaving the eggs in the solution for more than 4 min will reduce their viability. Pour the solution and roots over two stacked sieves with the 200 mesh above the 500 mesh. Most of the roots will be caught on 200-mesh sieve, and the nematode eggs will pass through and be retained on the 500-mesh sieve. Wash the sieves gently for several minutes with tap water to remove all traces of the bleach and wash most of the eggs onto the 500-mesh sieve. Discard the roots and gently rinse the eggs from the 500-mesh sieve into a beaker. A plastic wash-bottle works well for this step and allows back washing of the sieve to remove all the eggs. Wash eggs to one side of the 500-mesh sieve and gently tip the sieve to pour eggs into a 600-ml beaker. Use the wash-bottle to spray a stream of water on the backside of the sieve to wash eggs into the beaker. Bring the level of water in the beaker up to 500 ml. The resulting egg suspension must then be calibrated. Shake or stir the egg suspension and remove several 1-ml aliquots. Place the aliquots in a small counting dish or plastic tray, wait for the eggs to settle to the bottom (1 to 2 min) and count them with a dissecting stereomicroscope. When observing nematode eggs or juveniles, the light source must be transmitted from below the specimen using a mirrored stage. From the sample counts, determine the average number of eggs per milliliter in the stock solution and prepare dilutions to yield 2000, 1000, 500, and 250 eggs/ml. Remember to always shake or stir the egg suspensions before pouring or sampling because they sink very rapidly to the bottom. The prepared inoculum can be stored overnight at 5°C to 15°C. Susceptible and resistant tomato plants should have been transplanted to 15-cm diameter pots several weeks before and will be ready for inoculation.

Follow the protocols outlined in Procedure 9.1 to complete this part of the experiment.

Procedure 9.1 Inoculating Tomato Plants with Root Knot Nematodes

1. Inoculate young susceptible tomato plants with calibrated suspensions of fresh root knot nematode eggs prepared from galled roots as described under inoculum preparation. Working in groups of five students, each group should inoculate five pots. Each pot will be inoculated with a different inoculum level, including 0, 250, 500, 1000 and 2000 eggs/ml. The number of eggs added per pot equals 0 (10 ml of water only — this is the control), 2500 (10 × 250), 5000 (10 × 500), 10,000 (10 × 1000) and 20,000 (10 × 2000), respectively. Make sure the soil in the pots is evenly moist (not soaked). Nematode eggs will die immediately if exposed to dry soil.
2. Make three equally spaced holes, 3- to 5-cm deep and about 3 cm from the stem of each plant. A marker pen or large cork-borer makes a good diameter hole for the inoculation.
3. While swirling the suspension constantly, carefully pour equal amounts (a little more than 3 ml) of 10 ml of inoculum into each of the three holes. Squeeze the holes shut with your fingers and water lightly immediately. Do not over water the plants. Label each pot with your group number, date, and number of eggs per pot.
4. Repeat Steps 1 to 3 with a nematode-resistant tomato variety, but only use the 10,000 eggs (10 ml of 1000 eggs/ml) inoculum rate. Label this pot as resistant and use it for comparison only in the second experiment.

Follow the protocols outlined in Procedure 9.2 to finish this experiment.

Procedure 9.2 Assessing the Effects of Nematodes on Tomato Plants

1. Carefully wash the soil, causing as little damage as possible, from the roots of plants inoculated in Procedure 9.1. Turn the pot upside-down and gently tap the rim of the pot on a bench until the root ball separates from the pot. Slide the root mass out of the pot and immerse in a trashcan filled with water. Gently shake the root mass up and down until most of the soil has fallen out of the root ball. Wash the remaining soil from the roots under a stream of water in the sink. Keep the correct pot label with each plant to avoid confusion.

2. Remove the excess water from the plants by blotting with paper towels and bring the plants to the laboratory.

3. Separate the roots from the top by cutting the plant at the soil line.

4. Make the following measurements and record in Table 9.1:
   - Overall length of shoot (top) in centimeters
   - Overall length of root system in centimeters
   - Weight of shoot in grams
   - Weight of root system in grams
   - Estimated number of galls

   This estimate can be based on counts made on a part of the root system. The number found on a total of 30 linear centimeters of individual root segments is an acceptable estimate. A total of 30 cm of short segments of root should be collected from different parts of the root mass.

5. Visually compare the resistant tomato plant to the susceptible plant inoculated with the same number of eggs. The resistant plant should be larger, more vigorous, and have no galls on the roots. However, there may be dark and decaying lesions on the roots (necrosis). The mechanism of resistance to root knot nematodes in tomato is called hypersensitivity. As juveniles migrate into the roots and attempt to establish a feeding site, the plant cells die, thus preventing further feeding by the nematode. In response to a large inoculum level, this hypersensitive response may cause considerable root necrosis and damage to the plant, even though the nematodes do not complete their life cycle and reproduce.

6. Place several 2- to 3-cm segments of galled roots in a small beaker containing phloxine B stain. Remove segments after 15 min and observe the stained egg masses on the outer surface of the roots under the dissecting microscope. It is possible to estimate the total number of eggs on the root system by staining and counting egg masses on several 1-g aliquots of root and multiplying the number of egg masses by 300. This is a conservative estimate of the number of eggs per gram of root. Multiply again by the total weight of the root system to get the total number of eggs. This value can be compared to the initial inoculum levels to estimate the reproduction of the nematode.

Anticipated Results

The purpose of this exercise was to demonstrate the effects of root-knot nematodes on susceptible tomato plants at different inoculum levels. More galls should have been observed on the roots at higher inoculum levels than at lower inoculum levels. This difference may be difficult to see between the two highest inoculum levels as competition for infection sites would be quite high. The shoot weights and lengths should be similar between the controls and the lower two inoculum levels. At the lowest inoculum level, the shoots may actually be longer and weigh more. At the highest inoculum levels, the shoot weight and length should be lower. The root lengths should be shorter at the higher inoculum levels compared to those at the control because galling generally tends to reduce root growth. However, root weights may actually increase with increasing nematode inoculum levels up to the highest inoculum rates used. The galls generally make the roots heavier because the root-knot nematodes are also causing more photosynthate to be directed toward the roots than shoots as compared to uninfected plants. This effect is, in part, how the shoots become stunted as a result of nematode infection.

Questions

- Sketch the life cycle of the root-knot nematode beginning with the egg stage. What are vulnerable stages of the life cycle that could be targeted to keep nematode numbers below damaging levels?
- Are there other pathogens that cause root galling that might be confused with symptoms caused by root-knot nematodes?

EXPERIMENT 2. ISOLATION OF PLANT-PARASITIC AND FREE-LIVING NEMATODES FROM SOIL

The objectives of this experiment are to isolate plant-parasitic nematodes from soil and to demonstrate the diversity of nematode populations in row crop fields and ornamental plantings.

Materials

Each student or group of students will require the following:

- *Pictorial Key to Genera of Plant-Parasitic Nematodes* (Mai and Lyon, 1975)
- Soil sampling tubes, 1-quart plastic bags, plastic buckets, and ice chest
- Metal sieves for separating the nematodes from the soil. Two sieves are required, 250-pore openings (60 mesh) and 38-pore openings (400 mesh), each with a 20-cm diameter and 5-cm depth.
- Centrifuge equipped with a hanging bucket rotor and 50-ml, nonsterile centrifugation tubes
- Vortex mixer
- 2-l stainless-steel graduates, 150-ml beakers, and 1-l wash-bottles
- Sucrose solution (454 g sugar sucrose and water to make 1 l)

TABLE 9.1
Data Sheet for Experiment 1 (Procedure 9.2)

Group No:_____

| Plant No. | Number of Eggs Inoculated | Measurement | | | | Number of Galls |
| | | Shoot (top) | | Root | | |
		Length	Weight	Length	Weight	
1						
2						
3						
4						
5						

- Dissecting microscopes with light source
- Top-loading balance
- Plastic buckets and bags
- 2-l stainless steel graduated cylinder
- 150-ml beakers
- Petri dishes or Syracuse watchglasses
- Plastic squeeze water bottles

General Considerations

Locate a row crop or vegetable field that has been in continuous production for 5 to 10 growing seasons that can be sampled during the laboratory period. Also, find some 10- to 20-year old foundation plantings (shrubs) that can be sampled on the same day. Make the sucrose solution mentioned previously in the materials section and refrigerate (4°C) until needed.

Follow the protocols outlined in Procedure 9.3 to complete this experiment.

Procedure 9.3 Isolation of Parasitic and Free-Living Nematodes from Soil

1. Collect 20 soil cores (15- to 20-cm deep) in the root zone of an annual crop from the selected field, place in a plastic bucket, and mix thoroughly. Soil (500 to1000 cm³) should be sealed in a plastic bag, labeled, and placed in an insulated ice chest. Soil samples around the shrubs should be collected and handled in the same manner.
2. Add 250 cm³ of soil to 800 to 1200 ml of water in a 2-l stainless-steel graduated cylinder and mix vigorously for 30 sec. Allow soil to settle for 30 sec. Decant the suspension slowly onto a 60-mesh sieve nested on top of a 400-mesh sieve. Use a wash bottle to backwash nematodes and soil into a 150-

ml beaker. Nematode samples must be allowed to settle for 1 h before the centrifugation procedure.
3. Gently pour off most of the water from beakers containing soil and nematodes. Swirl contents and pour into 50-ml centrifugation tubes. Make sure tubes are balanced, using a top-loading balance — the weight of the tubes and contents should be within 1 g of each other. Weight may be adjusted by a squeeze bottle of water. Centrifuge at 420 *g* for 5 min (do not use brake). Gently pour water from tubes. Nematodes are in the soil pellet in the bottom of the tubes. Refill tubes with sucrose solution and mix with vortex mixer to resuspend nematodes. Centrifuge for 15 sec at 420 *g* (do not use brake). Decant sucrose solution/nematode suspension on to the 400-mesh sieve (pour very slowly). Gently rinse with water to remove sugar solution and backwash nematodes into 150-ml beaker. Beaker should contain no more than 25 ml of water.
4. Pour samples into petri dishes or Syracuse watchglasses and place on a dissecting scope. Nematodes are best observed if a light source from underneath is used. Use a pictorial key to identify nematode genera.

Anticipated Results

Students should be able to distinguish between free-living nematodes (nonparasitic types) and plant-parasitic nematodes. Free-living nematodes do not have stylets and are usually more active than plant-parasitic nematodes. Plant-parasitic nematodes have stylets that have knobs or flanges on the posterior end and are usually quite sluggish (see Figure 8.1). Predatory nematodes, which have rasp-like teeth or a stylet without knobs, may be present in samples. Because of the build-up of organic matter, samples from the shrubs should contain more free-living nematodes than should samples from the field. Plant-parasitic nematodes should be isolated from both samples. Some of the larger plant-parasitic nematodes such as dagger, ring, or lance (Chapter 8) may be found in the samples, and students should be able to identify them using the pictorial key.

Questions

- What characteristics and behaviors distinguish free-living nematodes (nonparasitic types) from plant-parasitic nematodes?
- Are any predatory nematodes present in the samples?
- Which sample has more free-living nematodes? Which sample has more plant-parasitic nematodes?
- Can any of the plant-parasitic nematodes be identified by the pictorial key?

ACKNOWLEDGMENT

The author and editors gratefully acknowledge the contributions of Dr. Gary Windham to this chapter.

REFERENCE

Mai, W.F. and H.H. Lyon. 1975. *Pictorial Key to Genera of Plant-Parasitic Nematodes.* Cornell University Press, Ithaca, NY, 219 pp.

SUGGESTED READINGS

Southey, J.F. (Ed.) 1986. *Laboratory Methods for Work with Plant and Soil Nematodes.* Her Majesty's Stationery Office, London, 202 pp.

Zuckerman, B.M., W.F. Mai and M.B. Harrison. 1985. *Plant Nematology Laboratory Manual.* University of Massachusetts Agricultural Experiment Station, Amherst, MA, 212 pp.

10 Plant Pathogenic Fungi

Ann B. Gould

CHAPTER 10 CONCEPTS

- Fungi are eukaryotic, heterotrophic, absorptive organisms with cell walls.

- Organisms that were collectively grouped as fungi have been placed into the following three different kingdoms: Stramenopila (Oomycota), Protista (Plasmodiophoromycota), and Fungi (Chrytridiomycota, Zycomycota, Ascomycota, and Basidiomycota).

- The Deuteromycota (*Fungi Imperfecti*) is an artificial group of true fungi that is placed in a form or artificial phylum because its teleomorphic stage is unknown or very rarely produced. Some members of this artificial group are referred to as *Mycelia sterilia* because they produce no spores at all.

- Chitin is a primary component of cell walls of true fungi and is not found in the cell walls of fungal-like organisms in the Oomycota. The Oomycota contain cellulose in their cell walls that is not found in the cell walls of true fungi.

- Most plant pathogenic fungi can live as saprophytes or parasites (facultative saprophytes). However, some plant pathogenic fungi require living plant cells for their nutrition (biotrophs).

- The sexual state of a fungus is known as the teleomorph and the asexual state is known as the anamorph.

- The anamorphic state of most fungi is the conidium (Ascomycota, Basidiomycota, and Deuteromycota) and amotile sporangiospores (Zygomycota). In Plasmodiophoromycota (kingdom Protista), the Chytridiomyctoa (kingdom Fungi) and most species of the Oomycota (kingdom Stramenopila), the anamorphic state is motile sporangiospores referred to as zoospores.

- Examples of teleomorphic spores include oospores (Oomycota), zygospores (Zygomycota), ascospores (Ascomycota), and basidiospores (Basidiomycota).

- Most fungi are dispersed as spores through air currents, water, and animals (primarily insects). Fungi may also be spread in or on infected plant parts, movement of soil, and on agricultural equipment.

Of the biotic plant disease agents known to humankind, fungi are the most prevalent, and the study of these organisms has some very rich history. Fungi were once considered to be plants and thus were in the domain of botanists. Fungi are actually microorganisms, however, and differ from plants such that they are now placed in their own kingdom (Table 10.1). With the development of techniques needed for studying microorganisms, the discipline of mycology (Greek for *Mycos* = fungus + *-logy* = study) developed also, and now the fungi encompasses organisms from many different groups. Indeed, not all fungi that cause plant disease are true fungi.

Although the formal, systematic study of fungi is a mere 250 years old, fungi have played important roles in the history of humankind for thousands of years. In their most important role, fungi are agents of decay, breaking down complex organic compounds into simpler ones for use by other organisms. Fungi are important in the production of food (wine, leavened bread, and cheese) and can also be a source of food. Fungi may be poisonous, may produce hallucinogens and have played a role in religious rites in many cultures. Fungi can attack wood products, leather goods, fabrics, petroleum products, and foods at any stage of their production, processing, or storage. They produce toxins (**mycotoxins**) and are a source of antibiotics, cyclosporin (an immunosuppressant), vaccines, hormones and enzymes, and can also cause diseases in both animals and plants. Those fungi found consistently in association with a particular plant disease are called fungal **pathogens**.

Fungi can destroy crops, and the economic consequences of this have been enormous throughout human

0-8493-1037-7/04/$0.00+$1.50
© 2004 by CRC Press LLC

75

TABLE 10.1
Some Differentiating Characteristics of Plants, Animals, and Fungi

	Animals	Plants	Fungi
Mode of obtaining food	Heterotrophy	Autotrophy	Heterotrophy
Chloroplasts	No	Yes	No
Chief components of cell wall	Cell wall lacking	Cellulose and lignin	Chitin, glucans, or cellulose (in some fungi-like organisms)
Chief sterols in cell membrane	Cholesterol	Phytosterols (ß-sistosterol, campesterol, stigmasterol)	Ergosterol
Food storage	Glycogen	Starch	Glycogen
Chromosome number in thallus	Diploid	Haploid, diploid, or polyploid	Haploid, diploid, or dikaryotic
Alternation of generations	No	Yes	Yes
Specialized vascular tissues for transport	Yes	Yes	No

history. Fungi reduce yield, destroy crops in the field and in storage, and produce toxins poisonous to humans and animals. Blights, blasts, mildews, rusts, and smuts of grains are mentioned in the Bible. The Greek philosopher Theophrastus (370–286 B.C.) recorded his speculative (but not experimental) studies of grain rusts and other plant diseases. Indeed, wheat rusts have been important from ancient times until the present wherever wheat is grown. The Roman religious ceremony, the Robigalia, appealed to the rust gods to protect grain crops from disease.

When societies depend on a single crop for a major degree of sustenance, plant diseases can have a devastating impact. Potato was the major food of Irish peasants in the 1800s; the disease late blight destroyed the potato crop in 1845–1846, resulting in massive starvation and emigration for years to come. The discipline of plant pathology was born over the scientific and political controversy caused by this disease. Until that time, fungi seen in plants were thought to be the result of plant disease, not the cause of plant disease. The famous German botanist Anton deBary worked with the late blight pathogen and proved experimentally that this was not the case. The causal agent of light blight of potato is now known as *Phytophthora infestans*. Other fungal diseases that played major roles in human history include chestnut blight, coffee rust, downy mildew of grape, and white pine blister rust.

So, what are these fascinating organisms called fungi, and what role do they play in plant disease?

FUNGAL CHARACTERISTICS

A **fungus** (pl. fungi) is a **eukaryotic**, **heterotrophic**, absorptive organism that develops a microscopic, diffuse, branched, tubular thread called a **hypha** (pl. hyphae). A group of hyphae is known collectively as a **mycelium** (pl. mycelia), which is often visible to the naked eye. The mycelium makes up the vegetative (nonreproductive) body or thallus of the fungus.

Fungi have definite cell walls, have no complex vascular system, and, except for a few groups, are not motile. The thallus of some fungi is single-celled (as in the yeasts) or appears as a **plasmodium** (as in the slime molds, Chapter 11). Many fungi can be grown in pure culture, facilitating their study, and they reproduce sexually or asexually by means of spores. As a group, the fungi encompass a way of life shared by organisms of different evolutionary backgrounds.

The hyphae of most fungi are microscopic and differ in diameter between species, from 3 to 4 μm to 30 or more μm wide. Septa (crosswalls) may or may not be present and usually contain small pores to maintain continuity with other cells. Those hyphae without crosswalls are called aseptate or **coenocytic**. Hyphae may branch to spread over a growing surface, but branching usually occurs only when nutrients near the tip become scarce. In culture, fungi form **colonies**, which can be discrete or diffuse, circular collections of hyphae or spores, or both, that arise from one cell or one grouping of cells.

Hyphal cells are bound by a cell envelope called the plasma membrane or **plasmalemma**. The plasmalemma of fungi differs from that of plants and animals in that it contains the sterol ergosterol, not cholesterol as in animals or a phytosterol as in plants. Outside the plasmalemma is the **glycocalyx**, which is manifest as a slimy sheath as in slime molds or as a firm cell wall found in most other fungi. This cell wall is composed chiefly of polysaccharides. In the fungal phyla Ascomycota, Basidiomycota, and Deuteromycota (mitosporic fungi) **chitin** [ß-(1→4) linkages of *n*-acetylglucosamine], and **glucans** (long chains of glucosyl residues) are major cell wall components. Zygomycota contain chitosan, chitin, and polyglucuronic acid, and Oomycota contain **cellulose** [ß-(1→4) linkages of glucose] and glucans. Chitin, glucans, and cellulose form strong fibers called microfibrils, which, embedded in a matrix of glycoprotein and polysaccharide, lend support to the hyphal wall. Cell walls also contain proteins and, in some fungi, dark pigments called **melanins**.

Fungal cells may contain one or many nuclei. A hyphal cell with genetically identical haploid nuclei is **monokaryotic**; cells with two genetically different but compatible haploid nuclei are **dikaryotic** (a characteristic of fungi in the Basidiomycota). Compared with animals and plants, the nuclei of fungi are small with fewer chromosomes or number of DNA base pairs. **Plasmids** (extrachromosomal pieces of DNA that are capable of independent replication) are found in fungi; the plasmid found in *Saccharomyces cerevisiae*, the common yeast fungus used to make wine and leavened bread, has been intensively studied. **Vacuoles** in hyphal cells act as storage vessels for water, nutrients or wastes, and may contain enzymes such as nucleases, phosphatases, proteases, and trehalase. Cells accumulate carbon reserve materials in the form of lipids, **glycogen**, or low-molecular-weight carbohydrates such as **trehalose**. **Mitochondria** (energy-producing organelles) in fungi vary in size, form, and number.

Fungal cells from different hyphal strands may often fuse in a process called **anastomosis**. Anastomosis, which is very common in some Ascomycota, Basidiomycota, and Deuteromycota fungi, results in the formation of a three-dimensional network of hyphae and permits the organization of some specialized structures such as **rhizomorphs**, **sclerotia**, and fruiting structures, also known as **sporocarps**.

HOW FUNGI INFECT PLANTS

NUTRITION

Fungi have the advantage over organisms in the plant kingdom in that they do not have chlorophyll and are not dependent on light to manufacture food. Thus, fungi can grow in the dark and in any direction as long as there is an external food source and water.

Fungal hyphae elongate by apical growth (from the tip). Unlike animals, fungi do not ingest their food and then digest it; they instead obtain their food through an absorptive mechanism. As the fungus grows through its food, hyphae secrete digestive **exoenzymes** into the external environment (Chapter 27). Nutrients are carried back through the fungal wall and stored in the cell as glycogen. Free water must be present to carry the nutrients back into fungal cells.

Fungi can grow rapidly over a surface stratum and may penetrate it or may produce an aerial mycelium. Fungi have the potential to utilize almost any carbon source as a food. This is restricted only by what exoenzymes the fungus produces and releases into the environment. Some of these enzymes are listed in Table 10.2.

Fungal growth may continue as long as the appropriate nutrients and environmental conditions are present. Growth ceases when the nutrient supply is exhausted or the environment is no longer favorable for development. Nutrients required by fungi include a source of carbon in the form of sugars, polysaccharides, lipids, amino acids, and proteins; nitrogen in the form of nitrate, ammonia, amino acids, polypeptides and proteins; sulfur, phosphorus, magnesium, and potassium in the form of salts; and trace elements such as iron, copper, calcium, manganese, zinc, and molybdenum.

TYPES OF FUNGAL PATHOGENS: HOW DO FUNGI GET THEIR FOOD?

Most fungi are **saprophytes** in that they use nonliving organic material as a source of food. These organisms are important scavengers and decay organisms, and along with bacteria recycle carbon, nitrogen, and essential mineral nutrients. Most plant pathogenic fungi are versatile organisms that can live as saprophytes or parasites. These fungi are called **facultative saprophytes**. They attack their hosts, grow, and reproduce as parasites and then act as a saprophytic member of the normal soil microflora between growing seasons. This makes them very difficult to control.

Many plant pathogenic fungi are classified as **necrotrophs**. These fungi are usually saprophytes and survive well as sclerotia, spores, or as mycelia in dead host material in the absence of a living host. Given the opportunity, however, these fungi become parasitic, kill, and then feed on dead plant tissues. Necrotrophs produce **secondary metabolites** that are toxic to susceptible host cells. Fungal enzymes degrade tissues killed by the toxins and the cell constituents are used as food. Disease symptoms caused by necrotrophs are manifested as small to very large patches of dead, blackened, or sunken tissue. A classic example of a necrotroph is the fungal pathogen *Monilinia fructicola*, which causes brown rot of peaches.

Biotrophic pathogens (sometimes called obligate parasites) grow or reproduce only on or within a suitable host. The biotrophic relationship is highly host specific. Biotrophic pathogens, especially fungi that attack foliar plant parts, derive all their nutrients from the host. They invade host tissues, but do not kill them (or may kill them gradually), thus ensuring a steady supply of nutrients. These fungi may produce special penetration and absorption structures called **haustoria**. Haustoria penetrate the host cell wall, but not the plasmalemma, essentially remaining outside the host cell while nutrients are transferred across the host plasmalemma into the fungus. Classic examples of pathogenic biotrophs include the powdery mildew fungi (Chapter 14) and rusts (Chapter 18).

Some biotrophs are part of a mutually beneficial symbiosis with other organisms. They are called **mutualists** and include the **mycorrhizal fungi** (fungi that grow in association with plant roots) and **endophytes**. Another

TABLE 10.2
Digestive Enzymes Produced by Fungi

Enzyme	Substrate	Utility
Cutinase	Cutin: A long-chain polymer of C_{16} and C_{18} hydroxy fatty acids, which, with waxes, forms the plant cuticle	Facilitates direct penetration of host cuticle by pathogenic fungi
Pectinase (pectin methyl esterase, polygalacturonase)	Pectin: Chains of galacturonan molecules [α-(1→4)-D-galacturonic acid] and other sugars; main components of the middle lamella and primary cell wall	Facilitates penetration and spread of pathogen in host; causes tissue maceration and cell death
Cellulase (cellulase C_1, C_2, C_x, and ß-glucosidase)	Cellulose: Insoluble, linear polymer of ß-(1→4) linkages of glucose; skeletal component of plant cell wall	Facilitates spread of pathogen in host by softening and disintegrating cell walls
Hemicellulase (e.g., xylanase, arabinase)	Hemicellulose: Mixture of amorphous polysaccharides such as xyloglucan and arabinoglucan that vary with plant species and tissues; a major component of primary cell wall	Role of these enzymes in pathogenesis is unclear
Ligninase	Lignin: complex, high-molecular weight-polymer (made of phenylpropanoid subunits); major component of secondary cell wall and middle lamella of xylem tissue	Few organisms (mostly Basidiomycetes) degrade lignin in nature; brown rot fungi degrade lignin, but cannot use it as food; white rot fungi can do both
Lipolytic enzymes (lipase, phospholipase)	Fats and oils (fatty acid molecules): Major components of plant cell membranes; also stored for energy in cells and seeds and found as wax lipids on epidermal cells	Fatty acid molecules used as a source of food by pathogen
Amylase	Starch: Main storage polysaccharide in plants; enzymes hydrolyze starch to glucose	Glucose readily used by pathogen as a source of food
Proteinase or protease	Protein: Major component of enzymes, cell walls and cell membrane; enzymes hydrolyze protein to smaller peptide fractions and amino acids	Role of these enzymes in pathogenesis unclear

group of organisms, the **hemibiotrophs**, function as both biotrophs and necrotrophs during their life cycle. An example of a hemibiotroph is the soybean anthracnose pathogen, *Colletotrichum lindemuthianum*. This fungus first lives as a biotroph by growing between the plasma membrane and cell wall of living cells. The fungus then suddenly switches to a necrotrophic phase and kills all the cells it has colonized.

DISEASE SYMPTOMS CAUSED BY FUNGI

Details of the infection process are described in Chapter 26. The visual manifestation of this process (or **symptoms**) varies depending on the infected plant part, the type of host, and the environment. Symptoms caused by fungal pathogens can be similar to those caused by other biotic and abiotic disease agents. Generally described, these symptoms are the following:

- **Necrosis** — Cell death; affected tissue may be brown or blackened, sunken, dry or slimy; in leaves, often preceded by yellowing (chlorosis, or a breakdown in chlorophyll).
- **Permanent wilting** — Blockage or destruction of vascular tissues by fungal growth, toxins, or host defense responses; affected tissue wilts and dies, associated leaves may scorch and prematurely abscise.

- Abnormal growth — **Hypertrophy** (excessive cell enlargement), **hyperplasia** (excessive cell division), **etiolation** (excessive elongation); affected tissue appears gall- or club-like, misshapen or curled.
- Leaf and fruit abscission — Premature defoliation and fruit drop.
- Replacement of host tissue — Plant reproductive structures replaced by fungal hyphae and spores.
- Mildew — Surfaces of aerial plant parts (leaves, fruit, stems, and flowers) covered with a white to gray mycelium and spores.

Common manifestations of these symptoms are listed in Table 10.3.

FUNGAL REPRODUCTION

The reproductive unit of most fungi is the **spore**, which is a small, microscopic unit consisting of one or more cells that provides a dispersal or survival function for the fungus. Spores are produced in asexual or sexual processes.

ASEXUAL REPRODUCTION

Asexual reproduction is the result of mitosis, and progeny are genetically identical to the parent. Most higher or true

TABLE 10.3
Symptoms of Plant Diseases Caused by Fungi

Symptom	Manifestation	Description	Example
Necrosis	Leaf spot	Discrete lesions of dead cells on leaf tissue between or on leaf veins, often with a light-colored center and a distinct dark-colored border and sometimes accompanied by a yellow halo; fruiting structures of fungus often evident in dead tissue	Strawberry leaf spot (*Mycosphaerella musicola*)
	Leaf blotch	Larger, more diffuse regions of deaf or discolored leaf tissue	Horse chestnut leaf blotch (*Guignardia aesculi*)
	Needle cast	Needles of conifers develop spots, turn brown at the tips, die, and prematurely fall (cast) to the ground	Rhabdocline needle cast of Douglas-fir (*Rhabdocline pseudotsugae*)
	Anthracnose	Sunken lesions on leaves, stems and fruit; on leaves, regions of dead tissue often follow leaf veins and/or margins	Anthracnose (or cane spot) of brambles (*Elsinoë veneta*)
	Scab	Discrete lesions, sunken or raised, on leaves, fruit and tubers	Apple scab (*Venturia inaequalis*)
	Dieback	Necrosis of twigs that begins at the tip and progresses toward the twig base	Sudden oak death (*Phytophthora ramorum*)
	Canker	Discrete, often elliptical lesions on branches and stems that destroy vascular tissue; can appear sunken, raised or cracked; causes wilt, dieback, and death of the branch; fruiting structures of fungus often evident in dead tissue	Cytospora canker of spruce (*Cytospora kunzei* or *Leucostoma kunzei*)
	Root and crown rot	Necrosis of feeder roots to death of the entire root system; often extends into the crown and can girdle the base of the stem; causes wilt, dieback, and death of the canopy	Rhododendron wilt (or Phytophthora root rot) (*Phytophthora cinnamomi*; *P. parasitica*)
	Cutting rot	Cuttings in propagation beds are affected by a blackened rot that begins at the cut end and travels up the stem, rapidly killing the cuttings	Blackleg of geranium (*Pythium* species)
	Damping-off	Seeds and seedlings are killed before (preemergent damping-off) or after (postemergent damping-off) they emerge from the ground	Damping-off caused by species of *Pythium*, *Rhizoctonia*, and *Fusarium*
	Soft rot	Fleshy plant organs such as bulbs, corms, rhizomes, tubers, and fruit are macerated and become water soaked and soft; tissues may eventually lose moisture, harden and shrivel into a mummy; common postharvest problem	Rhizopus soft rot of papaya (*Rhizopus stolonifer*)
	Dry rot	Dry, crumbly decay of fleshy plant organs	Fusarium dry rot of potato (*Fusarium sambucinum*)
Permanent wilting	Vascular wilt	Vascular tissue is attacked by fungi, resulting in discolored, non-functioning vessels and wilt of canopy	Dutch elm disease (*Ophiostoma ulmi*)
Abnormal growth	Leaf curl	Leaves become discolored, distorted and curled, often at the leaf edge	Peach leaf curl (*Taphrina deformans*)
	Club root	Swollen, spindle- or club-shaped roots	Clubroot of crucifers (*Plasmodiophora brassicae*)
	Galls	Enlarged growths, round or spindle-shaped, on leaves, stems, roots, or flowers	Cedar-apple rust (*Gymnosporangium juniperi-virginianae*)
	Warts	Wart-like outgrowths on stems and tubers	Potato wart disease (*Synchytrium endobioticum*)
	Witches' broom	Profuse branching of twigs that resembles a spindly broom	Witches' broom of cacao (*Crinipellis perniciosa*)
	Etiolation	Excessive shoot elongation and chlorosis, induced in poor light or by growth hormones	Foolish seedling disease of rice (*Gibberella fujikuroi*)

TABLE 10.3 (continued)
Symptoms of Plant Diseases Caused by Fungi

Symptom	Manifestation	Description	Example
Leaf and fruit abscission		Infection of petioles, leaves, and fruit cause tissues to drop prematurely	Anthracnose of shade trees (*Apiognomonia* species)
Replacement of host tissue		Plant reproductive structures are replaced by fungal hyphae or spores, or both	Common smut of corn (*Ustilago zeae*)
Mildew		Surfaces of aerial plant parts (leaves, fruit, stems, and flowers) are covered with white to gray mycelium and spores	Powdery mildew of rose (*Sphaerotheca pannosa* var. *rosa*)

fungi (Ascomycota, Basidiomycota, and Deuteromycota– the mitosporic fungi) produce nonmotile, asexual spores called **conidia** (Figure 10.1). Conidia are produced at the tip or side of a stalk or supporting structure known as a **conidiophore**. Conidiophores can be arranged singly or in fruiting structures. A **synnema** consists of a group of conidiophores that are fused together to form a stalk. An **acervulus** is a flat, saucer-shaped bed of short conidiophores that grow side by side within host tissue and beneath the epidermis or cuticle. A **pycnidium** is a globose or flask-shaped structure lined on the inside with conidiophores. A pycnidium has an opening called an **ostiole** through which spores are released.

Another type of asexual spore is the **chlamydospore**. Chlamydospores are thick-walled conidia that form when hyphal cells round up and separate. These spores function as resting spores and are found in many groups of fungi. Some members of the Oomycota (Chapter 20), or water mold fungi-like organisms, produce a motile spore called a **zoospore** in a sac-like structure called a **sporangium**. The stalk that supports the sporangium is called a **sporangiophore**. Zygomycota, best known for *Rhizopus stolonifer*, the fungus that causes bread mold and soft rot of fruit and vegetables, produce asexual spores called **sporangiospores**, also borne in a sporangium.

SEXUAL REPRODUCTION

Genetic recombination is the result of sexual reproduction and offspring are genetically different from either parent. The heart of the sexual reproductive process is fertilization. In fertilization, sex organs called **gametangia** produce special sex cells (gametes or gamete nuclei), which fuse to form a zygote. The fertilization process occurs in two steps: (1) **plasmogamy**, when the two nuclei, one from each parent, physically come together in the same cell; and (2) **karyogamy**, when these two nuclei fuse together to form a zygote.

In most fungi, the vegetative body, or thallus, has one set of chromosomes (haploid). The diploid (having two sets of chromosomes) zygote that results from fertilization therefore must undergo meiosis or a chromosome reduction division before it can develop into a haploid, multi-

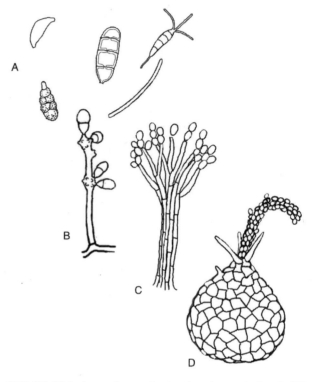

FIGURE 10.1 Asexual reproduction in mitosporic fungi. (A) Single conidia varying in shape, color (hyaline or brown) and number of cells. (B–D) Asexual fruiting structures. (B) Conidia produced on a distinct conidiophore. (C) Synnema. (D) Pycnidium. (Courtesy of N. Shishkoff.)

cellular organism. In contrast, most plants and animals, and some fungi-like species in the kingdom Stramenopila, possess a diploid thallus. Meiosis, therefore, occurs in the gametangia as the gametes are produced.

Homothallic fungi produce both male and female gametangia on a single mycelium that are capable of reproducing sexually (i.e., they are self-fertile). This is analogous to a monoecious plant, where both male and female flowers are found on the same plant. Homothallic gametangia may be obviously differentiated into male and female structures (**antheridia** and **oogonia**, respectively), or may be morphologically indistinguishable (sexually undifferentiated). Homothallic species occur in all phyla of fungi.

PLATE 1 (A) Fairy ring (*Chlorophyllum molybdites*) on lawn. (B) Bird's nest fungus (*Cyathus* species) on mulch. (C) Slime mold (*Physarum* species) on tall fescue (*Festuca arundinaceae*). (D) Slime mold (*Fuligo* species) on mulch.

PLATE 2 (A) Crown gall of *Euonymus* caused by *Agrobacterium tumefaciens*. (B) Fireblight of crabapple (*Malus* species) caused by *Erwinia amylovora*. (C) Bacterial leaf scorch of pin oak (*Quercus palustris*) caused by *Xylella fastidiosa*. (D) Aster yellows of common coneflower (*Echinacea purpurea*). (Courtesy of David Cook, University of Tennessee.)

PLATE 3 (A) *Wheat streak mosaic virus* on wheat (*Triticum aestivum*). H= healthy; I= infected (Courtesy of Marie Langham, South Dakota State University.) (B) Leaf symptomology of *Wheat streak mosaic virus*. (Courtesy of Marie Langham, South Dakota State University.) (C) *Bean pod mottle virus* of soybean (*Glycine max*). (D) *Tomato spotted wilt virus* of Hosta (*Hosta* species). (Courtesy of Anni Self, Tennessee Department of Agriculture.)

PLATE 4 (A) Foliar nematode (*Aphelenchoides* species) on Anemone (*Anemone* species). (B) Sting nematode *(Belonolaimus* species) damage on creeping bentgrass (*Agrostis palustris*). (C) Root knot nematode (*Meloidogyne incognita*) on tomato (*Lycopersicon esculentum*). (Courtesy of Steve Bost, University of Tennessee.) (D) Soybean cyst nematode (*Heterodera glycines*) on soybean (*Glycine max*). (Courtesy of Melvin Newman, University of Tennessee.)

PLATE 5 (A) Oak leaf blister caused by *Taphrina coerulescens*. (B) Powdery mildew of *Euonymus* caused by *Oidium* species. (C) Brown rot of peaches (*Prunus persica*) caused by *Monilinia fructicola*. (Courtesy of Steve Bost, University of Tennessee.) (D) Early blight of tomato (*Lycopersicon esculentum*) caused by *Alternaria solani*. (Courtesy of Steve Bost, University of Tennessee.)

PLATE 6 (A) Daylily (*Hemerocallis* species) caused by *Puccinia hemerocallidis*. (B) Brown patch of tall fescue (Festuca) caused by *Rhizoctonia solani*. (C) Butt rot of honey locust (*Gleditsia triacanthos*) caused by *Ganoderma lucidum*. (D) Leaf gall of azalea (*Azalea* species) caused by *Exobasidium vaccinii*.

PLATE 7 (A) Blue mold of tobacco (*Nicotiana tabcum*) caused by *Peronospora tabacina*. (B) Black rot of grape (*Vitis vinifera*) caused by *Guignardia* species. (C) Southern corn leaf (*Zea mays*) blight caused by *Cochliobolus heterostrophus* (*Helminthosporium maydis*). (D) Flower blight of geranium (*Pelargonium × hortorum*) caused by *Botrytis cinerea*.

PLATE 8 (A) Cold damage of new spring growth of boxwood (*Buxus* species). (B) Iron chlorosis of oak (*Quercus* species). (C) Phenoxy-class herbicide damage to Japanese maple (*Acer palmatum*). (D) Witch's broom of white pine (*Pinus strobus*).

Conversely, the sexes (or **mating types**) in **heterothallic** fungi are separated in two different individuals, which are self-sterile. This is analogous to dioecious plants, where male and female flowers are produced on different plants. The different mating types in heterothallic fungi are usually differentiated as plus (+) or minus (−), or by using letters (e.g., A and A′). Heterothallic fungi are not as common as homothallic ones.

Examples of sexual spores include **oospores**, **zygospores**, **ascospores**, and **basidiospores**. The name of a particular spore type usually reflects the parent structure that produced it. For example, ascospores are produced in a parent structure called an ascus and basidiospores are produced from a parent structure called a basidium.

ANAMORPH–TELEOMORPH RELATIONSHIPS

Most fungi are capable of reproducing both asexually and sexually. The asexual state of a fungus is called the **anamorph**, and the sexual state is the **teleomorph**. The **holomorph** encompasses the whole fungus in all its facets and forms (both anamorph and teleomorph). Some fungi have no known teleomorph, whereas others have no known anamorph.

Intriguingly, many fungi, particularly Ascomycota and some Basidiomycota, produce anamorphs and teleomorphs at different points in their life cycle. For this reason, many fungi have been given two different names, one for the anamorph when it was described, and one for the teleomorph when it was described. When the anamorph is finally associated with its corresponding teleomorph, the whole fungus (holomorph) is more properly referred to by its teleomorph name. For example, the sexual state or teleomorph of *Claviceps purpurea*, which causes ergot of rye and wheat, has an asexual state called *Sphacelia segetum,* a mitosporic fungus. The holomorph of this fungus is properly referred to as *Claviceps purpurea*. Molecular biology has been useful to assess the relationships between anamorphs and teleomorphs. Although a fungus may produce different taxonomic structures at different points in its life cycle, its genetics are the same. Examples of holomorphs include *Thanatephorus cucumeris* ana. *Rhizoctonia solani* (brown patch of turf), *Venturia inaequalis* ana. *Spilocaea pomi* (apple scab), and *Magnaporthe poae* ana. *Pyricularia grisea* (summer patch of turfgrass).

FUNGI AND THE ENVIRONMENT

Fungi are ubiquitous and grow in many different habitats and environmental conditions. Of these, a source of moisture is most critical for growth and reproduction and most fungi grow best in a damp environment. As stated previously, moisture is needed to move nutrients into hyphal cells. Indeed, some thin-walled species require a continuous flow of water to prevent desiccation. For foliar pathogenic fungi, relative humidity is important for spore germination and penetration of leaf tissue. For soil fungi, free moisture in the soil is important for dispersal as well.

Some fungi can adapt to very low moisture availability by regulating the concentration of solutes (or osmotic potential) in cells. A cellular osmotic potential higher than that of the environment will cause water to enter cells. When the environment becomes drier, concentration of solutes outside the cell becomes higher than that within the cells, and water leaves the cells. The net result is that the cell desiccates and the plasmalemma shrinks. Active release or uptake of solutes in hyphal cells helps certain fungi maintain adequate hydration under fluctuating environmental conditions. Other fungi produce resting structures with impermeable walls that withstand drying. These structures germinate when conditions become favorable for growth.

Since oxygen is required for respiration (generation of energy), most fungi do not grow well when submerged. Fungi that require oxygen for respiration are called **aerobic** or obligately oxidative. Those that do not need oxygen for respiration are **anaerobic** or fermentive. Fungi that can derive energy by oxidation or fermentation are **facultative fermentives**. Most plant pathogenic fungi are aerobic or facultatively fermentive.

Fungi must adapt to, or tolerate, considerable temperature fluctuations that occur daily or seasonally. Most fungi are **mesophilic** and grow well between 10°C and 40°C with the optimum temperature for most species between 25°C and 30°C. Some species are **thermophilic** (grow at 40°C or higher) or **psychrophilic** (grow well at less than 10°C). Fungi function best at a pH range of 4 to 7 and some produce melanins in cell walls to protect against damage from sunlight.

SURVIVAL AND DISPERSAL

With the exception of biotrophs, most plant pathogenic fungi spend part of their life cycle as parasites and the remainder as saprophytes in the soil or on plant debris. When a fungus encounters adverse environmental conditions, drains its environment of nutrients or kills its host, growth usually ceases. Inactive hyphae are subject to desiccation or attack by insects or other microbes, so pathogenic fungi must survive in a reduced metabolic or dormant state in plant debris or soil until conditions improve, or disperse to find other hosts.

Fungi produce spores to both survive adverse environmental conditions (e.g., chlamydospores) and for dispersal. Indeed, the different reproductive strategies exhibited by many fungi have different ecological advantages. Generally speaking, fungi reproduce asexually when food sources are abundant and dispersal and spread are of prime importance, and reproduce sexually at the end of the season when food is limited or in cases where dispersal is not as critical. For example, the biotrophic powdery mildew pathogens produce copious quantities of asexual spores (conidia) during the growing season when environmental conditions are favorable and there is abundant host material. These spores are easily dispersed in air and serve to spread disease rapidly over the summer months. At the end of the growing season, environmental and host factors trigger sexual reproduction of the fungus. The ascocarp produced (called a **cleistothecium** or **perithecium** — see Chapter 14) overwinters in plant parts or debris releasing ascospores (spores that result from meiosis) to initiate new infections the following spring.

Spore dispersal can be critical for fungal survival; spores released quickly from fruiting structures have a better chance of finding new hosts and initiating infections. Spores can be passively or actively liberated from the parent mycelium. Passive spore release mechanisms include rain-splash, mechanical disturbance (wind, animal activity and cultivating equipment) and electrostatic repulsion between a spore and its **sporophore** (supporting stalk). Spores can also be actively released; ascospores are forcibly discharged from the ascus in many members of the Ascomycota (e.g., *C. purpurea*) and sudden changes in cell shape may launch spores into the air (e.g., the aeciospore stage of the rust fungus *Puccinia*).

Most spores are passively dispersed (or **vectored**) through air or soil for short or long distances. The most common modes of dispersal for plant pathogenic fungi include air currents, water (via rain splash and dispersal through flowing water), and animals (notably insects). Insects have an intimate role in the life cycle and spread of fungi that cause many important diseases. For example, beetles that vector the Dutch elm disease pathogen *Ophiostoma ulmi* pick up fungal spores on their bodies and inoculate new hosts during feeding. Some plant pathogenic fungi such *Tilletia caries* (stinking smut of wheat) are dispersed on seed, which conveniently places them with their host when the time is appropriate for infection to commence. Soil pathogens that produce motile zoospores (some Oomycota and plasmodial slime molds) can more actively move through soil. Soil zoospores, attracted to roots by root exudates (**chemotaxis**), encyst and penetrate the root cortex to initiate new infections.

Some fungi, such as the basidiomycete *Armillaria mellea*, disperse through potentially inhospitable environments by producing rhizomorphs, root-like structures composed of thick strands of somatic hyphae. Rhizomorphs have active meristems that grow through soil from diseased plants to healthy roots, thus facilitating the dispersal of the fungus to new substrates. In cross section, a rhizomorph has a dark, outer rind and an inner cortex.

More commonly, however, fungi produce resting structures called sclerotia. Sclerotia are spherical structures 1 mm to 1 cm in diameter with a thick-walled rind and a central core of thin-walled cells that have abundant lipid and glycogen reserves. Sclerotia may remain viable in adverse conditions for months or years. Many plant pathogenic fungi in the Ascomycota, Basidiomycota, and mitosporic fungi produce sclerotia, especially those that infect herbaceous plants as a means of surviving between crops. Sclerotia may germinate to form sexual structures, as in *C. purpurea* or asexual conidia, as in *Botrytis cinerea*.

FUNGAL CLASSIFICATION

The classification of organisms (taxonomy) including fungi is based on criteria that change as more information becomes known about them. Taxonomy of fungi has been classically based on sexual and asexual spore morphology and how they are formed, followed by secondary considerations such as hyphal and colony characteristics. Molecular techniques that examine and compare the genetics of organisms are increasing in importance to taxonomists.

As previously mentioned, the fungi as a group used to be classified with the plants in kingdom Planta, but are now placed in three different kingdoms (Table 10.4). The endoparasitic slime molds belong to kingdom Protista. The water mold fungi (also called straminopiles) belong to the phylum Oomycota within kingdom Stramenopila (Chromista). These two fungal groups are often referred to in the literature as lower fungi or pseudofungi and are discussed further in Chapter 11 and Chapter 20. The other, true fungi (e.g., Chytridiomycota, Zygomycota, Ascomycota, and Basidiomycota) are members of kingdom Fungi (or Mycota). These fungi are also often referred to as the higher fungi. Interestingly, the Chytridiomycota, once placed in kingdom Protista, may be regarded as ancestors of the other true fungi.

Fungi are further divided into subcategories. The most basic classification of a fungal organism is the **genus** and **species** (or specific epithet). The generic name and specific epithet is constructed in Latin according to internationally recognized rules. Fungal names are also associated with an authority, which is the name of the person who discovered, described, or named the species. For example, the species *Claviceps purpurea* (Fr.:Fr.) Tul. was described by Elias Fries and modified by Tulanse. There are more than 70,000 described species of fungi.

In the classification scheme, related genera are grouped into families, families into orders, orders into classes, classes into phyla, and phyla into kingdoms. Each grouping has a standard suffix (underlined). For example, the Dutch elm disease pathogen, *Ophiostoma ulmi*, is classified as follows:

TABLE 10.4
Characteristics of Different Phyla of Plant Pathogenic Fungi

	Protista	Stramenopila	Fungi (Mycota)			
	Plasmodiophoromycota	Oomycota	Chytridiomycota	Zygomycota	Ascomycota	Basidiomycota
Habitat	Aquatic	Aquatic	Aquatic/terrestrial	Terrestrial	Terrestrial	Terrestrial
Form of thallus	Plasmodium (multinucleate mass of protoplasm without a cell wall)	Coenocytic hyphae	Globose or ovoid thallus (lacks a true mycelium)	Well-developed coenocytic hyphae; rhizoids, stolons	Well-developed septate hyphae; some are single celled (yeasts)	Well-developed septate hyphae
Chromosome number in thallus	Haploid (cruciform nuclear division)	Diploid	Diploid	Haploid	Haploid	Dikaryotic, haploid
Chief component of cell wall	Cell wall of thallus lacking	Cellulose/glucans	Chitin	Chitin/chitosan	Chitin/glucans	Chitin/glucans
Motile stage	Zoospores with two anterior, unequal, whiplash flagella	Zoospores with two flagella: one anterior, tinsel and one posterior, whiplash	Zoospores with one posterior whiplash flagellum	None	None	None
Pathogenic relationship	Obligate parasites	Facultative or obligate parasites	Obligate parasite	Facultative parasites; obligate symbionts (endomycorrhizal fungi)	Facultative or obligate parasites; endophytes	Facultative or obligate parasites
Asexual reproduction	Secondary zoospores in zoosporangia	Zoospores in sporangia; chlamydospores	Holocarpic (entire thallus matures to form thick-walled resting spores)	Sporangiospores in sporangia; chlamydospores (endomycorrhizal fungi)	Conidia on conidiophores produced on single hyphae or in fruiting bodies; budding (yeasts)	Budding, fragmentation, arthrospores, oidia, or formation of conidia
Sexual reproduction	Resting spores with chitin in cell walls, the result of fusion of zoospores to form a zygote	Oospores, the result of contact between male antheridia and female oogonia	Not confirmed	Fusion of gametangia to produce a zygospore	Formation of ascospores within an ascus	External formation of four basidiospores on a basidium; fusion of spermatia and receptive hyphae (rusts); fusion of compatible mycelia (smuts)
Major plant diseases (or groups of diseases)	Clubroot of crucifers, powdery scab of potato	Root, stem, crown, seed, and root rots, white rust, downy mildew	Potato wart, crown rot of alfalfa	Soft rot, fruit rot, bread mold, postharvest disease, endomycorrhizae	Powdery mildew, root, foot, crown, and corn rot, vascular wilt, canker, wood decay, anthracnose and other foliar diseases, needlecast, apple scab, postharvest disease	Root rot, heart rot, white and brown wood decay, fruit rot, smuts, rusts, snow mold of turfgrass, fairy ring

Kingdom **Fungi** (or **Mycota**)

   Phylum **Asco<u>mycota</u>**

      Class **Asco<u>mycetes</u>**

         Order **Ophiostoma<u>tales</u>**

            Family **Ophiostoma<u>taceae</u>**

               Genus *Ophiostoma*

                  Species (and authority) ***Ophiostoma ulmi*** **(Buisman) Nannf.**

Fungi are probably best identified by the phylum to which they belong. Fungi are placed in phyla based on the sexual phase of the life cycle. Below are brief descriptions of the different phyla, the kingdoms to which they belong, and examples of each.

KINGDOM PROTISTA

Plasmodiophoromycota (Chapter 11) is one of four groups of slime molds, including the endoparasitic slime molds that produce zoospores with two flagella and are obligate parasites. Sexual reproduction in this group includes fusion of zoospores to form a zygote. Some of these fungi vector plant viruses (Chapter 4).

Plasmodiophora brassicae (clubroot of crucifers) has a multinucleate, amoeboid thallus (plasmodium) that lacks a cell wall. The plasmodium invades root cells, inducing hypertrophy and hyperplasia (thus increasing its food supply). The "clubs" that result interfere with normal root function (absorption and translocation of water and nutrients). Plants affected by this disease wilt and stunt. *Plasmodiophora brassicae* survives in soil between suitable hosts as resting spores (product of meiosis) for many years (Chapter 11).

KINGDOM STRAMENOPILA (CHROMISTA)

Many members of the **Oomycota** (Chapter 20) produce motile, biflagellate zoospores that have a tinsel flagellum and a whiplash flagellum. The thallus is diploid and their cell walls are coenocytic and contain cellulose and glucans. Oomycetes reproduce asexually by motile zoospores in a zoosporangium (Figure 10.2) or a few species produce nonmotile sporangiospores. Chlamydospores occur in some species. Sexual reproduction results in an oospore (zygote) produced from contact between male (antheridia) and female (oogonia) gametangia (gametangial contact). Most members of the Oomycota attack roots, but some species of *Phytophthora*, the downy mildews and white rusts, also attack aerial plant parts. The Oomycota are often called water molds because most species have a spore stage that swims and thus require free water for dispersal.

The ubiquitous soil water mold *Pythium* causes seed rots, seedling damping-off, and root rots of all types of

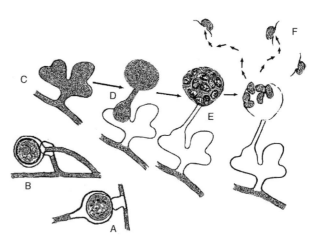

FIGURE 10.2 Reproduction in *Pythium aphanadermatum* (Oomycota), a root, stem, seed, and fruit rot pathogen with a very wide host range. (A–B) Sexual stage. Oogonium (female gamete) with an (A) intercalary or (B) terminal antheridium (male gamete). (C–F) Asexual stage. (C) Inflated, lobate sporangium. (D) Sporangium with vesicle. (E) Sporangium with vesicle containing zoospores. (F) Biflagellate zoospores. (Courtesy of N. Shishkoff.)

plants and is especially troublesome in greenhouse plant production and in turfgrasses. *Pythium* species are also responsible for many soft rots of fleshy vegetable fruit and other organs in contact with soil particles in the field, in storage, in transit, and at market.

Phytophthora species cause a variety of diseases on many hosts ranging from seedlings and annual plants to fully developed fruit and forest trees. They cause root rots; damping off; rots of lower stems, tubers, and corms; bud or fruit rots; and blights of foliage, twigs, and fruit. Some species of *Phytophthora* are host specific whereas others have a broad host range. The genus is best known for its root and crown rots (most *Phytophthora* species), late blight of potato and tomato (*P. infestans*), Rhododendron wilt (*P. cinnamomi*), and sudden oak death (*P. ramorum*).

The downy mildews (e.g., *Plasmopara* species) are obligate parasites that occur on many cultivated crops. They primarily cause foliage blights that attack and spread rapidly in young and tender green leaf, twig, and fruit tissues. These fungi produce sporangia (asexually) on distinctively dichotomously branched sporangiophores, which emerge through stomates on the lower surfaces of leaves, forming a visible mat with necrotic lesions on leaf surfaces. The downy mildews require a film of water on the plant and high relative humidity during cool or warm, but not hot, weather. Downy mildew of grape (*P. viticola*) destroyed the grape and wine industry in Europe after the disease was imported from the U.S. in 1875. It led to the discovery of one of the first fungicides, **Bordeaux mixture** (copper sulfate and lime; Chapter 33).

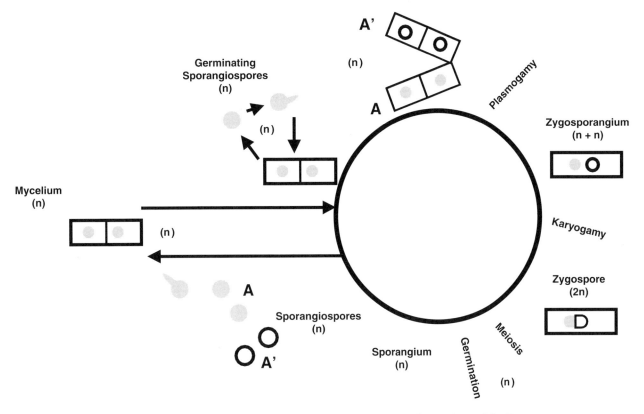

FIGURE 10.3 General life cycle of the Zygomycota. This life cycle is for heterothallis members of the Zygomycota.

KINGDOM FUNGI

Chytridiomycota

Chytridiomycota (Chapter 11), the chytrids, inhabit water or soil. They are coenocytic and lack a true mycelium. The thallus is irregular, globose, or ovoid; diploid; and its cell walls contain chitin and glucans. They have motile cells (zoospores) that possess a single, posterior, whiplash flagellum. The mature vegetative body transforms into thick-walled resting spores or a sporangium.

Olpidium brassicae is an **endobiotic** parasite of the roots of cabbage and other monocots and dicots. The entire thallus converts to an asexual reproductive structure (**holocarpic**); the sexual cycle in this organism, however, has not been confirmed. *Olpidium* is a known vector of plant viruses such as tobacco necrosis virus and lettuce big vein virus.

Zygomycota

The **Zygomycota** are true fungi with a well-developed, coenocytic, haploid mycelium with chitin in their cell walls. Zygomycetes produce nonmotile spores called sporangiospores. Zygospores (sexual spores) are produced by fusion of two morphologically similar gametangia gamet-

angial copulation of opposite mating type (heterothallic) (Figure 10.3). These fungi are terrestrial and are saprophytes or weak pathogens causing soft rots or molds. Members of this group also include parasites of insects (Entomophthoralean fungi), nematodes, mushrooms, and humans, as well as dung fungi and the endomycorrhizal fungi (see discussion on mutualists).

Rhizopus stolonifer is a ubiquitous saprophyte notably recognized as a common bread mold, which also causes a soft rot of many fleshy fruit, vegetables, flowers, bulbs, corms, and seeds. Black, spherical sporangia that produce asexual sporangiospores (also known as mitospores) appear on the swollen tip (**columella**) of a long, aerial sporangiophore. The fungus produces **rhizoids** (short branches of thallus that resemble roots) within its food substrate and hyphal **stolons** that skip over the surface of the substratum. *Rhizopus* does not penetrate host cells; it produces cellulases and pectinases that degrade cell walls, causing infected areas to become soft and water-soaked and absorbs the nutrients released by the enzymes. The fungus emerges through the broken epidermal tissue and asexually reproduces, forming a fluffy covering of fungal growth. When the food supply diminishes, gametangia combine to form a zygosporangium with a single zygospore inside.

Like *Rhizopus*, species of *Mucor* are common soil inhabitants and are often associated with decaying fruits and vegetables. They differ from *Rhizopus* fungi, however, in that they lack rhizoids. *Mucor amphibiorum* causes a dermal and systemic infection of toads, frogs, and platypus, and other species in this genus are human parasites and are implicated in dermal and respiratory allergies. Post-harvest rots of apple, peach, and pear are caused by *Mucor piriformis*. This species is a soil inhabitant that contaminates fruit that falls to the orchard floor. The fungus penetrates at the calyx or through wounds and within two months in cold storage causes complete decay of the fruit. Fungal mycelia can often be seen emerging in tufts from the surface of affected fruit. The disease is managed by destroying fallen fruit and by using sanitation practices where contaminated soil is not allowed to contact harvested fruit.

Ascomycota

The **Ascomycota** (Chapter 13 to Chapter 15) are true fungi with a well-developed, septate mycelium containing chitin in their cell walls. The sexual structure of the ascomycetes is an **ascus**, which is shaped like a sac and contains the products of meiosis, called **ascospores**, which normally number eight per ascus. Asci are either **unitunicate** (with a single wall) or **bitunicate** (with a double wall).

Ascospores are forcibly liberated from asci, which are produced in fruiting structures called **ascocarps**. Ascocarps produced by various ascomycetes include the following:

- **Cleistothecia** — Completely closed ascocarps lined with one or more asci.
- **Perithecia** — Flask-shaped ascocarps with an ostiole at the tip (Chapter 15)
- **Apothecia** — Open, cup-shaped ascocarps where asci are exposed (Chapter 15)
- Ascostroma or **pseudothecia** — Asci are produced in a cavity or **locule** buried within a matrix (stroma) of fungal mycelium (Chapter 15).

Asci not produced in an ascocarp are called **naked asci** (Chapter 13).

Asexual reproduction in the Ascomycota includes fission, budding, fragmentation, or formation of conidia or chlamydospores. Conidia are produced on conidiophores. The species of the Ascomycota are a diverse group of fungi that can use many different substrates and occupy a variety of niches.

One of the most infamous diseases caused by an already introduced ascomycete is ergot of rye and wheat (Chapter 1). The pathogen *C. purpurea* occurs worldwide. Grain in the seed head is replaced with fungal sclerotia (called ergots) that are poisonous to humans and animals. When ingested, fungal alkaloids restrict blood vessels and affect the central nervous system, causing gangrene in the extremities, hallucinations, and miscarriage. The resulting disorder in humans, known as ergotism or St. Anthony's fire, was common in the Middle Ages, especially in peasants who subsisted on bread made with contaminated rye flour.

Claviceps purpurea is homothallic. In the spring, sclerotia germinate on fallen seed heads to form perithecia at the periphery. Each perithecium produces many asci, which forcibly discharge eight ascospores that are wind-disseminated to infect the ovaries of young flowers. Within a week, droplets of conidia exude in a sticky liquid (honeydew) from the young florets of infected heads. Insects are drawn to the honeydew and carry the conidia on body parts to new flowers. Conidia are also spread by splashing rain. The ergots mature at the same time as seed and are harvested or fall to the ground where they overwinter. Grain containing as little as 0.3% of sclerotia by weight can cause ergotism.

Basidiomycota

The **Basidiomycota** (Chapter 18 and Chapter 19) are true fungi that often have large, macroscopic fruiting bodies. They have a well-developed, septate mycelium with chitin in their cell wall. The mycelium of basidiomycetes can be dikaryotic, each cell containing two haploid nuclei. The sexual structure is called a **basidium** (or club), on which four haploid basidiospores (the result of karyogamy and meiosis) are produced on stalks called sterigmata. Basidia are produced on **basidiocarps**. Basidiomycetes reproduce asexually by budding and fragmentation or by formation of conidia, **arthrospores** (formed when hyphal fragments break into unicellular sections), or **oidia** (produced by short branches, or oidiophores that cut off oidia in succession from the tip). Like ascomycetes, basidiomycetes utilize a variety of substrates and occupy many different niches. Some of the most important plant diseases, the rusts and the smuts (introduced in Chapter 18), are caused by basidiomycetes.

Some basidiomycetes are best known for the production of the classic mushroom or a basidiocarp. A mushroom basidiocarp consists of a **pileus** (cap) lined with **gills** and a **stipe** (stalk). A **volva** may remain at the base as a remnant of the universal veil that once enveloped the entire developing mushroom (button). The **annulus** is the ring around the stipe as a remnant of a partial or inner veil that once enveloped only the gills or pores on the underside of the mushroom. Basidia line the outer surface of gills.

Armillaria mellea affects hundreds of species of trees, shrubs, fruits, and vegetables worldwide. The pathogen is also an important decay fungus in forest ecosystems. *Armillaria* forms rhizomorphs and **mycelial fans** (fan-shaped hyphae) under the bark near the crown of infected trees. Rhizomorphs consist of cord-like "shoestring" threads of mycelium 1 to 3 mm in diameter. They have a compact outer layer (or rind) of black mycelium and a core of white mycelium, and form a branched network that develops over wood and into surrounding soil. Rhizomorphs initiate new infections when in contact with fresh substrate. Plants affected by the fungus show symptoms similar to those caused by other root diseases (reduced growth, small, yellow leaves, dieback, and gradual decline and death). *Armillaria* produces honey-colored mushrooms at the base of dead or dying trees in early fall.

DEUTEROMYCOTA (MITOSPORIC FUNGI)

Some fungi are classified in an artificial group (not a true phylum) called the Deuteromycota or the *Fungi Imperfecti* (Chapter 16). These fungi have no known sexual state in their life cycle and thus have not been classified in one of the other groups. Mitosporic fungi usually reproduce asexually (mitotically) via conidia and are considered to be anamorphic partners of associated teleomorphs in the Ascomycota or Basidiomycota, whose identity may or may not be known. They have hyphae that possess the characteristics of the teleomorph. Some mitosporic fungi have no known reproductive process and are called the sterile fungi or **Mycelia Sterilia**, although some of these fungi do produce sclerotia. Molecular techniques are very useful for placing mitosporic fungi with their associated teleomorphs, and, indeed, may one day eliminate the need for artificial names such as Mycelia Sterilia.

Mitosporic fungi occupy the same habitats as their teleomorph relatives; thus they are largely terrestrial. Most are saprophytes, some are symbionts of lichens, others are grass endophytes or mycorrhizal fungi, and many are plant pathogens. These fungi are divided into the following three informal classes: **Hyphomycetes** (spores are produced on separate conidiophores), **Agonomycetes** (Mycelia Sterilia), and **Coelomycetes** (spores are produced in fruiting structures called **conidiomata**). Classification within these groups is based on conidiophore, conidia, and hyphal characteristics.

MUTUALISTS

Certain biotrophs such as mycorrhizal fungi and endophytes live all or most of their life cycles in close association with their plant host. In many cases, this symbiosis is a mutualistic one because both the fungus and the plant host benefit from the association.

TURFGRASS ENDOPHYTES

Simply stated, endophytes are organisms that grow within plants. When searching the literature, the term is often used to describe a specific relationship between certain fungi and turfgrass hosts. Unlike turfgrass pathogens such as *Magnaporthe poae*, however, endophytic fungi do not harm the host; indeed, they appear to enhance stress tolerance and resistance to feeding by certain insects.

Endophytic fungi associated with turfgrass are ascomycetes in the family Claviciptaceae and include species of *Epichloë* [ana. *Neotyphodium* (= *Acremonium*)] and *Balansia*. Although these fungi colonize all above-ground plant parts, they are found most readily within the leaf sheath of the turfgrass host and do not colonize the roots. Turfgrass hosts include perennial ryegrass and several fescues (e.g., tall fescue, hard fescue, chewings fescue, and creeping red fescue).

The ability to reproduce sexually varies among the different types of endophytic fungi. In the heterothallic endophyte *Epichloë*, a stroma is composed of fungal mycelium embedded together with the flowers and seeds of the grass forms on the stem just below the leaf blade. The stroma is situated in such a way that nutrients flow from living host tissues into the fungus. Clusters of spores called **spermatia** are produced on the surface of the stroma. Flies (*Phorbia* species) visit the stroma and eat the spermatia, which remain undigested in the fly intestinal tract. When a female fly then visits another stroma of opposite mating type to lay eggs, she defecates and deposits spermatia all over the stroma. Fertilization follows and then perithecia form and ascospores are released to infect nearby florets. Interestingly, the fly larvae that hatch from the eggs then eat the stroma. Some endophytes such as *Neotyphodium* do not produce stroma and cannot sexually reproduce. These fungi grow into the seed at flowering time and are thus seed transmitted. Other endophytes exhibit both sexual and asexual reproductive characteristics.

Benefits conferred to the host by this endophytic association include enhanced nutrition, drought tolerance, and increased hardiness and resistance to disease. Notably, surface feeding by insect pests is deterred. Toxic alkaloids such as peramine are produced in infected hosts, and these compounds deter feeding by insects such as chinchbug, sod webworm, and billbug. *Neotyphodium* endophytes are used in turfgrass breeding programs to enhance these beneficial characteristics in turf intended for sports or landscape (not pasture) use.

Unfortunately, the endophyte association in pastures grasses can have disastrous impact on grazing livestock. Host plants containing *Neotyphodium* endophytes produce ergot alkaloids that reduce blood flow to extremities, causing tails and hooves to rot and fall off. In other associations, ingested lolitrems cause animals to spasm uncontrollably in a syndrome known as ryegrass staggers. Indeed, sleepy grass (*Achnatherum robustrum*) infected by a *Neotyphodium* endophyte produces lysergic acid, a relative of LSD; horses that ingest relatively small amounts of such compounds go to sleep for several days.

MYCORRHIZAL FUNGI

A mycorrhiza (or "fungus-root," coined by A. B. Frank in 1885) is a type of mutualism in which both partners benefit in terms of evolution and fitness (ability to survive and reproduce). This is primarily a nutritional relationship where the fungal partner increases the efficiency of nutrient uptake (notably phosphorus) by the host and in turn receives carbon made by the host during photosynthesis. The mycorrhizal association is intimate, diverse, and the fungi involved vary in taxonomy, physiology, and ecology. Mycorrhizal fungi infect more than 90% of vascular plants worldwide including many important crop plants, and thus the association is economically, ecologically, and agriculturally significant.

Based on the morphology of the association, mycorrhizae are classified into the following three major groups: ectomycorrhizal (outside of root), endomycorrhizal (within root) and ectendomycorrhizal (both inside and outside the root).

Ectomycorrhizae

In this group, a fungal **mantle** of septate hyphae forms over the entire root surface, replacing root hairs and the root cap. Hyphae proliferate between root cortex cells, forming an intercellular network, or **Hartig net**, and it is within this net that nutrients are exchanged between partners. The fungus does not penetrate cells of the cortex or stele (vascular bundle). The ectomycorrhizal relationship is very long lasting, lasting up to 3 years. The mycelium can extend and retrieve nutrients up to 4 m from the roots.

Ectomycorrhizal fungi are mostly members of the Basidiomycota and the Ascomycota and include tooth fungi, chanterelles, mushrooms, puffballs, club fungi, and truffles. Indeed, many fungal sporocarps seen beneath the tree canopy on the forest floor are ectomycorrhizal fungi. Approximately 5000 fungal species form associations with more than 2000 partners, including plants in the Betulaceae (birch), Fagaceae (beech and oak), Pinaceae (pine), and also with some pteridophytes (ferns and horsetail) and eucalyptus. The ectomycorrhizal association comprises about 20% of all mycorrhizal associations. The taxonomy of this group is based on identification of the fungal sporocarp.

Endomycorrhizae

The endomycorrhizae, in contrast, do not change the gross morphology of the root and hyphae proliferate both between and within the cells of the root cortex, but do not penetrate the stele. There are three different types of endomycorrhizae: glomalean, orchidaceous, and ericaceous.

Glomalean Endomycorrhizae

This mycorrhizal relationship, also known as arbuscular or vesicular-arbuscular, is worldwide in distribution and occurs in many different habitats. Glomalean mycorrhizal fungi have a very broad host range; a relatively small number of species (about 150) in the Zygomycota form endomycorrhiza, with more than 70% of them being angiosperms and gymnosperms.

In the infection process, chlamydospores resting in soil germinate in the vicinity of plant roots. They penetrate the epidermis, and an aseptate, irregular mycelium grows between the cells of the root cortex. Curious hyphal coils called **pelotons** may form between and within the cells of the outer cortex. **Arbuscules** (special, dichotomously branched haustoria) then develop intracellularly and serve as the site of nutrient and carbon exchange. These structures last only 4 to 15 days before hyphal content is withdrawn and the arbuscules disintegrate. **Vesicles** are produced intra- or intercellularly. These ovate to spherical structures contain storage lipids and may also serve as propagules. The fungus reproduces asexually via chlamydospores that form internally or on external hyphae that may extend as far as 1 cm from the root. As the sexual state of these fungi is unknown, the taxonomy of this group is based on characters such as chlamydospore morphology and content.

Although there are relatively few glomalean species, many host plants can accept more than one mycorrhizal endophyte. The ability of an endophyte to infect a certain host often depends on environmental conditions rather than host specificity. For example, a strain of a single species indigenous to a certain area may successfully compete with a nonindigenous strain of the same species introduced into the soil, suggesting that certain strains may adapt to local environmental conditions.

Orchidaceous Endomycorrhizae

This group includes those fungi that form associations with members of the orchid family. In nature, fungal hyphae penetrate the protocorms of orchids during the saprophytic stage, enabling the seedlings to continue development. Orchidaceous mycorrhizal fungi are primarily basidiomycetes, most of which form imperfect stages in the genus *Rhizoctonia*.

Ericaceous Endomycorrhizae

In this group, fungi in the Ascomycota and Basidiomycota form associations with plants in the Ericales (e.g., rhododendron, blueberry, heather, Pacific madrone, bearberry and, Indian pipe). In most associations, penetration of cortical cells by the fungus occurs, but arbuscules do not form. A mantle or sheath may form on the root surface and in some cases a Hartig-net may be present. An interesting example of an ericaceous mycorrhizal relationship is the one between several *Boletus* species (Basidiomycota) and *Monotropa*, or Indian pipe. This plant is a white, unifloral (produces a single flower), achlorophyllous (without chlorophyll), parasitic plant (Chapter 22) that receives all its carbon and nutrients from neighboring trees via a connecting mycorrhizal fungus.

Ectendomycorrhizae

In this last major group, typical ectendomycorrhizae form on forest trees. Under certain conditions, fungal hyphae penetrate the cortical cells close to the stele. The perfect stages of these fungi are basidiomycetes.

Significance of Mycorrhizal Association

Mycorrhizal fungi, in essence, act as root hairs. Fungal mycelia explore large volumes of soil and retrieve and translocate nutrients that may be otherwise unavailable to the host. The success of the mycorrhizal relationship is probably due to its long evolutionary history, its economy (fungal mycelium is less "expensive" for a host to maintain than an extensive system of root hairs), and its efficiency (fungi produce phosphatase enzymes that readily solubilize phosphorus from soil particles).

Partners associated with a mycorrhizal fungus often exhibit an increase in growth, which is attributed to an increase in phosphorus nutrition of the plant. External mycelia effectively increase the phosphorus depletion zone around each root, the phosphorus absorbed by the fungal mycelium is translocated to the site of nutrient exchange (such as arbuscules or the Hartig net), and is released to the host. Fertilizer high in phosphorus and nitrogen reduces mycorrhizal infectivity and sporulation in host plants.

The mycorrhizal relationship benefits plants in other ways as well. Ectomycorrhizae, for example, play a role in protecting plants from plant pathogenic organisms; the mantle acts as a physical barrier. Some ericoid mycorrhizae produce antimicrobial compounds that inhibit other competing organisms. In a broader sense, mycorrhizae serve to interconnect individuals within a plant community; thus mature plants can nurture seedlings in a community by means of a buffering effect.

Of course, the fungi too benefit from the association in that a high percentage of photosynthate manufactured by the host is translocated to the fungal symbiont. Host sugars (glucose and sucrose) obtained by the fungus are used to produce new hyphae or are converted to glycogen and stored in vesicles or other structures.

SUGGESTED READINGS

Agrios, G.N. 1997. *Plant Pathology*, 4th ed. Academic Press, San Diego.

Barr, D.J.S. 2001. Chytridiomycota, pp. 93–112 in *The Mycota VII. Part A: Systematics and Evolution*, D.J. McLaughlin, E.G. McLaughlin and P. A. Lemke (Eds.). Springer-Verlag, Berlin.

Braselton, J. 2001. Plasmodiophoromycota, pp. 81–91 in *The Mycota VII. Part A: Systematics and Evolution*, D. J. McLaughlin, E.G. McLaughlin and P. A. Lemke (Eds.). Springer-Verlag, Berlin.

Carlile, M.J., S.C. Watkinson and G.W. Gooday. 2001. *The Fungi*, 2nd ed. Academic Press, San Diego.

Hawksworth, D.L., B.C. Sutton and G.C. Ainsworth. (Eds.) 1983. *Ainsworth & Bisby's Dictionary of the Fungi.* Commonwealth Mycological Institute, Kew Surrey.

Kyde, M.M. and A.B Gould. 2000. Mycorrhizal endosymbiosis, pp. 161–198 in *Microbial Endophytes*, C.W. Bacon and J. F. White (Eds.). Marcel Dekker, New York.

Schumann, G.L. 1991. *Plant Diseases: Their Biology and Social Impact.* APS Press, St. Paul, MN.

Sinclair, W.A., H.H. Lyon and W.T. Johnston. 1987. *Diseases of Trees and Shrubs.* Cornell University Press, Ithaca.

Taylor, J.W., J. Spatafora and M. Berbee 2001. Ascomycota. Tree of Life Web Project. http://tolweb.org/tree?group=Ascomycota&contgroup=Fungi.

Ulloa, M. and R.T. Hanlin. 2000. *Illustrated Dictionary of Mycology.* APS Press, St. Paul, MN.

Volk, T.J. 2001. Fungi, pp. 141–163 in *Encyclopedia of Biodiversity*, Vol. 3. Academic Press, New York.

Volk, T.J. 2002. *Tom Volk's Fungi.* Department of Biology, University of Wisconsin-LaCrosse. http://botit.botany.wisc.edu/toms_fungi.

White, J.F., P.V. Reddy and C.W. Bacon. 2000. Biotrophic endophytes of grasses: a systemic appraisal, pp. 49–62 in *Microbial Endophytes*, C.W. Bacon and J.F. White. (Eds.).Marcel Dekker, New York.

11 Slime Molds and Zoosporic Fungi

Sharon E. Mozley, David Porter, and Marc A. Cubeta

CHAPTER 11 CONCEPTS

- Slime molds and zoosporic fungi represent several phylogenetically distinct groups of microorganisms.

- *Labyrinthula zosterae, Plamodiophora brassicae,* and *Synchytrium macrosporum* have been classified at various times with protists, slime molds, and zoosporic fungi.

- *Synchytrium macrosporum* is a zoosporic fungus, *L. zosterae* is a member of kingdom Stramenopila, and the taxonomic placement and phylogenetic relationship of *P. brassicae* to other microorganisms is less certain.

- Knowledge and understanding of the phylogeny and taxonomy of genetically diverse assemblages of plant pathogenic microorganisms can contribute to improved diagnosis and management of plant disease.

Three lower eukaryotic microorganisms *(Plasmodiophora brassicae, Labyrinthula zosterae,* and *Synchytrium macrosporum)* represent important pathogens of agricultural, aquatic, grassland, wetland, and woodland species of plants. The phylogeny (evolution of organisms over time) and taxonomy (classification of living organisms) of *L. zosterae, P. brassicae,* and *S. macrosporum* will be discussed in relation to their importance in plant pathology.

The scientific discipline of plant pathology is focused largely on the study of organisms that cause disease (i.e., pathogens) on economically important species of vascular plants. Plant pathologists must be familiar with a wide variety of pathogens to ensure accurate identification of the causal agent for successful deployment of economical and environmentally sound disease management strategies. In addition, plant pathologists must also be able to recognize a diverse range of plant-associated organisms typically studied in related disciplines outside of plant pathology (e.g., biology, botany, mycology, and zoology). Slime molds and certain groups of zoosporic fungi represent good examples of organisms that cause a relatively small number of diseases of economically important plants.

Zoosporic fungi are found in several phylogenetically unrelated groups of microorganisms that produce motile, flagellated **spores (zoospores)**, usually as a result of **asexual reproduction** during some stage of their life (Fuller and Jaworski, 1987). Thus, the term zoosporic fungi is descriptive and not of evolutionary or phylogenetic significance. Zoosporic fungi are found in two different kingdoms of organisms: the kingdom Fungi and the recently named kingdom Stramenopila (Dick, 2001). The oldest branch of the kingdom Fungi is the **phylum** Chytridiomycota, all of which are zoosporic fungi. Also in the

kingdom Fungi are the more familiar (but not zoosporic) Zygomycota, Ascomycota, and Basidiomycota. The zoosporic fungi in the kingdom Stramenopila include the hyphochytrids, labyrinthulids, and oomycetes. The kingdom Stramenopila represents an extremely diverse group of organisms that also includes nonzoosporic organisms such as brown algae, chrysophyte algae, and diatoms. Organisms classified in the Phylum Chytridiomycota produce zoospores with usually a single, posteriorly directed and "hair-less" (smooth) **flagellum**, whereas organisms classified in the kingdom Stramenopila (stramenopiles) are characterized by the production of zoospores with usually two flagella (biflagellate), but always with tripartite tubular hairs decorating one of them (Dick, 2001; Figure 11.1). A flagellum that possesses two rows of tripartite tubular hairs is also referred to as a heterokont or tinsel flagellum (Hawksworth et al., 1995).

The slime molds or Mycetozoans include three closely related groups — the protostelids, dictyostelids, and myxomycetes — and a fourth unrelated group, the acrasids, a polyphyletic (i.e., of multiple evolutionary origin) assemblage of organisms (Olive, 1975; Martin et al., 1983). Recent phylogenetic analyses based on protein sequence data support the three monophyletic groups (i.e., **clades**) of the Mycetozoa (Baldauf and Doolittle, 1997). Slime molds are heterotrophic organisms that form an amoeboid stage during their life and can be isolated from bark, decaying organic matter, dung, plants, and soil. Slime molds commonly ingest bacteria and fungal spores as a food source and can often be observed on grass or wood during periods of wet weather. In general, slime molds are not usually damaging to plants. However, slime molds may grow on the surfaces of plants or mulch and cause concerns to the public (see Plate 1C and Plate 1D follow-

0-8493-1037-7/04/$0.00+$1.50

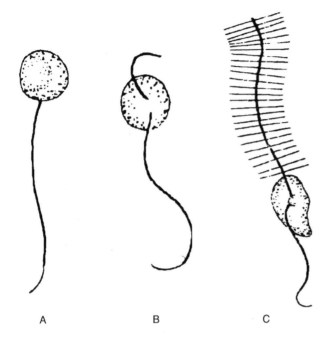

FIGURE 11.1 Zoospores of (A) *Synchytrium macrosporum* (B) *Plasmodiophora brassicae.* (C) *Labyrinthula zosterae.* The tripartite tubular hairs on the posterior flagellum of *L. zosterae* are not visible with a light microscope. (Parts (A) and (B) from Dick, M.W., 2001. *Straminipilous Fungi,* Kluwer Academic Publishers, Dordrecht, The Netherlands, and Part (C) from Sparrow, F.K., 1960. *Aquatic Phycimycetes,* 2nd ed., University of Michigan Press, Ann Arbor, MI. With permission. Redrawn by L.J. Gray.)

ing page 80). Although different species of slime molds may occur in a limited geographic area, as a group they are found in most temperate and tropical regions of the world (Figure 11.2).

Two groups of organisms often classified with both slime molds and zoosporic fungi are the plasmodiophorids and labyrinthulids (Braselton, 2001). The connection between slime molds and plasmodiophorids is tenuous and based only on the following two common characteristics: (1) the presence of a naked multinucleate protoplast similar to a myxomycete **plasmodium** and (2) the production of anteriorly directed, biflagellate zoospores with two hairless flagella. The plasmodiophorids have a distinct phylogenetic origin separate from the mycetozoans and the acrasids (Karling, 1964; Castlebury and Domier, 1998). The plasmodiophorids are usually considered as members of the zoosporic fungi because of their osmotrophic mode of nutrition and production of zoospores, but are distantly related to both the straminipiles and zoosporic fungi (Dick, 2001). As a group, the plasmodiophorids are obligate endoparasites (**biotrophs**) of algae, fungi, protists, and vascular plants. Some plasmodiophorids are important plant pathogens of agricultural crops and include the following: *Plasmodiophora brassicae* (clubroot of crucifers), *Polymxya graminis* (root diseases of cereals), and *Spon-*

gospora subterranean (powdery scab of potato). The last two pathogens can also transmit plant viruses (Chapter 4).

From a taxonomic perspective, labyrinthulids have also been associated with slime molds, fungi and various protists. Labyrinthulids produce a network of fine hyaline filaments through which the characteristic spindle-shaped cells move (called an ectoplasmic net), thus prompting some researchers to classify them with the slime molds. However, morphological, ultrastructural, and recent molecular data place the labyrinthulids into the kingdom Stramenopila as a sister group to the oomycetes (Chapter 20) and clearly separating them from both the plasmodiophorids and slime molds (Porter, 1990; Leander and Porter, 2001; Dick, 2001).

The labyrinthulids include organisms that are important decomposers and parasites of algae and vascular plants in coastal marine habitats. *Labyrinthula zosterae* is a pathogen of eelgrass (*Zostera marina*), an ecologically important seagrass that forms vast subtidal meadows in estuarine communities in temperate regions of the world. Between 1934 and 1935, most populations of eelgrass present in the North Atlantic were killed due to a pandemic "wasting disease" caused by *L. zosterae* (Porter, 1990). Since that time, Atlantic eelgrass beds have recovered, but are still not as extensive as before the pandemic. However, *L. zosterae* is still associated with localized die-back of eelgrass.

Chytrids are members of the phylum Chytridiomycota, the most basal clade within the kingdom Fungi (James et al., 2000). Because of their early evolutionary divergence, most chytrids exhibit certain characteristics not shared by the Ascomycota and Basidiomycota, such as determinate growth and production of zoospores via asexual reproduction. Chytrid sporangia are microscopic in size and much smaller than the fruiting bodies of most Basidiomycota and Ascomycota. Although a large number of chytrid species are parasitic on a wide variety of other organisms including fungi, only a few plant hosts are of economic importance. Chytrids often appear simple in form and structure on initial examination, but their morphological characteristics can exhibit considerable variation in shape and size. Chytrids are of great ecological importance as decomposers of a wide variety of biological substrates. For example, a large number of chytrids are known to degrade cellulose in the leaves and stems of plants, whereas others can degrade chitin, keratin, and sporopollenin, a biopolymer associated with pollen grains highly resistant to biological degradation. Chytrids present within the rumen of certain herbivores (e.g., cows and sheep) are also associated with the breakdown of cellulose substrates and provide much needed energy and nutrients for growth and development of animals.

Similar to other true fungi, chytrids have chitin in their cell walls, have flattened plate-like mitochondrial cristae, and exhibit an absorptive mode of nutrition. However,

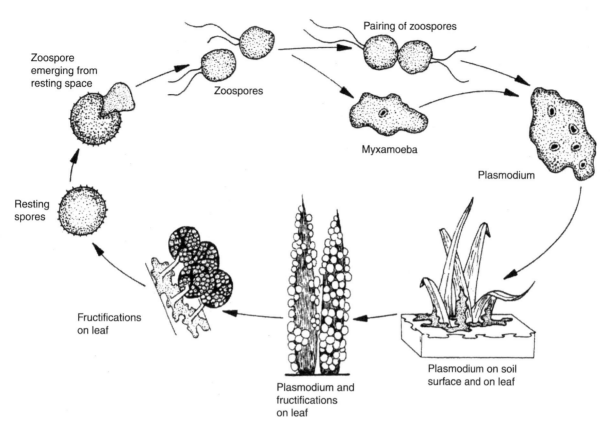

FIGURE 11.2 Life cycle of slime mold. (From Agrios, G.N., 1997. *Plant Pathology,* 4th ed., Academic Press, New York. With permission.)

unlike all other members of the kingdom Fungi, chytrids reproduce by the formation of zoospores and lack true mycelium. Chytrid zoospores differ from stramenopile zoospores in having only one posteriorly directed, smooth flagellum (Figure 11.1). Because chytrids are microscopic and usually determinant in their growth, they are often overlooked in the environment, but can be readily found and observed using baiting techniques and a simple dissecting microscope. Although chytrids are sometimes referred to as aquatic fungi or water molds, they are present and can be found nearly anywhere that water is available, including in soil. A large number of chytrids are parasites of algae, fungi, plants, and even frogs. Many chytrids can also exist as **saprophytes**.

The genus *Synchytrium* has over 200 species that are known to parasites of a wide range of algae and plants in fresh water and terrestrial habitats. *Synchytrium endobioticum* is the best-known species and is the causal agent of potato wart disease. The fungus can also attack tomato and noncultivated species of *Solanum*. Since the initial discovery of potato wart in Hungary in 1896, this disease has been identified on most continents. Because of the serious nature of potato wart, *S. endobioticum* was placed on the list of quarantined organisms established in 1912 by the United States Department of Agriculture (USDA) under the Plant Protection Act. Canada also has enacted

quarantine laws to prevent movement of potatoes and soil from Newfoundland, where the disease commonly occurs, to other Canadian provinces. A recent outbreak of potato wart in a single field in Prince Edward Island (PEI), Canada, in 2000 prompted the USDA to impose a quarantine that banned importation of seed potatoes from PEI into the U.S. for one year. Potato wart was also recently discovered in two additional fields on PEI in 2002. The **resting spores** of *S. endobioticum* can survive for 20 to 30 years in soil (Agrios, 1997).

Other chytrid pathogens of economically important species of agricultural plants include the following: *Olpidium* species (infect roots of many plants and a vector of at least six plant viruses — Chapter 4), *Urophlyctis* species (crown wart of alfalfa), and *Physoderma maydis* (brown spot of corn).

PLASMODIOPHORA BRASSICAE

Plasmodiophora brassicae is an important pathogen of cultivated agricultural crops that belong to the Brassicaceae (the mustard family). Members of this plant family are often referred to as crucifers (because of their cross-shaped flowers) or cole (which is German for stem). Crops and include the following: broccoli, Brussell sprouts, cabbage, Chinese cabbage, canola, cauliflower, collards, kale, kohlrabi, mustard, radish, rape, rutabaga, and turnip. Several additional genera of cultivated and weed species of plants in the genera *Alyssum, Amoracia, Brassica, Camelina, Capsella, Erysimum, Iberis, Lepidium, Lobularia, Lunaria, Matthiola, Nasturtium, Raphanus, Rorippa, Sinapis, Sisymbrium,* and *Thlaspi* are also hosts for *P. brassicae* (Farr et al., 1995). *Arabidopsis thaliana*, a mustard species (mouse-ear cress), widely employed as a model system in the genetic research of plant development and plant–microbe interactions, has been reported recently as a host for *P. brassicae*.

Plasmodiophora brassicae is a biotrophic (obligate) parasite that causes a devastating disease of crucifers known as clubroot. The disease occurs throughout the world in commercial crucifer production fields, but is also a serious problem in home gardens. Clubroot has been known since the 13th century in Western Europe and was first studied in detail by Woronin in Russia in the late 1870s. Woronin originally described *P. brassicae* as a slime mold. Despite the tremendous amount of research on clubroot, it still remains one of the most serious diseases of crucifers and is largely responsible for the disappearance of commercial cabbage production in many regions throughout the world.

Because *P. brassicae* can persist in soil for many years, traditional approaches such as crop rotation are of limited value in managing clubroot. Some success has been achieved in breeding plants for resistance to *P. brassicae*. However, field populations of *P. brassicae* are genetically diverse and composed of many races of the pathogen. Races are represented by genetically distinct individuals and typically identified by artificially inoculating a series of well-defined species of crucifers and observing them for disease symptoms (Williams, 1966; Cubeta et al., 1998). Although some resistant varieties of crucifers have been developed and are commercially available, none of them are resistant to all races of *P. brassicae*. One approach used for centuries to manage clubroot involves the modification of the soil environment by adding lime as either calcium carbonate or calcium oxide to increase soil pH to at least 7.2. This approach provides an economical and effective means of reducing the damaging effects of clubroot, and it is hypothesized that the increased soil pH interferes with the germination of resting spores, zoospore motility, and the initial root infection process. Unfortunately, a soil pH of 7.2 or higher may not be a favorable growth environment for many cultivated agricultural crop species.

LIFE HISTORY

Plasmodiophora brassicae can survive for at least 10 years in soil by forming resting spores (Agrios, 1997; Figure 11.3). During periods of cool, wet weather when the soil becomes saturated with water, resting spores (also referred to as cysts) germinate to produce usually one primary zoospore with two hairless flagella (Figure 11.1). These zoospores swim to the root and penetrate hairs of young roots or enter the plant through wounds in secondary roots (Williams, 1966). Once inside the root, *P. brassicae* produces a haploid amoebae-like structure called a plasmodium (pl. plasmodia) that passes through the cells and becomes established in them. In the plant cells, the nucleus of each plasmodium divides and becomes transformed into a multinucleate structure called a zoosporangium (pl. zoosporangia) that contains 4 to 8 secondary zoospores. Secondary zoospores are discharged through exit pores in the plant cell wall and usually fuse with each other to form zygotes, which, in turn, cause more infection of roots to form additional plasmodia. These diploid plasmodia divide by meiosis and produce clusters (sorus, pl. sori) of haploid resting spores with a single nucleus (uninucleate).

As plasmodia continue to grow and develop, they ingest proteins and sugars in the plant cells as a source of nutrients while stimulating the cells to divide (**hyperplasia**) and enlarge (**hypertrophy**). This abnormal plant growth results in the production of small, spindle-shaped swellings on roots that later develop into larger-sized galls or clubs. Root galls interfere with nutrient and water movement in the plant and initial symptoms on infected plants often appear as a yellowing and wilting of the lower leaves, particularly on warm, sunny days. Severely infected plants are often smaller than healthy plants. Eventually the galls become a food source for other soil-dwelling microorganisms that initiate their decay and release of resting spores into the soil.

SYNCHYTRIUM MACROSPORUM

Synchytrium macrosporum is a biotroph of more than 1400 different species of plants representing 185 families and 933 genera and ranging from liverworts (hepatophytes) to flowering plants (angiosperms), especially those in the families Asteraceae, Brassicaceae, Cucurbitaceae, Fabiaceae, and Solanaceae (Karling, 1964). *Synchytrium macrosporum* has the largest and widest host range of any biotrophic fungus in the kingdom Fungi (Karling, 1964). *Synchytrium macrosporum* is a weak pathogen that primarily attacks young seedlings.

Although most plants survive early infection and grow to maturity, some seedling death may occur in rare cases of severe infection, particularly if environmental conditions are favorable. As with *P. brassicae*, galls caused by the hypertrophy and hyperplasia of infected host epidermal cells is the most recognizable symptom produced by *S. macrosporum*. In general, galls form on leaves and stems of developing plants and range in size from 350 µm to 1.3 mm. However, galls can also form on the roots and underground fruits of certain legumes. The characteristic galls are composed of a central infected cell with a single

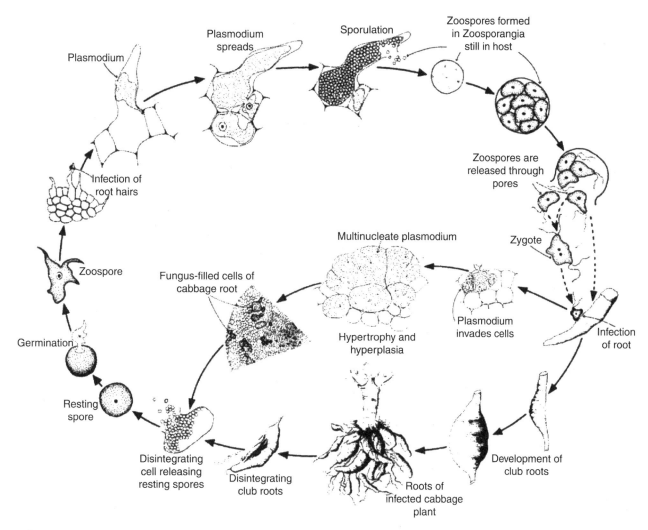

FIGURE 11.3 Life cycle of *Plasmodiophora brassicae* (From Agrios, G.N., 1997. *Plant Pathology,* 4th ed., Academic Press, New York. With permission.)

resting spore and a sheath of surrounding cells of increased size. Occasionally, some portion of the host cell cytoplasm is retained around the resting spore. Galls associated with infection by *S. macrosporum* containing resting spores only are referred to as monogallic, whereas galls produced by other species of *Synchytrium* containing either a resting spore or a zoosporangium are referred to as digallic.

LIFE HISTORY

Synchytrium macrosporum survives primarily as resting spores in soil and infected plant debris (Figure 11.4). Depending on the host and geographic location, resting spores usually germinate in the presence of moisture in late winter or early spring. When resting spores germinate, they function as a prosori (sing. prosorus), or containers for cellular contents that will later become sori. During germination, contents of the resting spore and prosorus exit the thick-walled casing through an opening or exit

pore that is eventually filled by a plug of dark pigmented material. The cytoplasm, surrounded by a plasma membrane, undergoes a number of mitotic divisions before partitioning into numerous sporangia. The number of sporangia within a single sorus can range from 120 to 800 µm; the diameter of an individual **sporangium** can range from 18 to 60 µm. Sporangia within a sorus can remain dormant for 1 to 2 months and infected plant tissue needs to dry out completely and then be rehydrated to induce resting spores to germinate (Karling, 1960). These conditions may simulate events that occur in nature where infected material senesce, dry out, and then become rehydrated with water from dew or rain.

The cytoplasm within each sporangium cleaves into individual zoospores, each with its own nucleus. In order for zoospores to be released, the sorus opens (dehisces) and releases individual sporangia from the soral membrane. The sporangia, in turn, release zoospores by a splitting of the sporangial inner membrane. Zoospores swim to new hosts through the thin film of water present either

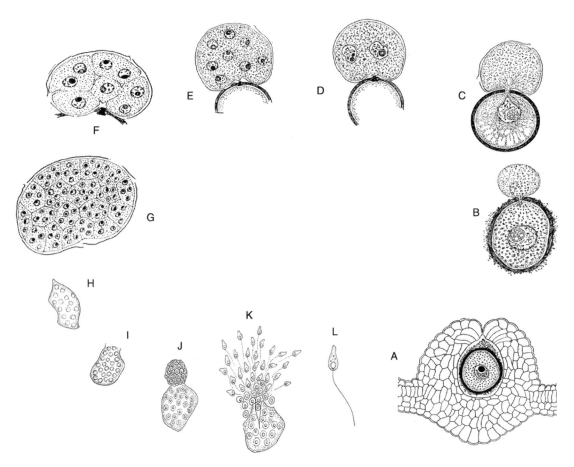

FIGURE 11.4 Life cycle of *Synchytrium macrosporum*. (A) Composite gall on leaf of ragweed (*Ambrosia trifida*). (B) Early germination stage of resting spore/prosorus. (C) Later stage of resting spore germination. (D) Binucleate incipient sorus. Notice the plug of dark material between the sorus and resting spore case. (E) Multinucleate incipient sorus. (F) Cleavage of soral contents into multiple uninucleate protospores, the precursor to sporangia. (G) Sorus of multinucleate sporangia. (H) Individual sporangium. (I) Individual sporangium with a ruptured sporangial wall showing an inner membrane prior to spore discharge. (J) Dehiscence of the sporangium. (K) Dispersal of the zoospores. (L) An individual zoospore. (From Karling, J.S., 1964. *Synchytrium*, Academic Press, New York. With permission.)

on the surface of a plant or in soil to infect young seedlings. Zoospores can also be dispersed in water splashed from one plant to another. The zoospores are ovoid to slightly elongate (3–3.8 μm × 4–4.5 μm, with a single yellowish-orange refractive lipid globule) and possess a single, posteriorly directed, hairless flagellum (12 to 14 μm in length). When zoospores are released from the sporangium in the spring, they alternate between swimming and moving in an amoeboid fashion for as long as 24 h before settling down on the surface of a host. Once settled, encystment occurs and the flagellum is either retracted into the zoospore, later referred to as a zoospore cyst, or cleaved, and a membrane is produced on the outside of the zoospore. A narrow germ tube develops from the encysted zoospore and penetrates the host cell wall. Cytoplasm flows into the host cell from the zoospore cyst and both the zoospore cyst and germ tube disintegrate. The cytoplasm can assume a variety of different shapes from round to amoeboid once inside the host and moves within the cell positioning itself near the host nucleus. The

cytoplasm, now called an initial cell or uninucleate thallus, increases in size and develops a thick wall as it matures. After the thick-walled initial cell goes into a state of dormancy, it becomes a resting spore.

Resting spores can be either spherical or ovoid in shape and range in diameter from 80 to 270 μm. The color of the resting spore wall varies from dark amber to reddish-brown and the wall is usually 4 to 6 μm thick. The walls of resting spores and sporangia are yellow-orange colored, as are the characteristic galls that form as a result of infection. Although Karling (1960) observed the production of zygotes from the fusion of two zoospores and their nuclei, no one has substantiated this observation or determined the role that **sexual reproduction** plays in resting spore formation.

LABYRINTHULA ZOSTERAE

The devastating eelgrass wasting disease of 1934 and 1935 brought attention to the obscure protist *Labyrinthula*,

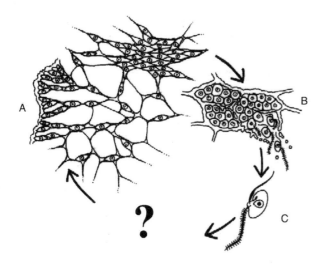

FIGURE 11.5 Life cycle of *Labyrinthula vitellina* (the sexual phase of the life cycle of *L. zosterae* has not been observed). (A) Colony of trophic cells within an ectoplasmic network. (B) Sorus within which occurs meiosis and the release of haploid zoospores (meiospores). (C) Biflagellate zoospore (which possibly develops into a new spindle-shaped trophic cell, although the developmental process is not known). (From Porter, D., 1990. Phylum Labyrinthulomycota, pp. 388–398 in: *Handbook of Protoctista*, Margulis, L. et al., Eds., Jones and Bartlett, Boston. With permission.)

which at the time was known only from a few 19th-century German publications. Although *Labyrinthula* was thought to be associated with the disease in the early 1930s (Porter, 1990), it was not until 1988 that by satisfying Koch's postulates, Muehlstein et al. (1991) demonstrated that *L. zosterae* was the causal agent for the necrotic lesions on eelgrass and die-back symptoms observed in seagrass meadows.

Ten species of *Labyrinthula* are recognized (Dick, 2001), but there are many yet to be described, particularly those associated with different species of seagrass. In addition to the eelgrass pathogen *L. zosterae*, a different species of *Labyrinthula* is associated with turtle grass (*Thalassia testudinum*) and has been implicated in a devastating die-off of seagrass in Florida Bay during the late 1980s (Porter, 1990). Some labyrinthulas are agents of disease (pathogens), whereas other species are not pathogens of seagrasses and exist primarily as saprophytes.

In contrast to *P. brassicae* and *S. macrosporum*, most species of *Labyrinthula* can be easily grown on nutrient medium in culture. In culture, labyrinthulas are most easily grown with a yeast or bacterium coinoculated on the nutrient medium as a food organism (Porter, 1990).

LIFE HISTORY

Labyrinthula species are known primarily from their trophic (or feeding) stage. This stage, which can be observed on nutrient medium in a petri dish, is composed

of a colony of cells within a network of interconnecting filaments (called the ectoplasmic network, Figure 11.5). The cells are spindle shaped and move with a gliding motion within the ectoplasmic network with speeds as high as 150 μm/min. The network plays an important role in the biology of the *Labyrinthula* colony by (1) providing a structure through which the cells move; (2) aiding in the attachment of the colony to the substrate; (3) housing digestive enzymes necessary for feeding by the cells; and (4) serving as a conduit for transmitting signals within the colony to coordinate communal motility (Porter, 1990).

In *Labyrinthula zosterae*, the trophic colony is the only stage of the life history that has been observed. In other closely related species such as *L. vitellina*, a sexual life cycle is observed and characterized by biflagellate, heterokont zoospores produced by meiosis (Margulis et al., 1990). However, the fusion of gametes (syngamy) has never been observed. The life cycle of *L. vitellina* is presented in Figure 11.5 (Porter, 1990).

Slime molds and zoosporic fungi represent phylogenetically and taxonomically distinct assemblages of microorganisms commonly associated with plants. The interdisciplinary scientific study of slime molds and zoosporic fungi has provided a foundation of knowledge for understanding their biology, ecology, genetics, and interactions with plants. In the following laboratory exercises, students will be provided with an opportunity to observe characteristic structures of three zoosporic plant pathogens — *P. brassicae, S. macrosporum,* and *L. zosterae* — and their parasitic interactions with different species of host plants. The biotrophic feeding behavior of *P. brassicae* and *S. macrosporum*, which prevents their culturing on a nutrient medium, also provides a unique opportunity for students to understand how Koch's postulates are modified to examine the disease causing activities of biotrophic plant pathogens.

REFERENCES

Agrios, G.N. 1997. *Plant Pathology,* 4th ed. Academic Press, New York.

Baldauf, S.L. and W.F. Doolittle. 1997. Origin and evolution of the slime molds (Mycetozoa). *Proc. Natl. Acad. Sci. USA,* 94: 12007–12012.

Braselton, J.P. 2001. Plasmodiophoromycota, pp. 81–91 in *The Mycota VII, Part A, Systematics and Evolution*, McLaughlin, D. J. et al., Eds. Springer-Verlag, Berlin.

Castlebury L.A. and L.L. Domier. 1998. Small subunit ribosomal RNA gene phylogeny of *Plasmodiophora brassicae. Mycologia* 90: 102–107.

Cubeta, M.A., B.R. Cody and P.H. Williams. 1998. First report of *Plasmodiophora brassicae* on cabbage in Eastern North Carolina. *Plant Dis.* 81: 129.

Dick, M.W. 2001. *Straminipilous Fungi.* Kluwer, Dordrecht, The Netherlands.

Farr, D.F., G.F. Bills, G.P. Chamuris and A.Y. Rossman. 1995. *Fungi on Plants and Plant Products in the United States.* APS Press, St. Paul, MN.

Fuller, M.S. and A. Jaworski. 1987. *Zoosporic Fungi in Teaching and Research.* Southeastern Publishing Corporation, Athens, GA.

James, T.Y., D. Porter, C.A. Leander, R. Vilgalys and J.E. Longcore. 2000. Molecular phylogenetics of the Chytridiomycota supports the utility of ultrastructural data in chytrid systematics. *Can. J. Bot.* 78: 336–350.

Hawksworth, D.L., P.M. Kirk, B.C. Sutton and D.N. Pegler. 1995. *Ainsworth & Bisby's Dictionary of the Fungi.* Cambridge University Press, Cambridge, U.K.

Karling, J.S. 1960. Inoculation experiments with *Synchytrium macrosporum. Sydowia* 14: 138–169.

Karling, J.S. 1964. *Synchytrium.* Academic Press, New York.

Leander, C.A. and D. Porter. 2001. The Labyrinthulomycota is comprised of three distinct lineages. *Mycologia* 93: 459–464.

Margulis, L., J.O. Corliss, M. Melkonian and D.J. Chapman. 1990. *Handbook of Protoctista.* Jones and Bartlett, Boston.

Martin, C.W., C.J. Alexopolus and M.L. Farr. 1983. *The Genera of Myxomycetes.* University of Iowa Press, Iowa City, IA.

Muehlstein, L.K., D. Porter and F.T. Short. 1991. Observations of *Labyrinthula zosterae* sp. nov., the causative agent of wasting disease of eelgrass, *Zostera marina. Mycologia* 83: 180–191.

Olive, L.S. 1975. *The Mycetozoans.* Academic Press, New York.

Porter, D. 1990. Phylum Labyrinthulomycota, pp. 388–398 in *Handbook of Protoctista,* Margulis, L., J.O. Corliss, M. Melkonian, and D.J. Chapman (Eds), Jones and Bartlett, Boston.

Sparrow, F.K. 1960. *Aquatic Phycomycetes,* 2nd ed. University of Michigan Press, Ann Arbor, MI.

Williams, P.H. 1966. A system for the determination of races of *Plasmodiophora brassicae* that infect cabbage and rutabaga. *Phytopathology* 56: 624–626.

12 Laboratory Exercises with Zoosporic Plant Pathogens

Marc A. Cubeta, David Porter, and Sharon E. Mozley

Plant diseases have been observed and recorded by humans for more than 2000 years (Agrios, 1997). Many plant diseases were initially described based on the observation of visible signs (vegetative and reproductive structures of the pathogen) and symptoms (reactions of the plant to infection) on fruits, leaves, roots, and stems. Since then, scientists in the discipline of plant pathology have continued to investigate the causal role that microorganisms play in plant disease and how their biology, ecology, and genetics influence pathogenesis (disease-causing ability of an organism). The majority of scientific studies conducted by plant pathologists have focused primarily on plant species of economic importance to agriculture and fostered the development of experimental methods to examine plant pathogens and their associated diseases. In general, most agricultural crops are subject to many diseases caused by a wide array of plant pathogens. Some of these pathogens have a narrow host range and can only infect a single species or variety of plant, whereas other pathogens have the ability to infect a wider range of hosts often in genetically different families of plants. The intensity and severity of disease can also vary depending on environmental conditions and genetic composition of the pathogen and plant.

In the following laboratory exercises, the biology of *Plasmodiophora brassicae,* a well-studied pathogen of crucifers in agricultural production systems (Sherf and Mac-Nab, 1986) and *Labyrinthula zosterae* and *Synchytrium macrosporum,* two important pathogens of plants in natural ecosystems, will be examined. These organisms produce motile spores (zoospores) that are an important component of their ecology and pathology (Agrios, 1997; Fuller and Jaworski, 1987; Karling, 1960; Porter, 1990). Differences in the life cycle, feeding (trophic) behavior, and symptom expression of each organism will provide the basis for determining differences in plant susceptibility and how these organisms cause plant disease. In each laboratory exercise, students will examine infected plant material to familiarize themselves with signs and symptoms of each organism. Various extraction and artificial inoculation methods (depending on the organism) will be employed to monitor and record disease development. Because *L. zosterae* can be readily cultured on nutrient medium, the students will initially isolate this organism from infected seagrass plants and then reisolate the organism from plants they have artificially inoculated to fulfill Koch's postulates and offer "proof of pathogenicity" (Chapter 36).

EXERCISES

EXPERIMENT 1. SUSCEPTIBILITY OF CRUCIFERS TO *PLASMODIOPHORA BRASSICAE*

The selection of species or varieties, or both, of crucifers with reduced susceptibility to *P. brassicae* can often be used to manage clubroot disease. However, because of the inherent genetic diversity that exists in field populations of *P. brassicae*, no variety is likely to be resistant to all genetic individuals of *P. brassicae*. In this laboratory exercise, a modification of the procedures developed by Williams (1966) and Castlebury et al. (1994) for isolation of resting spores and plant inoculation will be employed to examine the susceptibility of different crucifers to infection by *P. brassicae*. Each type of crucifer will be critically examined for the incidence and severity of disease symptoms and compared with the noninoculated control. Students will also have an opportunity to examine the characteristic microscopic structures of *P. brassicae* in infected plant tissue.

Materials

Each student or team of students will require the following laboratory items:

- 50- and 100-ml beakers
- Blender
- Centrifuge (table top or swinging bucket)
- Centrifuge tubes
- Cheesecloth
- Compound light microscope
- Dissecting microscope

- Distilled water
- 500- and 1000-ml Erlenmeyer flasks
- Forceps
- Funnel
- Glass slide and cover slips
- 1000-ml graduated cylinder
- Hemacytometer
- Mortar and pestle
- 6- to 8-cm deep paper cups
- Pasteur pipettes with rubber bulb
- 10-cm plastic petri dishes
- Plastic stakes and trays
- Potting soil (peat moss and vermiculite, 1:1 v/v)
- Seeds
- Staining solution (0.005% cotton blue in 50% acetic acid)
- Stir plate and stir bars
- 9-cm diameter Whatman # 1 filter paper
- Wooden petri dish holder

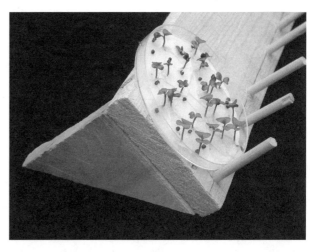

FIGURE 12.1 Wooden apparatus for holding petri plates at a 45° angle. (Courtesy of B. R. Cody North Carolina State University.)

Small packages of seed can usually be purchased at a local garden center, Asian market, or seed company. Try to include as many crucifers as possible in the laboratory exercise. Have each student or group of students select one or two species of plants for the inoculation experiments.

Follow the protocol in Procedure 12.1 to germinate seeds for the experiment.

FIGURE 12.2 Healthy (left) and infected (right) Chinese cabbage roots. (Courtesy of L. A. Castlebury.)

Procedure 12.1 Germination of Seeds

1. Prepare five petri dishes each for the following seeds: broccoli (*Brassica oleracea* var. *italica*), cabbage (*B. oleracea* var. *capitata)*, canola (*B. napus*), cauliflower (*B. oleracea* var. *botrytis*), Chinese cabbage (*B. pekinensis*), collard (*B. oleracea* var. *acephala*), kale (*B. oleracea* var. *acephala*), mustard (*B. nigra*), radish (*Raphanus sativus*), rutabaga (*B. napus*), and turnip (*B. rapa*).
2. Place a piece of Whatman #1 filter paper into a plastic petri dish (10-cm diameter) and moisten with distilled water. Arrange seeds (30 to 50 per dish) on filter paper 1 to 2 cm apart and gently press each seed into the filter paper with a pair of forceps. Offset seedling in each row to allow roots to grow straight and not become entangled with germinating seeds and seedlings below.
3. Incubate seeds at 20°C to 25°C with 12 h of supplemental lighting from two cool watt fluorescent bulbs placed 10-cm above dishes (250 µmol m⁻² s⁻¹). After seeds have germinated (usually 24 to 48 h), carefully place petri dishes on wooden holder at a 45° angle (Figure 12.1).
4. Check dishes daily to ensure filter paper remains moist. If filter paper begins to dry out, remove lid

and add 1 to 3 ml of distilled water to base of the plate.

Contact a local county agricultural extension agent or plant pathologist to obtain a sample (100 to 500 g fresh weight) of infected roots. Students can also collect infected roots from crucifers grown in a home garden or commercial field with a history of clubroot disease. Plants infected with *P. brassicae* are usually smaller and stunted compared to healthy plants and have characteristic spindle-shaped galls (Figure 12.2) on their roots. Infected plants should be placed in a paper bag and transported to the laboratory in a cooler for processing. Infected roots not used for the laboratory exercise can be stored for 2 to 3 years in a nondefrosting freezer at −20°C.

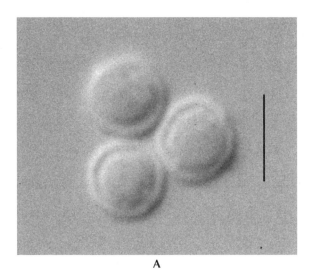
A

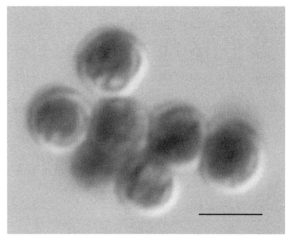
B

FIGURE 12.3 Unstained and stained resting spores of *Plasmodiophora brassicae*. Bar represents 4 μm. (Courtesy of L. A. Castlebury.)

Follow the protocol described in Procedure 12.2 to isolate and quantify resting spores of *P. brassicae*.

Procedure 12.2 Isolation and Quantification of *Plasmodiophora brassicae* Resting Spores

1. To release resting spores of *P. brassicae* from roots, place 100 g of diseased roots (either fresh or frozen) in 400 ml of sterile distilled water and macerate in a blender for 2 min at high speed.

2. Place a glass funnel with five layers of cheesecloth on top of a 1000-ml Erlenmeyer flask and collect filtrate from blended solution. If cheesecloth becomes clogged with plant material and does not filter properly, remove plant material and liquid from cheesecloth and repeat blending and filtering procedure.

3. Remove plant debris from cheesecloth and place in a mortar with 5 to 10 ml of distilled water. Grind debris with a pestle for 1 to 2 min and filter the solution as described previously and combine with previously collected filtrate.

4. Place equal volumes of filtrate into centrifuge tubes (10- or 50-ml tubes depending on size of rotor) and adjust their weight by adding appropriate amounts of distilled water to each tube. Once centrifuge tubes have a similar weight and are balanced, place them in the rotor and centrifuge for 10 to 15 min at 2000 *g* at room temperature.

5. Gently remove centrifuge tubes and place in a rack. Carefully remove the top, gray-colored fraction with a Pasteur pipette (this fraction contains the resting spores and often will appear above a whitish-colored layer in the middle of the tube) and place into another centrifuge tube. Repeat the centrifuge process as needed, particularly if filtrate is cloudy or contaminated with excessive plant material. Place the collected filtrate containing the resting spores in a glass beaker with a stir bar for quantification.

6. Gently mix collected filtrate with a stir plate to evenly distribute resting spores. Pipette a small drop of the collected filtrate on a clean glass slide, add a cover slip, and examine under a microscope at 400× to 1000×. Resting spores are round and approximately 4 μm in diameter (Figure 12.3). For better resolution of resting spores, add one drop of staining solution (0.005% cotton blue in 50% acetic acid) to the slide preparation prior to examination (Figure 12.3). If resting spores are observed, determine their concentration in collected filtrate with the following procedure. Spores without cytoplasm are not viable and should not be counted.

7. Measure the volume of collected filtrate in a graduated cylinder, pour collected filtrate into a glass beaker, and gently swirl the mixture to evenly distribute resting spores. Place one drop of spore solution at the edge of a cover slip on a hemacytometer (counting chamber with a ruler) and allow solution to be drawn into area between cover slip and glass slide. Let slide sit for 2 to 3 min to allow spores to settle on the slide.

8. Examine center of slide and locate the area of the ruler with 25 cells (5 rows and 5 columns) and consisting of 16 smaller squares each. Count the number of resting spores in each corner and center squares (5 squares total). Calculate mean number of spores per square, apply correction factor for area (2.5×10^5) and determine concentration of resting spores per ml of collected filtrate.

Example: Upper left square = 74; upper right square = 56; lower left square = 45; lower right square =

60; and center square = 80; (74 + 56 + 45 + 60 + 80)/5 = 63; $63 \times (2.5 \times 10^5) = 1.575 \times 10^7$ resting spores per ml of collected filtrate.

9. After determining concentration of resting spores, adjust spore concentration to 1×10^7 spores per ml and pour equal amount of stock solution into 50-ml beakers. The number of beakers required is determined by the total number of plant species and varieties to be inoculated. Freshly extracted resting spores are preferred for inoculation. However, resting spore solutions may be stored for 3 to 5 days at 4°C or 3 to 4 months at −20°C prior to use.

Follow the protocol outlined in Procedure 12.3 to inoculate seedlings and collect data.

Procedure 12.3 Inoculation of Seedlings and Collection of Data

1. For each crucifer examined, fill 20 paper cups (6 to 8 cm deep) with a potting mixture consisting of 1 part peat moss and 1 part vermiculite. Poke several small holes in the bottom of cup to allow for adequate drainage and water potting mixture until moist. Place one set of 10 cups into a plastic tray and repeat this process for a second set of cups (two replicates).

2. Remove 20 seedlings from petri dishes and dip their roots into quantified resting spore solution of *P. brassicae* for 10 sec. Create a 5-cm deep hole with a pencil and transplant one seedling into each individual paper cup. In a separate set of 20 cups, also include seedlings dipped in sterile distilled water as a control. Incubate seedlings at 18°C to 28°C for 6 to 8 weeks. During first week of incubation, keep potting mixture saturated with water. Thereafter, water seedlings as needed and fertilize after 3 weeks (1 g of 20-20-20 fertilizer in 3.78 l of water).

3. After incubation, remove inoculated seedlings and gently wash soil from roots in running tap water. Examine roots of each seedling for galls. Determine percentage of infected seedlings based on gall symptoms (disease incidence) and severity of infection with the 0–4 rating scale (Figure 12.4). Follow the example presented in Table 12.1. Repeat the procedure above for each of two replicates of each plant species examined and then calculate a separate average for disease incidence and severity. Also, examine seedlings in each control treatment for the incidence and severity of clubroot disease.

4. With a sharp scalpel or razor blade cut a thin section of healthy and diseased roots. Place section in distilled water on a glass slide and observe cells for the presence of resting spores (Figure 12.5).

5. Cut infected roots from the remaining seedlings, place in a plastic freezer bag, and store at −20°C for future laboratory exercises.

6. At the completion of the experiment, place all paper cups, plant material, and soil in an autoclave for 1 h at 121°C.

Anticipated Results

Crucifers will vary in their susceptibility to infection by *P. brassicae*, and galls of various sizes will be produced on infected roots. No disease symptoms (i.e., galls) should be observed on roots of crucifer plants in the control treatment. Resting spores should be readily observed inside of root cells with the microscope after sectioning and staining.

Questions

- Why is it important to include a noninoculated control and replicates of each crucifer crop in the experiment?
- What was the response of each crucifer species and variety to infection? Were there differences in disease incidence and severity?
- What are some differences in morphology of infected and healthy roots based on macroscopic and microscopic examination?

EXPERIMENT 2. CHARACTERIZATION AND COMPARISON OF PLANT INFECTION BY THE GALL-INDUCING CHYTRID *SYNCHYTRIUM MACROSPORUM*

The plant pathogen *Synchytrium macrosporum* has the widest host range of any known biotrophic fungus (Karling, 1960). The organism can infect more than 1400 different species of plants, most of which grow in natural rather than agricultural ecosystems. Plants infected with *S. macrosporum* can also exhibit considerable variation in symptom expression. In this laboratory exercise, students will examine diseased chickweed plants for resting spores, sporangia, and sori of *S. macrosporum* and observe the unique swimming pattern of zoospores associated with this organism. Subsequent experiments will be conducted to compare and contrast disease symptoms of asparagus, bean, corn, and turnip artificially inoculated with *S. macrosporum*. The inability to culture *S. macrosporum* on nutrient medium also requires the use of a modified method for isolation of resting spores and inoculation of plants.

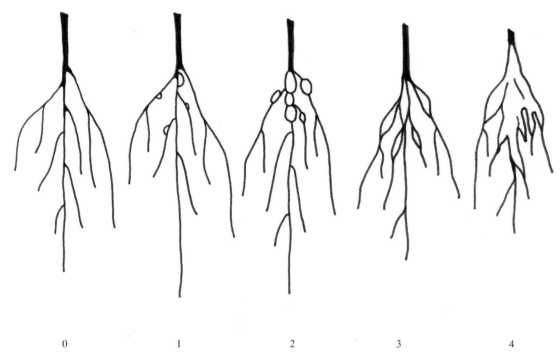

0   1   2   3   4

FIGURE 12.4 Clubroot disease rating scale (0 to 4) for crucifer seedlings. (Modified by L. J. Gray from Williams, P.H., 1966. *Phytopathology* 56: 624–626.)

TABLE 12.1
Example Data Sheet to Calculate Disease Incidence and Severity

Crucifer Tested	No.ofHealthy Plants with Rating = 0	No. of Diseased Plants with Rating = 1	No. of Diseased Plants with Rating = 2	No. of Diseased Plants with Rating = 3	No. of Diseased Plants with Rating = 4
Cabbage	2	1	3	2	2

Note: Disease incidence = (number of diseased plants/total number of plants) × 100
  [(1 + 3 + 2 + 2)/(2 + 1 + 3 + 2 + 2)] × 100 = 80%

Disease severity = No. of diseased plants in each rating category × correction factor)/total number of plants
  [(1 × 1) + (3 × 2) + (2 × 3) + (2 × 4)] / (2 + 1 + 3 + 2 + 2) = 2.1

Materials

Each student or team of students will require the following laboratory items:

- 50-ml beakers
- Commercial peat based soil mix
- Commercial soluble fertilizer
- Compound light microscope
- Dissecting microscope
- Dissecting needles
- Glass slides and cover slips
- Kimwipes®
- Pasteur pipettes with rubber bulb
- 10-cm diameter petri dishes
- Plastic bags with twist ties
- Plastic pots (10- or 15-cm diameter)
- Plastic or wooden stakes
- Seeds (asparagus, bean, corn, and turnip)
- Tape (double-sided)
- Tooth picks (wooden)
- Tween 88 (0.1% solution)
- Whatman #1 filter paper

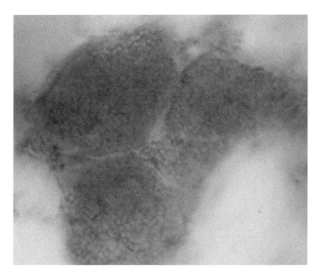

FIGURE 12.5 Resting spores of *Plasmodiophora brassicae* in root cells of Chinese cabbage. (Courtesy of L. A. Castlebury.)

Follow the protocol in Procedure 12.4 to prepare squash mounts.

Procedure 12.4 Squash Mount Procedure

1. Collect several chickweed (*Stellaria media*) plants infected with *Synchytrium macrosporum*. Diseased plants will have galls with dark amber to reddish-brown resting spores or bright orange-yellow sori from germinated resting spores on aboveground stems (Figure 12.6). Infected chickweed plants can be collected from February to March in the Southeastern U.S. (e.g., Alabama and Georgia). This may be a later date for areas farther north depending on when chickweed seeds germinate.

2. Place a piece of infected chickweed tissue in a clean petri dish, secure with double-sided tape, and then affix petri dish on the stage of a dissecting microscope with double-sided tape.

3. Add six separated drops of sterile double-distilled water on an inverted petri dish lid and place lid next to dissecting scope. Flame sterilize tips of two dissecting needles, cool for 10 sec, and carefully remove orange sori and resting spores from tissue.

4. Rinse sori and resting spores by serially running them through the six drops of sterile water before placing on a clean glass slide in a small drop of sterile water.

5. Carefully place a clean cover slip over the drop of water and gently tap it with the eraser end of a pencil to break open sori and release the individual sporangia. Examine the prepared slide with a compound light microscope.

6. Place slide in a petri dish with a moist Kimwipe folded into the bottom. Take a small, wooden toothpick, break in half, and use the two halves to hold slide above moistened Kimwipe.

7. Examine slide after 1 h and again after every 2 h. Sporangia from mature sori are usually ready to release zoospores after 1 to 2 h. Periodically moisten Kimwipe by adding a small amount of water to ensure that the slide does not dry out.

Anticipated Results

The students should observe the multicellular sheath produced from host epidermal cells surrounding both sporangial and resting spore galls. The students should also observe zoospore discharge from individual sporangia, be able to distinguish different parts of the life cycle, and observe the jerky swimming pattern of the zoospores, which is typical for chytrid fungi.

Questions/Activities

- Draw an infected chickweed plant.
- Draw resting spores, sori and sporangia observed with the microscope. Indicate magnification of each drawing.
- What kind of swimming pattern do the zoospores have and how are they released from the sporangia?
- Using the life cycle diagram in Chapter 11 (Fifure 11.4), label the structures observed.

To prepare host plants for the inoculation experiments, the following list of germination times for each plant species will be useful: bean (*Phaseolus vulgaris*) 6 to 10 days; corn (*Zea mays*) 5 to 7 days; turnip (*Brassica rapa*) 7 to 10 days; and asparagus (*Asparagus officinalis*) 14 to 21 days. Asparagus seeds should be soaked in warm water for 48 h prior to planting. Be sure to replenish warm water when it has cooled down at least two or three times during the 48-h soaking period. Asparagus seed germinate slowly and should be planted 2 to 3 weeks before planting bean, corn, and turnip seeds. This will ensure that all species of plants can be inoculated with *S. macrosporum* at the appropriate growth stage.

Follow the protocol in Procedure 12.5 to prepare inoculum of *S. macrosporum*.

FIGURE 12.6 Sporangia, sori, and resting spores of *Synchytrium macrosporum*. (From Powell, M.J., 1993. *Mycologia* 85: 1–20. With permission.)

Procedure 12.5 Preparation of Inoculum of *Synchytrium macrosporum*

1. Soak fresh or dried leaves of chickweed infected with *S. macrosporum* in sterile distilled water for 5 to 7 days to soften tissues around prosori with resting spores of the fungus.
2. With dissecting needles and a dissecting microscope, tease prosori from tissue and rinse gently in distilled water.
3. Transfer prosori to a petri dish lined with two layers of moistened, Whatman #1 filter paper and incubate in the laboratory for 2 to 3 weeks or until prosori germinate and sori mature.
4. Transfer 25 to 50 mature sori individually with a fine dissecting needle into a drop of distilled water on a clean glass slide.
5. Cover the suspension of sori with a cover slip and press gently with a pencil eraser to break open the sori to release zoosporangia.
6. Zoospores are released from zoosporangia within 1 to 2 h and can be collected by gently washing slide with distilled water into a 10-ml beaker 2 to 2.5 h after zoospores have formed.

A well-drained, commercial peat-based soil mix should be used to propagate each host plant. The soil mix should be kept moist, but not saturated with water. Host plants can be grown in a greenhouse or in the laboratory at room temperature with supplemental fluorescent lighting (14-h photoperiod and 0.3 m from the light source). When plants are 21 days old, fertilize once a week with a half-recommended rate of a complete soluble fertilizer. Plant six seeds of each host plant in a pot of diameter 10 or 15 cm and thin to three plants after seedlings have emerged from the soil.

Because *S. macrosporum* is not available commercially, cultures for use in the laboratory exercise will be prepared from infected chickweed plants collected from the field. Infected plant tissue can be dried and stored at room temperature in a low humidity environment to provide a source of viable resting spores for future experiments.

Follow the protocol in Procedure 12.6 to inoculate plants with *S. macrosporum*. The inoculation procedure is a modification of the method originally developed by J.S. Karling (1960).

Procedure 12.6 Inoculation of Plants with *Synchytrium macrosporum*

1. Swab the emerging leaves of each seedling with Tween 88 (0.1% v/v) and rinse with sterile distilled water. This procedure provides a wet surface for zoospores to swim on the leaf. The Tween solution should be applied to leaves of similar age 1 to 2 weeks prior to conducting the experiment to determine whether it is toxic to each species of plant. If phytotoxicity is observed, dilute the Tween solution

with water to a concentration that does not damage the leaves.

2. Dilute the zoospore mount from Steps 5 and 6 of Procedure 12.5 with 5 ml of sterile distilled water and place a drop of the zoospore mount onto the treated emerging leaf. Inoculate nine plants (three plants in three separate pots). Also, include another pot of three seedlings on which a few drops of sterile distilled water are placed on emerging leaves (control treatment).

3. Place a bell jar or plastic container over the seedling to maintain high humidity. If seedlings are too large to cover, then place a wet pad of absorbent cotton around the inoculated leaf. For corn and asparagus, several pots can be covered with a plastic dome or a large plastic bag with a twist tie.

4. Each seedling should be inoculated once a day for six consecutive days.

5. Once the inoculations are completed, remove the bell jar or plastic container, bag, or dome and maintain plants in a greenhouse or laboratory for 3 to 4 weeks after inoculation. Galls should be apparent 2 weeks after infection and will mature in 2 to 4 months.

Anticipated Results

Synchytrium macrosporum should not infect asparagus, but will usually cause moderate to severe infection of bean, corn, and turnip. The type, size, and structure of galls produced by *S. macrosporum* will vary on chickweed, corn, bean, and turnip.

Questions

- What differences can you note in appearance of the galls on asparagus, bean, corn, and turnip? Are these galls similar in appearance to the galls observed on chickweed?
- Do you think these differences are attributable to the plant, the fungus, or both? Explain your answer.
- Complete Table 12.2. Answer yes or no in the "Infected" column. Put "Yes" in the appropriate box corresponding to each species of plant if galls, resting spores, or prosori are present on the plant tissue and "No" if no symptoms (i.e. galls) or signs of the fungus are evident. For the "Degree of Infection" column, indicate the severity of disease with the following scale: sparse = 2 to 10 galls per plant, moderate = 10 to 20 galls per plant, heavy = 20 to 100 galls

per plant, and severe > 100 galls per plant. For the "Type of Gall" column, indicate whether galls were single-celled (e.g., simple) or multi-celled (e.g., composite). If the degree of infection was severe, then place the term "Confluent" in the type of gall column. Confluent galls occur when several epidermal cells adjacent to one another become infected and the individual sheaths of hypertrophied host cells that form around each developing gall merge together (Karling, 1960).

- Summarize the results of your observations in Table 12.1. What can you conclude from these observations?
- Karling advocated the use of host range for identification of species of *Synchytrium*. What modern techniques could be used to aid in the identification of a fungus rather than conducting a host range study? Do you think host range studies are still important? Explain your answer.

Exercise 3. Observations of Eelgrass Wasting Disease

Labyrinthula zosterae is an important pathogen of seagrass in estuarine environments. The organism produces zoospores that are involved in the disease cycle and can be grown in pure culture by using yeast cells to supplement their nutrition. In this laboratory exercise, students will observe eelgrass (*Zostera marina*) wasting disease and isolate *L. zosterae* from diseased eelgrass exhibiting typical symptoms. To demonstrate that *L. zosterae* is the causal agent of eelgrass wasting disease, students will follow Koch's postulates to establish "proof of pathogenicity."

Materials

Each student or team of students will require the following laboratory items:

- Agar
- Air pump and tubing
- Alcohol lamp
- Antibiotics (ampicillin, penicillin G, and streptomycin sulfate)
- Artificial or natural seawater
- Buchner funnel with rubber cork
- Carboy (25 l)
- Clorox® (sodium hypochlorite, NaOCl)
- Compound light microscope
- Culture of yeast (any nonfilamentous, nonmucoid yeast, preferably one cultured from non-

TABLE 12.2
Susceptibility of Asparagus, Bean, Corn, and Turnip to *Synchytrium macrosporum*

Species	Infected	Degree of Infection	Type of Gall
Asparagus officinialis			
Brassica rapa			
Phaseolus vulgaris			
Zea mays			

surface disinfested and decaying eelgrass leaves)
- Dissecting microscope
- Dissecting needles and scissors
- Erlenmeyer flasks (2 l) with cotton stoppers and glass tube for aeration
- Ethanol (80% EtOH)
- Filters (0.4 μm pore diameter)
- Flasks (2 l, side-arm)
- Forceps
- Germanium dioxide (GeO$_2$)
- Glass pipettes (5 ml)
- Glass slides and cover slips
- Paper towels (sterile)
- Plastic petri dishes (4.5- and 10-cm diameter)
- Plastic screw top tubes (50 ml)
- Plastic ziplock bags (3.78 l)
- Scalpel
- Serum seawater plus medium (see procedure later)
- Sterile distilled water
- Tygon tubing
- Vacuum pump with rubber tubing
- Whatman #1 filter paper (9 cm)

Approximately 20 to 30 l of artificial or natural seawater (28 to 32 parts per thousand salinity) is required for this laboratory experiment. The seawater will be used to prepare serum seawater plus medium to isolate *L. zosterae* and for seagrass inoculation experiments. Various pre-packaged mixtures that approximate seawater chemistry are commercially available and can be easily prepared by mixing these ingredients with distilled water (e.g., Instant Ocean, etc.).

Follow the protocol outlined in Procedure 12.7 to prepare and filter seawater from natural sources.

Procedure 12.7 Preparation and Filtering of Seawater from Natural Sources

1. Place a Buchner funnel containing two pieces of Whatman #1 filter paper into a 2-l side-arm flask connected to a vacuum source with flexible rubber tubing.
2. Once the vacuum has been established, slowly pour the seawater into the Buchner funnel and continue until the flask is full.
3. Filtered seawater should be stored at 4°C to 10°C in clean plastic carboys and covered with black plastic. For the production of sterile seawater use a filtration system (0.4 μm pore diameter) or autoclave for 20 min at 121°C.

Follow the protocol outlined in Procedure 12.8 to prepare serum seawater plus (SSA+) isolation medium.

Procedure 12.8 Preparation of Serum Seawater Plus (SSA+) Isolation Medium

1. Add 12 g of agar (Difco) and 3 mg of germanium dioxide (a diatom inhibitor) to 1 l of filtered seawater (Procedure 12.7). Autoclave medium for 20 min at 121°C.
2. After medium has cooled to 50°C, add 4 to 10 ml of sterile horse serum (1% v/v, BBL or Gibco) and 250 mg each of the antibiotics ampicillin, penicillin G, and streptomycin sulfate.

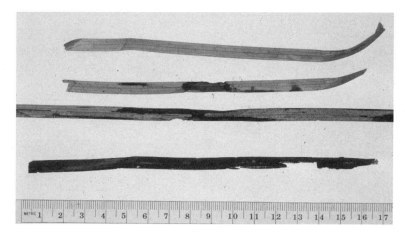

FIGURE 12.7 Eelgrass leaves with necrotic lesions symptomatic of wasting disease.

3. Gently swirl to mix the medium and pour into 10-cm plastic petri dishes. *Caution:* All SSA+ medium should be used poured into petri dishes at this time. Reheating (remelting) this medium after it has solidified will coagulate the horse serum protein and render it useless as an isolation medium.

Eelgrass is found in coastal estuaries throughout the Northern Hemisphere. In North America, eelgrass is distributed in seagrass meadows along the Atlantic coast from Labrador to North Carolina and along the Pacific Coast from Alaska to Baja California. Since most seagrass meadows are protected by law, check with local authorities to identify appropriate eelgrass beds where plants can be collected. Eelgrass shoots can be collected at the lowest tides by wading or swimming a short distance into the water.

Follow the protocol outlined in Procedure 12.9 to collect healthy and diseased samples of *Z. marina*.

Procedure 12.9 Collecting Healthy and Diseased Samples of *Zostera marina*

1. Collect approximately 20 healthy plants (without disease symptoms manifested by the appearance of blackened or dead necrotic areas present on leaves) with at least three inner (youngest) green leaves from the eelgrass bed.
2. While collecting, also include a portion of the rhizome and roots with each collected shoot. Place shoots in a 3.78-l plastic ziplock bag with a seawater dampened paper towel.
3. Do not submerge collected plants directly in seawater. For transport, keep plants cool but not directly on ice.

4. After collecting healthy plants, carefully examine plants in the eelgrass beds for wasting disease symptoms. Plants infected with *L. zosterae* will have black streaks and patches of necrotic tissue (lesions) on older leaves (Figure 12.7).
5. Collect approximately 30 leaves with disease symptoms by selecting leaves with some areas of healthy green tissue adjacent to the necrotic lesions. Place diseased leaves into a new 3.78-l plastic ziplock bag and keep moist and cool until ready to isolate the pathogen as described previously.
6. If you are unable to collect healthy and diseased eelgrass plants, request them from a colleague in a coastal area. Eelgrass plants can be shipped overnight in an insulated container and successfully used for this laboratory experiment.

One week before beginning the isolation procedure, streak several plates of SSA+ medium with a culture of yeast (see materials and supplies given previously). The yeast will serve as a food source for *Labyrinthula*. Monthly transfer on SSA without antibiotics can maintain dual axenic cultures of *Labyrinthula* and yeast.

Follow the protocol outlined in Procedure 12.10 to isolate *L. zosterae* from seagrass.

Procedure 12.10 Isolation of *Labyrinthula zosterae* from Seagrass

1. Dip forceps and scissors in 50-ml tube containing 80% EtOH and flame to disinfest.
2. Cut eelgrass leaves into small pieces (5 to 10 mm²) with the disinfested forceps and scissors. Use leaf pieces from the edge of the blackened, necrotic lesions where *L. zosterae* is likely to be most active.

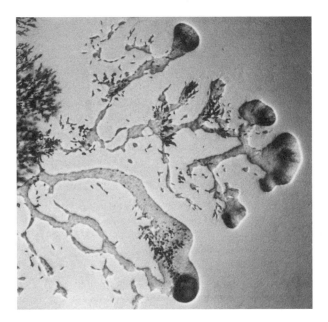

FIGURE 12.8 *Labyrinthula zosterae* growing on SSA+ medium from diseased eelgrass leaf.

3. Place cut leaf pieces to a sterile petri dish of 4.5-cm diameter, add 0.5% sodium hypochlorite to cover leaf pieces, and gently swirl them for 1 min. Aseptically transfer each disinfested leaf piece to new petri dish of 4.5-cm diameter, add sterile distilled water, and gently swirl for 2 min.

4. Repeat the rinsing process with sterile, filtered seawater. Transfer leaf pieces to a sterile paper towel or filter paper to remove excess water and place 4 to 5 disinfested pieces of eelgrass leaf tissue on a 10-cm diameter petri dish containing SSA+ medium.

5. Observe each SSA+ dish daily for growth of *L. zosterae* from each piece of eelgrass tissue (Figure 12.8).

6. Continue to observe dishes for at least 1 week or until they become overgrown by bacteria or fungi, making observation and isolation of *Labyrinthula* difficult.

7. When an appropriate *Labyrinthula* colony is located and ready for transfer, first streak a clean SSA+ dish with a small amount of yeast from an actively growing culture.

8. Then transfer a portion of the actively growing *Labyrinthula* colony to the yeast streak on the new SSA+ dish.

Follow the protocol outlined in Procedure 12.11 to demonstrate that *L. zosterae* is the causal agent of eelgrass wasting disease, using Koch's postulates (Muehlstein et al., 1991).

Procedure 12.11 Demonstrating that *Labyrinthula zosterae* is the Causal Agent of Eelgrass Wasting Disease, Using Koch's Postulates

1. Fill 2-l Erlenmeyer flasks to the neck with artificial or natural sterilized seawater. Place a single healthy eelgrass shoot in each flask and weigh down with a short piece of heavy rubber tubing slipped over the rhizome. Stopper flasks with cotton through which a glass tube extends to near the bottom of the flask (a sterile 5-ml pipette is a good substitute). Attach an air pump or airline with a cotton plug filter to the glass tube and adjust the flow rate to deliver about one bubble per second.

2. Place several 1-cm pieces of green eelgrass leaf in distilled water and autoclave for 20 min at 121°C. Place some of these sterilized leaf pieces in cultures of *Labyrinthula* isolated for Procedure 12.10. Place them on the agar surface adjacent to the spreading colonies of *Labyrinthula*, but not on top of the yeast cells. Allow the *Labyrinthula* cells to grow into the leaf piece for 24 to 48 h.

3. Cut 0.5-cm pieces of thin Tygon tubing and slit the tube wall along one radius to create a small clip to attach the inoculated leaf pieces to healthy eelgrass plants. Sterilize clips before use.

4. Remove a green shoot of eelgrass from a flask and place on a sterile paper towel. With a sterile forceps, pick up a piece of inoculated carrier leaf and clip it to a green leaf of the eelgrass shoot with a Tygon clip. Replace the inoculated shoot into the flask.

5. As an experimental control, clip a sterilized piece of leaf tissue to a green leaf of an eelgrass shoot in a separate flask.

6. Repeat the experimental and control inoculations for at least five flasks each.

7. Place the flasks on a lighted bench or greenhouse where the plants will receive at least 20% full sunlight.

8. Observe plants for wasting disease symptoms 1 to 7 days after inoculation (Figure 12.9).

9. Complete Koch's postulates by reisolating *L. zosterae* on SSA+ medium from diseased leaves with necrotic lesions in the flasks.

Anticipated Results

Wasting disease should be readily observed on eelgrass, and *L. zosterae* should be easily cultured from infected eelgrass leaves on SSA+ isolation medium. Eelgrass plants inoculated with *L. zosterae* will produce typical wasting disease symptoms of leaves, followed by isolation

FIGURE 12.9 Eelgrass shoot with necrotic lesion 60 h after inoculation with *L. zosterae*.

and observation of the microorganism from diseased leaves on SSA+ isolation medium. No symptoms should be observed and *L. zosterae* should not be isolated from eelgrass plants in the control treatment.

Questions

- Why is it necessary to clip a noninoculated, disinfested leaf piece to a green eelgrass plant as an experimental control?
- Low salinity has been reported to inhibit wasting disease. How could you test this hypothesis with the inoculation apparatus set up for this laboratory exercise?
- How will you determine whether necrotic lesions on leaves of other seagrasses are caused

by *L. zosterae* or another plant pathogenic microorganism(s)?

ACKNOWLEDGMENTS

Paul H. Williams, emeritus professor, University of Wisconsin, for his many useful suggestions and clubroot resistance screening protocol that served as a template for developing Laboratory Exercise 1. Timothy James (Duke University) and Lisa Castlebury (USDA Systematic Botany and Mycology Laboratory) for presubmission review and to LC for providing photographs of resting spores and diseased cabbage. Celeste A. Leander for providing comments and suggestion for Laboratory Exercise 3. Andy Tull (University of Georgia) for advice on growing plants for the laboratory exercises. Lynnette J. Gray for illustrations and Bryan R. Cody for images. Pamela E. Puryear (North Carolina State University) for assistance in literature searches.

REFERENCES

Agrios, G.N. 1997. *Plant Pathology,* 4th ed. Academic Press, New York.

Castlebury, L.A., J.V. Maddox and D.A. Glawe. 1994. A technique for the extraction and purification of viable *Plasmodiophora brassicae* resting spores from host root tissue. *Mycologia* 86: 458–460.

Fuller, M.S. and A. Jaworski. 1987. *Zoosporic Fungi in Teaching and Research.* Southeastern Publishing Corporation, Athens, GA.

Karling, J.S. 1960. Inoculation experiments with *Synchytrium macrosporum. Sydowia* 14: 138–169.

Muehlstein, L.K., D. Porter and F.T. Short. 1991. Observations of *Labyrinthula zosterae* sp. nov., the causative agent of wasting disease of eelgrass, *Zostera marina. Mycologia* 83: 180–191.

Porter, D. 1990. Phylum Labyrinthulomycota, pp. 388–398 in *Handbook of Protoctista,* Margulis, L., J.O. Corliss, M. Melkonian and D.J. Chapman (Eds), Jones and Bartlett, Boston.

Powell, M.J. 1993. Looking at mycology with a Janus face: A glimpse at Chytridiomycetes active in the environment. *Mycologia* 85: 1–20.

Sherf, A.F. and A.A. MacNab. 1986. *Vegetable Diseases and Their Control,* 2nd ed. John Wiley & Sons, New York.

Williams, P.H. 1966. A system for the determination of races of *Plasmodiophora brassicae* that infect cabbage and rutabaga. *Phytopathology* 56: 624–626.

13 Archiascomycete and Hemiascomycete Pathogens

Margery L. Daughtrey, Kathie T. Hodge, and Nina Shishkoff

CHAPTER 13 CONCEPTS

- *Taphrina* and *Protomyces* are two genera of Archiascomycetes, an early diverging clade of Ascomycota.

- Both *Taphrina* and *Protomyces* have a saprobic yeast stage and a parasitic mycelial stage during which asci are formed.

- The asci of *Taphrina* and *Protomyces* are naked (not contained in an ascoma).

- Typical symptoms of diseases caused by *Taphrina* or *Protomyces* are galls, leaf curls, and leaf spots.

- Peach leaf curl is a common disease caused by *Taphrina deformans*.

- *Eremothecium* is unusual in that it is a plant pathogen within the Hemiascomycetes (ascomycetous yeasts).

Archiascomycete and hemiascomycete **pathogens** cause a number of fairly obscure diseases as well as some commonly recognized problems such as peach leaf curl and oak leaf blister. Many of the diseases discussed in this chapter affect weeds or native plants rather than cultivated species, which tend to receive the most attention from plant pathologists. The plant–pathogen interactions in these groups are unique and quite fascinating.

Phylogenetic studies by Nishida and Sugiyama (1994) and others reveal the following three major lineages within the Ascomycota: Archiascomycetes, Hemiascomycetes, and Euascomycetes (Sugiyama, 1998). The Archiascomycetes appear to have diverged before the Hemiascomycetes (ascomycetous yeasts) and Euascomycetes (filamentous ascomycetes).

HEMIASCOMYCETE PATHOGENS

Most ascomycetous yeasts (Hemiascomycetes) are not plant pathogens, but the genus *Eremothecium* (syn. *Ashbya*, *Holleya*, and *Nematospora*) is an exception (Kurtzman and Sugiyama, 2001). *Eremothecium* is a filamentous fungus that also produces yeast cells within the host (Batra, 1973). **Asci,** which are formed directly from the **mycelium**, contain needle-like **ascospores**.

Sucking insects, especially the true bugs, often vector this genus. *Eremothecium coryli* fruit rot associated with stink bug feeding was responsible for losses of more than 30% in field tomatoes in California (Miyao et al., 2000).

In addition to tomatoes, *E. coryli* infects cotton, citrus, hazelnuts, and soybeans, and a related species, *E. sinecaudii,* infects the seeds of mustards. The three species *E. cymbalariae, E. ashbyi,* and *E. gossypii* all infect *Hibiscus* species and coffee, and also cause surface lesions on citrus fruits and cotton bolls. *Eremothecium ashbyi* is also used for the industrial production of riboflavin (vitamin B$_2$) via a fermentation process.

ARCHIASCOMYCETE PATHOGENS

The Archiascomycetes undergo sexual reproduction that is **ascogenous**, but lack **ascogenous hyphae** and **ascomata**. The following four orders are currently grouped within the newly proposed class Archiascomycetes: the Taphrinales, the Protomycetales, the Pneumocystidales, and the Schizosaccharomycetales (Kurtzman and Sugiyama, 2001). The last two are animal pathogens and fission yeasts, respectively. The Protomycetales and Taphrinales are exclusively plant pathogens and each contains a single family, the Protomycetaceae and Taphrinaceae, respectively. Members of these families produce both yeast-like and mycelial states; **asexual reproduction** occurs by budding.

PROTOMYCETACEAE

The Protomycetaceae includes five genera (*Burenia*, *Protomyces*, *Protomycopsis*, *Taphridium*, and *Volkartia*) com-

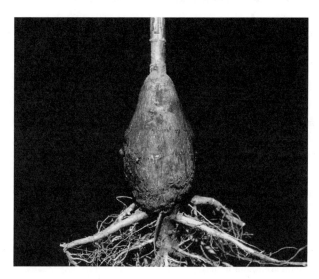

FIGURE 13.1 Stem gall of giant ragweed (*Ambrosia trifida*) caused by *Protomyces gravidus*. (From Holcomb, G. E., 1995. *Plant Disease* 79 (cover). With permission.)

prising 20 species (Alexopoulos et al., 1996). The genus *Protomyces* causes **galls** on leaves, stems, flowers and fruit of plants in the Apiaceae (dill family) and Asteraceae (composite family). *Protomyces macrosporus*, for example, causes a leaf gall of wild carrot (*Daucus carota*); it is reported to occur on 26 other genera, including 14 genera in the Asteraceae. *Protomyces gravidus* causes a stem gall on giant ragweed (*Ambrosia trifida*) in the U.S. (Figure 13.1) and also affects tickseed (*Bidens)*. Of the other members of the family, Protomycetaceae, *Burenia,* and *Taphridium* occur only on Apiaceae, whereas *Protomycopsis* and *Volkartia* occur only on Asteraceae. These three genera cause color changes or galls and spots on their hosts.

Protomyces produces an intercellular **diploid** mycelium and thick-walled **intercalary** ascogenous cells that form in the host tissue. These cells are sometimes mistaken for the spores of **smuts**, such as the **white smut**, *Entyloma* (Preece and Hick, 2001). The ascogenous cells of *Protomyces* overwinter and germinate to form asci in the spring. Inside each multinucleate ascus, four spore mother cells formed by **mitosis** continue to divide to form many ascospores. Hundreds of these spores are forcibly released en masse and they continue budding after release. Fusion of two of these ascospores forms a diploid cell that can reinfect leaves.

TAPHRINACEAE

The genus *Taphrina* was created in 1832 by Fries; it is the only genus of the Taphrinaceae and includes some 95 species (Mix, 1949). *Taphrina* species are **parasites** of members of many plant families. Those on ferns generally differ from species that attack **angiosperms** in having thin,

clavate asci. The fern hosts of *Taphrina* show some very elaborate symptoms: *T. cornu-cervi* causes antler-shaped galls on *Polystichum aristatum* and *T. laurencia* causes highly branched bushy protuberances on the fronds of *Pteris quadriaurita* in Sri Lanka. Although Taphrinas affect many herbaceous and woody plants, the symptoms they cause are often overlooked. The woody plant hosts occurring in the U.S. and Canada are shown in Table 13.1. Some of the most frequently encountered *Taphrina* diseases are those affecting the genus *Prunus* — a number of species occur on plums, apricots, and cherries. *Taphrina pruni* causes deformed fruits called "plum pockets" or "bladder plums" on plums and related species. The **witches' brooms** caused by *T. wiesneri* on Japanese flowering cherries caused some concern in Washington, D.C. in the 1920s, but these perennial infections were successfully eradicated by pruning (Sinclair et al., 1987).

Taphrina species often cause small, yellow leaf spots that may or may not be thickened or blistered into concave or convex areas; these often turn brown with age (Figure 13.2). Spots on maples (*Acer* species) caused by a number of different *Taphrina* species may be brown to black. Asci generally appear as a white bloom on one or both leaf surfaces within the areas colonized by *Taphrina*. Twig deformation, galls, witches' brooms and distorted **inflorescences** or fruit are also possible. Alder cones infected by *T. robinsoniana* show antler-shaped outgrowths (Figure 13.3).

Within the host, *Taphrina* forms a mycelium that is intercellular, **subcuticular**, or stays within the epidermal wall. There are no ascomata. The naked asci develop from ascogenous cells either in a subcuticular layer or within a wall **locule** (Figure 13.4). Ascospores in many species bud within the ascus, packing the ascus with **blastospores**. This budding may continue after spores are released from the ascus. Much of the *Taphrina* life cycle is spent in the **saprotrophic** yeast state, during which the fungus is haploid and uninucleate. The yeast-like **anamorph** has been named *Lalaria*, after a pebbled beach on a Greek island (Moore, 1990). A dikaryotic mycelium develops only within the parasitized host tissue and asci are produced. *Taphrina* species generally overwinter as blastospores in buds or on bark, although in a few cases overwintering is accomplished by means of perennial mycelium within the host.

Taxonomic understanding of *Taphrina* is being revolutionized by modern molecular research. Assignment of *Taphrina* to the early diverging Archiascomycete **clade** based on molecular traits fits with earlier knowledge of the unique morphological and physiological traits of this genus, some of which are more typical of **Basidiomycota** than of **Ascomycota**.

Cultures made from ascospores or blastospores of *Taphrina* grow slowly on artificial media as pale pink yeast colonies. Without knowledge of the organism's parasitic

TABLE 13.1
Taphrina **Species and Their Woody Plant Hosts Reported in the U.S. and Canada**[a]

Host	Species
Acer species	*Taphrina aceris*, UT
	T. carveri, c, s USA, ONT
	T. darkeri, nw USA, w Can
	T. dearnessii, c, e NA
	T. letifera, e, nc USA, NS
	T. sacchari, c, e USA, QUE
Aesculus californicus	*T. aesculi*, CA, TX
Alnus species	*T. alni* (catkin hypertrophy), AK, GA, c Can, Eur, Japan
	T. japonica, nw NA, Japan
	T. occidentalis (catkin hypertrophy), nw NA, QUE
	T. robinsoniana (catkin hypertrophy), c, e NA
	T. tosquinetii (catkin hypertrophy), NH, NS, Eur
Amelanchier alnifolia	*T. amelanchieri* (witches' broom), CA
Amelanchier species	*T. japonica*, CA
Betula species	*T. americana* (witches' broom), NA
	T. bacteriosperma, n hemis
	T. boycei, NV
	T. carnea, ne USA, e, w Can, Eur
	T. flava, ne USA, e Can
	T. nana, WY, e, w Can, Eur
	T. robinsoniana, NF
Carpinus caroliniana	*T. australis*, e USA, ONT
Castanopsis species	*T. castanopsidis*, CA, OR
Corylus	*T. coryli*, USA, Japan
Malus	*T. bullata*, WA, Eur
Ostrya virginiana	*T. virginica*, c, e US, e Can
Populus species	*T. johansonii* (catkin hypertrophy), USA, e Can, Eur, Japan
	T. populi-salicis, nw NA
	T. populina, n hemis
	T. rhizophora NY, WI, Eur
Potentilla species	*T. potentillae*, USA, Eur
Prunus species	*T. armeniacae*, USA
	T. communis (plum fruit hypertrophy), NA
	T. confusa (chokecherry fruit hypertrophy, witches' broom), USA, e Can
	T. farlowii (cherry leaf curl, fruit and shoot hypertrophy) c, e USA
	T. deformans, cosmopolitan
	T. flavorubra (cherry and plum fruit and shoot hypertrophy)
	T. flectans (cherry witches' broom), w USA, BC
	T. jenkinsoniana, NV
	T. pruni (plum pockets), n hemis
	T. pruni-subcordatae, (plum witches' broom), w USA
	T. thomasii (witches' broom on holly-leaf cherry), CA

TABLE 13.1 (continued)
***Taphrina* Species and Their Woody Plant Hosts Reported in the U.S. and Canada**[a]

Host	Species
	T. wiesneri (cherry witches' broom) cosmopolitan
Pyrus species	*T. bullata*, WA, BC, Eur
Quercus species	*T. caerulescens*, NA, Eur, n Africa
Rhus species	*T. purpurescens*, c, e USA, Eur
Salix laevigata	*T. populi-salicis*, CA
Sorbopyrus auricularia	*T. bullata*, BC
Ulmus americana	*T. ulmi*, c, e USA, UE, Eur

[a] Diseases caused by the fungi listed for each host genus are leaf blisters or curls unless otherwise noted. Geographic distributions are identified with postal abbreviations of states and provinces or more broadly as follows: c, central; e, east; n, north; s, south; w, west; Can, Canada; Eur, Europe; hemis, hemisphere; NA, North America.
Source: Courtesy of W. A. Sinclair, Cornell University.

FIGURE 13.2 *Taphrina purpurescens* leaf blister of dwarf sumac (*Rhus copallina*). Affected leaves display red-brown blisters.

FIGURE 13.3 Catkin hypertrophy on hazel alder, *Alnus serrulata,* caused by *Taphrina robinsoniana.*

phase, the fungus in culture might be mistaken for one of the pink basidiomycetous yeasts that are common on plant tissue.

Some Important Diseases Caused by *Taphrina*

Most diseases caused by *Taphrina* attract the attention of only the most curious observer. The following two diseases stand out as more obvious and damaging to cultivated plants: oak leaf blister caused by *T. caerulescens* (see Plate 5A following page 80) and peach leaf curl caused by *T. deformans.*

Oak leaf blister is a minor disease in most climates, but is relatively damaging in the southern U.S. on shade trees in the red oak group. About 50 oak (*Quercus*) species are susceptible, including members of both the white oak and red oak groups. Even in areas where the disease is not especially harmful to the oaks, the peculiar leaf symptoms attract comment. The symptoms may include yellowish convex blisters on the leaves, leaf curling, and leaf drop (Figure 13.5). Sporulation usually occurs on the concave side of the blisters. A layer of asci forms between the cuticle and **epidermis**; the asci swell, break the cuticle, and forcibly discharge their ascospores. The leaf blisters typically turn brown after ascospore production. Control measures are not usually necessary for oak leaf blister.

Taphrina deformans causes the best-known disease in this group, peach leaf curl. This disease is notorious because

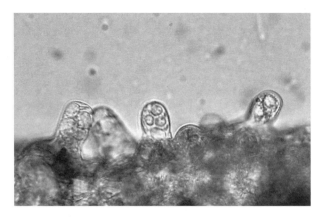

FIGURE 13.4 Asci of *Taphrina purpurescens*. The asci are "naked," that is, they are produced directly on the host tissues and not inside an ascoma. Several ascospores can be seen in the ascus at center.

FIGURE 13.6 Peach leaf curl caused by *Taphrina deformans*. (Photographed by W. R. Fisher (CUP 36520). Courtesy of Cornell Plant Pathology Herbarium.)

FIGURE 13.5 Leaf blister of red oak (*Quercus* species) caused by *Taphrina caerulescens*. (Photographed by H. H. Whetzel and H. S. Jackson (CUP 1938a). Courtesy of Cornell Plant Pathology Herbarium.)

it causes striking symptoms and significant damage on economically important hosts. Peaches, nectarines, and almonds are affected; the disease occurs all over the world where peaches are grown, but has a more limited geographic occurrence on its other hosts. Peach leaves that are infected when they are still undifferentiated tissue later show yellow, pink or reddish areas that are buckled and curl

into weird shapes; entire leaves may be distorted (Figure 13.6). Cynthia Westcott, noted ornamentals pathologist, described peach leaves as sometimes looking "as if a gathering string had been run along the midrib and pulled tight" (Westcott, 1953). The leaf distortion results from growth-regulating chemicals (Chapter 29) secreted by the fungus: **cytokinins** and **auxins** are known to be produced by some *Taphrina* species. Leaf drop may lead to poor fruit quality and weaken trees. The young peach fruits may be distorted or show reddish warty spots lacking the usual fuzz; infected twigs are swollen. The blistered portions of leaves or fruit develop a powdery gray coating of naked asci on the upper surface. After the ascospores are released, the diseased leaves, sometimes turn brown, wither and drop.

Ascospores that land on the surface of the peach tree will bud to produce a saprobic, **epiphytic** yeast phase. During the summer, the fungus lives invisibly and harmoniously on the peach tree. This haploid inoculum can persist on the host plant for several years. In the spring, the expanding leaf buds are subject to fungal invasion. The fusion of two yeast cells is thought to form the mycelium that infects the plant. Asci form beneath the cuticle of the host and push to the outside of the plant to release ascospores once again.

There is a single infection period for *T. deformans* in the spring, with no secondary cycles later in the growing season. Thus, by the time the symptoms are noticed, it is too late to achieve any disease control in that same year. The timing of spring rains and the severity of the preceding winter influence how extensive the disease symptoms are from one year to the next. Infection of immature peach leaves is optimum at temperatures from 50°F to 70°F and requires rain. Because the fungus overwinters on the surface of the host, this is one of the few diseases that can be controlled with spray treatments at the close of the previous growing season; dormant sprays in late fall or treatments before bud swell in the spring can curb infection. Lime-sulfur and bordeaux mixture have traditionally been used in this fashion. Other effective fungicides include chlorothalonil, fixed coppers, ferbam, or ziram. Chemical control is used for only a few other diseases caused by *Taphrina* species, primarily those affecting plums and cherries.

So, although most of the curious swellings, discolorations and spots due to fungal pathogens in the Hemiascomycetes and Ascomycetes remain little-studied and largely ignored, the interaction of *T. deformans* with peach trees is considered a major disease because it can cause significant economic loss. Societal value placed on the host plant is often what determines whether a disease is of minor or major import — it is not just a matter of how many plants are killed or disfigured, but which plant species are affected that determines whether control efforts and research dollars will be focused on the problem.

REFERENCES

Alexopoulos, C. J., C. W. Mims and M. Blackwell. 1996. *Introductory Mycology*. Wiley & Sons, New York.

Batra, L. B. 1973. Nematosporaceae (Hemiascomycetidae): Taxonomy, pathogenicity, distribution, and vector relations. *USDA Agric. Res. Ser. Tech. Bull.* 1469: 1–71.

Kurtzman, C. P. and J. Sugiyama. 2001. Ascomycetous yeasts and yeast-like taxa, pp. 179–200 in *The Mycota VII, Part A, Systematics and Evolution*, McLaughlin, D. J., E. G. McLaughlin and P. A. Lemke (Eds.), Springer-Verlag, Berlin.

Mix, A. J. 1949. A monograph of the genus *Taphrina*. *Univ. Kan. Sci. Bull.* 33: 1–167.

Miyao, G. M., R. M. Davis and H. J. Phaff. 2000. Outbreak of *Eremothecium coryli* fruit rot of tomato in California. *Plant Dis.* 84: 594.

Moore, R. T. 1990. The genus *Lalaria* gen. nov.: Taphrinales anamorphosum. *Mycotaxon* 38: 315–330.

Nishida, H. and J. Sugiyama. 1994. Archiascomycetes: Detection of a major new lineage within the Ascomycota. *Mycoscience* 35: 361–366.

Preece, T. E. and A. J. Hick. 2001. An introduction to the Protomycetales: *Burenia inundata* on *Apium nodiflorum* and *Protomyces macrosporus* on *Anthriscus sylvestris*. *Mycologist* 15: 119–125.

Sinclair, W. A., H. H. Lyon and W. A. Johnson. 1987. *Diseases of Trees and Shrubs*. Cornell University Press, Ithaca, NY.

Sugiyama, J. 1998. Relatedness, phylogeny, and evolution of the fungi. *Mycoscience* 39: 487–511.

Westcott, C. 1953. *Garden Enemies*. D. Van Nostrand, New York.

14 The Powdery Mildews

Margery L. Daughtrey, Kathie T. Hodge, and Nina Shishkoff

CHAPTER 14 CONCEPTS

- Powdery mildews are obligate parasites that show interesting morphological adaptations to herbaceous vs. woody hosts.

- Powdery mildews have specialized feeding cells called haustoria that absorb nutrients from their hosts.

- Molecular genetic studies and scanning electron microscope studies have led to a recent reversal in taxonomic thought and a new paradigm: The anamorphs reflect phylogeny much better than the teleomorphs.

- Many powdery mildew fungi have recently been renamed to reflect the newly apparent relationships.

- Powdery mildews injure many ornamental crops and garden plants because of the highly conspicuous colonies that cause aesthetic injury; vegetable, field and fruit crops suffer from yield and quality reduction.

The **powdery mildew** fungi in the phylum Ascomycota cause easy to recognize diseases. The fungus grows across the surface of the host in conspicuous colonies, creating whitish circular patches that sometimes coalesce until the entire leaf surface is white (see Plate 5B following page 80). The colonies may form on either the upper or lower surface of leaves, as well as on stems, flower parts, and fruits. Because energy from **photosynthesis** is diverted into growth of the pathogen, infected plants may be stunted and produce fewer or smaller leaves, fruits or grain. The impacts of powdery mildews on their hosts may be mainly aesthetic or may reduce yield or quality. Many floral and nursery crops (Figure 14.1) and also fruit, vegetable, and field crops are affected.

Powdery mildews are spread over great distances by wind and can also be moved about the world on plants with inconspicuous or latent infections. Within the powdery mildew family, Erysiphaceae, some members have very narrow host ranges, whereas others have broad host ranges and affect plants in multiple plant families. Powdery mildews differ from the vast majority of other fungal pathogens in that, with few exceptions, the mycelium grows superficially over host tissues. Only specialized feeding cells called **haustoria** (Chapter 10) penetrate the host epidermis (Figure 14.2). Powdery mildew life cycles (Figure 14.3) are entirely **biotrophic**. No species has been grown in **axenic** culture apart from its host for any significant duration and none can grow on dead plant material.

Conidiophores develop directly from the mycelium on the host surface throughout the growing season; they produce infective **conidia** either one at a time (Figure 14.4) or in short chains. In tropical climates or green-houses, this may be the only spore stage in the life cycle. Conidia are wind-borne or splash-dispersed and serve as secondary inocula during the growing season. Crops grown in greenhouses may suffer repeated cycles of infection via conidia year-round. Spore release typically follows a diurnal pattern, with the highest number of conidia being released around midday in response to a decrease in relative humidity. Outdoors, rain may serve as a trigger for the release of conidia. Powdery mildews are unusual among fungi in that their conidia do not require large amounts of free water to germinate. Significant disease spread can thus occur even during dry weather.

Most species of the Erysiphales are **heterothallic**; two compatible individuals must mate before asci can be formed. Compatibility is genetically determined by the mating type locus. Two different **alleles** confer two possible mating types. Some homothallic species are known, but the genetic basis of homothallism has not been determined.

After mating, ascomata are produced superficially on the host, typically toward the end of summer (Figure 14.5). Ascomata can be an important overwintering stage in temperate climates as they provide the primary inoculum at the beginning of each growing season for a number of powdery mildews. In arid climates, they aid in survival during hot, dry periods. In direct contrast to conidia, free water is required to stimulate ejection of the ascospores and allow their germination. In some species that attack perennial hosts, such as *Podosphaera pannosa* (sect. *Sphaerotheca*) f. sp. *rosae* on rose, the fungus can also survive as mycelium inside infected buds. In powdery mildews of tropical climates, the ascomata are of little

FIGURE 14.1 Powdery mildew of poinsettia caused by *Oidium* species. White colonies on the red bracts of this Christmas favorite make the plants unsaleable. The sexual state of this pathogen is as yet unknown.

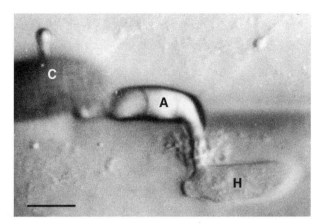

FIGURE 14.2 Penetration structures of *Blumeria graminis* f. sp. *hordei*, powdery mildew of barley. A conidium (C, upper left) that is slightly out of focus on the leaf surface has germinated to produce a primary (top) and a secondary germ tube. The secondary germ tube has produced a large appressorium (A: center) and penetrated the epidermal cell wall. A haustorium (H: bottom right) has been formed inside the epidermal cell. This basic unit of infection may be replicated in hundreds of epidermal cells underlying a single mildew colony. (Photo courtesy of J. R. Aist, Cornell University. With permission.)

importance and have apparently been lost altogether in some species.

The appendages on the ascocarps function in dispersal by helping the ascocarps to adhere to plant surfaces and **trichomes**. For powdery mildews on woody plants, the elaborate appendages allow the ascomata to cling to rough bark surfaces. In *Phyllactinia*, which causes the powdery mildew of alders and numerous other hardwoods, the moisture-sensitive appendages play a unique role in dispersal (Figure 14.5). On maturity, the appendages press down on the substrate, breaking the ascocarp away and releasing it into the wind. Sticky secretions produced from a second set of appendages on top of the ascocarp promote its adherence when it lands. The ascocarp passes the winter upside-down. In spring, the ascocarp splits around the circumference and flips open completely, like a hinged jewel box, so that the formerly upside-down asci in the "lid" now point upward to discharge their infective spores. In other powdery mildews, the appendages are less complex and the ascocarps typically open from the pressure of asci against the upper surface.

HOST RELATIONSHIPS

Powdery mildews engage in fascinating interactions with their host. Unlike other plant pathogens, a single individual penetrates its host at many different sites and does not proliferate within the plant tissues. A few powdery mildews are atypical in that they do penetrate the stomatal space and form a limited **hemiendophytic** mycelium. The complex organs of infection formed by powdery mildews are among the best-studied host–pathogen interfaces in plant pathology (Figure 14.2).

THE INFECTION PROCESS

Powdery mildew conidia deposited on a hydrophobic leaf surface excrete an adhesive matrix within minutes of contact (Nicholson and Kunoh, 1995). This matrix is believed to mediate host recognition. In a compatible interaction, conidia rapidly germinate by producing one or more germ tubes that follow the contours of the host surface (Figure 14.6). At the germ tube's apex, an **appressorium** (Figure 14.2, Figure 14.6, and Figure 14.7) is produced at the site where penetration will occur. In different species, the appressoria can be undifferentiated, simple, forked, or lobed, and these morphological variations can be useful in identification (Figure 14.8). The fungus also continues to grow across the host's surface forming the distinctive superficial mycelium.

The difficult traverse of the plant cell wall is the next step in infection. At this stage, a hypha with narrow diameter, known as a **penetration peg**, grows from the bottom surface of the appressorium. Through a combination of enzymatic and mechanical action, it penetrates the wall of an epidermal cell. Having gained entry, the fungus produces a haustorium (Figure 14.2), a special feeding cell that assimilates nutrients from the plant. Haustoria can be simple, lobed, or digitate. The haustorium is not in direct contact with the host cytoplasm; rather, it is enfolded by the cell membrane of its host cell. The fungus induces changes in the plant membrane surrounding the haustorium that results in leakage of nutrients that are then taken up by the fungus. The invaded cell remains alive and functioning for some time and receives nutrients from

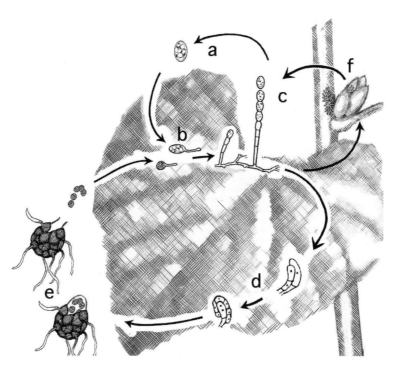

FIGURE 14.3 Life cycle of a powdery mildew (a) Conidium. (b) Condium germinating and forming an appressorium on the leaf surface. (c) Conideophores and conidia. (d) Sexual reproduction between compatible mating types. (e) Ascocarps, ascus, and ascospores. Note germinating ascospore at (b). (f) Mycellium overwintering in bud. (Drawing by N. Shishkoff.)

surrounding cells that are then leaked to the fungus. This basic pathogenic interaction is repeated in many different cells, and the superficial mycelium spreads across the leaf surface and occasionally develops appressoria, each leading to penetration and parasitism of a single cell. Although the host is seldom killed outright, the continual drain of nutrients often depresses growth and sometimes causes puckering or chlorotic or necrotic spotting of the leaf surface.

HOST RESPONSES

Host plants resist powdery mildew attack through passive and active defense systems (Chapter 28). Leaf surface features, such as the thick waxes that protect the undersurfaces of some grass leaves, may inhibit germination. The cuticle and cell wall present formidable barriers to penetration because their thickness and durability affect the penetration ability of fungal germ tubes. Many plants actively respond to penetration attempts by forming **papillae** (sing. papilla), tiny, thick deposits of callose and other materials that are deposited on the inner surface of the cell wall. Papillae are a physical barrier to invasion and also serve as foci for the release of antimicrobial compounds. The **hypersensitive response** may be induced in incompatible interactions, resulting in death of an invaded host cell and starvation of the pathogen. All these mechanisms are genetically controlled and provide important sources of resistance for plant breeders.

GENERAL TAXONOMY

Powdery mildews are classified in the Class Leotiomycetes (with primitive cup-fungi), the Order Erysiphales, and the Family Erysiphaceae. For years, the taxonomy of the powdery mildews relied largely on the morphology of the ascomata and their appendages, because they showed more obvious differences (Braun 1987; 1995). Recent advances in molecular phylogeny have shown the morphology of the anamorph to be a better indicator of relationships (Saenz and Taylor, 1999; Takamatsu et al., 1999). The characteristics of the conidial chains and the ornamentation of the conidial surface are particularly useful (Cook et al., 1997). Now it is thought that differences in the ascomata merely reflect adaptation to particular hosts. Braun and Takamatsu (2000) and Braun et al. (2002) have renamed many powdery mildews based on new genetic and morphological data.

THE ANAMORPH

Conidia (meristemic arthrospores) are produced on unbranched conidiophores, singly or in chains. The name *Oidium* is often used informally to refer to any **anamorph** of a powdery mildew. The genus *Oidium* in the more formal sense refers to powdery mildews producing ovoid conidia singly or in chains. *Oidium* species are divided into subgenera based on morphological features, and these neatly correlate with particular **teleomorphs**, so that *Oid-*

FIGURE 14.4 Conidiophores and conidia of *Oidium* sp., which causes powdery mildew of poinsettia. (Photo courtesy of G. Celio, University of Georgia. With permission.)

ium subgen. *Oidium* is associated with the teleomorph *Blumeria*, the subgen. *Pseudoidium* with the genus *Erysiphe*, and so on. Other powdery mildew anamorphs have conidia that are not ovoid or are borne singly (Figure 14.8).

THE TELEOMORPH

The ascomata of powdery mildews are initially fully closed and they have often been referred to as **cleistothecia**. They differ from other cleistothecial fungi, however, in that the ascomata do eventually rupture and the ascospores are forcibly discharged. Developmental features also suggest that the ascomata are more similar to perithecia than cleistothecia.

Mycologists once assumed that primitive powdery mildews were *Erysiphe*-like, with undifferentiated ascomatal appendages (Figure 14.8). New genetic evidence suggests that the primitive powdery mildews lived on woody hosts and possessed curled appendages (Takamatsu et al., 2000). Species on herbaceous hosts typically lack elaborate appendages. This suggests that appendages may be "costly" to maintain, valuable in anchoring ascomata to persistent plant parts, but lost quickly when the plant host has no persistent parts on which to anchor. Related species thus, display conserved anamorph morphology, whereas teleomorph morphology may be plastic and strongly influenced by host and environment.

IDENTIFICATION OF POWDERY MILDEWS

The new nomenclature (Braun et al., 2002) is sure to be confusing for a time because many familiar names have been changed in order to reflect our new understanding of relationships among the powdery mildew fungi. In this system, the Family Erysiphaceae is divided into five tribes

including, in all, 13 current genera. A simple overview of modern powdery mildew genera is presented in Figure 14.8. It is easy enough to recognize that a disease is caused by a powdery mildew fungus by noting the white powdery colonies on infected plant parts. Morphological differences separating the species can be subtle, however, so the best route to diagnose powdery mildew diseases is through the use of a thorough host index, such as Farr et al., 1980 (also available online).

SOME IMPORTANT DISEASES CAUSED BY POWDERY MILDEWS

POWDERY MILDEW OF GRAPE

Powdery mildew of grape is caused by *Erysiphe* (sect. *Uncinula*) *necator*. The disease can reduce yield and fruit quality, as well as stunt vines and decrease their winter hardiness. *Vitis vinifera* and hybrids such as Chardonnay and Cabernet Sauvignon are more susceptible to powdery mildew than are varieties derived from American species. Wine quality can be affected when as few as 3% of berries are infected. Leaves, green shoots, and fruits are susceptible to infection. White mildew colonies appear on upper and lower surfaces of leaves, and may coalesce. Immature leaves may be distorted and stunted. Shoot infections appear as dark patches. Infected berries can display white colonies (Figure 14.9), abnormal shape, or rust-colored spots and may split open, rendering them vulnerable to fruit rot by *Botrytis cinerea*. Infected leaves may be cupped-up or scorched and may fall prematurely.

Erysiphe necator overwinters as ascomata in crevices on the surface of the bark of vines or as mycelium in buds (Gadoury and Pearson, 1988). Ascospore release is triggered by even minuscule amounts of rain in the spring when the temperature is warm enough. The ascospores are spread by wind to the green tissues of the grapevine from bud break through bloom. The short time required from infection until the production of new inoculum gives this powdery mildew powerful epidemic potential (Chapter 30). Because little free moisture is needed for germination of conidia, powdery mildew can be a serious problem even in years when black rot and downy mildew are hampered by dry conditions. Relative humidity of 40% to 100% is all that is required for the formation of microdroplets of water used by conidia for germination and infection. Temperature is optimum for disease development from 20°C to 25°C, but infection is possible over a wide temperature range of 15°C to 32°C.

All cultural practices that promote the drying of plant surfaces will reduce powdery mildew. Full sun exposure is desirable; alignment of rows with the direction of the prevailing wind is recommended. Vines should be pruned and trained to reduce shading and allow air circulation

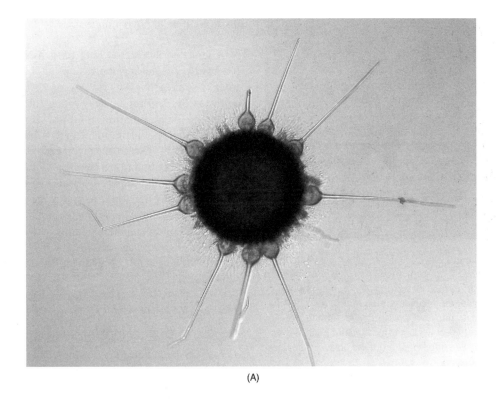

(A)

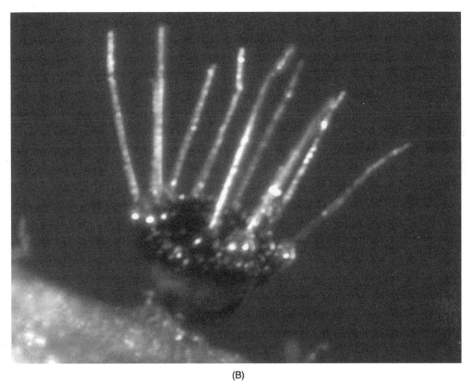

(B)

FIGURE 14.5 The ascoma of *Phyllactinia guttata*, the fungus causing one of the powdery mildew diseases of oak. (A) As it appears when first formed, with the bulbous-based appendages lying flat against the oak leaf surface. (B) After liberation. The appendages have successfully flexed to lift the ascoma from the leaf undersurface. After a brief tumble, the top of the liberated fruiting body has adhered by its sticky cushion to nearby plant material. It will overwinter in this upside-down position. In spring, the ascoma will split open around its equator, revealing asci in its "lid" in the perfect position to discharge infective spores into the air. (Courtesy of K. Hodge, Cornell University. With permission.)

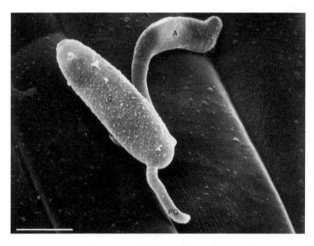

FIGURE 14.6 Scanning electron micrograph of a conidium (C) of *Blumeria graminis* f. sp. *hordei* (causes of powdery mildew of barley) on a leaf surface. A primary (PG) and a secondary germ tube are visible; the secondary germ tube (top, right) has started to differentiate into an appressorium (A) In this species, the primary germ tube serves a sensing function, and the secondary germ tube leads to penetration of the host. (Photo courtesy of J. R. Aist, Cornell University.)

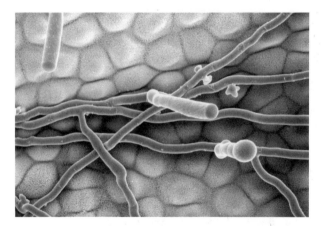

FIGURE 14.7 Scanning electron micrograph of a powdery mildew colony on a leaf surface. The small, lobed appressoria mark points of penetration into the leaf. Several conidiophores can be seen projecting upward off the leaf surface. (Photo courtesy of G. Celio, University of Georgia. With permission.)

within the canopy. Excessive nitrogen should be avoided. Irrigation should be done with trickle systems rather than over-the-vine impact sprinklers. Control of infections during the immediate prebloom through early postbloom period is stressed for grapes because this is when fruit is most susceptible to the disease. The more susceptible *V. vinifera* and hybrid cultivars require continued suppression of powdery mildew on the foliage until at least **veraison** in order to maintain a functional leaf canopy to ripen the crop and avoid premature defoliation. Controlling later-season infection also helps to reduce the overwinter-

ing **inoculum** and may improve winter hardiness in colder regions.

Elemental sulfur has long been used internationally for powdery mildew suppression, although it is **phytotoxic** to some native American (e.g., Concord) and hybrid cultivars. Modern fungicides such as the sterol inhibitors [both the demethylation inhibitor (DMI) and, more recently, morpholine groups] and strobilurins are also widely used in disease management programs (Chapter 33). Certain nontraditional products (e.g., oils, potassium bicarbonate and monopotassium phosphate salts and dilute hydrogen peroxide) also provide suppression and are used to variable extents where they are labeled for this purpose. It appears that much of the activity of such materials is eradicative, presumably due to the susceptibility of the exposed fungal colony to topical treatments of these substances.

POWDERY MILDEW OF CEREALS

Powdery mildew diseases affect **monocots** as well as **dicots**. The pathogen *Blumeria graminis* has several host-specialized **formae speciales** that infect different cereal crops. An example is *Blumeria graminis* f. sp. *tritici*, a pathogen of wheat (*Triticum aestivum*) (Wiese, 1977).

White colonies of powdery mildew are most common on the upper surface of older wheat leaves. The haustoria of *Blumeria* are unusual because they have long finger-like projections. The conidiophores generate long chains of conidia that can germinate over a wide temperature range (1°C to 30°C), with an optimum temperature of 15°C to 22°C and an optimum relative humidity of 85% to 100%. A complete disease cycle takes place in 7 to 10 days, so epidemics can develop quickly. Wheat plants are most susceptible when they are heavily fertilized and rapidly growing.

As wheat plants mature, ascomata begin to appear on the leaf surface as long as both mating types are present. The ascomata are important, because sexual recombination allows new races of powdery mildew to be produced. These new races may be able to grow on wheat cultivars that were not susceptible to the original powdery mildew population. Ascomata on wheat stubble also function as overwintering structures, and in mild climates the mycelium itself can survive. Infections by ascospores require rain and may occur in midsummer or fall. Powdery mildew of wheat prospers in both humid and semiarid climates. Yield is decreased by powdery mildew, because photosynthesis is reduced whereas **transpiration** and **respiration** are increased. Wheat yields may be reduced by as much as 40%.

In spite of the strong impact of this disease, it is not always cost-effective to use fungicides to control powdery mildew on wheat. **Systemic fungicides** are used for powdery mildew control in Europe, primarily. Host-plant resistance is used worldwide for powdery mildew man-

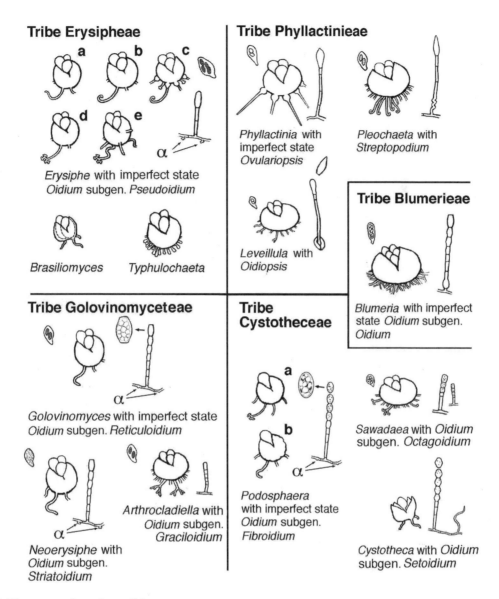

FIGURE 14.8 The genera of powdery mildews.

- Tribe Erysipheae: *Erysiphe* is shown here with its anamorph *Oidium* subgenus *Pseudoidium*, and its characteristic lobed appressoria (α). (This genus now embraces several former genera: (a) *Erysiphe*, (b) *Uncinula*, (c) *Bulbouncinula*, (d) *Microsphaera* and (e) *Medusosphaera*). *Brasilomyces* has a one-layered transparent outer ascomatal wall and no known anamorph. *Typhulochaeta* has club-shaped appendages and no known anamorph.
- Tribe Golovinomycetinae: *Golovinomyces* has two-spored asci, the anamorph *Oidium* subgenus *Reticuloidium* and mycelium with simple appressoria (α). *Neoerysiphe* has asci that only mature after overwintering and anamorph *Oidium* subgenus *Striatoidium* with lobed apressoria (α). *Arthrocladiella* has dichotomously branched appendages and anamorph *Oidium* subgenus *Graciloidium*.
- Tribe Blumerieae: *Blumeria,* sole genus in this tribe, is found on grasses; its anamorph is *Oidium* subgenus *Oidium*.
- Tribe Phyllactinieae: *Phyllactinia* has bulbous-based appendages and anamorph *Ovulariopsis*. *Pleochaeta* has curled appendages and the anamorph *Streptopodium*. *Leveillula*, which is unusual in having internal mycelium and conidiophores that grow out of stomatal pores, has undifferentiated appendages and the anamorph *Oidiopsis*.
- Tribe Cystotheceae: *Podosphaera* has simple appressoria (α) and the anamorph *Oidium* subgenus *fibroidium*. Conidia contain fibrosin bodies. (This genus now embraces the former genera (a) *Podosphaera* and (b) *Sphaerotheca*.) *Sawadaea* has curled appendages (that can be uni-, bi-, or trifurcate) and anamorph *Oidium* subgenus *Octagoidium*. *Cystotheca*, which is unusual in having thick-walled aerial hyphae, has an ascomatal outer wall that easily splits in two and anamorph *Oidium* subgenus *Setoidium*. (Drawing by N. Shishkoff.)

agement. Plant breeders work to supply wheat lines that are not susceptible to the powdery mildew **races** in a given geographic area; new lines are needed frequently to keep up with adaptations of the pathogen. Rotating crops and eliminating "volunteer" wheat plants and crop debris aid in disease management (Chapter 32).

FIGURE 14.9 Powdery mildew of grape, caused by *Erysiphe* (sect. *Uncinula*) *necator*. The powdery white bloom on the fruit makes these grapes unfit for consumption or processing. (Photo by W.R. Fisher, courtesy of the Cornell University Herbarium (CUP 11853a). With permission.)

FIGURE 14.10 Powdery mildew of rose. The arrow indicates a severe infection of the flower pedicel that may affect the longevity and saleability of the flowers.

POWDERY MILDEW OF ROSE

Powdery mildew diseases are especially important on ornamental plants, on which any visible infection may lower the aesthetic appeal. Powdery mildew of rose (*Rosa × hybrida*) affects roses in gardens as well as in field or greenhouse production. It is the most common disease of roses grown in greenhouses for cut flowers and also on potted miniature-roses. Young foliage and pedicels are especially susceptible and may be completely covered with mildew (Figure 14.10). Young shoots and flower petals may also be infected. The rose powdery mildew is *Podosphaera* (sect. *Sphaerotheca*) *pannosa* f. sp. *rosae*.

Infections may develop quickly and conidia may start to germinate within just a few hours of landing on the host (Horst, 1983). Chains of conidia are produced only 3 to 7 days after the initial infection. Disease development is optimum at 22°C. The fungus overwinters in buds unless the climate is so mild that conidia are continually produced year-round, as in greenhouses. Control of rose pow-

dery mildew in greenhouses is achieved by air circulation with fans plus heating and ventilation to reduce humidity, coupled with the use of fungicides. Strobilurins, DMIs, and morpholine fungicides are commonly used for disease control in greenhouses. Products featuring potassium bicarbonate, botanical extracts, and biocontrols, including the fungus *Pseudozyma flocculosa*, have recently been developed. Gardeners wishing to circumvent the problem of powdery mildew on outdoor roses can seek disease-resistant cultivars and species.

POWDERY MILDEW ON CUCURBITS

Numerous vegetable crops are susceptible to powdery mildew, but cucurbits are arguably the group most severely affected. The powdery mildew on cucurbit crops (*Cucumis* species, including squash, pumpkin, and cucumber) is *Podosphaera* (sect. *Sphaerotheca*) *xanthii* (referred to in earlier literature as *Sphaerotheca fuliginea* or *S. fusca*). Infections on both upper and lower leaf surfaces occur and these reduce yield by lowering plant vigor and increasing the number of sun-scorched fruits (Figure 14.11).

FIGURE 14.11 Powdery mildew of cucumber, showing colonies on the upper and lower leaf surfaces. (Photo by M.T. McGrath. From McGrath, M.T., 2001. *Plant Dis.* 85: 236–245. With permission.)

Resistant cultivars of cucumbers and melons are available, but most squash varieties are susceptible. Susceptible melons may have significantly poorer fruit quality due to low sugar content. Even the flavor of winter squash can be harmed because the fruits of mildew-infected plants have fewer stored soluble solids that affect taste. The color and handle quality of pumpkins can be ruined by powdery mildew. The disease develops on cucurbits during warm summer weather (a mean temperature of 68°F to 80°F is most favorable). Conidia are thought to be airborne over long distances to initiate infections.

The fungicides used for disease management in cucurbits often are systemic materials with specific, single-site modes of action that make them vulnerable to the development of resistance in the pathogen (McGrath, 2001). The fungicide benomyl is no longer effective against cucurbit powdery mildew because of resistance development. Due to the rapid epidemic development of powdery mildews, they are more capable of developing resistance to fungicides than are many other fungi. The pathogen population can shift quickly to resistant individuals even when these are initially at an undetectable level in the population. Growers of any crop with high susceptibility to a powdery mildew are encouraged to rotate among chemical classes with different modes of action to delay the development of resistance. The use of materials with multisite modes of action, such as chlorothalonil, copper, sulfur, horticultural oil, and potassium bicarbonate, is also a good strategy to avoid resistance.

An **integrated pest management** (IPM) program for cucurbit powdery mildew can be very effective (McGrath and Staniszewska, 1996; McGrath, 2001, Chapter 35). Elements of such a program include using resistant varieties when available, scouting for the first colonies beginning at the time of fruit initiation, using air-assist sprayers to maximize spray coverage on difficult-to-cover lower

leaves and using fungicides according to a resistance-management strategy. For greenhouse-grown cucumbers, silicon amendment of the nutrient solution has reduced disease. A number of organisms including the fungi *Pseudozyma flocculosa*, *Sporothrix rugulosa*, *Tilletiopsis* species, *Ampelomyces quisqualis* and *Verticillium lecanii* have been used for biocontrol (Chapter 34) of cucumber powdery mildew in greenhouses (Dik et al., 1998). Mechanisms of antagonism used by these fungi include **hyperparasitism** or **antibiosis**. As powdery mildews may thrive at lower relative humidity than the antagonistic fungi that might be deployed against them, environmental conditions in a given situation may determine whether biocontrol is effective.

POWDERY MILDEW ON FLOWERING DOGWOOD

Many woody plant species are subject to one or more powdery mildews, but these diseases rarely have a significant economic impact. Lilac foliage, for example, frequently shows conspicuous powdery mildew in late summer, but the shrubs are attractive again during bloom the following spring. Powdery mildew has been noted on dogwoods (*Cornus* species) since the 1800s, but a serious powdery mildew disease on *C. florida*, the flowering dogwood, became apparent in the Eastern U.S. only in the mid-1990s (Figure 14.12). *Erysiphe* (sect. *Microsphaera*) *pulchra* is the powdery mildew that now affects flowering dogwood. Aesthetic injury is caused on landscape trees by whitening and twisting of the leaves at the tips of branches in mid-summer, whereas economic injury is seen during nursery production when young trees with powdery mildew are severely stunted. In many cases, infection of landscape trees is subtle, such that the thin coating of mycelium on the leaves is often overlooked. Symptoms of reddening or leaf scorch are often seen on infected leaves in dry summers, and it is thought that the mildew may contribute to the drought stress on the host tree.

Ascomata on leaf debris are the only documented overwintering mechanism for this powdery mildew. The abundance of ascomata varies from year to year, perhaps due to variation in weather conditions during the fall when they are maturing. The nonnative Kousa dogwood, *C. kousa*, is not appreciably affected by powdery mildew. Several selections of powdery mildew-resistant cultivars of *C. florida* are under development. Fungicides such as propiconazole and chlorothalonil are used for disease suppression during nursery crop production.

In some ways, powdery mildews seem to be mild diseases, because the fungi only rarely kill the plants that they parasitize. Powdery mildew fungi and their hosts come closely together in a carefully balanced parasitic relationship that results in nutrient flow to the powdery mildew without extensive death of plant host cells. Perennial plants may be infected by powdery mildew summer

FIGURE 14.12 Dogwood powdery mildew causes curling and deformation of leaves at the tips of branches; the fungus may entirely coat the leaves. The eastern flowering dogwood (*Cornus florida*) is quite susceptible, but most cultivars of the Kousa dogwood (*Cornus kousa*) are quite resistant. (Photo by K. Britton, USDA-USFS.)

after summer and still leaf out vigorously every spring. Annual plants usually bear fruit and set seed in spite of powdery mildew. Yet to the horticultural producers aiming to have high yield, high-quality crops, or to gardeners wanting attractive and vigorous plants, powdery mildews are formidable diseases. Curbing powdery mildew epidemics on economic hosts is essential to maximize the growth potential of the plants and maintain the marketability of the products harvested from them.

REFERENCES

Braun, U. 1987. *A Monograph of the Erysiphales (Powdery Mildews)*. J. Cramer, Berlin-Stuttgart.

Braun, U. 1995. *The Powdery Mildews (Erysiphales) of Europe*. G. Fischer Verlag, Jena.

Braun, U., R.T.A. Cook, A.J. Inman and H.-D. Shin. 2002, in *The Powdery Mildews: A Comprehensive Treatise*, Belanger, R.R., A.J. Dik and W.R. Bushnell (Eds.) APS Press, St. Paul, MN.

Braun, U. and S. Takamatsu. 2000. Phylogeny of Erysiphe, Microsphaera, Uncinula (Erysipheae) and Cystotheca, Podosphaera, Sphaerotheca (Cystotheceae) inferred from rDNA ITS sequences — some taxonomic consequences. *Schlechtendalia* 4: 1–33.

Cook, R.T.A., A.J. Inman and C. Billings. 1997. Identification and classification of powdery mildew anamorphs using light and scanning electron microscopy and host range data. *Mycol. Res.* 101: 975–1002.

Dik, A.J., M.A. Verhaar and R.R. Belanger. 1998. Comparison of three biocontrol agents against cucumber powdery mildew (*Sphaerotheca fuliginea*) in semi-commercial scale glasshouse trials. *Eur. J. Plant Path.* 104: 413–423.

Farr, D.F., G.F. Bills, G.P. Chamuris and A.Y. Rossman. 1980. *Fungi on Plants and Plant Products in the United States.* APS Press, St. Paul, MN.

Gadoury, D.M. and R.C. Pearson. 1988. Initiation, development, dispersal and survival of cleistothecia of *Uncinula necator* in New York vineyards. *Phytopathology* 78: 1413–1421.

Horst, R.K. 1983. *Compendium of Rose Diseases.* APS Press, St. Paul, MN.

McGrath, M.T. 2001. Fungicide resistance in cucurbit powdery mildew: Experiences and challenges. *Plant Dis.* 85: 236–245.

McGrath, M.T. and H. Staniszewska. 1996. Management of powdery mildew in summer squash with host resistance, disease threshold-based fungicide programs, or an integrated program. *Plant Dis.* 80: 1044–1052.

Saenz, G.S. and J.W. Taylor. 1999. Phylogeny of the Erysiphales (powdery mildews) inferred from internal transcribed spacer (ITS) ribosomal DNA sequences. *Can. J. Bot.* 77: 150–169.

Takamatsu, S., T. Hirata, Y. Sato and Y. Nomura. 1999. Phylogenetic relationship of *Microsphaera* and *Erysiphe* sect. *Erysiphe* (powdery mildews) inferred from the rDNA ITS sequences. *Mycoscience* 40: 259–268.

Takamatsu, S., T. Hirata, and Y. Sato. 2000: A parasitic transition from trees to herbs occurred at least two times in tribe Cystotheceae (Erysiphaceae): Evidence from nuclear ribosomal DNA. *Mycol. Res.* 104: 1304–1311.

Wiese, M.V. 1977. *Compendium of Wheat Diseases.* APS Press, St. Paul, MN.

15 Ascomycota: The Filamentous Fungi Forming Perithecia, Apothecia, and Ascostromata

Kenneth J. Curry and Richard E. Baird

CHAPTER 15 CONCEPTS

- Recognition of ascomycetes is based on sexual reproduction culminating in formation of an ascus.

- Asci develop within a fruiting body or ascoma.

- Ascoma may be a perithecium, an apothecia, or an ascostroma, each of which corresponds with a major fungal evolutionary group.

- Some ascomycetes have asexual states that are classified as deuteromycetes.

The fungi considered in this chapter are commonly called the ascomycetes and are the most diverse group of fungi. They are characterized by a specialized unicellular sporangium, the **ascus** (pl. asci), in which sexually derived spores, ascospores, are formed. The ascomycetes considered here are further characterized by the development of a protective structure, the **ascoma** (pl. ascomata), in which asci develop in a single layer or plane, the hymenium. The basic construction of the ascomycetes (and most other fungi) is filamentous and gives them tremendous surface area to absorb nutrients from their substrate.

The ascomycetes may be divided into two major types based on lifestyle. About 40% of the ascomycetes are lichenized (Hawksworth et al., 1995). These lichenized ascomycetes have an obligate symbiotic relationship with a photosynthetic symbiont (symbiotic partner) that can be a unicellular green alga or one of the cyanobacteria (blue-green alga). The ascomycete member is never found without its photosynthetic partner under natural conditions. The photosynthetic partner, however, is sometimes found free-living in nature.

The remaining 60% of the ascomycetes are either saprophytes or are engaged in symbiotic relationships that range from parasitism through commensalism to mutualism. Most of these ascomycetes, in sharp contrast to lichenized ascomycetes, grow with their hyphae immersed in the substrate from which they obtain nutrients. Their reproductive structures are usually the only parts exposed. Hyphae grow through a substrate by exploiting fresh nutrient sources. Older hyphae gather nutrients and transport materials to growing tips. The oldest hyphae die as the fungus grows into new areas. Thus, the fungus is constantly replacing its senescent parts.

The majority of the ascomycetes are saprophytes. Their carbon and energy come from the organic remains of dead organisms. They grow in dead plant material where they degrade, among other things, cellulose and lignin (Chapter 27). Fungi excel at degrading these two tremendously abundant molecules. They grow in soil associated with organic remains composed primarily of plant material.

A minority of the ascomycetes are parasites. Their carbon and energy come from living organisms. A small number of these ascomycetes are animal parasites. A few grow in or on vertebrates, including humans. Some produce structures to trap nematodes and rotifers. Most ascomycete parasites have plant hosts with which they exhibit a wide range of relationships. Some ascomycetes grow within plant tissues as endophytes. The relationship is parasitic in that the endophyte takes nutrients from its host, but no disease symptoms are manifested. A parasite that causes disease symptoms is a pathogen (Bos and Parlevliet, 1995). Some ascomycete pathogens are obligate biotrophs. They invade living plant cells, keeping the host cells alive while withdrawing nutrients. Some ascomycete pathogens are necrotrophs, killing host cells to obtain nutrients. Between the extremes of **biotroph** and **necrotroph** is the **hemibiotroph** (Luttrell, 1974). Hemibiotrophs invade and maintain living host cells for a period of hours or days, and then revert to necrotrophy. Hemibiotrophs and necrotrophs can also exist in nature as saprophytes. Their ability to function as parasites allows them to create their own saprophytic substrates.

REPRODUCTION

We recognize the following three reproductive categories for the ascomycetes: (1) ascomycetes that only reproduce sexually, (2) ascomycetes known to reproduce sexually, but also observed in asexual reproduction, and (3) fungi (deuteromycetes) thought to be ascomycetes, but only observed to reproduce asexually.

Sexual reproductive structures all function with one end to bring together two compatible nuclei, where compatible means the nuclei can fuse. Typically for ascomycetes this results in two haploid nuclei fusing to form a diploid nucleus. The act of nuclear fusion, called karyogamy, usually does not follow immediately on bringing the two nuclei into a single cell. Bringing the nuclei together typically first involves bringing the cytoplasm of two cells together or plasmogamy (literally the marriage of cytoplasm). The nuclear pair resulting from plasmogamy is called a dikaryon. The two haploid nuclei, though physically separate, function in some ways as if they were diploid. Typically the nuclear pair undergoes a synchronized division, with the resulting new nuclear pair migrating into a growing hyphal tip to be delimited from the first pair by a septum. The successive growth of dikaryotic hyphae and synchronized division of the dikaryotic nuclei can continue for some time. Ultimately, particular dikaryotic cells differentiate into asci, karyogamy or nuclear fusion occurs, meiosis occurs, and the sexual process culminates in the production of ascospores. Thus, sexual reproduction is defined as comprising two steps, nuclear fusion and meiosis. Other events might be closely associated with the sexual process, such as the occurrence of gametangia to effect nuclear transfer, but they are not necessary to the definition of sexual reproduction.

The transfer of nuclei in many ascomycetes can only occur between two compatible organisms. Ascomycetes that require a partner for sexual reproduction are referred to as heterothallic. Both members of the pair may produce both male and female gametangia, so gender is not ascribed to one organism; thus, ascomycetes, with a few exceptions, are hermaphroditic. A number of ascomycetes do not require a partner to complete sexual reproduction. Formation of ascogonia and antheridia, dikaryotic hyphae, nuclear fusion, and meiosis occur within a single organism and is referred to as homothallic. Note that although homothallic fungi have met the definition for sexual reproduction, they will not participate at the same level of genetic variation through gene exchange as heterothallic ascomycetes.

The principal propagule of dissemination of ascomycetes is a spore. Asexual reproduction, in contrast with sexual reproduction, involves the production of propagules of dissemination via mitosis. Nuclear fusion and meiosis are both notably absent. The most common asexual propagule of dissemination for ascomycetes is a spore called a conidium. Think of a spore as any propagule, whether derived sexually through meiosis or asexually through mitosis, which from its inception is designed for dissemination. A spore may comprise one cell or two or more cells. Typically, the number of cells is regulated genetically and precisely.

NOMENCLATURAL CONVENTIONS

Identifying ascomycetes is sometimes complicated when only the asexual state is found. The connection between sexual and asexual states within a single species is often unknown. Three pragmatic terms describe the combinations of sexual and asexual states (Reynolds and Taylor, 1993). The **holomorph** refers to both the sexual and asexual states of the fungus, even if one or both states are not present at the time of observation (Figure 15.1). The **teleomorph** indicates only a sexual state of the fungus even if the asexual state is present. The anamorph refers to only the asexual aspect of the fungus even if the sexual state is present. The **anamorph** of ascomycetes is usually given a separate binomial name from the teleomorph out of pragmatic convenience. The asexual state is often seen without the sexual state. The practice of assigning two names to a holomorphic fungal species is sanctioned in the International Code of Botanical Nomenclature and by long tradition. The anamorphic name is assigned to the Deuteromycota or Fungi Imperfecti (Chapter 16). The Deuteromycota are considered to be a form-group with members referred to form-species, form-genera, etc. The designation of *form* is applied to indicate a less formal taxonomic status to the asexual state than to the sexual state, while recognizing the ubiquity of the asexual state in nature. Formal taxonomy is tied to the sexual state or the teleomorph, indicating our belief that sexual reproduction is somehow more important in an evolutionary model then asexual reproduction. For example, if teleomorphic fungi meet certain morphological criteria, the binomial name *Glomerella cingulatum* is applied. If members of this fungus are seen expressing only the asexual or anamorphic state, the name *Colletotrichum gloeosporioides* is used. The anamorph and teleomorph stages of the fungus are not commonly seen together, but if so, the teleomorph name is used as the holomorph name because of its formal taxonomic priority over the anamorph designation. Thus, the holomorph and teleomorph share one name, in this example *G. cingulatum*, and the anamorph of the same fungus has a separate name, in this example *C. gloeosporioides* (Sutton, 1992).

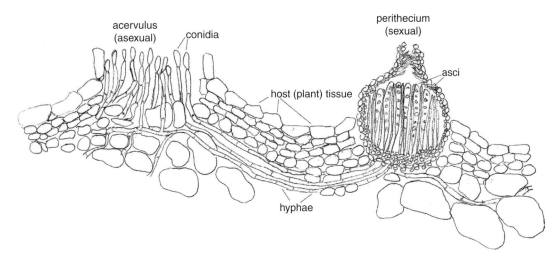

FIGURE 15.1 Nomenclatural conventions. The plant pathogen, *Glomerella* species, is shown here developing in host plant tissue. The anamorph or asexual aspect of the fungus, *Cercosporella* species, is depicted on the left developing as an acervulus. The teleomorph or sexual aspect of the fungus, *Glomerella* species, is shown on the right. The holomorph name is also *Glomerella* species.

GENERALIZED LIFE CYCLE OF THE ASCOMYCETES

Considerable value will be gained in understanding a generalized ascomycete life cycle that may serve as a starting point for understanding specific details of development of individual organisms among the species of ascomycetes. For this generalized life cycle, the presence of male and female gametangia, the development of dikaryotic hyphae, the formation of asci in a single layer or plane called the hymenium, and the formation of ascospores is included (Figure 15.2). A modern account is found in Alexopoulos et al. (1996).

The life cycle begins with the differentiation of hyphae to form an ascogonium (female gametangium) with a trichogyne and to form an **antheridium** (male gametangium). Nuclei from the antheridium migrate through the trichogyne into the ascogonium, where they are thought to pair with ascogonial nuclei. Hyphae develop from the walls of the ascogonium, and the paired nuclei (dikaryons) migrate into these dikaryotic hyphae. The dikaryotic hyphae grow along with monokaryotic hyphae to form a protective fruiting body or ascoma in which asci will develop. The dikaryons undergo synchronized divisions, distributing themselves in pairs through the growing dikaryotic hyphae. The dikaryotic hyphae form a hymenial layer within the developing ascoma. The tip of each dikaryotic hypha bends over 180° to form a crozier (resembles a crochet hook). The nuclear pair (dikaryon) in the tip of the crozier undergoes nuclear fusion to form a diploid nucleus while the crozier grows into the ascus. The diploid nucleus undergoes meiosis to yield four haploid nuclei. These four nuclei typically undergo a synchronized mitotic division to produce eight haploid nuclei. Double membranes are elaborated within the ascus and

surround each of the eight nuclei. A spore wall is laid down between these two membranes, and thus eight ascospores are formed. Ascospores can be disseminated passively by disintegration of the ascus, or, more com-

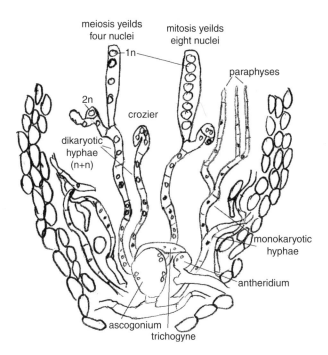

FIGURE 15.2 A diagrammatic apothecium is shown with the gametangia (ascogonium and antheridium) producing dikaryotic (*N* + *N*) hyphae from which croziers develop. The dikaryon fuses to form a diploid nucleus in a cell called the ascus mother cell. The ascus mother cell develops into the mature ascus while the diploid nucleus undergoes meiosis to yield four nuclei, each of which typically divide by mitosis to give eight nuclei in the ascus. A spore wall develops around each nucleus, resulting in eight ascospores.

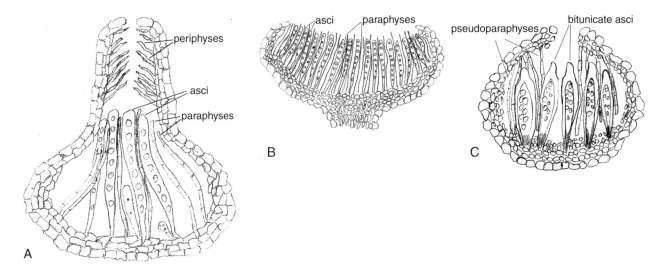

FIGURE 15.3 Diagrammatic representations of ascomata. (A) A perithecium with periphyses and paraphyses. (B) An apothecium with paraphyses. (C) A pseudothecium with pseudoparaphyses and bitunicate asci.

monly, disseminated actively by forcible discharge from the ascus. In the latter case, the ascus develops internal turgor pressure and forcibly ejects the ascospores one at a time through a preformed opening at the tip of the ascus. Ascospores may be disseminated by wind, water, or animal vectors, usually insects. Spores landing on suitable substrates germinate and grow into the substrate to exploit the nutrient sources. Environmental conditions that stimulate reproduction are not well understood, but exhaustion of the nutrient source is commonly associated with reproduction.

ASCOMATA

All the taxa addressed in this chapter produce asci in structures that, in the most general terminology, are fruiting bodies or ascomata (sing. ascoma; Figure 15.3). These can be further distinguished developmentally as ascocarps, which include perithecia and apothecia, and ascostromata (sing. ascostroma), which include pseudothecia (Alexopoulos et al., 1996). The distinction between ascocarp and ascostroma hinges on developmental studies that indicate the ascocarp as developing from sexual stimulation of the activity of an ascogonium and antheridium. In contrast, the ascostroma develops as a stroma first with gametangia and asci developing later in locules in the stromatic tissue.

Two major architectural types of ascocarps are represented in this chapter. The **perithecium** is typically flask-shaped enclosing asci in a single locule with a single pore or ostiole through which ascospores are discharged at maturity. A few of the species represented here produce ascocarps without ostioles. These ascocarps might be called cleistothecia (introduced in Chapter 14), but differ from the typical cleistothecium in that

their asci are organized in a hymenium within the ascocarp instead of the scattered arrangement that characterizes cleistothecia. The morphological situation may be complicated if the perithecium develops embedded in a stroma. The stroma develops first, just as an ascostroma, but the sexual structures stimulate the development of an ascocarp wall around the developing asci; therefore, the final structure is a perithecium encompassing asci and surrounded by stroma.

The basic plan for the **apothecium** is an open cup with or without a stipe (stalk) and with asci organized in a hymenium across the surface of the cup. Convolutions of the basic cup lead to a variety of morphologies and corresponding taxa.

The ascostroma may express a predictable or an irregular morphology. Asci may develop in multiple, adjacent locules formed in the stroma or they may develop in one locule. Some taxa produce a flask-shaped stroma, a **pseudothecium**, which resembles a perithecium.

TAXONOMY

The ascomycetes addressed in this chapter are organized into the following three major groups characterized principally by type of ascoma: the pyrenomycetes, the discomycetes, and the loculoascomycetes. These groups have had both formal taxonomic names with positions in the Linnaean hierarchy and trivial names of convenience. Trivial names will be used here in deference to the state of flux in higher ascomycete taxonomy. Taxonomy within the three major groups is organized around major centrum types (i.e., the ascoma and its internal organization). Various types of molecular information are currently being used to support, refine, and refute morphological and developmental morphological data on which past taxonomic schemes have been constructed (Taylor et al., 1994; Samuels and Seifert, 1995). The morphological groups described here are generally supported by molecular data as being monophyletic, that is, each group represents a single phylogenetic lineage and all of its descendants.

PYRENOMYCETES

This group typically includes fungi producing ovoid to cylindrical asci that develop in a hymenium within a perithecium. Two plant pathogens representing this group are described here.

Glomerella species are responsible for anthracnose diseases, commonly of annual plants, and fruit rots, including bitter rot of apple and ripe rot of grape. The asexual state (anamorph), *Colletotrichum* species, is commonly encountered in these diseases. A conidium landing on a host surface produces a short germ tube terminating in a specialized swelling, the appressorium. A penetration peg from the appressorium grows through the host cuticle, thus allowing the fungus to gain entry into the plant. The fungus grows usually as a necrotroph, destroying host tissue. Eventually, the fungus reemerges at the host surface, developing a tissue called stroma just below the host cuticle. Expansion of the stroma (acervulus; Chapter 16) and the developing conidiophores and conidia rupture the host cuticle, allowing conidia to disperse to reinfect the same plant or infect other hosts. The fungus overwinters at the end of the temperate growing season in various ways, not all known to us. Taking bitter rot of apple as an example, the infected fruit dries and hardens, a process called mummification. The mummy may remain on the tree or fall to the ground. Perithecia form on the surface at the beginning of the following spring, where they discharge their ascospores to initiate infection on new host material. Controls for *Glomerella* species are generally through sanitation or removal of mummified fruits and dead or diseased tissues that were invaded by the pathogen. Fungicides are also effective, but require full season applications when environmental conditions are suitable for fungal growth and sporulation.

Claviceps purpurea causes ergot of rye and wheat and is occasionally found on oats, barley, corn, and some wild grasses. The fungus overwinters as compact hyphal masses (sclerotia; Chapter 16) and produces stalked stromata in the spring. Perithecia form within the enlarged heads of the stromata. The ascus contains eight thread-like, multicellular ascospores that are forcibly discharged. Spores that happen to reach host flowers germinate, and hyphae invade and destroy the ovary. Conidia are produced in a sticky matrix and are dispersed to other host flowers by insects attracted to a sticky exudate or honey dew. Honey dew production gradually diminishes toward the end of the growing season and the mycelium develops into a hard mass, the sclerotium. This sclerotium may fall

to the ground to overwinter or be harvested with healthy seed to be sown the following year. The sclerotium is the source of ergot, a group of alkaloids including secoergolenes, ergolines, and lysergic acid derivatives that protect it from herbivores. Control practices include use of sclerotium-free seed, rotation with resistant crops, and deep tillage practices for burial of sclerotia.

DISCOMYCETES

This group includes fungi that produce asci in a hymenium within an apothecium. Two major phylogenetic groups of the discomycetes are recognized based on asci that are operculate or inoperculate. Operculate asci have an apical apparatus that is a hinged lid opening under positive pressure during spore discharge. Inoperculate asci have a variety of pore types that direct ascospores through the ascus tip during spore discharge. The basic cupulate architecture of the ascocarp is expressed in a broad variety of modifications, including a closed cup that develops underground (truffles). One representative plant pathogen in this group is described here.

Species of *Monilinia* are responsible for brown rot of stone fruits (e.g., peach; see Plate 5C following page 80). Young peaches are generally resistant to brown rot caused by *Monilinia fruticola* and its anamorph *Monilia*, but become increasingly susceptible as they approach maturity. Fungal invasion occurs through hair sockets, insect punctures, and other wounds. The mycelium spreads through the host tissue rapidly and destroys cell walls by pectinases and other extracellular enzymes (Chapter 27). Mycelium reemerges on the surface of the fruit to produce conidia that are spread variously by wind, water, and insects to other fruit. Some fruit falls to the ground where it quickly disintegrates under the action of saprophytic fungi. Heavily infected fruit remaining on the tree lose water content and becomes a shriveled mass of plant tissue and mycelium — a mummy. The mummy may persist on the tree through the winter or it may fall to the ground and last in the soil for up to three years. In either case, the fungus overwinters in the mummy. The following spring, mummies on trees express the anamorph and begin a new infection cycle with conidial production. Mummies on the ground express the teleomorph and begin the infection cycle by producing stipitate brown apothecia with inoperculate asci. The ascospores are forcibly discharged and wind dispersed. Ascospores landing on susceptible host tissue are thought to infect in a manner similar to that of conidia. Control of the brown rot pathogen is primarily by the use of select fungicides on a time schedule. Cultural practices such as sanitation of mummies and pruning and removal of infected or dead material reduce disease levels. Also, insect control can significantly reduce disease incidence.

LOCULOASCOMYCETES

Fungi producing asci in an ascostroma are included in this group. The asci are constructed with two functionally distinct walls. The inner wall is thick and expands rapidly during ascospore discharge, rupturing the thin outer wall. In contrast to the unitunicate (one coat) ascus characteristic of the pyrenomycetes and discomycetes, the loculoascomyctes have a bitunicate (two coats) ascus. Diseases caused by fungi in this group include apple scab (*Venturia inaequalis*), black rot of grapes (*Guignardia bidwellii*; see Plate 7B following page 80), and citrus, scab (*Elisone fawcetii*). One representative plant pathogen is described here.

The genus *Venturia* contains several species of plant pathogens including *V. inaequalis*, which causes apple scab and attacks some related plants. Both leaves and fruit can be infected. Infection begins as a velvety, olive-green lesion that gets darker and dries as it matures. *Venturia inaequalis* overwinters in leaves on the ground. A stroma, which in this species is a pseudothecium, develops slowly within the leaf tissue. An ascogonium develops within the pseudothecium and projects its trichogyne into the surrounding plant tissue. The fungus requires an opposite meeting strain to proceed (i.e., heterothallic). If that strain exists in the leaf, an antheridium will develop near the trichogyne and transfer nuclei to it. Dikaryotic hyphae will develop from the ascogonium and form asci through the crozier mechanism described earlier in this chapter. This very slow development culminates in spring with asci forcibly discharging ascospores, each consisting of two cells of unequal size, hence the name "inaequalis." Spores are dispersed by air currents. Those landing on suitable host surfaces produce short germ tubes with appressoria by which they penetrate the host cuticle and invade the tissue. Hyphae re-emerge at the surface to form acervuli and conidia that continue spreading infection throughout the growing season. This anamorphic state is *Spilocaea pomi*. Conidia infect via appressoria and penetration of host cuticle just as with ascospores. Late in the season, hyphae grow deep into leaf tissue to start the overwintering phase of the life cycle. Control of the apple scab pathogen is primarily with fungicides on a timed schedule. Removal of infected or dead plant tissues and use of resistant varieties are other recommended practices.

The ascomycetes described in this chapter include a significant number of plant pathogens (although this is still a small percent of ascomycete diversity) and some of the most agriculturally devastating pathogens are represented here. These pathogens are not confined to a few families within the larger group, but are scattered throughout many families of the group. They represent the widest range of relationships from the necrotrophs (that quickly kill their host cells for nutrients), through the hemibiotrophs to the genetically intimate biotrophs, carefully balanced to keep their hosts alive while obtaining nutrients from them. Our understanding of the taxonomy of the group has been confounded by the plethora of life cycles and reproductive strategies and the extraordinary adaptability of fungi. We know so much about these fungi and yet so little, that students working on this group can expect a lifetime of challenges, frustrations, and satisfactions. The key to controlling these fungi with respect to agriculture lies in understanding much more about them.

REFERENCES

Alexopoulos, C.J., C.W. Mims and M. Blackwell. 1996. *Introductory Mycology*, 4th ed. John Wiley & Sons, Inc., New York.

Bos, L. and J.E. Parlevliet. 1995. Concepts and terminology on plant/pest relationships: Toward a consensus in plant pathology and crop protection. *Annu. Rev. Phytopathol.* 33: 69–102.

Hawksworth, D.L., P.M. Kirk, B.C. Sutton and D.N. Pegler. 1995. *Ainsworth & Bisby's Dictionary of the Fungi*, 8th ed. CAB International, University Press, Cambridge, UK.

Luttrell, E.S. 1974. The parasitism of vascular plants. *Mycologia* 66: 1–15.

Reynolds, D.R. and J.W. Taylor (Eds.). 1993. *The Fungal Holomorph: Mitotic, Meiotic and Pleomorphic Speciation in Fungal Systematics*. CAB International, Wallingford, UK.

Samuels, G.J. and K.A. Seifert. 1995. The impact of molecular characters on systematics of filamentous ascomycetes. *Annu. Rev. Phytopathol*, 33: 37–67.

Sutton, B.C. 1992. The genus *Glomerella* and its anamorph *Collectotrichum.*, pp. 1–26. in *Colletotrichum: Biology, Pathology and Control*, Bailey, J.A. and M.J. Jeger (Eds.), CAB International, Redwood Press, Ltd., Melksham, UK.

Taylor, J.W., E. Swann and M.L. Berbee. 1994. Molecular evolution of ascomycete fungi: Phylogeny and conflict., in *Ascomycete Systematics: Problems and Perspectives in the Nineties*, Hawksworth, D.L. (Ed.), Plenum Press, New York.

16 Deuteromycota: The Imperfect Fungi

Richard E. Baird

CHAPTER 16 CONCEPTS

- Taxonomy of the Deuteromycota is based on **asexual** spore formation or the absence of spores.

- Sexual stages are primarily in the **Ascomycota**, but there are a few in the **Basidiomycota**.

- Species can be parasitic or saprophytic.

- Asexual spores called **conidia** are nonmotile.

- Conidia are formed on conidiophores either singly or grouped in sporodochia, pycnidia, acervuli, or synnemata.

Species of the Deuteromycota (deuteromycetes), also known as the imperfect fungi or mistosporie fungi, are among the most economically destructive group of fungi. These fungi cause leaf, stem, root, fruit, and seed rots; blights; and other diseases. The Southern corn leaf blight epidemic in the early 1970s, which caused damage of one billion dollars, was incited by *Helminthosporium maydis*, the anamorph or the asexual form of the ascomycete *Cochliobolus heterostrophus.* Other deuteromycetes, such as *Aspergillus flavus*, produce mycotoxins (aflatoxins) in infected corn kernels. Mycotoxins when ingested by humans or animals can cause cancer of the digestive tract or other serious illnesses or death.

The deuteromycetes were called imperfect fungi in the early literature because they were thought not to produce sexual spores like those by species of the Ascomycota (Chapter 13 and Chapter 15) and Basidiomycota (Chapter 19). Descriptions and classifications of these fungi were based solely on production of **conidia** or on mycelial characteristics, or both. Deuteromycetes are now known to be the anamorphic stage of members of the Ascomycota and Basidiomycota. For example, *Fusarium graminearum* is the imperfect (asexual) stage of *Gibberella zeae*. The ubiquitous pathogen *Rhizoctonia solani*, which does not produce asexual spores, is the anamorph of the basidiomycete *Thanatephorus cucumeris*.

DEUTEROMYCOTA OR FUNGI IMPERFECTI

A brief history of this group may be helpful in understanding why a fungus may be known by two scientific names. During the 1800s, fungal identification was based strictly on morphological characters. The object of these studies was to identify pathogenic fungi when very little was known about their anamorph–teleomorph (sexual spore) relationship. The asexual fruiting bodies were often the only structures present on infected host tissues, and scientists were unaware that sexual reproductive states existed. Early mycologists, such as Persoon (1801), Link (1809), and Fries (1821), described genera and species of imperfect (lacking a known sexual stage) fungi that were later classified as Fungi Imperfecti. These studies initiated the description of many deuteromycetes and other fungi and are considered the starting point for fungal classification. Saccardo (1899) compiled descriptions of the known fungi into one unified source in his *Sylloge Fungorum* series. Taxonomic keys and descriptions to the genera and species of the Deuteromycota using spore shape, size, presence of cross-walls (septa) in the hyphae, and fruiting body type were developed by numerous scientists over the next half century. Conidiospore development (ontogeny) was used as a basis to identify genera and species after the 1950s. Asexual spore development on conidiophores and within fruiting bodies was considered a more natural classification for these fungi. Although considered a more natural classification scheme than those previously based on spore and conidiophore morphology, the earlier systems continue to be used by plant pathologists and disease diagnosticians because of the ease in identifying the genera and species. Since these early works, hundreds of papers and monographs were developed that include additional genera and species with information on their associated sexual stages.

Many of the imperfect fungi do not readily form a sexual stage in culture or on host tissue. Therefore, artificial systems for identification were developed and are currently being used. Recognizing that the majority of deuteromycetous fungi with a teleomorphic stage also have a second name for the sexual stage may be critical in understanding how to identify these fungi.

LIFE HISTORY

The deuteromycetes are primarily terrestrial in distribution, but can occur in salt or fresh water. They survive by deriving nutrients as saprophytes on plant debris or as parasites on living hosts. The degree of parasitism or pathogenicity varies depending on the fungus, the isolate, and host they invade. Many species of deuteromycetes are not only parasitic on plants but also infect animal cells.

Deuteromycetes may cause damage directly by infecting a host or indirectly by producing toxins. Direct infections by deuteromycetes to living hosts, other than plants, are termed mycoses and approximately 15 types are known. The more common mycoses include candidiosis (*Candida albicans*), and superficial infections called dermatophytosis, which include athlete's foot or dandruff caused by several *Tinea* species. A common species of deuteromycete associated with animal, avian, and human disease is *Aspergillus,* which causes aspergillosis. Infections may induce lesions on the surface of the skin or may damage internal organs such as lungs and liver. Infection frequently occurs when hosts are under stress, and immunosuppressive situations result from the poor health (e.g., AIDS virus).

Indirect damage from some of these fungi results from ingestion of infected food (feed) or through inhalation of particles containing mycotoxins that are either carcinogenic or cause other health problems. Mycotoxins are produced by fungi during the growth of the crop, or during transportation, processing, and storage. For example, aflatoxin levels in corn increase during drought and when high levels of nitrogen fertilizers are applied. Different types of mycotoxins are produced by species of *Aspergillus, Fusarium,* and *Penicillium*. One of the most important groups of toxins is the aflatoxins produced by *Aspergillus* species and occur primarily on crops such as corn, cotton seed, and peanuts.

Another group of mycotoxins are fumonisins that are produced by *F. verticillioides* (= *moniliforme*) and *F. proliferatum* (Gelderblom et al., 1988). Fumonisin is primarily associated with *F. verticillioides*, which can routinely be cultured or identified from corn tissues (Bacon and Nelson, 1994). Several forms of the toxin exist, but FB_1, FB_2, and FB_3 are the most common and important that are typically associated with food and feed (Gelderblom et al., 1988). If ingested, fumonisin can cause a neurological disorder in horses, called leukoencephomalacia, pulmonary edema in pigs, and esophageal cancer in humans.

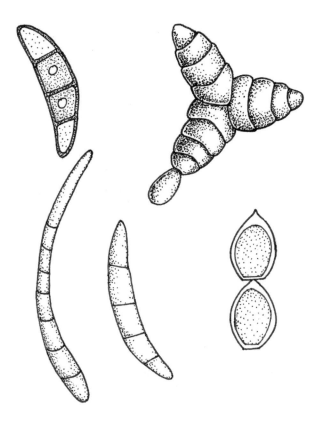

FIGURE 16.1 Conidia or asexual spores can be simple or complex and can be single or multicellular. (Drawing courtesy of J. MacGowen, Mississippi State University).

TAXONOMY OF THE DEUTEROMYCOTA

The size, shape, and septation pattern of conidia are used as the primary characters for the practical and working identification of deuteromycetes genera and species. Conidia are defined as asexual, nonmotile spores that belong to the anamorphic stage of a fungus life cycle. The Saccardian system, which used spore type to identify the deuteromycetes, was the first major tool employed by mycologists to identify genera (Saccardo 1899). The Saccardian system was incorporated into other systems that included more natural classification schemes based on conidial ontogeny or development. This more advanced system is usually referred to as "The Hughes-Tubaki-Barron System of Classification." The work by Barnett and Hunter (1986) includes keys and descriptions for identification of genera. Finally, Hennebert and Sutton (1994) identified subtle differences in spore development on conidiophores for identifying genera and species.

MORPHOLOGICAL STRUCTURES

Conidia (Figure 16.1) are produced on specialized hyphae called **conidiophores**. Because there are a large number of deuteromycete fungi, much variation can occur in their

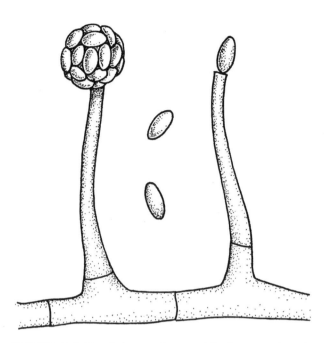

FIGURE 16.2 Fungi that produce conidia borne on loosely spaced conidiophores are called Hyphomycetes. (Drawing courtesy of J. MacGowen, Mississippi State University).

reproductive structures. Conidia vary in shape and size and can be one- or two-celled or multicellular, depending on the number of septa present. Septa within the spores vary, from transverse (across) to longitudinally oblique. Shapes range from filiform (thread-like), ovoid (egg-shaped), clavate (club-shaped), cylindrical (cylinder-shaped), stellate (star-like), or branched. Conidia can be ornamented with appendages and appear hyaline to colored.

Conidiophores are specialized hyphae, branched or unbranched, bearing specialized conidiogenous cells at the points where conidia are produced. Conidiophores may occur singly (separate) or in organized groups or clusters. If conidiophores are formed individually (Figure 16.2) and not enclosed in specialized structures, then the fungi that produce these forms are called Hyphomycetes. The hyphomycetes are subdivided into two groups based on color of hyphae and spores. The dematiaceous group has dark hyphae and spores, whereas the moniliaceous species possess light or pale-colored hyphae and spores. Dematiaceous genera of deuteromycetes include plant pathogenic species such as *Alternaria, Aspergillus, Bipolaris,* and *Penicillium.*

An example of an economically important dematiaceous hyphomycete is *Alternaria solani*, the causal organism of early blight of tomato (see Plate 5D following page 80). The disease caused by *A. solani* is considered by many to be the most economically damaging to tomatoes in the U.S. The disease occurs each year because the pathogen overwinters in plant debris in soil as chlamydospores (thick-walled survival spores). As the temperature warms in the spring, the chlamydospores germinate within the

plant debris or soil, and hyphae continue to grow saprophytically, forming conidia on individual conidiophores. Conidia produced during the saprophytic stage are disseminated by wind, rain, and insects or are transported in soil on farm machinery. For long-distance dissemination, infected seed is the primary source. Under wet and warm conditions, the conidia present on the tomato plant tissue germinate and hyphae can either penetrate the host through stomata or directly through the cuticle. Infections usually occur first on the mature foliage. Lesions develop and conidia are produced within the necrotic areas of the tomato foliage or stems. Secondary infections can occur from the local dissemination of new conidia formed on the host and plants can become completely defoliated. The fungus can develop chlamydospores that remain dormant in the dead plant tissues to start the cycle over for the following season. Symptoms of early blight can be observed on all above-ground tomato parts. Following germination, pre- and postemergence damping-off of plants can occur. Because lesions are generally first observed on mature leaves, the disease appears to progress from the lower portion, moving upward to the top of the plant. Infections increase forming circular lesions up to 4 to 5 mm in diameter, and the lesions become brown, with concentric rings giving the necrotic area a target-shaped appearance. Leaves that are infected often are observed with yellowing areas. As multiple lesions occur from secondary infections, the leaves turn brown and die. The entire plant can become defoliated and die at this stage. Controls for the disease include avoiding purchase of infected seed and soil for transplants that harbor the fungus; crop rotation with other solanaceous plants, such as potatoes, eggplant, and peppers; removal or burial of crop residue; and use of disease-free plants, fungicides, and resistant varieties.

DEUTEROMYCOTA CONIDIOMATA

If conidiophores are grouped together into organized clusters, then they are formed within specialized structures called conidiomata. The different types of conidiomata include **acervuli, pycnidia, sporodochia,** and **synnemata.** Modern references assign acervuli and pycnidia to the group Coelomycetes. Sporodochia-forming species are considered under the dermatiaceous or moniliaceous subgroups of Hyphomycetes. Synnemata have been placed under the Hyphomycetes in the subgroup stilbaceous fungi (Alexopoulus et al. 1996).

Conidiophores and conidiomata on hosts or in culture are used for identification. The ability of the deuteromycetes to form these structures in culture vs. that on a host varies per genus and species. A structure that is routinely formed on a host may not often be observed when grown in culture media. An example is setae (sterile hyphae or hairs) that are associated with acervuli of *Col-*

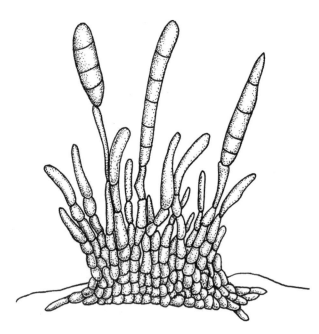

FIGURE 16.3 Sporodochia form on the surface of the host plant containing clusters or groups of conidiophores. (Drawing courtesy of J. MacGowen, Mississippi State University.)

letotrichum species. Setae (sterile-like appendages) routinely form in the acervuli produced on living hosts, but may be absent when growing on selective medium. As most identification keys are based partially on the morphology of the pathogen on host tissue, proper identification may require direct observation of conidiomata development on plant tissue rather than on artificial media.

Sporodochia are similar to acervuli except that the cluster or rosettes of conidiophores form on a layer or cushion hyphae on the host surface (Figure 16.3), whereas acervuli are imbedded in epidermal tissue or the plant cuticle. Sporodochia also appear as mat-like cottony structures due to the clustering of condiophores. However, in culture, sporodochia that resemble those on host materials are rarely observed, making species identification difficult. Examples of specific genera of fungi forming sporodochia include *Epicoccum, Fusarium,* and *Strumella.*

Fusarium oxysporum f. species. *vasinfectum,* causal agent of Fusarium wilt of cotton, is a good example of a sporodochia-forming fungus. The conidia of this pathogen overwinter in plant debris or can be introduced into fields by infected seed or in soil transported by farm equipment. The fungus forms chlamydospores in the soil or in plant debris. Under optimum weather conditions, the conidia or chlamydospores germinate and the fungus grows saprophytically, producing conidia on conidiophore rosettes. Germinating conidia or hyphae that come in contact with host root tissue can invade by direct penetration or indirectly through wounded areas on the root. Root wounding is often increased by the root-knot nematode, *Meloidogyne incognita* (Chapter 8), and the occurrence of this pest has been directly associated with fields with increased levels of *Fusarium* wilt. Following invasion into

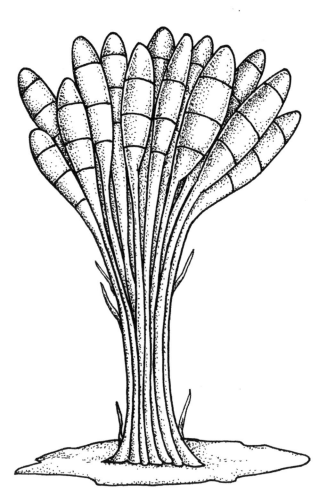

FIGURE 16.4 Synnemata consist of fused conidiophores at the base, forming conidia at the apex or on the sides of the structure. (Drawing courtesy of J. MacGowen, Mississippi State University.)

the root, the hyphae then grow inter- and intracellularly through the cortex and endodermis. The fungus penetrates the vascular system, and conidia are rapidly produced and distributed systemically into the transpiration stream of the cotton plants. The fungus physically obstructs the lumens of the xylem tissue, preventing water movement, which eventually results in wilting and death of the plants. Chlamydospores form in dead host tissues and overwinter in the plant residue. Control practices include use of clean seed from uninfested fields, use of resistant varieties, and reduction of root-knot nematode levels through chemical control, rotation, or with nematode-resistant varieties.

Synnemata conidiomata form conidiophores that are fused and the conidia often form at or near the apex (Figure 16.4). Species that produce conidia in this fashion belong to the Stilbellaceae group of deuteromycetes. Conidiophores of the synnemata-forming species are usually elongate and easy to identify from cultures because of their upright, whisker-like appearance. Synnemata are generally formed in culture unlike the sporodochial-forming species. Common genera that form synnemata are *Graphium, Arthrosporium, Isaria,* and *Harpographium.*

Graphium ulmi, the anamorph for the causal agent of Dutch elm disease, produces synnemata during the asexual or conidial stage of the life cycle. In the spring, the pathogen, which overwinters in plant debris, grows saprophytically and forms mycelia where the synnemata are produced. The spores of the pathogen can then infect healthy trees by an insect vector, and if roots of healthy plants are grafted to an infected tree, the pathogen can be transmitted to the healthy tree. The conidia are primarily disseminated by the elm bark beetle (*Scolytidae*) carried on their body parts. As the insects feed on the tree tissues, conidia are deposited into the feeding wound sites. The spores germinate, become established, and grow into the xylem tissue. Blockage of the vascular system occurs from the fungus and by defense responses of the host tree. Symptoms of Dutch elm disease include yellowing and wilting on one to many branches early in the season, depending on when infection occurs. Leaves of infected trees turn brown and die in portions of the tree or the entire tree may be affected. If the tree survives the first years following invasion, death will occur sometime during the second year of infection. If stems of the tree are sectioned, a brown discoloration is observed in the outer xylem of twigs, branches, and sometimes roots. Also, in dead and dying trees, insect larval galleries from the elm bark beetle can be observed under the bark of the tree trunk. Controls for Dutch elm disease include methods to eliminate the vector and pathogen by removing dying and dead wood (sanitation; Chapter 32). Fungicides injected into the trunk will stop the spread of the pathogen, but treatments must be repeated continuously from one year to the next.

COELOMYCETES CONIDIOMATA

Acervuli (sing. acervulus) contain a defined layer of conidiophores and conidia formed just below the epidermal or cuticle layer of plant tissues (Figure 16.5). Conidiophores and conidia erupt through the host epidermis or cuticle exposing the acervulus. Conidiophores within these structures are generally short and simple compared to the Hyphomycetes, such as *Aspergillus* and *Penicillium.* Once exposed, acervuli are usually saucer-shaped in appearance (Barnett and Hunter, 1986). In culture, fungi that typically form acervuli on host tissue can often appear to produce sporodochia. The eruption of host cuticle or epidermis, which defines an acervulus, cannot be observed in culture.

An example of an acervulus-forming pathogen is *Colletotrichum lagenarium,* the causal agent of anthracnose of cucurbits. *Colletotrichum lagenarium* overwinters on plant debris and in seeds from infected fruits. The fungus grows saprophytically in the soil and produces conidia on dead tissue. The conidia are disseminated locally by rain-

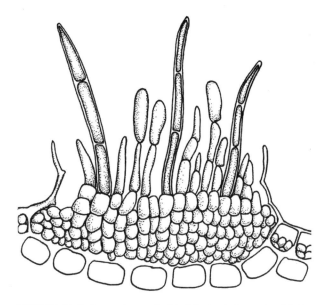

FIGURE 16.5 Acervulus embedded in host tissue containing clusters or groups of conidiophores with setae. (Drawing courtesy of J. MacGowen, Mississippi State University.)

fall and soil transported on farm equipment and over longer distances by infested seed. When weather conditions are suitable for infection, conidia germinate on the host. The germ tubes (hyphae) form appressoria at the point of contact with the host cell, quickly followed by formation of infection pegs, which allow direct entry into the host. Hyphae then grow intercellularly, killing host cells. As the fungus continues to grow, angular light to dark brown or black lesions are formed between the leaf veins. The lesions are elongate, narrow, and water-soaked in appearance and become sunken and yellowish to brown. When conditions are favorable, acervuli form on stromal tissue and conidiophores containing conidia erupt through the cuticle of the host. Spores are exposed to the environment and disseminated. Girdling of the stems or petioles can occur and defoliation results. Control practices include the use of disease-free seed, crop rotation with resistant varieties, cultural practices that remove or bury plant debris, fungicide sprays, and use of resistant cucurbit varieties when available.

Pycnidia (sing. pycnidium) differ from acervuli by the formation of the flask-shaped structures composed of fungal tissue that enclose the conidia and conidiophores (Figure 16.6). Pycnidia shapes described by Alexopoulus et al. (1996) include the following: papillate, beaked, setose, uniloculate, and labyrinthiform. Conidiophores that form within the pycnidium can be extremely short as in *Phoma* or larger as in *Septoria* or *Macrophoma.* Pycnidia resemble perithecia, which are sexual reproductive structures of some species of the Ascomycota (Chapter 15). If observed microscopically, the sexual spores of the ascomycetes are borne in asci and not on conidiophores. Other important considerations when identifying pycnidia-producing deuteromycetes is the presence or absence of an ostiole, which

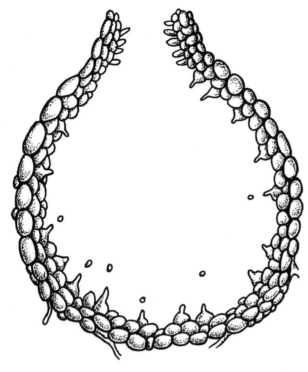

FIGURE 16.6 Pycnidium, a flask-shaped structure composed of fungal tissue, can be either embedded in or superficial on host tissue. Large numbers of conidiophores and conidia formed within the structure. (Drawing courtesy of J. MacGowen, Mississippi State University.)

is located at the apex and where conidia are exuded in a thick slimy layer or cirrhus. Keys to the identification of deuteromycetes often refer to imbedded pycnidia within the host material, but pycnidia of the same fungal species will occur superficially on artificial media.

Septoria glycines, the causal agent of brown spot of soybean, produces conidia within pycnidia. The fungus can reinfect plants within the same field during subsequent years since the conidia or mycelium overwinter on debris of host stem and leaf tissues. During warm and moist weather, sporulation occurs as the fungus is growing saprophytically and the conidia are disseminated by wind or rain. The spores germinate and hyphae invade by passing through stomata. The pathogen grows intercellularly, killing adjacent cells. New conidia form in lesions that can serve as a source for secondary infections on the same host plants. The pathogen primarily invades foliage, causing a flecking appearance on mature leaves, but infections can also occur on stems and seed. If environmental conditions are optimum, secondary infections can occur, resulting in defoliation that generally moves from the lower leaves and progresses upward. Lesions are usually irregular, becoming dark brown and can be up to 4 mm

in diameter. During initial development of the pathogen, lesions often coalesce to form irregular-shaped spots. On young plants, leaves turn yellow and abscise, but late in the growing season, infected foliage can be rusty brown before falling off. Control methods are limited to the use of resistant varieties, rotation to nonhost crops, and fungicides. The latter methods are generally not considered to be economically feasible.

Species in the **Mycelia Sterilia** are traditionally included in the Deuteromycota as a group that does not form asexual spores. Identification of these fungi is based on hyphal characteristics, absence or presence of sclerotia (survival structures) and number of nuclei per hyphal cell. *Rhizoctonia solani* and *Sclerotium rolfsii* have teleomorphs that place them into the Basidiomycetes (Chapter 19). Both of these fungi are important plant pathogens that occur worldwide and attack agronomic, vegetable, and ornamental crops. *Rhizoctonia solani* has many morphological and pathogenicity forms called anastomosis groups.

Rhizoctonia species, which are responsible for brown patch of turfgrass, survive as sclerotia in plant debris in the soil. Over a wide range of temperatures, the sclerotia germinate and the fungus grows saprophytically until a suitable host becomes available. Once the fungus becomes established in the host, circular lesions develop. Leaves and sheaths lose their integrity and appear water-soaked. The damaged tissue at first has a purplish-green cast, which then becomes various shades of brown depending on weather conditions and the type of grass. Often, dark-purplish or grayish-brown borders can be observed around the infected areas. Fungicide applications are effective in preventing or reducing severity of the brown patches. The disease may also be controlled by the following cultural practices: avoiding excessive nitrogen applications that enhance fungal growth, increasing surface and subsurface drainage, removing any sources of shade that reduce direct sunlight and increase drying of the leaf surface, and reducing thatch build-up when possible.

In summary, the deuteromycota is a very diverse group based on the presence of conidiophores and conidia, except for the fungi that do not produce asexually. The reader should keep in mind that when mycologists first created this artificial group, the sexual reproductive stages were unknown. Although the teleomorphs have now been determined for many of the deuteromycetes, these sexual stages are often rare or almost never seen on a host or in culture. As many deuteromycetes are plant pathogens, the group has been maintained for identification based on their asexual reproductive structures and cultural characteristics when identifying members of the Mycelia Sterilia.

REFERENCES

Alexopoulus, C.J., C.W. Mims and M. Blackwell. 1996. *Introductory Mycology*, 4th ed. John Wiley & Sons, New York, 869 pp.

Bacon, C.W. and P.E. Nelson. 1994. Fumonisin production in corn by toxigenic strains of *Fusarium moniliforme* and *Fusarium proliferatum. J. Food Prot.* 57: 514–521.

Barnett, H.L. and B.B. Hunter. 1986. *Illustrated Genera of Imperfect Fungi*, 4th ed. Burgess, Minneapolis, MN, 218 pp.

Fries, E.M. 1821. *Systema Mycologicum 1.* Gryphiswaldiae, Lund, Sweden, 520 pp.

Gelderblom, W.C.A., K. Jaskiewicz, W.F.O. Marasas and P.G. Thiel. 1988. Novel mycotoxins with cancer-promoting activity produced by *Fusarium moniliforme. Appl. Environ. Microbiol.* 54: 1806–1811.

Hennebert, G.L. and B.C. Sutton. 1994. Unitary parameters in conidiogenesis, pp. 65–76 in *Ascomycetes Systematics: Problems and Perspectives in the Nineties,* Plenum, New York.

Link, J.H.F. 1809. Observationes in ordines plantarum naturales. *Mag. Ges. Naturf. Freunde*, Berlin, 3: 3–42.

Persoon, D.C.H. 1801. *Synopsis Methodica Fungorum.* H. Dieterich, Göttingen, 706 pp.

Saccardo, P.A. 1899. *Sylloge Fungorum Omnium Hueusque Cognitorum*, Vol. 14. Self-published, Pavia, Italy.

17 Laboratory Exercises Illustrating Some Fungi in the Deuteromycota

Richard E. Baird

The Deuteromycota (deuteromycetes) is a heterogeneous and artificial (does not reflect phylogeny) phylum erected on the production of asexual spores called conidia that are formed on conidiophores. The conidiophores develop either free on mycelium or are enclosed in structures called conidiomata (fruiting structures). Laboratory exercises for the deuteromycetes are designed to teach students how to recognize different asexual reproductive structures and the role of environmental conditions or nutrition on the shapes and forms of the conidiomata. The Mycelia Sterilia includes genera of fungi within the deuteromycetes that do not produce asexual spores at all. Many of the fungi within the Mycelia Sterilia are economically important pathogens (e.g., *Rhizoctonia* and *Sclerotium* species) and are included in this chapter to teach students important characteristics used for their identification.

EXERCISES

EXPERIMENT 1. IDENTIFICATION OF DEUTEROMYCETOUS FUNGI BY OBSERVING ASEXUAL REPRODUCTION STRUCTURES

When attempting to identify deuteromycetes, it is important to note morphological features of the fungus under the dissection microscope. Fungi placed on microscope slides often fragment or spores do not remain attached to conidiophores, making identification difficult. Morphological features to observe under the dissection microscope include the following: conidia borne singularly, in chains, clusters or clumps; conidia borne in slime droplets; conidia borne at apex or laterally on conidiophores; and presence of pycnidia, sporodocia, synnemata, or acervuli (Chapter 16). The purpose of this exercise is to demonstrate variations in conidiomata that are used to identify the deuteromycetes in host material and on artificial media. This exercise will train participants on how to recognize the different asexual structures involved in spore production that are important to identify the deuteromycetes. Table 17.1 suggests the use of fungi representatives of each conidiomata type that can be compared during this exercise. Depending on availability, other fungi can be used.

Materials

Each student or team of students will require the following materials:

- Cultures of at least two species of fungi showing each of the four types of asexual development in Table 17.1
- Fresh living materials with diseased tissue — Table 17.2
- Dissection and light microscopes
- Dissecting needles or probes
- Slides, cover slips and distilled water for mounting hyphae and spores
- A stain such as aqueous analine blue (0.05% or 5 mg/l)
- Eye dropper with bottle
- Cellophane tape
- Single edge razor blade or scalpel with #11 blade
- Pencil eraser
- Alcohol lamp for surface-sterilizing instruments

Follow the protocols listed in Procedure 17.1 to complete this exercise.

Procedure 17.1 Observation of Asexual Structures of Deuteromycetous Fungi

1. Observe the fungal culture with a stereo dissection microscope at low power. Note the shape and form of the fruiting structures. Fruiting structures can be observed more easily on the hyphae nearest to the margin of active growth in culture. Conidia in mass may obscure fruiting structures. The mycelium (hypha) becomes much thicker and often obscures the asexual structures near the center of the dishes.
2. Viewing asexual fruiting structures of deuteromycetes in plant material is often essential to identify the pathogen. Transition areas in the zone between healthy and necrotic plant tissue are usu-

0-8493-1037-7/04/$0.00+$1.50
© 2004 by CRC Press LLC

TABLE 17.1
Species of Fungi Used to Demonstrate Deuteromycete Conidiophores and Conidiomata

Conidiophore Location	Fungi Species
Naked hyphae	*Alternaria, Bipolaris, Cladosporium, Penicillium*
Sporodochia	*Epicoccum, Fusarium, Tubercularia, Volutella*
Pycnidia	*Diplodia, Macrophoma, Phomopsis, Septoria*
Synnemata	*Briosia, Graphium, Harpographium, Trichurus*

TABLE 17.2
Fungi for Cultures and Host Material for Comparison

Fungus	Host
Colletotrichum lindemuthianum	Green beans
Colletotrichum gloeosporiodes	Apple fruit
Colletotrichum circinans	Onions
Cytospora species	Spruce
Entomosporium mespili	Red tip photinia
Melanconium species	Grapefruit
Phomopsis species	Soybeans

ally good areas to look for structures. If fruiting structures are not evident in this area, move the tissue piece so that areas with greater tissue damage are viewed. If fruiting structures are not observed within the lesions, place the plant tissue overnight in a moist chamber. To obtain fruiting bodies, place a moistened tissue paper in bottom of a plastic food storage container (a plastic bag will also work), set the plant tissue onto tissues, and close box. Store at room temperature and check in 12 to 24 h for sporulation. Adding a drop of water directly to the lesion may also stimulate sporulation.

3. After viewing the diseased plant material with a dissection microscope, closer examination with a compound or light microscope is necessary to identify the fungus. Slice thin sections (1 mm) of a leaf lesion with a single edge razor blade or scalpel and mount on a microscope slide with a drop of analine blue or water. Look for conidia or fruiting bodies on the plant tissue and floating in the mounting medium.

4. To observe sporulation of a fungal colony growing in a petri dish, place a drop of water or analine blue to the area of interest, and add a cover slip directly onto the agar medium. Fruiting structures can be observed with minimal damage to the fruiting structures. For closer observation, mount fruiting structures on a slide in a drop of distilled water and a cover slip. If fruiting bodies cannot be seen on the agar plate, simply remove a tiny portion of the mycelium (#1 cm) and place it into the water on the slide. Flame sterilize the dissection needles and allow to cool before removing mycelium from the petri dish. Sometimes the cover slip is not appressed directly to the microscope slide when a mass of mycelium and agar is mounted. If this happens, use a pencil eraser to gently press on the cover slip and flatten the preparation. First use the lowest power (generally a 4× or 10× objective lens) on a specific area containing the fruiting structures and proceed to high, dry objective lens (40×). As with the stereo microscope, observe the hyphae nearest to the growing margin and move inward until light can no longer penetrate through the mycelium. Observe and draw the sporulating structures in detail.

5. A third method of viewing the deuteromycetes is with cellophane tape. This method is particularly useful for obtaining fruiting structures and spores from fresh material. Place a piece (1 to 2 cm) of cellophane tape with the adhesive side downward onto the plant material. Press slightly and remove from the tissue. Take the piece of cellophane, and place the side that was in contact with plant tissue downward onto a microscope slide containing water. Observe spores or fruiting structures with the 10× and 40× objectives lens and record your observations.

Anticipated Results

Students should be able to differentiate the various ways conidia are produced either naked as individual conidiophores or clustered together into synnemata, pycnidia, acervuli, and sporodochia.

Questions/Activities

- Describe and draw the following structures: individual conidiophores (hyphomycetes), single or multiple branched, spore or conidia attachments; sporodochia, conidiophore arrangements, locations on host tissues (living plant tissues).
- How do fruiting structures vary among Hyphomycetes?
- How do pycnidia differ from sporodochia or acervuli?
- Where are the conidiophores located in a pycnidium?
- Does the pycnidium have a pore (ostiole) in the neck (when present)? What is the purpose of an ostiole?
- Are the pycnidia embedded in or on the surface of host tissue?
- What is a synnema?
- How do the conidiophores in synnemata differ from the conidiophores in sporodochia?
- Where are the spores borne on synnemata and how are they attached?
- In detail, discuss the strategies for observing fungi on host material and in culture.
- Why is a dissection microscope important for fungal identification?

EXPERIMENT 2. THE EFFECTS OF LIGHT AND TEMPERATURE ON GROWTH AND SPORULATION OF FUNGI

The purposes of this experiment are to show how temperatures affect growth of fungi and to compare sporulation and growth potential of fungi when placed under different lighting regimens. Through direct observation, students will learn that growth conditions will affect spore production, cultural characters, and growth rates of fungi.

Common fungi that can be used in this exercise include *Alternaria* species (single conidiophores), *Epicoccum nigrum*, or *Fusarium solani* (sporodochia types); *Colletotrichum graminicola* or *Pestalotia* species (acervuli types); and *Phoma* or *Phomopsis* (pycnidia types). Many other fungi may be substituted in this experiment. Experiments typically require up to 14

days to complete, but will vary depending on the fungi included in the study.

The fungi should be grown on potato dextrose agar (PDA, Difco) for several days, perhaps a week, at room temperature to provide sufficient inoculum. The number of conditions (treatments) may vary depending on the available space and number of replicate plates (V-8 medium) used. We suggest that the class be divided into four groups, each with their own fungal species.

Materials

Each team of students will require the following materials:

- Four isolates of common deuteromycetes (one for each group). Inoculate each fungus onto five petri dishes (10-cm diameter) containing potato dextrose agar (PDA, Difco).
- A minimum of 21 10-cm diameter petri dishes containing V-8 agar medium (Diener, 1955): 200 ml of V-8 juice (Campbell Soup Co.), 3 g $CaCO_3$, 15 g Difco agar, and 800 ml of water
- Aluminum foil
- Parafilm
- Sterile cork borer, dissection needle or spatula for subculturing of fungi
- Alcohol burner, matches, and container of 95% ethanol for sterilizing subculturing tools
- Incubators set for different temperatures and lighting conditions
- Millimeter ruler for measuring growth of colonies
- Light source for observing cultures

Follow the protocol outlined in Procedure 17.2 to complete the experiment.

Procedure 17.2. Effects of Temperature and Light on Growth and Sporulation of Fungi

1. Use a flame-sterilized cork borer to cut agar pieces from cultures growing on Difco PDA medium. If possible, remove plugs from actively growing margins (see Figure 27.2). Mycelium closest to the original inoculation plugs can differ physiologically from mycelium nearest the margin, resulting in different growth rates.
2. Each group should transfer a single plug (mycelium side down) of its fungus onto the center of a petri dish containing V-8 agar. Inoculate a minimum of 21 dishes and wrap with parafilm. Label dishes

with the name of the fungus and numbered from 1 to 21.

3. Each group should place at least three of the inoculated petri dishes in incubators set at 10°C, 20°C, 30°C, or 40°C. Record the number of the petri dishes at each temperature. If insufficient incubator space is available, the 20°C and either 10°C or 30°C treatments should be used. These cultures should be exposed to light for a minimum of at least 8 h per day. Otherwise, sporulation could be adversely affected.

4. Each group should place at least three of the inoculated petri dishes in incubators set for the following light treatments: continuous artificial light, alternating light and darkness (12 h each), and continuous darkness. The continuous dark treatment may also be achieved by wrapping the cultures in aluminum foil. All incubators should be maintained at the same temperature (e.g., 20°C to 25°C).

5. Record growth data every two days for 14 days (maximum) or until the mycelium reaches the edge of the dishes for both temperature and light treatments. Using a millimeter ruler, measure the radial growth from inoculation plug to the perimeter of the colony. The mycelial growing edge can be easily viewed by placing the dishes toward a light source. Repeat in three locations for each dish and find the average to the nearest millimeter for each dish. Now average the radial growth for all three replicates for each fungus at each temperature. Plot this data on a graph with the Y axis as growth (mm) and the X axis as days in culture. Calculate the standard deviation (SD) for each point by the following formula:

$$SD = \sqrt{\sum (x - \bar{x})^2 / (n-1)} = \sqrt{x^2 / (n-1)}$$

where x is the sample measurement, $\bar{x}$ the calculate mean of data, $X = x - \bar{x}$, and n the number of samples used to calculate the mean = 3.

6. Sporulation should also be recorded when growth is measured. A rating scale for sporulation can be developed, such as none = 1, slight = 2, moderate = 3, and heavy = 4. Find the average rating for each fungus at each treatment. Plot sporulation on the Y axis and days on the X axis.

7. Prepare a final report from the data. Include graphs showing the growth and sporulation of species under the various temperatures and lighting conditions.

Anticipated Results

Students should observe that changes in temperatures affect growth and sporulation of fungi. However, some fungal species used in the experiment may not respond with the same growth and sporulation patterns as other species when compared under similar growth conditions.

Questions/Activities

- How did growth vary between species? *Hint*: Graph the data and calculate a growth rate.
- Did the fungi exhibit the same growth rate throughout the experiment? If not, what may have caused the changes in growth rates?
- When was sporulation first observed for each fungus? How did it vary per light regime?
- How did the effects of different light regimes compare between species (consult with other groups)?
- Compare different ways you might be able to collect data on sporulation (e.g., number of conidiophores per unit area, spore counts, etc.).

EXPERIMENT 3. IDENTIFICATION AND VARIABILITY OF FUNGI GROWING ON HOST MATERIAL AND IN CULTURE

The purpose of this exercise is to compare and contrast the morphological variability of "imperfect fungi" on plant material compared to isolates grown in nutrient cultures. Asexually reproducing fungi, like other fungi, are affected by environmental conditions, such as light, temperature, moisture, or nutrition found in the different growth substrates. Different nutrient agars may affect the morphology of the fruiting bodies and can complicate identification of the organism. Fruiting bodies often look different in culture compared to host tissue (saprophytic or parasitic). Keys to the deuteromycetes are based on their appearance on plant tissue; therefore, correct identifications made strictly from growth on culture media can be difficult, if not impossible, at times.

Materials

Each group of students will require the following materials:

- Dissection and light microscopes
- Slides, cover slips, and distilled water or 10% KOH (10 g KOH in 90 ml water)
- Several cultures of each of six fungi growing in 10-cm diameter petri dishes with representative samples of the same six species on plant

material (see Table 17.2 for suggestions). Herbarium material may be substituted for fresh
- Dissection needle, sectioning knife for woody tissue, and forceps

Follow the protocol outlined in Procedure 17.3 to complete the exercise.

Procedure 17.3. Comparison of Fungal Structures from Host Plant and Culture

1. Observe with a dissecting scope the corresponding cultures and plant material for a specific fungus. Cut very thin sections of diseased host tissue with a razor blade or scalpel and mount on microscope slide in either water or 10% KOH for dried specimens. Make a corresponding slide of the fungus growing on nutrient agar. *Note*: The dissection microscope may be the best method to examine any variation of the fruiting bodies from nutrient agar medium and on plant tissue.
2. Make drawings and record similarities and differences in structures.

Anticipated Results

Fungi that grow in culture often appear morphologically different on nutrient agar than in host materials. The differences may be slight or extreme depending on the species used. The differences, however, may be enough to affect correct identification of the pathogen based on growth and sporulation on artificial media.

Questions/Activities

- Detail the fruiting structure shapes (morphology) for each fungus observed in this exercise. How are they similar when compared on the plant tissue and the agar medium? How do they differ when compared on the plant tissue and the agar medium?
- How might the variability of the fruiting bodies on the different growth sources affect their identification?

EXPERIMENT 4. IDENTIFICATION OF *RHIZOCTONIA* SPECIES (MYCELIA STERILIA) USING HYPHAL CHARACTERISTICS

Rhizoctonia species do not form conidia in culture (or in nature), but are identified based on cultural characters such as mycelial shape and color, sclerotia formation, and

hyphal morphology. The hyphae are brown, elongate, inflated, and branch at right angles. This exercise will familiarize students with general cultural characteristics of this important genus.

Materials

Each student or group of students will require the following materials:

- Several 10-cm diameter petri dishes of *R. solani* growing on PDA
- Several 10-cm diameter petri dishes of a binucleate *Rhizoctonia*-like fungus (BNRLF) growing on PDA
- Dissection needle or microspatula to remove the fungal tissue to be observed
- Dissection and light microscopes, slides, cover slips, and distilled water to mount the hyphae

Follow the protocol outlined in Procedure 17.4 to complete the experiment.

Procedure 17.4. Identification of *Rhizoctonia solani*

1. Compare *R. solani* and BNRLF cultures for the following characteristics (see Step 2):
   - Hyphae are generally darker for *R. solani*.
   - Cells of the hyphae are elongate and often slightly inflated.
   - Hyphae branch at right angles and constrict at the area of branching, followed by septum (cross wall) formation.
   - Sclerotia are common in culture, light at first becoming dark brown to black, and are variable in size.
   - Asexual fruiting bodies are absent.
2. Cut a small plug (5 mm²) of agar containing fungal mycelium and place on microscope slide in a drop of water. Cover with a plastic cover slip and flatten.
3. Observe the preparation with a compound microscope at 100× and 400×.

Anticipated Results

Students should observe the morphological differences between the two groups of *Rhizoctonia*, including thickness and color of the hyphae (mycelium), presence of sclerotia, and patterns of sclerotial occurrence in the petri dishes. Hyphae observed microscopically should appear as described previously.

Questions/Activities

- What do the various *R. solani* isolates look like in culture?
- Discuss how sclerotia varied in shape and size, and the patterns formed in the petri dishes.
- What distinguished this group of fungi from others?

EXPERIMENT 5. DETERMINE THE NUMBER OF NUCLEI IN HYPHAL CELLS OF *RHIZOCTONIA* SPECIES

The genus *Rhizoctonia* is divided into groups based on number of nuclei (binucleate or multinucleate) per hyphal cell. The multinucleate isolates are *R. solani* (anastomosis groups) and the binucleate isolates are either *R. cerealis* or binucleate *Rhizoctonia*-like fungi (BNRLF) belong to the *Ceratobasidium* anastomosis group (CAG). Nuclear stains can be used to differentiate the groups.

Materials

Each student or team of students will require the following materials:

- 10-cm diameter petri dishes containing 4 to 6 different anastomosis groups (AG and CAG) of *Rhizoctonia* species growing on PDA
- 10-cm diameter petri dishes containing 5 ml of water agar
- Dissection needle or spatula to remove the fungal tissue to be observed
- Dissection and light microscopes
- Slides and cover slips
- 0.5% Aniline blue in lactophenol or 0.05% trypan blue in lactophenol in dropper bottles
- Alcohol lamps

Follow the protocol outline in Procedure 17.5 to complete the experiment.

Procedure 17.5 Determining the Nuclear Number in Cells of *Rhizoctonia* Species

1. Inoculate known isolates of *Rhizoctonia* species onto thin layers of water agar in 10-cm diameter petri dishes and incubate at room temperature for 48 to 72 h. Label dishes with the species name.
2. After the fungi have grown at least 5 cm from the inoculation plug, place a drop of 0.5% analine blue in lactophenol onto the hyphae at the edge of the colony. Place a cover slip over colored area, lightly press the cover slip down and onto the mycelium, and observe the hyphal cells at 400× under a com-

pound microscope. Be careful not to move the cover slip so that the agar comes in contact with the objectives lens. Count the number of nuclei in each hyphal cell and for each isolate.
3. Another method for nuclear staining is to remove a section (1 cm²) of the fungal mycelium from the thin-layer water agar plates and place onto a microscope slide. Then add the dye as in Step 2. Place a cover slip over the material, and heat slightly with a small burner until the agar begins to melt. Place the slide onto the microscope stage, view at 400× and count the nuclei as before.

Anticipated Results

The nuclear condition of the hyphal cells should be easily determined for the different isolates. From these observations, CAG or AG isolates can then be determined for the unknowns.

Questions

- How many nuclei did you observe in hyphal cells of *R. solani*?
- How many nuclei did you observe in hyphal cells of CAG types?

EXPERIMENT 6. DETERMINATION OF ANASTOMOSIS GROUPING FOR ISOLATES OF *RHIZOCTONIA* SPECIES

The genus *Rhizoctonia* is unusual because the asexual stage does not produce spores and the sexual stage is rarely observed. Because identifying these fungi is very difficult by cultural characters alone, alternative procedures were developed to identify the various anastomosis groups (AG) by a method known as **anastomosis** pairing reactions.

Anastomosis reactions that may occur when *Rhizoctonia* isolates are paired include the following: (1) Perfect fusion is when the cell wall membranes combine and a mixture of continuous living cytoplasm occurs (Figure 17.1). This reaction happens only when pairing the same isolates or ones from same anastomosis group and **vegetative compatibility** group. (2) Imperfect fusion is when the connection of cell walls and membrane contact are not clearly seen (Figure 17.2). Hyphal anastomosis occurs, but isolates are from different vegetative compatibility groups. Often the hyphae at the point of contact are smaller in diameter than the rest of the hyphae (Carling, 1996). Those isolates showing imperfect fusion are related, such as two AG-4 isolates identified from different sources. (3) Contact fusion occurs when hyphal cell walls of *Rhizoctonia* isolates come into contact, but no fusion, lysis, wall penetration, or membrane contact occurs (Figure 17.3).

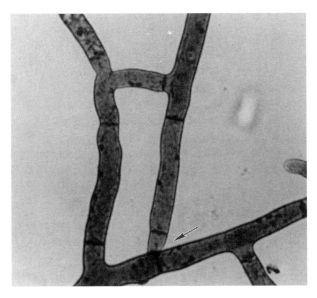

FIGURE 17.1 Perfect fusion during the anastomosis pairing experiment for *Rhizoctonia* species isolates. This reaction occurs with closely related isolates from the same genetic source or population. (Photo courtesy of Dr. D. Carling, University of Alaska.)

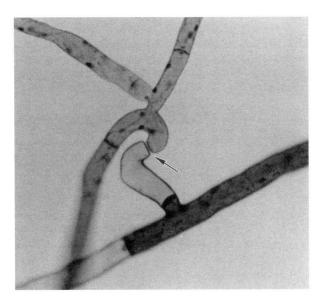

FIGURE 17.3 Photograph showing contact fusion during the anastomosis pairing experiment for isolates of *Rhizoctonia* species. Isolates come into contact, but cell wall fusion does not occur. Isolates exhibiting this pairing reaction are distantly related. (Photo courtesy of Dr. D. Carling, University of Alaska.)

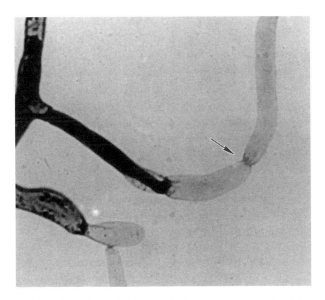

FIGURE 17.2 Imperfect fusion during the anastomosis pairing experiment for isolates of *Rhizoctonia* species. Isolates causing this reaction are related, but from different genetic sources or populations. (Photo courtesy of Dr. D. Carling, University of Alaska.)

Materials

Each student or team or students will require the following materials:

- Four known *Rhizoctonia* AG (CAG) cultures with four AG (CAG) grown on PDA. Selection should include isolates capable of showing the different types or degrees of anastomosis that occur during pairing.
- Cultures of isolates from unknown anastomosis groups
- Compound microscope and cover slips
- 12 10-cm diameter petri dishes containing 1.5% water agar
- 12 microscope slides
- Forceps, dissection needles, or microspatula.
- Alcohol burner and jar of 95% ethanol for flame sterilization of tools

Follow the protocol outlined in Procedure 17.6 to complete the experiment.

Hyphal cells at the contact point and adjacent cells may occasionally die, but do not anastomose. No fusion occurs among totally unrelated isolates. The unknown and tester isolates show no reaction to the presence of the others. The following experiment is intended for graduate students or advanced undergraduate special topics.

Procedure 17.6 Determination of Anastomosis Grouping of Isolates of *Rhizoctonia* Species

1. Tester isolates will be used to compare morphological (growth form, color, and shape in culture) or visible differences between the isolates. In this

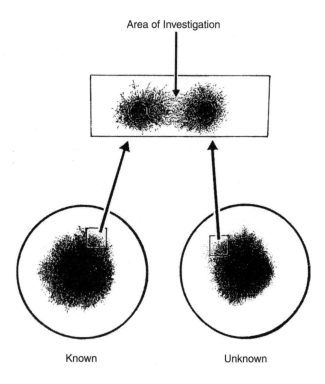

Area of Investigation

Known   Unknown

Rhizoctonia solani

FIGURE 17.4 Diagrammatic representation of pairing of *Rhizoctonia* isolates on an agar slide. (Drawing by J. MacGowen, Mississippi State University.)

exercise, unknown AG (*R. solani*) isolates will be compared to known (tester) AG isolates. Selection of tester isolates is based on similar growth forms, shapes, and colors of the unknown *Rhizoctonia*. All isolates should be inoculated onto Difco potato-dextrose agar (PDA) dishes at the same time and grown for a minimum of 10 to 14 days at room temperature before the comparisons can be made.

2. *Rhizoctonia* isolates may be separated by anastomosis reactions by using the water agar dish method or water agar slide method.

3. Water agar dish method: Place a mycelial plug of the unknown *Rhizoctonia* isolate off-center in the 10-cm diameter petri dish containing 1.5% water agar. Transfer a plug of the unknown tester to the dish, opposite and off-center so that the two plugs are approximately 3 to 5 cm from each other (Figure 17.4).

4. Anastomosis pairing reaction is made immediately after the hyphae from the two isolates meet (slightly touch), as seen visually with a dissection microscope. Place a drop of 0.5% analine blue in lactophenol at the point where the two meet, add a

cover slip, and view at 400×. The following four types of hyphal interactions are possible: perfect fusion (C3), imperfect fusion (C2), contact fusion (C1), and no fusion (C0) (Matsumoto et al., 1932; Carling, 1996).

5. Water agar slide method: With forceps, surface sterilize a microscope slide by dipping in 95% ethanol and flaming. Alternatively, slides may be placed in a beaker covered with aluminum foil or in a glass petri dish and autoclaved for 20 min. Holding the slide with the forceps, dip the sterile slide into 1.5% PDA and place onto petri dishes containing 1.5% water agar. The water agar in the dish prevents the agar on the slide from drying out during the pairing studies.

6. After the PDA coated slides have hardened, place the mycelial plugs 3 to 5 cm apart on the slides. One side of the slide will have the unknown and the other side the tester isolate of *R. solani*.

7. Incubate the petri dish containing the inoculated slides at room temperature for 1 to 2 days and observed regularly for hyphal growth of the two isolates. View the hyphae of the two isolates microscopically when the hyphae come into contact. Avoid overlapping of hyphal growth. Too much overlap makes it extremely difficult to locate the hyphae of the two isolates.

8. Remove the slide from the petri dish and stain the zone where the hyphae of the two isolates intersect with 0.5% analine blue and lactophenol. Place a cover slip over the top of the stained area and determine the anastomosis pairing reaction by using the compound microscope.

Anticipated Results

The four types of fusion ranging from perfect, imperfect, contact, to no fusion should be observed. Students will observe perfect fusion reactions and cellular death associated with true anastomosis pairing vs. the contact types. From these observations, students will be able to determine which unknown anastomosis groups they have been screening.

Questions

• How are tester isolates chosen to use for comparison with an unknown *Rhizoctonia*?

• What are the major types of anastomosis pairing reactions that can occur?

REFERENCES

Carling, D.E. 1996. Grouping in *Rhizoctonia solani* by hyphal anastomosis reaction, pp. 37–47 in *Rhizoctonia Species: Taxonomy, Molecular Biology, Ecology, Pathology and Disease Control*, Sneh, B., S. Jabaji-Hare, S. Neate and G. Dijst (Eds.), Kluwer, Boston, MA.

Diener, U.L. 1955. Sporulation in pure culture by *Stemphylium solani*. *Phytopathology* 45: 141–145.

Matsumoto, T., Yumamoto, W. and Hirane, S. 1932. Physiology and parasitology of the fungi generally referred to as *Hypochnus sasakii* Shirai. I. Differentiation of the strains by means of hyphal fusion and culture in differential media. *J. Soc. Trop. Agric.* 4: 370–388.

18 Smut and Rust Diseases

Larry J. Littlefield

CHAPTER 18 CONCEPTS

- Smut and rust diseases are important; rusts are far more important worldwide and attack a wider range of host plants than smuts do.

- Smut diseases are important primarily on cereals; many can be managed effectively by chemical seed treatment or resistant cultivars. Three major types of smut infection, each type requiring different management strategies, are (1) seedling infection, (2) floral infection, and (3) meristem infection.

- Rust diseases cause millions of dollars damage annually to numerous agronomic, horticultural, and timber species.

- Rust fungi are highly specific, obligate parasites, often attacking limited numbers of host species or cultivars. The degree of host specificity is expressed as *formae speciales* or physiological races in many rust fungi.

- The life cycle of rusts may contain up to five different spore stages; some rust species (autoecious forms) complete their life cycle on one host species; others (heteroecious forms) require two.

- Many rusts are disseminated by wind for long distances, which contributes greatly to their widespread occurrence.

- Several management practices are used to reduce severity and economic impact of rust diseases, depending on the particular host and rust species. They include genetic resistance, foliar fungicide sprays, elimination of alternative host species, and other cultural practices.

The pathogens responsible for smut and rust diseases are members of the kingdom Fungi and the phylum Basidiomycota, and are placed in the orders Ustilaginales and Uredinales, respectively. Those orders differ from most other Basidiomycota in not having their sexual reproductive spores (**basidiospores**) borne on or in fruiting bodies, as in mushrooms and puffballs. Rather, their basidiospores are borne on specialized germ tubes (basidia) produced on the germination of thick-walled, typically dormant, resting spores (**teliospores**). Smut fungi can be cultured on artificial media much more easily than the rusts. However, as pathogens in nature, both the smuts and rusts function as obligate parasites, that is, they require living host tissue as a nutrient source. Both types of pathogens cause important plant diseases, but rust fungi attack a much broader host range and worldwide are far more important economically than are the smuts.

SMUT DISEASES

Smuts are most important on cereal crops (e.g., barley, wheat, oats, grain sorghum, and corn), although violets,

onions, sugarcane, forage grasses, date palms, and several other crops can also be infected. The economic impact of smut infection is most commonly evident as individual kernels or entire heads of wheat and barley (Figure 18.1a and Figure 18.1b), or ears of corn (Figure 18.1d and Figure 18.1e), being largely replaced by a dark-brown, dry, powdery mass of teliospores. The most common smut diseases in North America include corn smut (Figure 18.1d through Figure 18.1g), stinking smut or common bunt of wheat (Figure 18.1a), and loose smut of wheat (Figure 18.1b), barley, and oats. With the development of highly effective seed treatment chemicals for small grains and many smut-resistant corn varieties, smuts are now far less damaging economically than as recently as 50 years ago. However, smuts, as well as many other plant pathogens, can spread easily over great distances by wind or modern transportation, and such pathogens, once controlled, have a history of reappearing sometimes in yet more aggressive forms. Thus, it is important to understand the biology of smut diseases and the fundamentals of their management. As recently as 1996, Karnal bunt first appeared in the U.S., coming from Mexico, where it had previously been intro-

FIGURE 18.1 (a) Common bunt of wheat (*Tilletia tritici*); diseased kernels, left, with partially to completely ruptured pericarp (arrows), revealing dense teliospore masses (tm), which eventually disintegrate into individual teliospores; healthy kernels, right. (b) Loose smut of wheat (*Ustilago tritici*); progressive emergence, left to right, of the diseased head and transformation of kernels to masses of dry teliospores, providing windborne inoculum for immature ovaries in healthy heads. (c) Excised, hydrated, loose smut-infected embryo of barley, containing *U. nuda* hyphae (h) that can later infect apical meristem of the young seedling. (Original magnification × 25.) (From North Dakota State University. With permission.) (d) Corn smut (*U. maydis*) converts kernels into large gall-like masses of teliospores (g) initially surrounded by an outer, white membrane that eventually disintegrates. (From University of Illinois. With permission.) (e) Corn smut-infected ear (left), healthy ear (right). Smut galls (g) are mature, with a partially ruptured outer membrane exposing masses of black teliospores within. (From University of Illinois. With permission.) (f) Corn smut galls on infected tassels. (g) Mature corn smut gall on an infected stalk. Millions of teliospores (t) are exposed upon disintegration of the gall's outer membrane. (h) Label from can of corn smut galls (*Cuitlacoche*) sold in Mexico. (From Hongos de Mexico, S.A. de C.V., Mexico City. With permission.)

duced inadvertently from India on seed used in international wheat breeding programs. Although Karnal bunt has little effect on yield, it has had major economic impacts on export sales of U.S. wheat.

Smut fungi, as do the rusts, reproduce **sexually**, but can do so only once per growing season. However, smuts lack the highly prolific **asexual** reproductive stage characteristic of many rust fungi. That latter character precludes secondary infection during a growing season, thus limiting their geographic distribution as compared to many rusts.

Although **life cycles** of most smut fungi are similar, their **disease cycles** can be very different, resulting in different approaches to disease management for various smut diseases. The most important differences among smut disease cycles and hence the major differences in their management are related to the stage of host development at which primary infection occurs. Although there are three major stages of host plant growth at which different smut fungi can initiate infection, they all require the presence of meristematic host tissue for infection to actually occur. Understanding those differences is essential to implementing an appropriate disease management strategy. Different smuts can infect plants in the seedling stage (where infection typically occurs in below-ground cotyledon tissue), in the flowering stage (where infection of ovaries occurs early in their development), or in juvenile to mature stages (where infection can occur in meristems of any above-ground parts of the plant). Different types of smut diseases include the following examples.

SEEDLING INFECTION

Examples of such smut diseases in wheat include common bunt or stinking smut caused by *Tilletia tritici* and *T. laevis*; dwarf bunt by *T. controversa*; Karnal bunt by *T. indica*; and flag smut by *Urocystis agropyri*. Symptoms of common bunt or stinking smut (Figure 18.1a) are expressed in mature kernels resulting from infection in the seedling stage earlier that same growing season (Figure 18.2a). That infection resulted from teliospores present in soil from previous years' crops, particularly in drier soils, or from teliospores deposited onto the surface of seed during previous seasons' harvest or subsequent shipping or storage. Such teliospores can remain viable in soil or on the seed surface for up to 10 years or more, until moisture and temperature conditions are satisfactory for germination, and then proceed to infect the cotyledon of germinating seedlings. **Mycelium** within a newly infected seedling invades the apical meristem of the plant; as the stem differentiates and elongates the mycelium is carried upward, eventually residing in the head and the developing kernels. Smut mycelium that differentiates to form teliospores eventually replaces all internal tissues

of the kernel and is surrounded by the **pericarp**. The pericarp breaks down during harvesting or handling, releasing teliospores that can be windborne or infest otherwise healthy seed, thus completing the disease cycle. The most economical, widely used management procedure for common bunt is fungicide seed treatment to kill surface-contaminating teliospores and to protect seedlings from soilborne teliospores. Other management strategies include planting only pathogen-free seed or smut-resistant cultivars, and planting when soil temperature is greater than 26°C (68°F), the optimum for teliospore germination.

FLORAL INFECTION

Examples in wheat and barley include loose smut of wheat, caused by *Ustilago tritici*, and loose smut of barley, by *U. nuda*. The major symptom of loose smut (Figure 18.1b) is the often naked rachis bearing only the remnants of kernels that had been totally replaced by brown to black, dry, powdery masses of teliospores. That condition results from infection that occurred one or more years earlier during the flowering process that led to production of the current year's planted seed (Figure 18.2b). Infected heads emerge slightly before the uninfected heads; the former are transformed into masses of teliospores that provide **inoculum** for the current season's crop as the new, uninfected heads begin to emerge. Infection occurs in the young florets, where mycelium grows into the young embryos (Figure 18.1c) and where it can remain viable for several years. When that infected seed is planted and germinates, the mycelium in the ovary invades the apical meristem and is eventually carried passively into the developing head. There it converts the entire kernels, including the pericarp, into the mass of dry, black teliospores, thus completing the disease cycle. Those spores can then function as primary inoculum for that year's crop as new heads emerge and begin to form kernels. Visual evidence of that infection will not be seen until heading time of the crop produced from those infected kernels. Successful chemical control of loose smut requires seed treatment with one of several systemic fungicide seed treatments available. Such materials must enter into the seed where they can kill smut mycelium within the embryo. Surface, contact fungicide treatments are ineffective for controlling floral infecting smuts. Other recommendations include planting pathogen-free seed, as determined by laboratory analysis, or resistant cultivars, if available.

MERISTEM INFECTION

Corn smut, caused by *U. maydis,* is the most common example of this type of smut infection. It is also more easily visible to the untrained eye than any of the small grain smuts due to large galls (up to 2 to 3 inches in diameter) that often occur in clusters, sometimes occupying nearly the entire volume of

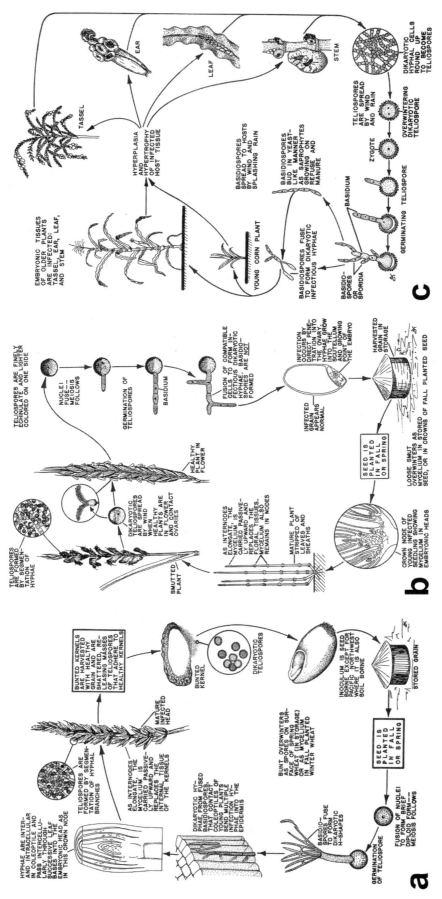

FIGURE 18.2 (a) Disease cycle of wheat common bunt disease (*Tilletia tritici*). (b) Disease cycle of wheat loose smut disease (*Ustilago tritici*). (c) Disease cycle of corn smut disease (*U. maydis*) (Parts (a)–(c) from Kenaga, C.B., E.B. Williams and R.J. Green, 1971. *Plant Disease Syllabus*, Balt Publishers, Lafayette. With permission.)

an ear of corn while still attached to the stalk (Figure 18.1d and Figure 18.1e). Galls of various sizes are also common on tassels (Figure 18.1f), stalks (Figure 18.1g), brace roots, and leaves. Before maturity, galls appear as somewhat white, translucent, rather brittle masses (Figure 18.2c). As galls mature, their internal portions become progressively dark, with the white exterior membrane initially remaining intact. Later, the outer membrane disintegrates and the entire gall becomes a dark-brown, dry, powdery mass consisting of millions of teliospores formed from hyphae that originally comprised the gall. Teliospores become windborne and contaminate soil, or the galls can fall to the ground where they overwinter in soil as crop debris (Figure 18.2c). The following season, soilborne teliospores from galls germinate, producing basidiospores that may land on susceptible host tissue (i.e., *any* meristematic, above-ground corn tissue); following fusion of germ tubes from sexually compatible basidiospores, the newly fused infection hyphae can then penetrate the corn plant (e.g., into individual silks), where they grow toward and into the developing kernels, or they may penetrate and infect leaf, stem, or tassel meristems. Once established in the host, smut mycelium can continue to invade host tissue, eventually resulting in gall formation.

Corn smut is no longer a serious disease in hybrid dent corn production because of the many resistant varieties available. However, smut remains a significant problem in sweet corn as resistance in those cultivars is limited. Management of smut in sweet corn is facilitated by maintaining well-balanced fertility, without excess nitrogen. Fungicide sprays or seed treatments provide no benefit in control of corn smut.

A major difference in the epidemiology of smut diseases compared to rusts is the absence of asexually produced **repeating spores** in the life cycle of smuts. Consequently, smuts, lacking the ability to form billions of asexual reproductive spores easily disseminated over hundreds of miles by wind, cannot build up major epidemics compared to many rust diseases.

Although considered only a plant disease in the U.S., corn smut is regarded as a choice food in Mexico, where young smut galls are harvested prior to turning black and powdery. Commercially the product is sold under its Spanish name *Cuitlacoche* (often spelled *Huitlacoche*; pronounced "wheat-la-coach-e," Figure 18.1h). The 19th-century Hidatsa Indians of present southwestern North Dakota harvested, boiled, dried, and later mixed corn smut galls ("Mapë′di") into boiled corn as a flavor additive.

RUST DISEASES

Currently and historically rusts are among the most damaging of all plant diseases worldwide. The devastation they caused was justification for annual sacrifices to ancient Roman gods and for 17th- and 18th-century legislation in France and the American colonies, respectively (see elim-

ination of alternative hosts later in this chapter). Rusts continue to cost untold millions of dollars annually to producers of cereals, vegetables, ornamentals, timber, sugarcane, coffee, and other crops, with those costs being passed on inevitably to processors and consumers. In this highly abbreviated coverage of rust diseases and their pathogens, it is essential to choose only those most important examples which illustrate particular concepts of pathogen biology, host–pathogen relationships, and disease management. Suggested readings are provided for greater breadth and depth of information.

Common symptoms and signs of cereal and other herbaceous rusts are shriveled seed (Figure 18.4a) and erumpent lesions (pustules) on leaves and stems (Figure 18.4b through Figure 18.4d); also see Plate 6A following page 80, respectively. Rust fungi have a highly complex life cycle that often contains up to five different spore stages; some rusts require two unrelated, alternative hosts to complete their life cycle (Figure 18.3a through Figure 18.3c). The five sequential stages of the complete rust life cycle, beginning and ending with the teliospores, are shown in microscopic view in Figure 18.4e to Figure 18.5e. The overwintering stage (or oversummering stage, depending on the rust and the climate in which it exist) is that of teliospores (Figure 18.4e and Figure 18.5e) borne in telia (Figure 18.3d and Figure 18.4d). Teliospores germinate to form basidiospores (Figure 18.3a and Figure 18.4e); if basidiospores land on a susceptible host they can initiate infection, resulting in production of **pycniospores** (Figure 18.4g), produced in pycnia (Figure 18.4f). Pycniospores function as spermatia (Figure 18.3a) and accomplish fertilization of the female flexuous hyphae, thus establishing the $N + N$ (i.e., dikaryotic) nuclear condition (Figure 18.3b). The next stage is that of **aeciospores** (Figure 18.4i), borne in aecia (Figure 18.4h and Figure 18.4i), that is followed by production of **urediniospores** (Figure 18.5a through Figure 18.5c) in uredinia (Figure 18.4b and Figure 18.5a), and eventually by teliospores (Figure 18.5d and Figure 18.5e) in telia to complete the cycle. The life cycle, showing the nuclear condition of different hyphal and spore stages is summarized in Figure 18.3b. Rusts that complete their entire life cycle on one host species (not necessarily on one host plant) are termed autoecious rusts, whereas those requiring two alternate host species for that process are heteroceious rusts. Terminology used to identify primary vs. alternate hosts of autoecious rusts lacks consistency and can be rather confusing. Some define the term primary host as that host which produces the telial stage, and the alternate host as the one that produces the pycnial/aecial stage. Others define primary host as the host of greater economic importance, with alternate host being the less important host economically. Figure 18.3a illustrates the classic hetereocious disease cycle of *Puccinia graminis* (wheat stem rust), which requires wheat or some related

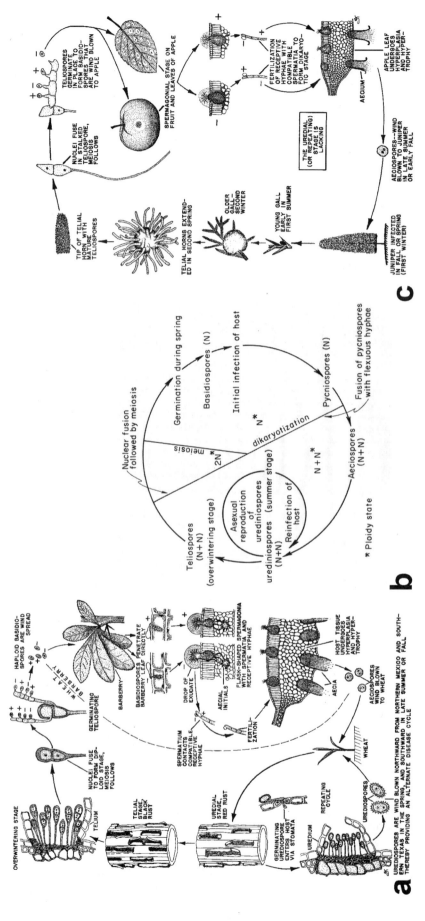

FIGURE 18.3 (a) Disease cycle of wheat stem rust disease (*Puccinia graminis* f. species tritici). (From Kenaga, C.B., E.B. Williams and R.J. Green, 1971. *Plant Disease Syllabus*, Balt Publishers, Lafayette. With permission.) (b) Generalized life cycle of rust fungi, showing the sexual and asexual reproductive stages and ploidy number of those stages. (From Littlefield, L.J., 1981. *Biology of the Plant Rusts: An Introduction*, Iowa State University Press, Ames. With permission.) (c) Disease cycle of cedar-apple rust (*Gymnosporangium juniperi-virginianae*). (From Kenaga, C.B., E.B. Williams and R.J. Green, 1971. *Plant Disease Syllabus*, Balt Publishers, Lafayette. With permission.)

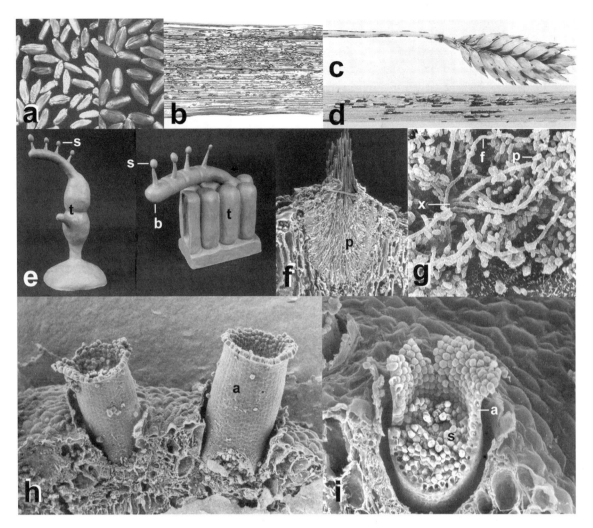

FIGURE 18.4 (a) Stem rust (*Puccinia graminis* f. species *tritici)* infection induces shriveling and reduced kernel size (left) compared to kernels from healthy plants (right). (From United States Department of Agriculture, Agricultural Research Service. With permission.) (b) Uredinial pustules of *P. graminis* f. species *tritici* on wheat leaf. (c) Telial pustules of *P. graminis* f. species *tritici* on stem and glumes. (d) Telial pustules of *P. graminis* f. species *tritici* on stem. (Parts (c) and (d) from United States Department of Agriculture, Agricultural Research Service. With permission.) (e) Models of germinating teliospores of the stalked, two-celled teliospore (t) of *Puccinia* species (left) and the sessile, single-celled teliospores (t) of *Melampsora* species (right). Each teliospore cell can potentially germinate to form an elongated septate basidium (b) that produces four basidiospores (s) borne on thin, peg-like sterigmata (From Littlefield, L.J., 1981. *Biology of the Plant Rusts: An Introduction*, Iowa State University Press, Ames. With permission.) (f) Longitudinal fracture through a pycnium (p) (= spermagonium) of *P. triticina*. Pycniospores (= spermatia), not present in this micrograph, are produced inside pycnia and belong to one of two possible mating types. They exude from the pycnial opening and are carried passively by leaf contact, splashing water or insects to the receptive ("female") hyphae (see Figure 18.4g) extending from pycnia of a second, compatible mating type, thus accomplishing spermatization. (Original magnification × 200.) (From Brown, M.F. and H.G. Brotzman, 1979. *Phytopathogenic Fungi: A Scanning Electron Stereoscopic Survey*, University of Missouri-Columbia Extension Division, Columbia. With permission.) (g) Leaf surface of pycnia/aecial host of *P. triticina*, showing elongated, flexuous hyphae (f) (the receptive "female" hyphae) that extend from the apical opening of an underlying pycnium (x). Pycniospores (p) (= spermatia) have exuded from that same underlying pycnium, but pycniospores and flexuous hyphae produced from the same pycnium are sexually incompatible. (Original magnification × 750.) (h) Two aecia (a) of *P. graminis* f. species *tritici* extending beyond the surface of the barberry host leaf. (Original magnification × 100) (From Littlefield, L.J. and M.C. Heath, 1979. *Ultrastructure of Rust Fungi*, Academic Press, New York. With permission.) (i) Diagonal fracture through an aecium (a) of *P. graminis* f. species *tritici,* exposing aeciospores (s) contained within the tubular aecium. (Original magnification × 200).

species, and common barberry for completion of its life cycle. In wheat stem rust disease, wheat is most commonly referred to as the primary host, with barberry being the alternate host, but not always. Similar cycles requiring two unrelated hosts occur in many other heteroecious cereal rusts, for example, wheat leaf rust (*P. triticina* = *P. recon-*

dita f. species *tritici)* and oat crown rust *(P. coronata).* Autoecious rusts, such as those on bean, asparagus, flax, or rose, complete their entire life cycle on only their one respective host. Many rusts, known as short-cycle rusts, lack one or more spore stage(s) in their life cycle. Cedar-apple rust (Figure 18.3c) lacks the uredinial stage; consequently, it has no repeating urediniospores, the most important spores epidemiologically in rust diseases of most agronomic or horticultural crops. As a result, no major "explosions" of cedar-apple rust occur, as is possible in cereal, bean, or coffee rusts. However, significant damage can occur in apple orchards due to rust resulting from massive primary infection by untold numbers of basidiospores originating from the cedar (botanically, a juniper) alternative host, not from repeating generations of urediniospores as in cereal rusts. Hollyhock rust (Figure 18.5d and Figure 18.5e), which produces only teliospores that overwinter and germinate to form basidiospores capable of initiating new infections the following year, exemplifies the greatest degree of shortened life cycle possible (i.e., the microcyclic form), yet containing the essential steps for sexual reproduction. Other abbreviated forms of rusts lack the sexual stage completely, or it is yet to be discovered. Many such rusts, for example, coffee rust (*Hemileia vastatrix*), survive very successfully in the uredinial stage only.

The economically important stage of a rust pathogen's life cycle can be different, depending on the rust. A few heteroceious, herbaceous rusts (e.g., hops rust), produce pycnia and aecia rather than uredia and telia on their more economically important host. More commonly, as in rusts of cereals, bean, sunflower, flax, alfalfa, asparagus, and many other herbaceous angiosperm hosts, the most damaging effect on yield results from the uredinial stage (Figure 18.4b and Figure 18.5a), irrespective of their being either autoecious or heteroecious. The subsequent telial stage on the same host species is essential to complete the life cycle, but typically it develops too late in the growing season to seriously reduce yield. The uredinial stage is not only the major cause of severe yield reduction in herbaceous hosts, but is the spore stage most commonly and most widely spread by wind, potentially resulting in epidemics spanning hundreds of miles during a growing season. The annual south-to-north dispersal of both wheat leaf and stem rust urediniospores occurs along the well-documented **Puccinia Path**. Both those pathogens can overwinter as uredinial stage infections on winter wheat in southern Texas and Mexico; urediniospores produced there in early spring are blown northward and have been trapped at elevations up to 14,000 ft. (ca. 4300 m). Progressive infections of susceptible winter wheats occur from northern Texas through South Dakota, and eventually into susceptible spring wheats in Minnesota, North Dakota, and central Canada by mid-summer. Similarly, the rapid spread of coffee rust in Asia and Africa resulted largely from windborne urediniospores, as in much of South America following the rust's introduction into the Western Hemisphere, possibly carried there on plants from Africa in 1970 (see Figure 18.5k).

In contrast to rusts of most agronomic crops, rusts of conifers cause the greatest damage in the pycnial and aecial stages of the life cycle. Trunk and branch cankers (Figure 18.5f) and galls (Figure 18.5g) resulting from infections two or more years old initiated by basidiospores from an alternative host can eventually girdle infected regions, resulting in death of all tissue beyond the canker. Several species of pine rusts cause losses of millions of dollars annually in both northern and southern forests of the U.S., as well as in pine, spruce, fir, larch, and other timber species worldwide. In contrast, an important example of a uredinal stage rust in commercial trees is leaf rust of poplars, which causes severe defoliation and reduced pulp wood production.

Rust pathogens are highly host specific, with most species being limited to one or only a few closely related hosts. Within some rust species there are also "specialized forms" (Latin plural, *formae speciales*; abbreviated "f. species ") that are limited to certain host species, for example, *P. graminis* f. species *tritici* on wheat and barley, *P. graminis* f. species *avenae* on oats, *P. graminis* f. species *secalis* on rye, and several others, as demonstrated by Eriksson in Sweden in 1894. There are also physiologic races within many species of several rusts, for example, *Melampsora lini* on flax, and in many *formae speciales* of rusts, such as *P. graminis* f.species *tritici* that can attack only certain cultivars of host species, as described by Stakman in 1914. The identity, and sometimes the actual genotypes, of physiologic races are determined by inoculating defined sets of differential varieties with urediniospores, and approximately 10 to 14 days later observing the various degrees or types of immune, resistant, or susceptible responses of those varieties. The distribution of reaction types across the set of differential varieties tested determines the identity of the rust race(s) so tested.

Parasitically, rust fungi obtain nutrition through specialized branches (**haustoria**) (Figure 18.5h and Figure 18.5i) that extend from intercellular hyphae into living host cells. Nutrients absorbed by haustoria are transported through the intercellular hyphae, supporting further expansion of the fungus colony and the production of reproductive spores. Nutrients diverted from the host into the pathogen contribute to reduced yield or quality of the plant's economic product (e.g., Figure 18.4a).

Resistance to rusts is often expressed as hypersensitivity or the hypersensitive reaction (HR; Chapter 28 and Chapter 31), a very localized reaction in which only a few host cells are killed (Figure 18.5j), thus depriving the obligate parasite of a source of nutrition. Other mechanisms, such as production of toxic compounds (phytoalexins) associated with HR, can contribute to death of the patho-

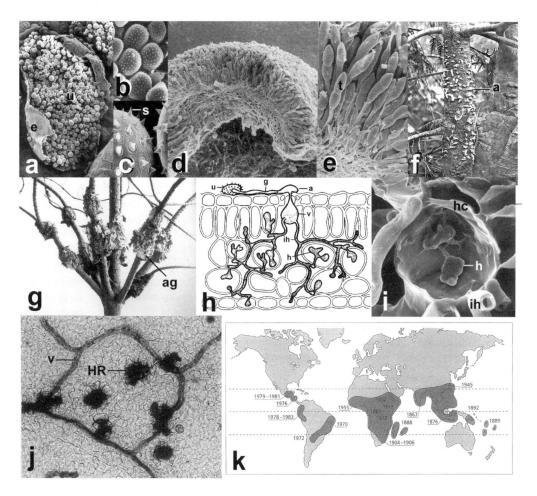

FIGURE 18.5 (a) Surface view of a uredinium of flax rust (*Melampsora lini*) containing several hundred urediniospores. Host epidermis (e) around the periphery of the uredinium is ruptured on expansion of that structure. (Original magnification × 100.) (From Littlefield, L.J., 1981. *Biology of the Plant Rusts: An Introduction*, Iowa State University Press, Ames. With permission.) (b) Higher magnification of urediniospores shown in Figure 18.5a. (original magnification × 750.) (From Littlefield, L.J., 1981. *Biology of the Plant Rusts: An Introduction*, Iowa State University Press, Ames. With permission.) (c) Spines (s) cover the surface of those uredin-iospores. (Original magnification × 2200) (From Littlefield, L.J. and M.C. Heath, 1979. *Ultrastructure of Rust Fungi*, Academic Press, New York. With permission.) (d) Longitudinal fracture of a telium of hollyhock rust (*Puccinia malvacearum*) on surface of host leaf. (Original magnification × 50.) (e) Higher magnification of a portion of Figure 5d reveals two-celled teliospores (t) characteristic of the genus *Puccinia*. (Original magnification × 130.) (Parts (d) and (e) from Littlefield, L.J., 1981. *Biology of the Plant Rusts: An Introduction*, Iowa State University Press, Ames. With permission.) (f) Aecial canker of white pine blister rust (*Cronartium ribicola*) on trunk of a young tree. Aecia (a), containing aeciospores, are the white, blister-like lesions on the canker surface. All portions of the trunk and branches beyond the canker will likely die. (g) Multiple aecial galls (ag) of western gall rust (*Endocronartium harknessii*) on lodgepole pine. (Parts (f) and (g) from Ziller, W.G., 1974. *The Tree Rusts of Western Canada*, Canadian Forestry Service Publication No. 1329, Department of the Environment, Victoria, B.C. With permission.) (h) Diagrammatic illustration of a typical uredinial stage rust infection in a host leaf. The urediniospore (u) germinates to form a germ tube (g), which bears at its apex an appressorium (a) over a host stoma. From that, infection occurs through the stoma, forming a substomatal vesicle (v), intercellular hyphae (ih), and ultimately numerous haustoria (h) that absorb host nutrients and transport them to other portions of the expanding or spore producing regions of the colony. (i) Internal view of a rust-infected coffee leaf showing intercellular hyphae (ih) and several haustoria (h) of *Hemileia vastatrix* that extend into the host cell (hc). (Original magnification × 1400.) (Parts (h) and (i) from Littlefield, L.J., 1981. *Biology of the Plant Rusts: An Introduction,* Iowa State University Press, Ames. With permission.) (j) Cleared, stained leaf of flax expressing immune, necrotic, hypersensitive reactions (HR), each resulting from an unsuccessful infection by an incompatible race of *Melampsora lini*. v = leaf vein. (Original magnification × 100.) (From Littlefield, L.J., 1973. *Physiol. Plant Pathol.* 3: 241–247, Academic Press, London. With permission.) (k) Current distribution of coffee rust (*Hemileia vastatrix*), in outlined, dark-shaded areas and the years in which infections were first noted in those areas. Mode of its introduction into the Western Hemisphere is uncertain (wind or inadvertent human activity). Its widespread distribution within much of Asia, Africa, and South America is thought to have resulted primarily from wind dispersal of urediniospore inoculum. (From Lucas, J.A., 1997. *Plant Pathology and Plant Pathogens*, 3rd ed., Blackwell Science, Oxford. With permission.)

gen in infected tissue. Other types of resistance include tolerance, where the host is capable of producing an acceptable yield in spite of rust infection, and slow rusting, when rust development in host tissues is significantly retarded rather than being halted (killed), as in hypersensitivity. These latter types of resistance are thought to result from more widely based, multigenic responses than is hypersensitivity, which often results from the action of a single resistance gene(s) in the host. Rapid development and spread of new physiologic races of rusts results in part from selection pressure placed on rusts by the planting of resistant cultivars over large regions of crop production. The genetic control of resistance and susceptibility of many hosts and the pathogenicity of many rust pathogens, respectively, is governed by the gene-for-gene relationship first demonstrated by Flor in the 1950s with flax rust.

A wide array of management techniques for rust diseases is dictated by the many variations in life cycles of rust fungi, the disease cycles of the different diseases, the diversity of host plants and differences in their cultural methods, the variably important significance of wind dispersal of spores, different economic impacts of rust diseases on different hosts and differences in breeding techniques for herbaceous vs. timber species. Examples of some more commonly used methods include the following.

- Genetic resistance (Chapter 31) — Wherever genetically resistant cultivars are available they are generally the most effective and least expensive means of managing rust diseases. Many cultivars of rust-resistant agronomic, orchard, and vegetable crops are available to commercial growers and homeowners. If they are not available or when new races of rusts arise, different means of disease management must be used, including the following.
- Foliar fungicidal sprays (Chapter 33) — Both protective and eradicant sprays are widely recommended for control of many rust diseases of herbaceous and orchard crops and ornamentals; many are quite effective. Unfortunately, the widespread, continued use of such fungicides sometimes provides selection pressure for evolution of new, fungicide-resistant or fungicide-tolerant strains of rust pathogens. Judicious use of foliar sprays is thus essential.
- Cultural practices (Chapter 32) — Practices such as planting susceptible conifers in low-risk environments, based on environmental evaluations, is recommended for some conifer rusts and is suggested even for some cereal rusts. Sanitation procedures, such as inspecting and destroying infected seedlings and pruning out

infected branches well below cankers or galls, another type of cultural practice, are recommended for some pine rusts. Other examples include providing adequate greenhouse ventilation to reduce high humidity favorable for geranium rust; similarly, for producers of organic asparagus, planting crop rows parallel to the prevailing wind is recommended to facilitate air movement and to reduce humidity in asparagus fields for management of rust disease. Removing or plowing under crop debris, including overwintering teliospores, is recommended for some vegetable diseases. However, such efforts may be futile if other growers in the region fail to do the same, thus providing the region an ample supply of windborne inoculum, similar to the near impossibility of controlling corn smut by sanitation.

- Elimination of alternate hosts — Elimination of alternate hosts has long been a recommended practice for many heteroecious rusts, dating back to the 1660 French legislation and similar laws in the American colonies spanning several decades in the 1700s that prohibited growing barberry bushes in close proximity to wheat fields. Those laws, based on empirical observation, far predated scientific proof for the linkage of wheat and barberry by DeBary in Germany in 1865. Influenced by a severe rust epidemic in 1916 and crucial demands for North American wheat during World War I, massive barberry eradication campaigns were implemented in 18 northerly, wheat-producing states of the U.S. and three adjacent Canadian provinces. The campaign lasted for some 50 years, although less in some states and provinces than others, but resulted in over 100 million barberry bushes being destroyed. The practice greatly reduced localized infections of wheat rust early in the growing season, which definitely increased yields in the participating states and provinces, and reduced the potential for developing new races by genetic recombination on the barberry host. However, barberry eradication did little to reduce the annual northward advance of urediniospores and the consequent wheat stem rust infections along the Puccinia Path, as the latter was more a result of meteorological than biological factors. Removal of junipers near apple orchards was formerly recommended as a control for cedar-apple rust, but the often impracticality of that, given the widespread distribution of junipers and the windborne distribution of basidiospores from that host, was later realized. Similarly, only partial success was achieved by eradicating wild currant bushes in western U.S. states for control of white pine blister rust, although the practice was more successful in eastern states.

LABORATORY EXERCISES

Unfortunately, rusts do not lend themselves easily to classroom laboratories situated in departments or universities in which rust research is not being conducted. It is best if one can provide the necessary supply of urediniospore inoculum personally or obtain it from colleagues working with rust fungi. Having to purchase an initial supply of spores and increase that sufficiently for class use is difficult to accomplish if one is not already familiar with basic methods of rust culture. Also, because rusts are plant pathogens, they cannot be legally shipped across state lines without obtaining proper permission from USDA/APHIS, which can require several weeks. Two alternative rust disease laboratory exercises are suggested, however, depending on local availability of host plant seeds and rust urediniospores. Both exercises have similar objectives and demonstrate similar methods. The following are the educational objective of the experiments: to learn methods of maintaining rust cultures on living host plants; to demonstrate different manifestations of susceptible and resistant responses, depending on the availability of host cultivars having different levels of resistance; to demonstrate effects of different temperatures on rate of rust development and production of urediniospores, depending on the availability of growth chambers or greenhouses of different temperatures; and to observe anatomical properties of urediniospores, such as surface ornamentation, germ pores (numbers and distribution), and normal anatomy, that is, size and shape of partially dehydrated spores under normal, ambient air conditions vs. textbook anatomy of hydrated, liquid-mounted spores typically used for microscopic examination (Littlefield and Schimming, 1989).

EXPERIMENT 1: WHEAT LEAF RUST

Depending on the local availability of differentially resistant or susceptible cultivars of wheat and differentially virulent races of *P. triticina*, one can demonstrate different manifestations of resistance and susceptibility to leaf rust infection. Total time required to complete the experiment is about 3 to 4 weeks.

Materials

The following materials are needed per student or group of 2 or 3 students and are based on supplies needed for two pots per student group (the minimum recommended). Numbers of items needed may increase depending on availability of growth chambers in which to observe effect of temperature on disease development:

- Two plastic pots of 3- to 4-inch diameter
- Sufficient commercial potting mix for two pots
- 20 to 30 seeds of a rust susceptible wheat cultivar
- 4 to 5 ml dry talc
- Microspatula
- 0.5 to 1.0 ml fresh urediniospores of leaf rust fungus (or one pot of young, leaf rust infected wheat plants to use for "brush-on" inoculation alternative described)
- A test tube or other small vial
- A piece of cheesecloth, approximately 6 inches × 6 inches
- A household water-mist applicator bottle (one bottle sufficient for many students)
- A 1 gallon ziplock or similar food storage bag
- Sufficient pot labels, one per pot, depending on total number of pots used

Follow the protocol outlined in Procedure 18.1 to complete this experiment.

Procedure 18.1 Production of Wheat Leaf Rust in the Laboratory

1. Grow wheat seedlings in 3- to 4-inch pots containing commercial potting mix (10 to 15 plants per pot) until 2 to 3 weeks old.
2. Inoculate plants with a mixture of urediniospores and dry talc. (Mix thoroughly ca. 4 to 5 ml dry talc and less than 1 ml fresh or stored urediniospores — about the volume of dry spores that can be picked up on the terminal centimeter of a microspatula.) Place that inoculum mixture into a test tube, a screw cap vial, or similar vessel and affix 4 to 5 layers of cheesecloth over the open top of that vessel with a rubber band. The inoculum mixture can then be applied to the wheat leaves by gently shaking the inverted test tube above the leaves, allowing the talc and spores to settle onto the leaves. Only a faint layer of talc-spore deposit on the leaves is sufficient for the experiment to succeed. Alternatively, inoculate healthy wheat

plants by simply "brushing" them with an inverted pot of rust-infected wheat inoculated about 14 days ago. Be sure that pustules are evident. This method will transfer sufficient urediniospore inoculum to the healthy plants and result in infection.

3. Very lightly atomize water onto the plants, being careful not to wash off the inoculum–talc mixture. Place pots into a 1-gal plastic bag containing a water soaked paper towel, seal the bags, allow inoculated plants to incubate at room temperature for 18 to 24 h. Partially open bags to allow plants to dry gradually over the next 3 to 4 h and then remove the pots to a greenhouse or growth chamber.

4. By maintaining inoculated plants at different temperatures, ranging from about 15°C to 30°C, the effect of temperature on rust development can be observed over the next 7 to 14 days.

5. Examine spores microscopically in aqueous mounts to observe and record spore size, ornamentation, number, and distribution of germ pores. Examine dry urediniospores on microscope slides, not mounted in liquid, to observe their normal size and partially collapsed shape under typical ambient atmospheric conditions compared to those spores mounted in liquid.

Anticipated Results

If favorable infection conditions are created, urediniospores should germinate and infect cells within 12 to 24 h. Symptoms and signs should develop over 7 to 14 days after inoculation. Initial development of small chlorotic spots ("flecks") on leaves, which in immune reactions will remain as such or in intermediate and susceptible reactions will continue to develop into larger infections, eventually producing uredinial pustules that rupture the leaf epidermis and bear urediniospores.

Questions

- What results would you expect if you had used the urediniospores to inoculate the pycnia/aecial (alternate) host (*Thallictrum* species) rather than wheat?
- What anatomical stages of rust fungus developed, that could have been observed microscopically, during the 18- to 24-h period in the moist plastic bags?
- If you inoculated host varieties having different levels of rust resistance, describe and compare development of symptoms and visible signs of the pathogen in the different host varieties. How was hypersensitivity expressed, if it occurred?

EXPERIMENT 2: BEAN RUST

The basic objectives and procedures are similar to those for the wheat leaf rust exercise. This alternative is included for those who might have more immediate access to the bean rust pathogen, *Uromyces appendiculatus*, or for classes comprising students more interested in horticultural than in cereal crops. Total time required to complete the experiment is about 3 to 5 weeks.

Materials

The following materials are needed per student or group of two or three students. This exercise requires basically the same number and kinds of supplies needed for the wheat leaf rust study (Procedure 18.1), except that plant only three bean seeds per pot:

- Edible dry pinto beans available in grocery stores provide inexpensive, susceptible seeds for this experiment
- Tween 20 solution if mist inoculation method is preferred
- Bean rust fungus, *U. appendiculatus*

Follow the instructions provided in Procedure 18.2 to complete this experiment.

Procedure 18.2 Production of Bean Rust in the Teaching Laboratory

1. Grow seedlings in commercial potting mix in 4-inch pots until 2 to 3 weeks old (three plants per pot, thin to two plants prior before inoculation with rust). Soaking seeds in wet paper towels or disposable diapers 24 h prior to planting will hasten germination. For best disease development, inoculate plants when leaves are about 50% to 70% fully expanded. (If leaves are expanding too rapidly before the scheduled class time, place pots into refrigerator. Cold temperatures will reduce the rate of leaf expansion without hurting the plant).

2. Inoculate plants with a mixture of *U. appendiculatus* urediniospores and dry talc; subsequently incubate plants and examine for development of symptoms, signs, and urediniospore anatomy as described for wheat leaf rust in Procedure 18.1. Alternatively, mix dry spore inoculum in water containing Tween 20 (two drops/liter), using about 20 to 30 ml of the water/Tween solution. Use atomizer to moist-inoculate plants with urediniospores suspended in the water/Tween solution. Lightly mist plants with water, moist-incubate for 24 h at room

temperature, gradually dry plants, and place into greenhouse or growth chamber (see Procedure 18.1). Observe infection development and urediniospore morphology as described for wheat leaf rust in Procedure 18.1.

Questions

- With this rust, what spore stage would have been produced as a result of your inoculation had you used aeciospores rather than urediniospores as inoculum?
- Describe anatomical properties of urediniospores produced as a result of the inoculation, including pigmentation, spore ornamentation, and number and distribution of germ pores.
- What effect did incubation temperature have on development of the rust? What appeared to be the optimum temperature for this rust's development?

The following are some variations in the above experiments, depending on availability of supplies, growth chambers, or greenhouse space:

- Include, incubate, and maintain controls in each exercise. Mock-inoculate other plants with water mist only to show that rust will not develop on the plants without having been inoculated with the pathogen.
- Cross-inoculate plants, that is, bean plants with *P. triticina* and wheat plants with *U. appendiculatus*, to show restricted host range of each pathogen. Nonhost plants will not get rust infection.

REFERENCE

Littlefield, L.J. and W.K. Schimming. 1989. Size and shape of urediniospores as influenced by ambient relative humidity. *Mycotaxon* 36: 187–204.

SUGGESTED READINGS

Agrios, G.N. 1997. *Plant Pathology*, 4th ed. Academic Press, San Diego.

Alexopoulos, C.J., C.W. Mims and M. Blackwell. 1996. *Introductory Mycology*, 4th ed. John Wiley & Sons, New York.

Buller, A.H.R. 1950. *Researches on Fungi, VII: The Sexual Process in the Uredinales*. University of Toronto Press, Toronto.

Bushnell, W.R. and A.P. Roelfs. 1984. *The Cereal Rusts*, Vol. 1. Academic Press, New York.

Littlefield, L.J. 1981. *Biology of the Plant Rusts: An Introduction*. Iowa State University Press, Ames.

Roelfs, A.P. and W.R. Bushnell. 1985. *The Cereal Rusts*, Vol. 2. Academic Press, New York.

Schieber, E. and G.A. Zentmyer. 1984. Coffee rust in the Western hemisphere. *Plant Dis.* 68: 89–93.

Schumann, G.L. 1991. *Plant Diseases: Their Biology and Social Impact*. APS Press, St. Paul, MN.

19 Fleshy and Other Basidiomycetes

Richard E. Baird

CHAPTER 19 CONCEPTS

- Fungi in this group all produce haploid **basidiospores** on **basidia**. Basidia can be found lining the surface of gills (typical mushroom), tubes (boletes), pores and teeth (shelf and conks), and internally (puffballs).

- Some fleshy basidiomycetes cause heart, butt, and root rots of trees.

- The hyphae of many basidiomycetes have **clamp connections**.

- Hyphal cells are separated by a very complex **dolipore septum**.

- Many fleshy basidiomycetes are associated with **wood decay**; some form **ectomycorrhizae** with plant partners; and relatively few are **parasitic**.

Approximately 30% of all known fungi are classified into the phylum Basidiomycota (basidiomycetes) (Alexopoulus et al., 1996). A large percentage of this total belongs to the group known as the fleshy basidiomycetous fungi that produce soft to flexible fruiting bodies that decay or decompose rapidly. Commonly called mushrooms (Figure 19.1), fleshy basidiomycetes include gilled and pored fungi, the puffballs, bird's nest fungi, jelly fungi, and the stinkhorns (Christensen, 1950). Species that form structures associated with wood decay are generally shelf to conk-like (lacking a stalk; Figure 19.2; see Plate 6C following page 80) or resupinate (appressed to the plant tissue; Figure 19.3).

In nature, these fungi exist as somatic mycelium (hyphae grown in a mass or layer) on wood, litter, and in soil. When environmental conditions are favorable, they form fruiting bodies of various sizes and shapes. Fleshy basidiomycetes occur worldwide in diverse habitats in the cold, temperate, and tropical regions. Individual species, however, are restricted in distribution based on habitats or flora constituents.

Although the majority of basidiomycetous fungi are saprophytic, certain mushroom-forming species are necrotrophic and still others, the rusts and smuts (Chapter 18), are obligate biotrophic plant pathogens. Saprophyte fungi, such as bird's nest fungi (see Plate 1B following page 80) and stinkhorns, may be brought to the attention of plant pathologists by the public when these fungi are found associated with landscape plantings. Many species form mycorrhizal associations with woody plants and forest trees whereas some species of fleshy basidiomycetes are associated with aquatic or wetland ecosystems such as bogs (Alexopoulus et al., 1996). In addition to the types of basidiomycetous fungi mentioned, the genera *Ceratobasidium* and *Thanatephorus*, the teleomorphs of *Rhizoctonia* species (Chapter 16 and Chapter 17), produce basidia on extremely small mycelial mats on soil or plant surfaces. A discussion of *Ceratobasidium* and *Thanatephorus/Rhizoctonia* is included later in this chapter due to the economic importance of this genus to agriculture.

MORPHOLOGY

Although species of fleshy basidiomycetes range from saprophytic to parasitic, the morphological structures of the fruiting bodies within this group are classified by several important characteristics. One of the most important is the presence of **basidia** (sing. basidium), of which several forms or types occur. The basidium is the site where karyogamy and meiosis occurs. The resulting nuclei that form following meiosis migrate through sterigmata, small stalk-like outgrowths, on which spores are formed externally. The **basidiospores** (haploid or *N*) are formed from the products of meiosis and are also called meiospores.

Hyphae of the basidiomycetes appear to be similar to other phyla such as the Ascomycota except that the hyphae contain a specialized septum called the **dolipore septum**. The dolipore septum (Figure 19.4) is unique to most members of the Basidiomycota and restricts movement of nuclei and other organelles between the hyphal cells. The exception is the rust fungi (Chapter 18); these basidiomycetes have cross walls similar to species of the Ascomycota. Another important characteristic that distinguishes the basidiomycetous fungi from other fungal phyla is the presence of **clamp connections** (Figure 19.5)

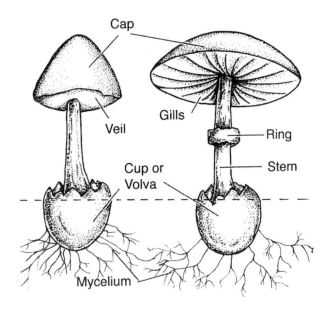

FIGURE 19.1 Diagrammatic representation of a mushroom. (Courtesy of J. MacGowen, Mississippi State University.)

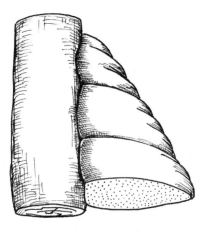

FIGURE 19.2 Woody mushrooms called conks are shelf-like structures where basidiospores are formed. Conks grow on living or dead woody plants or trees. (Courtesy of J. MacGowen, Mississippi State University.)

on the hyphae where septa have formed. This structure occurs in less than 50% of all species of basidiomycetes. Clamp connections help ensure the dikaryotic (N + N) condition of new cells during compartmentalization (septa formation) of the mycelium. This enables the basidiomycete fungi to reproduce sexually.

Although most fleshy basidiomycetes are primarily associated with decay of organic material, other species form unions with plants and forest trees in what is called a mycorrhizal or symbiotic association. However, several mushroom-forming genera can be pathogenic on economically important crops such as corn, vegetables, grasses, and forage crops. Members included in the genus *Marasmius, Xerompholina*, and *Armillaria* have also been

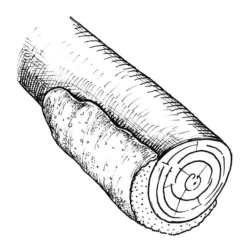

FIGURE 19.3 Resupinate basidiocarp growing on a dead branch. (Courtesy of J. MacGowen, Mississippi State University.)

reported to be pathogenic on agricultural and horticultural crops.

BASIDIOMYCETE TAXONOMY

The Basidiomycota is a diverse group of fungi. Molecular studies have caused mycologists to subdivide the group into the Hymenomycetes, Ustilaginomycetes, and Urediniomycetes (Swann and Taylor, 1993; McLaughlin et al., 1995; Berres et al., 1995). The Hymenomycetes, as indicated by the name, produce aggregates or clusters of basidia on a layer of specialized hyphae called a **hymenium**. All are members of the fleshy basidiomycetes. This group consists of the Agaricales (mushrooms), gastromycetes (e.g., puffballs and bird's nest fungi), Aphyllophorales (e.g., wood decay fungi), the groups of jelly fungi, and the Ceratobasidiales. Ceratobasidiales is an order in which the economical important genus *Rhizoctonia* (asexual state) belongs. Members of the deuteromycete genus *Rhizoctonia* may belong to either of the two sexual or basidia-forming states. These include the genera *Ceratobasidium*, which has two nuclei per cell, and *Thanatephorus,* which has more than two nuclei per cells.

The Agaricales or gilled fungi belong to an order referred to as toadstools or mushrooms, which are the fruiting bodies. Although there is extreme variation in morphology, mushrooms generally consist of a stalk (stipe) and a cap (pileus). The cap and sometimes the stalk support gills (lamellae) or pores where the basidia and basidiospores are produced. When a stalk is absent, the fruiting bodies are called woody conks, resupinate, or

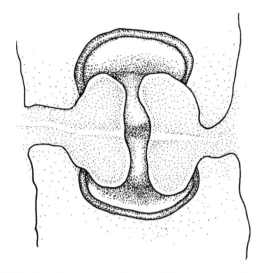

FIGURE 19.4 Hyphae containing a dolipore septum that occur only in certain genera and species of basidiomycetes. The structure of the septum restricts movement of nuclei between hyphal cells and maintains nuclear condition. (Courtesy of J. MacGowen, Mississippi State University.)

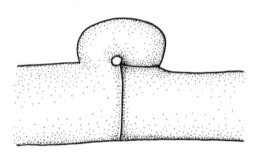

FIGURE 19.5 Hyphae with clamp connections that occur only in some basidiomycetes. (Courtesy of J. MacGowen, Mississippi State University.)

bracket fungi. The gills or pores have a hymenial layer where basidia and basidiospores are formed (Figure 19.1).

HABITAT

Members of the Agaricales are found on a broad range of plants and are commonly associated with decay of dead plants and woody tissues (lignicolous). Others types are more specific or occur in select ecosystems such as the coprophilous (dung inhabiting) fungi or the hallucinogenic mushroom species. *Psilocybe cubensis* is a very common coprophilus species found in pastures in the southern U.S. and is often collected for its hallucinogenic properties.

Certain mushroom-forming species are extremely host specific. Examples of this specific host–fungi relationship is the saprophytic mushroom-forming species *Strobilurus esculentus*, which forms mushrooms only on fallen spruce cones, or *Auriscalpium* species (tooth-forming fungus) observed only on cones of pine and firs. The mycorrhizal-associated mushroom species *Suillus pictus* and *S. americanus* are primarily found where white pine trees grow (Weber et al., 1988).

MYCORRHIZAE

Almost all vegetation and trees require a beneficial symbiotic relationship with certain genera and species of fungi. In a mycorrhizal relationship, fungal hyphae envelope the thin rootlets of trees, but do not harm the roots. The symbiotic relationship occurs when carbohydrates or sugars from photosynthesis are passed from the plant to the fungal associate, whereas the fungus acts as a secondary root system providing water and nutrient uptake to the plant or tree host (Figure 19.6). The transport or movement of the materials to either partner in the symbiotic association involves complex chemical pathways not discussed here. There are generally considered to be two types of mycorrhizal fungal associations including endomycorrhizae and ectomycorrhizae. Hyphae of endomycorrhizal fungi grow into the cortical cells of the root and do not form a thick layer of tissue around the root hairs (Figure 19.7). These fungi do not belong to the fleshy basidiomycetes, but are primarily members of the Zygomycota (Chapter 10). Alexopolous et al. (1996) estimated that there were over 2000 species of ectomycorrhizal fungi and the majority were fleshy basidiomycete mushroom-forming species. The hyphae of these fungi attach to the feeder rootlets of plants, forming a fungal mantle or thick layer of mycelium around the surface of the roots. In a true ectomycorrhizal association, the fungus initially forms a mantle (Figure 19.7) around the roots before growing between the outer cortical cell layer to form what is called the Hartig net (Wilcox, 1982). This network of fungi is the main distinguishing feature that defines ectomycorrhizae, because a number of fungi might be able to form what could be interpreted as a mantle, but do not have intercellular growth. The root hairs in an ectomycorrhizal association become forked or multiforked or are irregularly shaped. The color of the mantle varies depending on the fungal species. Other fungi that form ectomycorrhizal associations are members of the phylum Ascomycota, including the cup fungi and truffles, but the number of species and genera are fewer in number than for the Basidiomycota.

Ectomycorrhizal fungi are associated with many plant and tree species. Important tree partners include members of the Pinaceae (pines, true firs, hemlock, spruce, and others), Fagaceae (oaks, beech, and others), Betulaceae (birch, alders, and others), Salicaceae (poplars, willows, and others), and Tiliaceae (basswood). On a given tree

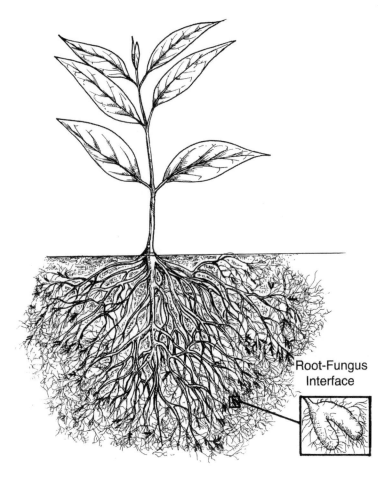

Root-Fungus
Interface

FIGURE 19.6 Ectomycorrhizal associations between a fungus and a tree. Fungus acts as a "secondary root system" to provide nutrients and water for the host tree species. (Courtesy of J. MacGowen, Mississippi State University.)

root, many ectomycorrhizal-forming species can occur. A specific tree species can partner with many different fungal species at one time. Douglas fir in the northwestern U.S. has more than 200 ectomycorrhizal associates. In turn, a single fungus can associate with a broad range of plants or be specific to select tree species. An example is the mushroom-forming species *S. americanus* that primarily occurs under white pine. In contrast, *Amanita vaginata* occurs under many species of pines and hardwoods.

Pathogenic genera and species of fleshy basidiomycetous fungi can infect agricultural crops and forest tree species. For agricultural crops, a complex of fungi called the sterile white basidiomycetes, containing genera that produce white mycelium with clamp connections in culture, have been identified as mushroomforming species, including *Marasmius* and *Xeromphalina*. Cultures of these fungi have been reported to infect roots and lower stems of corn, sorghum, millet, and vegetable crops such as snap beans.

Mushroom-, conk-, or bracket-forming basidiomycetous fungi have been shown to be pathogenic on forest trees. A classic example of pathogenicity by a mushroomforming species is *Armillaria mellea* or *A. tabescens*

(shoestring fungus-honey mushrooms). The fungus becomes established on trees or woody plants by forming rhizomorphs, black shoestring-like structures that transport nutrition and water, extending out from the mycelial mat. Rhizomorphs are a clustering of hyphae into strands, forming a complex mycelial structure (Figure 19.8). *Armillaria mellea* is one of the most common, best-known and most damaging fleshy fungal disease of forest, shade and ornamental trees, and shrubs. It flourishes on an extraordinary broad host range and under diverse environmental conditions. *Armillaria mellea* becomes established when host plants are under stress (e.g., nutrients, water), whereas *A. tabescens* can be highly virulent. The latter fungus invades bark and cambium tissues of the roots and the root collar of coniferous tree species. The extent and speed of invasion depends on host species and degree of stress to the plants. Plant stress increases by plant competition and by mechanical injury, resulting in declining root systems. After the host dies, the fungus continues to decay the host tissue and may survive in the soil for decades.

Another fleshy basidiomycete species pathogenic to forest trees is *Heterobasidium annosum* (the causal agent

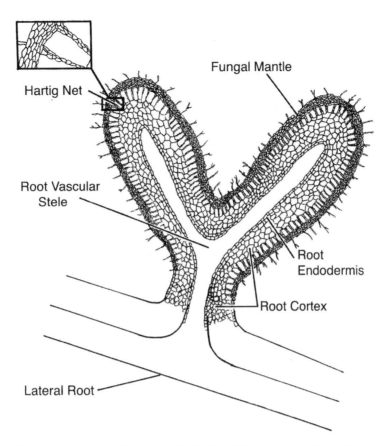

FIGURE 19.7 Endomycorrhizal fungus growing into the cortical cells of the tree roots. Hartig Net (box) from an ectomycorrhizal association. Hyphae from the fungus form an outer layer around the root and grow between the cortical cells. (Courtesy of J. MacGowen, Mississippi State University.)

of annosum root rot). This basidiomycetous fungus occurs in coniferous forest ecosystems and can invade tree roots when under stress from environmental conditions such as drought or when management practices are minimal. The fungus occurs throughout the U.S. and Canada, but causes severe economic damage in the southeastern U.S. Young seedlings can be killed by *H. annosum*, but in older trees the heartwood is invaded and chronic infections can continue for years. When roots of different trees graft, the fungus can invade the uninfected tree. The conks or shelf-like woody mushrooms can develop on infected living trees or on dead hosts. The conks generally are found slightly buried in the humus or can be seen just above the ground surface on the butt of the tree (Figure 19.2). Recognition of the basidiocarps or conks of *H. annosum* is useful for diagnosis.

WOOD ROTS CAUSED BY FLESHY BASIDIOMYCETES

The majority of fleshy basidiomycetes are saprophytic and often have the ability to decay wood. Fungi in this group derive nutrients from dead plant or woody tissues by enzymatically and chemically degrading cellulose, hemicellu-

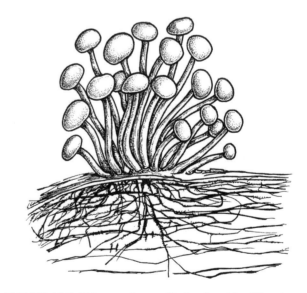

FIGURE 19.8 Rhizomorphs are black, shoestring-like structures deriving nutrients and water for fungal growth and sporulation. (Courtesy of J. MacGowen, Mississippi State University.)

lose, or lignin (Fergus, 1966; see Chapter 10 and Chapter 27). Some of these fungi occur on living trees, but they generally decay only the dead tissues of the heartwood.

Two groups of fleshy basidiomycetous fungi are most important in deriving nutrition in their role as wood rotters: the resupinate fungi (form flat basidiomata on host, Figure 19.3) placed into the Thelephoraceae or Corticiaceae (Alexopoulus et al., 1996), and the remainder traditionally placed into the group polypores. Within these groups, the fungi can either be referred to as white rot or brown rot decay species.

WHITE ROTS

White rot fungi can generally use all components of the wood, including lignin, cellulose, and hemicellulose. The wood degraded by these fungi is light in color due to the loss of lignin (brown color). Because lignin gives strength to the wood, the tissue becomes spongy, stringy, or mottled. Genera in the Agaricales, such as *Agaricus, Coprinus,* and *Pleuteus,* are common plant and wood decay fungi. Many genera in the Aphyllophorales, such as polypores (pored fungi), chanterelles, tooth and coral fungi, and the corticiod-forming fungal species, also decay wood. An example of a white rot fungus is *Oxyporus populinus* that attacks numerous hardwood tree species and causes trunk and limb rots. *Trametes versicolor* attacks the sapwood of hardwood trees such as sweet gum, *Liquidambar styraciflua.*

BROWN ROTS

Brown rot fungi rots, also known as dry rot, generally degrade cellulose and hemicellulose and have either limited or lack the ability utilize lignin. Lignin is more or less brown and remains after other wood polymers have been decayed. Wood decomposed by these fungi appears brown, cubicle, and crumbly. There is a defined margin between healthy and decayed wood. This form of rot is more common in coniferous trees than hardwood tree species. *Poria placenta* and *Phaeolus schweinitzii* are often associated with brown rot of timber and lumber.

FAIRY RINGS

Fairy rings are fleshy mushroom-forming basidiomycetes that grow in circular zones (see Plate 1A following page 80). Although many species form fairy rings, some of the more common genera include *Coprinus, Chlorophyllum, Lepiota,* and *Marasmius.* The rings are readily observed in lawns and golf courses, but can be found in forests and other ecosystems. The rings occur from the expansion or outward circular growth of the mycelium that initially originated from one central point. The mushrooms, however, almost always form near the outer edge of the active growth zones. Some fairy rings have been found to be hundreds of years old and extend over very long distances. A ring can be disrupted due to physical barriers or loss of the nutrient base, forming arcs or partial zones. In golf course greens or home lawns, fairy rings may be considered unsightly or may affect growth of the grass, causing obvious discolored patches and dead tissues within the zones. The mycelium from the fungus can act as physical barrier preventing adequate moisture and reduced vigor of the invaded grass. Control or prevention of rings includes reducing thatch and maintaining adequate moisture to the thatch layers. Labeled fungicides are effective for control of these fungi.

OTHER BASIDIOMYCETES (NONFLESHY)

An unusual plant pathogenic basidiomycete genus, *Exobasidium,* forms basidia directly from dikaryotic mycelium. A layer of hyphae forms between the epidermal cells of the plant tissue, causing tissue distortion, including swelling, blistering, and galling. Infections can occur on host species such as azalea, rhododendron, camellia, blueberry, ledum and leucothoe, and other ornamental and forest plants (see Plate 6D following page 80). Hyphae of *Exobasidium* grow intercellularly and invade cells by forming haustoria. Controls include removal or sanitation of galls in landscape situations or small greenhouses. Fungicides can be used, but success depends on applications prior to bud break and follow-up sprays. Resistant host plants are generally not available.

The genus *Rhizoctonia* (Chapter 16 and Chapter 17) is one of the most economically important groups of pathogenic species in the world. Billions of dollars are lost annually from seed decay, root rots, seedling disease, and aerial or foliar blights (see Plate 6B following page 80). Although *Rhizoctonia* is known primarily for disease of agronomic, vegetable, and ornamental crops, several species were reported to form mycorrhizal associations with plants. *Rhizoctonia* species do not produce conidia in culture and can be difficult to identify. Isolates of *Rhizoctonia* species may vary greatly in the number and size of sclerotia (survival structures), hyphal growth (appressed or aerial) and color in culture, and presence of the dolipore septum (Chapter 17). Without conidia, these cultural characters alone make it difficult to identify *Rhizoctonia* species.

Identification is further complicated by different mating types, called anastomosis groups (AG) (Carling, 1996). These types only mate by hyphal fusion to compatible anastomosis types. During complete anastomosis, the hyphal cell walls of the two compatible types fuse, lysis occurs, and the nuclei are exchanged between cells. Until recently, mating studies were the only techniques available for isolate identification. Isolates of *Rhizoctonia* that do not mate are called vegetatively incompatible.

The hyphae of the different groups have two levels of nuclei per cell. The binucleate (two nuclei per cell) group is known as the *Ceratobasidium* anastomosis group (CAG) and is named after the teleomophic state for this group of *Rhizoctonia* species Approximately 20 different mating types have been identified in this group. The second group, which includes *Rhizoctonia solani*, has three or more nuclei per cell (multinucleate) and isolates are divided among 14 anastomosis groups (AG).

The number of nuclei per cell and anastomosis pairing studies have been the most useful and effective methods to identify *Rhizoctonia* species. Recently developed molecular techniques are more accurate in identifying CAG and AG forms of *Rhizoctonia*. The anastomosis pairing groups have been correlated with groups defined by the newer molecular techniques (Vilgalys and Cubeta, 1994).

The teleomorph for the AG forms of *Rhizoctonia* has been identified as *Thanatephorus cucumeris*. At least two other teleomorphic genera have been identified for *Rhizoctonia*, but will not be mentioned here. The sexual stage for the CAG and AG types of *Rhizoctonia* species is generally nondescript. A hymenium layer of hyphae may form on the soil surface or occasionally in culture with the basidia and basidiospores being produced directly from that area (Sneh et al., 1991). Sexual structures are either rare or extremely difficult to find in nature.

Target spot of tobacco, also called Rhizoctonia leaf spot, is caused by basidiopore infections from *Thanatephorus cucumeris*. The fungus overwinters in soil or plant debris as sclerotia. The sclerotia germinate and hyphae grow saprophytically in the soil. Under favorable environmental conditions, basidiospores are formed on tiny resupinate mycelial tissues that form a hymenial surface containing basidia and basidiospores. The spores are windborne and disseminated to host tobacco plant foliage. The spores germinate and appressoria form from the growing hyphae, allowing direct entry of the fungus and subsequent colonization of the tissue. Once established in the host, target-shaped lesions develop from these initial infections. New basidiospores are produced within the lesions, causing secondary infections and ultimately resulting in defoliation of the tobacco plant. New sclerotia form and overwinter in the dead host tissue or soil, completing the disease cycle.

The fleshy basidiomycetes include a very large and diverse group of fungi. Species within the group occur in different ecosystems, including agricultural fields, forests, and ornamental landscapes. These fungi survive and reproduce either as saprophytes or parasites and are involved in mycorrhizal associations. Fruiting bodies of basidiomycetes can vary in shape and size, ranging from mushroom-like to those that form woody conks on dead wood, or species can be very tiny and wispy with only a hymenial layer present in soil or on plant tissues. The key morphological structures used by taxonomists to group basidiomycetes together include the basidium, dolipore septum, and, when present, clamp connections. The basidium is where meiosis and subsequent basidiospore formation occur.

REFERENCES

Alexopoulus, C.J., C.W. Mims and M. Blackwell. 1996. *Introductory Mycology*, 4th ed. John Wiley & Sons, New York, 869 pp.

Berres, M.E., L.J. Szabo and D.J. McLaughlin. 1995. Phylogenetic relationships in auriculariaceaous basidiomycetes based on 25S ribosomal DNA sequences. *Mycologia* 87: 821–840.

Carling, D.E. 1996. Grouping in *Rhizoctonia solani* by hyphal anastomosis reaction, pp. 37–47 in *Rhizoctonia Species: Taxonomy, Molecular Biology, Ecology, Pathology and Disease Control*, Sneh, B. et al. (Eds.), Kluwer, Dordrecht, The Netherlands.

Christensen, C.M. 1950. *Keys to the Common Fleshy Fungi*, 4th ed. Burgess, Minneapolis, MN, 45 pp.

Fergus, C.L. 1966. *Illustrated Genera of Wood Decay Fungi*. Burgess, Minneapolis, MN, 132 pp.

McLaughlin, D.J., M.E. Berres and L.J. Szaba. 1995. Molecules and morphology in basidiomycetes phylogeny. *Can. J. Bot.* 73: 684–692.

Sneh, B., L. Burpee and A. Ogoshi. 1991. *Identification of Rhizoctonia Species*. APS Press, St. Paul, MN, 133 pp.

Swann, E.C. and J.W. Taylor. 1993. Higher taxa of basidiomycetes: An 18S rRNA gene perspective. *Mycologia* 85: 923–936.

Vilgalys, R. and M.A. Cubeta. 1994. Molecular systematic and population biology of *Rhizoctonia*. *Annu. Rev. Phytopathol.* 32: 132–155.

Weber, N.S. and A.H. Smith. 1988. *A Field Guide to Southern Mushrooms*. University of Michigan Press, Ann Arbor, MI, 280 pp.

Wilcox, H.E. 1982. Morphology and development of ecto- and ectendomycorrhizae, pp. 103–113 in *Methods and Principles of Mycorrhizal Research*, Schenk, N.C. et al. (Eds.), APS Press, St. Paul, MN.

20 Oomycota: The Fungi-Like Organisms

Malissa H. Ament and Robert N. Trigiano

CHAPTER 20 CONCEPTS

- Organisms in the Oomycota were once classified in the kingdom Fungi, but are now included in the kingdom Stramenopila (Chromista).

- The group is characterized by production of oospores, thick-walled resting spores usually resulting from sexual reproduction and capable of surviving adverse environmental conditions.

- Sexual reproduction is by gametangial contact between oogonia and antheridia (meiosis occurs in the gametangia). Asexual reproduction includes production of sporangia that can produce motile, biflagellate sporangiospores (zoospores); some species do not produce motile cells.

- Hyphae are coenocytic (solid septa may be present at the base of reproductive structures or in old and damage hyphae), are relatively large (6- to 10-μm diameter), contain cellulose in the wall, and have diploid (2N) nuclei.

- Species in this group ecologically range from saprophytes to obligate plant parasites. Most of the more important plant pathogens are classified in the Peronosporales.

The phylum Oomycota (oomycetes) encompasses an incredibly diverse group of organisms that were for many years considered to be fungi or at least were always included in the study of the fungi. Most mycologists and other biologists recognized that the oomycetes, although being achlorophyllous and having mycelium-like colonies, were somewhat different and did not fit the conventional and prevailing concept of true fungi. The oomycetes have been variously categorized into different kingdoms and other groups in the past and some confusion in their taxonomic status still exists today. For the purposes of this chapter, we will follow Alexopoulus et al. (1996) and place them in the kingdom Stramenopila.

The oomycetes include organisms that range from saprophytic through obligately parasitic. Some are strictly aquatic whereas others are considered terrestrial — they occupy widely divergent habitats. Whatever their mode of nutrition or ecological niche, all species share some morphological and physiological characteristics. The following is not an exhaustive list, but these primary features readily distinguish the oomycetes from all other plant pathogens.

- The life cycles of oomycetes can be best described as diplontic; that is, for the vast majority of their life history, the ploidy level of the nuclei is diploid (2N). Haploid nuclei are only found in the gametangia where meiosis occurs. The zygote is the only diploid structure in the true fungi, which have typical haplontic life cycles where nuclei are either haploid (N) or exist as a **dikaryon** ($N + N$).

- The hyphae are generally **coenocytic**, but septa or crosswalls do occur at the bases of reproductive structures and in older portions of the mycelium (Figure 20.1). Hyphal walls of the oomycetes contain cellulose, which is not found in true fungi; instead true fungi have some amount of chitin in their walls.

- Sexual reproduction in the oomycetes is accomplished by **gametangial contact** between **oogonia** and **antheridia** (Figure 3.2). The resulting thick-walled **oospore**, which is diploid, is considered to be the resting or overwintering spore for many species (Figure 20.2).

- Asexual reproduction by the vast majority of oomycete species is completed by **zoospores** produced in sporangia. Typically many, more-or-less kidney-shaped zoospores are cleaved from the cytoplasm in the **sporangium** or vesicle. Each zoospore has one tinsel flagellum (see E in Figure 20.6) directed forward and one whiplash flagellum facing behind.

- A number of biochemical synthetic pathways and storage compounds in the oomycetes are very different than those present in true fungi.

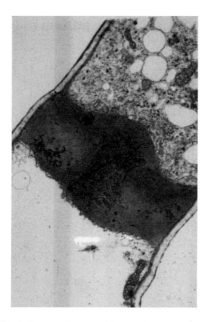

FIGURE 20.1 Septum in an old intracellular hyphae of *Perono-spora tabacina*. (From Trigiano, R.N., C.G. Van Dyke, H.W. Spurr, Jr. and C.E. Main, 1985. *Tob. Sci.* 29: 116–121. With permission.)

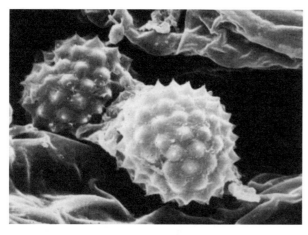

FIGURE 20.2 Oospore of *Peronospora tabacina*.

Just as there is some confusion regarding the placement of members of the Oomycota in a kingdom, there is also some discussion on how the phylum should be divided into orders. Most mycologists and plant pathologists agree that there should be either five or six orders, which may or may not include the order Pythiales. We will again follow the example of Alexopoulos et al. (1996) and recognize six orders (does not include the Pythiales). Most of the important plant pathogens fall into a single order, the Peronosporales. The species in Peronosporales are very diverse and include soil-inhabiting facultative parasites, such as *Pythium* and some *Phytophthora* species that cause seedling diseases, wilts, and root and seed rots.

Other *Phytophthora* species are facultative saprophytes that cause aerial blights, root rots, etc. The downy mildews and white rust organisms are obligate parasites that attack only the aerial portions of plants. Note that in the Saprolegniales (Saprolegniaceae), there are several species of *Aphanomyces* that can cause root rots of field crops such as peas and sugar beets. These organisms will not be discussed in this chapter.

The Peronosporales can be subdivided rather easily into the following three major families based primarily on characteristics of sporangia and sporangiophores: the Pythiaceae, the Peronosporaceae, and the Albuginaceae. Species in the Pythiaceae have sporangia that are borne variously, but not in chains, on somatic hyphae or on sporangiophores of indeterminate growth (grows continuously). This family includes the genera *Pythium* and *Phytophthora*. The seven genera classified in the Peronosporaceae (the downy mildews) form sporangia on well-defined, branched sporangiophores of determinate growth (Hawksworth et al., 1995). Some common genera are *Peronospora*, *Plasmopara*, *Bremia*, and *Pseudoperonospora*. The Albuginaceae contains a single genus *Albugo* that produces chains of sporangia on club-shaped sporangiophores of indeterminate growth. The species of *Albugo* are collectively known as the white rusts (Alexopolous et al., 1996).

PYTHIACEAE

The Pythiaceae encompasses ten genera (Hawksworth et al., 1995) and includes *Pythium* and *Phytophthora* species. These two genera include some of the most destructive and notorious plant pathogens. The Irish potato famine of 1845 and 1846 was caused by *Phytophthora infestans*. This organism was largely responsible for the starvation of many who were overly dependent on the potato crop and for the subsequent mass emigration to the U.S. A more detailed account of this disease and its influence on human affairs appears in Chapter 1 and Chapter 30.

PYTHIUM

There are about 120 species of *Pythium* according to Hawksworth et al. (1995) and many cause seed and root rots and pre- and postemergence damping-off of many greenhouse and field crops. *Pythium* root diseases can also severely damage established plants in the greenhouse and field (Figure 20.3). Roots of infected plants often lack small feeder roots and the secondary roots are discolored (brown or black). Infected plants often wilt and some may exhibit symptoms that mimic mineral deficiencies. Older plants infected with *Pythium* can also exhibit necrotic, watery-appearing lesions on the stem (Figure 20.4). *Pythium* blight has also become a serious problem of turfgrass,

FIGURE 20.3 Root rot of chrysanthemum caused by *Pythium* species. (Courtesy of A.S. Windham, University of Tennessee.)

FIGURE 20.5 Pythium blight of grass. (Courtesy of A.S. Windham, University of Tennessee.)

FIGURE 20.4 Blackleg of geranium caused by *Pythium ultimum*. (Courtesy of A.S. Windham, University of Tennessee.)

especially on high-value sports fields and golf courses (Figure 20.5).

In preemergence damping-off, the germinating seedling is attacked, colonized, and killed before breaking through to the soil surface. In postemergence damping-off, seedlings emerge from the soil and soon begin to wilt. In this scenario, the young feeder roots are destroyed and cortical tissues of the primary root are invaded by the pathogen. Dark, water-soaked lesions develop on stem at the soil line. Extracellular hydrolytic enzymes (pectinase, cellulases, and hemicellulases; Chapter 26 and Chapter 27) produced by the pathogen degrade cell walls and cause plant tissues to lose their structural integrity. When the seedlings can no longer support themselves, they collapse and fall onto the soil surface (damping-off). Often and especially in high-humidity environments such as greenhouses, white, watery-appearing mycelia can be seen growing on the rotting remains of the shoots.

The life history of *P. debaryanum* (Figure 20.6) is generally representative of most *Pythium* and other oomycete species in that meiosis occurs in gametangia, diploid oospores are the resting or overwintering spore and biflagellated zoospores are produced. *Pythium* species survive in soil as somatic hyphae or oospores. Zoospores from sporangia or oospores are attracted to roots where they encyst and germinate to infect tissue. Somatic hyphae may also directly infect roots. Hyphae grow intercellularly within host tissue and do not produce **haustoria**. The asexual portion of the life cycle is completed with formation of sporangia. Sporangia may be indistinguishable from somatic hyphae or quite distinct, depending on the species. The contents of sporangia migrate into **vesicles** where zoospores are differentiated. Vesicles burst and zoospores are liberated to the surrounding environment.

Sexual reproduction is initiated with the differentiation of gametangia, antheridia, and oogonia from somatic hyphae. A few **heterothallic** species of *Pythium* require different mating types, but most are **homothallic**, where both gametangia involved in the sexual process may arise from the same hyphae or may be formed on different hyphae that lie close together. One or more antheridia contact a single oogonium and meiosis occurs, reducing the ploidy of the nuclei in the gametangia to haploid. A nucleus from an antheridium passes into the oosphere (plasmogamy) and eventually fuses with a nucleus in the oosphere (karyogamy) to restore the diploid condition. The oosphere develops a smooth thick wall, which is capable of germinating after a resting period. Oospores may germinate directly via a germ tube or by producing a sporangia and oospores, depending on prevailing environmental conditions (primarily moisture and temperature). Some of the more important *Pythium* species include *P. ultimum, P. aphanidermatum,* and *P. debaryanum.*

Unfortunately, host resistance to *Pythium* species is not available for field and greenhouse crops. Management

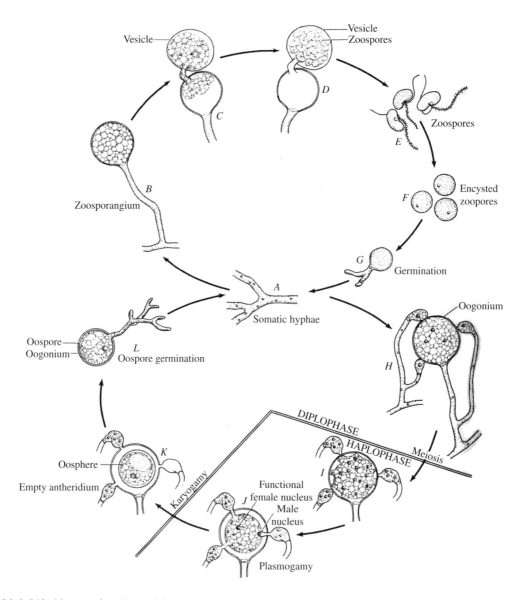

FIGURE 20.6 Life history of *Pythium debaryanum*. (Drawing by R.W. Scheetz. From Alexopoulos, C.J., C.W. Mims and M. Blackwell, 1996, *Introductory Mycology*, 4th ed., John Wiley & Sons, New York, 868 pp. With permission.)

of these diseases is primarily through cultural practices and fungicides, although some biocontrol tactics are used to control some oomycetes species (Chapter 34). Good soil drainage and management of soil moisture are important cultural practices used to control diseases caused by *Pythium* species in both field and greenhouse situations. Sound nutritional management of the crop, especially in regard to avoiding excess nitrogen fertilization (Chapter 32), can help to avoid diseases caused by these organisms. Crop rotation using nonhosts may also be an effective disease management tactic. In nursery crops, the adoption of pine bark as a container medium instead of peat has helped manage root rot diseases. Seeds may be treated with captan or other contact fungicides to prevent root rots. Plants may be drenched or sprayed with one of sev-

eral systemic acylalanine fungicides (metalaxyl or mefenoxam) either to prevent or control infection. Acylalanine class of fungicides is only active against species of the oomycetes; the true fungi are not affected by these compounds.

PHYTOPHTHORA

There are approximately 50 species of *Phytophthora* according to Alexopoulos et al. (1996); however, with the advent of molecular techniques in recent years, this number is subject to change. *Phytophthora* species cause many of the same type of diseases as *Pythium* species on the same plants (Table 20.1) and thus the two may be easily confused with one another when considering symptomol-

TABLE 20.1
Some Diseases Caused by *Phytophthora* Species

Phytophthora Species	Disease
P. fragariae	Red stele of strawberry
P. megasperma	Root and stem rot of soybean
P. parasitica	Blackshank of tobacco
P. citrophthora	Foot rot of citrus
P. cinnamoni	Root rots of forest and nursery crops and avocado
P. ramorum	Sudden oak death
P. infestans	Late blight of potato and tomato
P. palmivora	Black pod of cacao

FIGURE 20.7 Foliar blight of vinca by *Phytophthora* species. (Courtesy of A.S. Windham, University of Tennessee.)

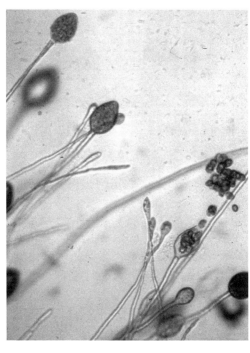

FIGURE 20.8 Sporangiophores and sporangia of a *Phytophthora* species. (Courtesy of A.S. Windham, University of Tennessee.)

ogy only. In addition, *Phytophthora* species can cause a number of other diseases, including foliar blights of field, nursery, forest, and greenhouse crops (Figure 20.7).

The life history of *Phytophthora* species differs in detail from that of most *Pythium* species. The most notable difference between the two genera is that zoospores of *Phytophthora* species are delimited and functional within the sporangium, whereas in *Pythium* species, zoospores are formed from the cytoplasm of sporangia that have migrated into a vesicle. *Phytophthora* species also have indeterminate sporangia, but many species have very differentiated sporangiophores or sporangia (Figure 20.8). For example, *P. infestans* produces sympodially branched sporangiophores, which have swollen nodes, and produce lemon-shaped, papillate sporangia. Some *Phytophthora* species produce haustoria unlike closely related *Pythium* species.

Phytophthora infestans

Phytophthora infestans causes the late blight of potato (and tomato). *Phytophthora infestans* is probably the best-known member of the genus because of its economic impact on agriculture and history. This organism is heterothallic and requires two mating types for sexual reproduction by gametangia contact. Sexual reproduction is not necessary for the survival of the organism as the mycelium may survive in infected tubers. The disease can be initiated when infected tubers are planted or when volunteer potatoes sprout in the spring. Under favorable environmental conditions (cool and moist), the pathogen grows into the shoot and after sufficient time, a matter of days sometimes, sporangiophores emerge through stomata on the bottom surface of leaves and sporangia are produced. Sporangia are dispersed by air currents and rain to other plants and the soil surface. Several (typically less than ten) biflagellated zoospores form and are liberated from each sporangium. Zoospores on leaves encyst and germinate via germ tubes, which form **appressoria**. An infection peg penetrates develops from the appressorium and penetrates the leaf surface either directly or indirectly through stomata. Sporangia that are in soil also produce zoospores, which may infect existing or developing tubers. Under favorable conditions for disease development, many "generations" or cycles of sporangia may be produced to infect additional plants — late blight is a **polycyclic** disease (Chapter 2 and Chapter 30) and may quickly devastate entire fields. As the pathogen grows and sporulates, host tissue is destroyed creating lesions that coalesce. Often the entire shoot portion of the plant is destroyed and yield and tuber quality are dramatically reduced. Infected tubers are small with sunken lesions, stained purple or brown, and may have an offensive odor due to secondary bacterial and fungal invaders.

Control of late blight of potatoes is accomplished by several tactics. Seed potatoes should be free of the pathogen, volunteer plants should be rouged from fields, and infected debris from any previous potato crop should be destroyed (sanitation; Chapter 32). Disease-resistant varieties should also be used when possible, although resistance may be broken by environmental conditions favorable to pathogen development or when new strains of the pathogen develop. Contact fungicides may be applied at prescribed intervals according to disease development models based on temperature and moisture (Chapter 34). Acylalanine fungicides have been extensively used to protect crops.

Control of other diseases caused by *Phytophthora* species is similar to control tactics used for *Pythium* species. Well-drained soils and careful management of soil moisture will help minimize damage to many crops by these pathogens. For greenhouse crops, growing media and containers should be decontaminated to ensure pathogen propagules are killed. Pine bark medium has been used to control disease in container nursery crops and a number of *Phytophthora* resistant varieties of crop species (fruits, etc.) are available. Systemic fungicides such as acylanaline fungicides and Fosetyl-al are commonly used to protect various crops from infection.

PERONOSPORACEAE — THE DOWNY MILDEWS

The **downy mildews** belong to the Peronosporaceae and are obligate, biotrophic parasites of aerial portions of flowering plants. There are seven genera in the family including *Basidiophora, Bremia, Bremiella, Peronospora, Plasmopara, Pseudoperonospora,* and *Sclerospora.* The genera are distinguished by the highly differentiated branching patterns of sporangiophores that always exhibit determinate growth (Figure 20.9). Sporangiophores emerge through stomata on lower leaf surfaces and sporangiophore development is completed before any sporangia are formed. Formation of sporangia on individual sporangiophores is synchronous (Figure 20.10) and mature sporangia are disseminated by air currents. The sporangia of many species germinate by producing biflagellated zoospores characteristic of species in the Oomycota. However, the sporangia of *Pseudoperonospora, Peronospora,* and some species of *Bremia* germinate via germ tubes and for this reason have been mistakenly termed conidia by some authors. Regardless of the mode of sporangial germination, these propagules serve to infect either healthy plants or healthy tissue remaining on an infected plant. Leaf surfaces are penetrated either directly through the epidermal wall or through stomata. Intercellular hyphae grow throughout the mesophyll and establish haustoria within host cells (Figure 20.11A and Figure 20.11B). The infection may be localized in a leaf (see Plate 7A following page 80) or become systemic in the plant. When environmental conditions are conducive, usually cool temperatures with high humidity and darkness, sporangiophores and sporangia are produced and start the secondary disease cycle. Many "crops" of sporangia can be produced during growing season and thus diseases caused by these organisms can be considered polycyclic. Sexual reproduction is by gametangial contact and ornamented oospores are usually produced within host tissue and serve as overwintering structures (Figure 20.2). Oospores usually germinate directly via germ tubes, but sometimes germinate indirectly by zoospores.

Members of the Peronsporaceae infect and cause diseases on many plants (Table 20.2). Probably the most well-known disease in this group is downy mildew of grapes caused by *Plasmopara vitacola.* Briefly, a root aphid native to North America was introduced to France in the mid-1860s with devastating consequences for the wine industry. In an attempt to control the aphid, the wine producers imported American grape rootstock, which had good tolerance to the insects. Unfortunately, the plants were infected with *P. vitacola* to which the French grapes varieties were extremely susceptible. The disease threatened the continued existence of the industry. In 1882, Alexis Millardet noticed that some vines sprayed with an

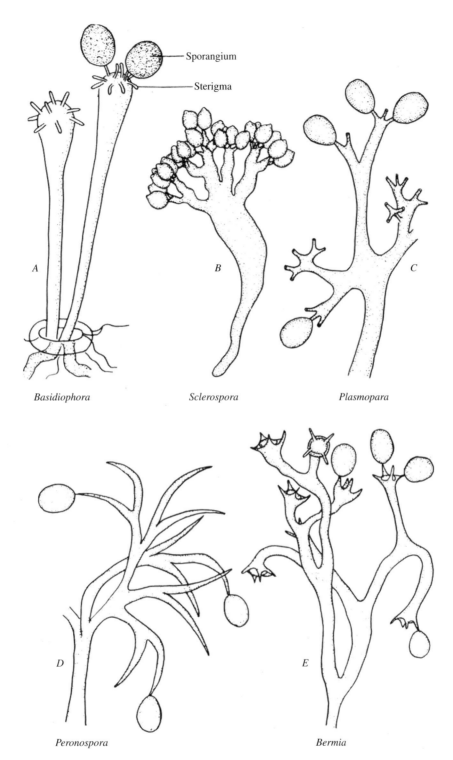

Basidiophora   Sclerospora   Plasmopara

Peronospora   Bermia

FIGURE 20.9 Sporangiophores characteristic of five genera of Peronosporaceae. (From Alexopoulos, C.J., C.W. Mims and M. Blackwell, 1996, *Introductory Mycology*, 4th ed., John Wiley & Sons, New York, 868 pp. With permission.)

unknown substance did not have the disease. As it turns out, the owner of the vines had applied a mixture of copper sulfate and lime to discourage passers-by from eating his

crop. Millardet, being a professor at the university, seized upon his observation and developed the first fungicide — Bordeaux mixture (Chapter 33). Thus, the French wine

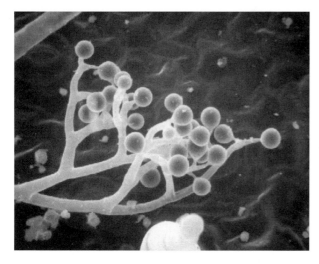

FIGURE 20.10 Synchronous development of sporangia on a sporangiophore of *Peronospora tabacina.* (From Trigiano, R.N., C.G. Van Dyke, H.W. Spurr, Jr. and C.E. Main, 1985, *Tob. Sci.* 29: 116–121. With permission.)

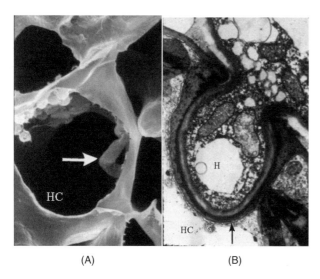

(A)   (B)

FIGURE 20.11 Haustoria of *Peronospora tabacina* in tobacco. (A) Scanning electron micrograph of a haustorium (arrow) in a host cell (HC). (B) Transmission electron micrograph of a haustorium (H) in a host cell (HC). Notice that the plasmalemma (cell membrane: arrow) is intact. (From Trigiano, R.N., C.G. Van Dyke and H.W. Spurr, Jr., 1983, Haustorial development of *Peronospora tabacina* infecting *Nicotiana tabacum*, *Can. J. Bot.* 61: 3444–3453. With permission.)

industry had solutions to the aphid and downy mildew problems.

Some varieties of some crops are resistant to downy mildews, but contact fungicides are still used to control diseases. More recently, systemic fungicides, including acylalanine compounds and Fosetyl-al, which may or may

TABLE 20.2
Some Common Downy Mildew Diseases

Species	Crop
Bremia lactucae	Lettuce
Peronospora tabacina (syn. *hyoscyami*)	Tobacco (blue mold)
P. destructor	Onion
P. antirrhini	Snapdragon
Peronosclerospora sorghi	Sorghum
Plasmopara vitacola	Grape
P. halstedii	Sunflower
Pseudoperonospora cubensis	Many cucurbit species
Sclerophthora macrospora	Cereal and grass species

FIGURE 20.12 *Albugo ipomoeae-pandurancae* infecting morning glory. (Left) Top of leaf showing chlorotic tissue. (Right) Bottom surface of leaf with white crusty pustules containing sporangia. (Courtesy of A.S. Windham, University of Tennessee.)

not be combined with contact fungicides, have been used successfully to control downy mildew diseases. Fungicides may be applied at prescribed intervals according to disease development models based on temperature and moisture (Chapter 34) for blue mold of tobacco (*Peronospora tabacina*) and diseases of other crops.

ALBUGINACEAE — THE WHITE RUSTS

The **white rusts** are obligate parasites on flowering plants and all belong to the genus *Albugo*. Species are differentiated by host and oospore ornamentation. *Albugo candida* can be a serious pathogen of crucifers. Another common species is *A. ipomoeae-panduranae* and is found growing in sweet potatoes and, more frequently, in wild morning glory (Figure 20.12).

The life history of *Albugo* species is similar to that exhibited by many other oomycetes except in the production of sporangiophores and sporangia. The club-shaped, indeterminate sporangiophores are contained within the host tissue and each produces several sporangia under the epidermis. The pressure exerted by the production of sporangia and mycelium against the host epidermis distorts the shape of the sporangia from globose to box- or cube-like. The pressure also ruptures the epidermis, liberating the sporangia and forming a white crust. Old pustules on morning glory may appear orange to pink. Sporangia are disseminated by wind and rain and germinate via zoospores that have typical oomycete morphology. Zoospores encyst, form a germ tubes, and initiate infections on a suitable host plants. Oospores are formed by gametangial contact and serve as resting spores.

The plant pathogens classified in the Oomycota are very interesting and form a distinct group of organisms. They cause a number of economically important diseases and have influenced the history of humans. There is no doubt that with additional research we will better understand their taxonomic position and learn more about how to control the diseases they cause.

REFERENCES

Alexopoulos, C.J., C.W. Mims and M. Blackwell. 1996. *Introductory Mycology.* 4th ed., John Wiley & Sons, New York, 868 pp.

Hawksworth, D.L., P.M. Kirk, B.C. Sutton and D.N. Pegler. 1995. *Ainsworth & Bisby's Dictionary of the Fungi*, 8th ed. CAB International, Wallingford, UK, 616 pp.

Trigiano, R.N., C.G. Van Dyke and H.W. Spurr, Jr. 1983. Haustorial development of *Peronospora tabacina* infecting *Nicotiana tabacum. Can. J. Bot.* 61: 3444–3453.

Trigiano, R.N., C.G. Van Dyke, H.W. Spurr, Jr. and C.E. Main. 1985. Ultrastructure of sporangiophore and sporangium ontogeny of *Peronospora tabacina. Tob. Sci.* 29: 116–121.

SUGGESTED READING

Agrios, G.N. 1997. *Plant Pathology*, 4th ed. Academic Press, New York, 635 pp.

21 Laboratory Exercises with the Oomycetes

Robert N. Trigiano, Richard E. Baird, and Steven N. Jeffers

The water molds or oomycetes occur worldwide and can be found in diverse ecosystems including estuaries, lakes, oceans, rivers, and streams. Taxa within the group, however, can also occur in agricultural fields and in forest habitats growing as saprophytes or parasites. Because of the diversity of oomycetes and their occurrence in many different environments, several representative experiments have been included in this chapter to enable students to learn more about methods to study this unique group.

Many fascinating and intriguing experiments can be conducted with the members of the oomycetes, including those that are not plant pathogens; however, we limit consideration to some of the more common species that cause plant diseases such as *Pythium, Phytophthora, Peronospora,* and *Albugo.* Soil-inhabiting species of *Pythium* and *Phytophthora* are easy to grow and manipulate in axenic cultures, whereas species of *Peronospora* and *Albugo* are obligate parasites that require a living host to complete their life cycles. The obligate pathogens are difficult to work with in most experimental systems. Although these obligate pathogens are very challenging to maintain, some educational and research exercises can be completed with relative ease. As has been emphasized throughout this book, please secure the proper permits to obtain and transport pathogens and then destroy all experimental materials by autoclaving or by other means when the experiments are completed.

The following experiments are designed to provide hands-on experiences for students working with *Pythium* species that cause root rots and damping-off of peas and beans, isolation of *Phytophthora* species from diseased plant tissues and directly from soil, formation of sporangia and oospore by *Pythium* and *Phytophthora* in culture, *in vitro* coculture of *Peronospora* species, and microscopic observation of *Peronospora* and *Albugo* species in host materials. The last two experiments are designed as special topics for advanced undergraduate and graduate students.

EXERCISES

EXPERIMENT 1. ROOT ROT OF BEAN AND PEAS CAUSED BY *PYTHIUM* SPECIES

A number of *Pythium* species cause root and seed rots of various crops. These diseases can devastate both field-grown and greenhouse crops. This experiment is designed to demonstrate symptoms and signs of the diseases. For an alternative exercise that uses a floricultural crop, see Procedure 32.1.

Materials

Each student or team of students will require the following items:

- Culture of *Pythium ultimum* or *P. aphanidermatum.* May obtain cultures from ATCC with permit.
- Untreated (no fungicides) seeds of any cultivar(s) of common edible beans (*Phaseolus vulgaris*) and peas (*Pisum sativum*)
- Six 10-cm diameter plastic pots, pot labels, and permanent marker
- Sand for plastic pots and pencil or large glass rod for making holes in sand
- Two plastic flats with Promix® or other soilless medium and paper towels
- Laboratory blender, scissors, and long transfer forceps
- 250-ml flask containing 125 ml of sterile cornmeal (CM) broth cornmeal broth (Difco, Detroit, MI) and two 1000-ml beakers
- Four 10-cm diameter petri dishes containing corn meal (CM) agar (add 15 g agar to above formulation)
- Compound microscope and microscope slides

Follow the protocol listed in Procedure 21.1 to complete this experiment.

Procedure 21.1 Root and Seed Rots Caused by *Pythium* species

1. Seven to ten days before the laboratory, plant seeds in flats filled with soilless medium. Grow in a cool greenhouse or laboratory. Each team of students will require at least 12 germinated seeds of each bean and pea. Space seeds so that the roots of individual plants do not grow together and seedlings can be easily separated.

2. One week before the laboratory, inoculate the sterile CM broth in 250-ml flasks with several plugs from the margin of 5-day old colonies of either *P. ultimum* or *P. aphanidermatum* growing on CM agar dishes. Incubate the liquid cultures at 18°C to 22°C either on a slow (30 rpm) shaker or on a shelf in an incubator or laboratory. Prepare an equal number of uninoculated flasks (medium without *Pythium* species).

3. Autoclave the sand the day before the experiment. This step is not absolutely essential for the success of the exercise.

4. Gently remove the seedlings from the flats and wash the particles of soilless medium from the roots using tap water. Store plants with roots wrapped in moist paper towels on the laboratory bench.

5. Swirl the liquid culture to dislodge hyphal growth from the glass and empty the contents of the flasks into the blender. Add 375 ml of sterile distilled water and homogenize the mixture with short bursts (high speed) of the blender. Pour the suspension into a 1000-ml beaker.

6. Dip the roots of three bean plants into the *Pythium* species suspension. Make three large holes in the sand in each plastic pot using a pencil or glass rod. Be sure that sand is moist. Very gently plant the beans with as little damage to the roots as possible. Label each pot: *Pythium* intact roots. Repeat the procedure with pea seedlings and label.

7. Trim about 25% of the root length from another group of three bean plants. Dip the remaining roots in the suspension and plant as in Step 6. Label the pots: *Pythium* cut roots. Repeat the procedure with pea seedlings and label.

8. Plant three bean seeds about 2-cm (0.5- to 1.0-inch) deep in a pot. Pour about half of the remaining *Pythium* species homogenate onto the surface of the sand in the pot. Label pot: *Pythium* bean seeds. Repeat the procedure with pea seeds and label.

9. Repeat Step 6 to Step 8 by using the contents of an uninoculated flask mixed with 375 ml of sterile water for the dip and drench treatments. Label pots according to the treatment.

10. Set the plants on a laboratory bench near a window and observe for symptom development. Water with distilled water and do not allow the sand to dry out.

Anticipated Results

Depending on the temperature at which the inoculated plants are grown and the amount of inoculum applied, symptoms of damping-off should be evident between three and seven days with both bean and pea plants. The inoculated plants should appear wilted at first, and then necrotic and water-soaked lesions will occur on the stem at the soil line. Infected plants will not be able to maintain stature and will fall over. Those plants whose roots were cut will typically display symptoms one or two days earlier than those with uncut roots. Roots of infected plants should appear dark and very soft compared to white and firm roots from uninoculated plants. Seeds treated with *Pythium* will either fail to germinate and rot or germinate poorly with the seedling succumbing to the disease very quickly. All plants that were not treated with *Pythium*, including those with cut roots, should grow normally unless a contaminating pathogenic *Pythium* species is present in the potting medium. Uninoculated seeds should germinate and grow normally.

Questions

- What are the controls in this experiment and why are they necessary?
- How would Koch's postulates be completed for this experiment?

EXPERIMENT 2. ISOLATION OF *PHYTOPHTHORA* SPECIES FROM PLANT TISSUES AND SOIL

Many members of the Pythiaceae, including *Pythium* and *Phytophthora* species, are widely distributed in soils. They may survive for long periods without a host. This exercise is designed to provide experience in isolating *Phytophthora* species from diseased plants and infested soil by using a selective medium.

Materials

Each student or group of students will need the following items:

- Roots, stems, or leaves from diseased plants (rhododendron, azalea, soybean, tomato, pepper, and tobacco work well)
- Soil sampling tool (2.5-cm diameter works well)
- Ice chest
- Paper towels and plastic bags
- Scalpel with #10 blade (*Caution*: Very sharp)

TABLE 21.1
Growth/Isolation Media for *Phytophthora* Species

Medium	Quantity
V8 Agar (V8A) Growth and Sporulation Medium	
V8 juice[a]	200 ml
$CaCO_3$	2 g
Distilled water	800 ml
Agar	15 g
PARP(H) Isolation Medium	
Delvocid (50% pimaricin): P	10 mg
Sodium ampicillin: A	250 mg
Rifamycin-SV (sodium salt): R	10 mg
75% PCNB (Terraclor): P	67 mg
Hymexazol[b]: H	50 mg
Clarified V8 juice[c]	50 ml
Agar[c]	15 g
Distilled water	950 ml

Note: All antimicrobial amendments should be added after the base agar medium has been autoclaved and cooled to 50°C to 55°C (autoclaved vessel can be held in hand without much discomfort).

[a] Clarified V8A (cV8A) can be made by stirring 200 ml V8 juice with 2 g $CaCO_3$ for 15 min and then centrifuging for 10 min at 4000 *g*. Supernatant may be stored frozen at −20°C and use 100 ml with 900 ml water and 15 g agar to make cV8A.

[b] Inclusion is optional: hymexazol inhibits *Pythium* species.

[c] Corn meal agar (CMA) may be substituted for clarified V8 juice and agar.

- Twenty plastic 10-cm diameter petri dishes containing PARPH medium (Jeffers and Martin, 1986; Ferguson and Jeffers, 1999; Table 21.1)
- Incubator set at 20°C without light
- Sieves with 4-mm and 2-mm openings
- Several aliquots of 100 ml of 0.3% water agar (3 g agar in 1 l of water) in 250-ml beakers
- Magnetic stirrer and stir bars
- 1-ml wide-bore pipette and bulb or pump
- Top-loading balance and plastic weigh boats
- Compound microscope, glass slides, and coverslips

Follow the protocol outlined in Procedure 21.2 to complete this exercise.

Procedure 21.2 Isolation of *Phytophthora* Species from Infected Plant Tissue and Infested Soil

- Isolation of *Phytophthora* species from infected plant tissues
  1. Place roots and stem segments from diseased plants into plastic bags and keep moist by damp towels. Healthy plants should also be sampled.

Transport the samples to the laboratory in a cool ice chest. Keep samples in the dark.
  2. In the laboratory, gently wash the tissues under running tap water for 5 to 10 min and blot excess moisture with a paper towel.
  3. Cut samples into 1-cm segments and place four segments on each of five PARPH petri dishes. Push each piece into the agar so that the tissues are surrounded by agar. Incubate the cultures at 20°C in the dark.
  4. Observe colony growth on PARPH agar dishes after 48 to 72 h. Continue to examine dishes for up to one week. Transfer small pieces of mycelium from individual whitish-colored *Phytophthora* colonies to fresh dishes containing PARPH medium and incubate at 20°C in the dark.
- Isolation of *Phytophthora* species from soil
  1. Collect soil cores up to 20-cm deep with a sampling tool approximately 2-cm in diameter. Collect 10 core samples within 20 cm of target symptomatic plants and 10 samples of noninfested soil from around healthy appearing plants of the same type.
  2. Place each set of 10 soil cores into separate and individually labeled plastic bags to make one infested and one noninfested composite soil

sample. Return the composite soil samples to the laboratory in a cool ice chest. All samples should be stored in a dark, cool place until the assay begins.

3. Prepare each composite soil for isolation by breaking up the clods. Remove rocks and plant debris by first using a coarse (4 mm) screen followed by a smaller screen (2 mm). Throughly mix the soil and return to plastic bags.

4. Add 50 ml of the infested soil to 100 ml 0.3% water agar contained in a 400-ml beaker with a stir bar. Place on a magnetic stirrer at high speed for about 3 min. Pipette 1-ml aliquots of the suspension onto PARP(H) dishes. Evenly spread the suspension across the surface of the agar. Use up to 5 dishes per composite soil sample Repeat for the noninfested soil.

5. Incubate the dishes at 20°C in the dark for 48 to 72 h. Do not enclose dishes in plastic bags or boxes. Wash soil from dishes under running tap water. Examine dishes with a dissecting microscope (30× to 50×) and count colonies. Subculture from these colonies to fresh PARP(H) medium if desired. Save subcultured colonies for later identification. Compare number of colonies from the different locations and soil types.

Anticipated Results

Colonies of *Phytophthora* species should develop on PARPH from both soil and diseased tissues between 24 and 72 h. Colonies of *Pythium* species may also develop if hymexazol is omitted from the isolation medium (PARP) or if there are hymexazol-tolerant *Pythium* species present in the samples. Soil from sources without infected plants may or may not contain *Phytophthora* and *Pythium* species. *Phytophthora* colonies should not develop on PARPH from healthy plant tissues.

Questions

- What morphological characteristics are used to recognize and identify species of *Pythium* and *Phytophthora*?
- What characteristics are used to distinguish *Pythium* and *Phytophthora* species from each other?
- Can a plant be infected with *Phytophthora* and *Pythium* at the same time? Can soil be infested with both genera?

EXPERIMENT 3. PRODUCTION OF SPORANGIA AND OOSPORES BY *PHYTOPHTHORA* AND *PYTHIUM* SPECIES

Many species of *Phytophthora* and *Pythium* form sporangia and oospores in culture if provided the proper environmental conditions. These simple experiments are designed to allow students to observe asexual and sexual reproduction in the Pythiaceae.

Materials

Each team of students or class will require the following materials:

- Agar (CM) cultures of several homothallic species of *Pythium* and *Phytophthora* (e.g., *P. cactorum* or *P. citricola*)
- Freshly gathered grass (tall fescue, blue grass, etc.) clippings autoclaved for 20 min on two successive days
- Sterile distilled water. [To enhance sexual reproduction in *Pythium* cultures, a drop of chloroform containing cholesterol (1 mg in 10 ml chloroform) may be added to 8 ml of distilled water. Allow the chloroform to evaporate under a transfer hood before adding grass or *Pythium*.]
- Sharp scissors to cut grass
- Fine-tipped forceps
- Several 60-mm diameter plastic petri dishes
- V8A and clarified V8A (Table 21.1)
- Centrifuge and Whatman #1 filter paper
- Incubator or laboratory bench equipped with fluorescent lights
- Sterile plastic drinking straws
- Soil
- Nonsterile soil extract solution (NS-SES). Stir 15 g of soil in 1 l of distilled water for at least 4 h and allow the suspension to settle overnight. Decant the water and centrifuge for 10 min at 4000 g followed by filtration through Whatman #1 filter paper (Jeffers and Aldwinckle, 1987). Store in the refrigerator
- Dissecting and compound microscopes
- Microscope slides and coverslips
- Lactoglycerol solution (1:1 lactic acid:glycerol v:v) with 0.1% acid fuchsin
- Brightly-colored nail polish

Follow the protocols in Procedure 21.3 to complete this experiment.

Procedure 21.3 Production of Sporangia and Oospores by *Pythium* and *Phytophthora* Species

1. Grow cultures of *Pythium* species on CM agar in 60-mm diameter plastic petri dishes at 20°C in the dark for three days.

2. Add 8 ml of sterile distilled water to a number of empty 60-mm diameter petri dishes. Separate and add 5 to 10 autoclaved blades of grass to the water. Remove several plugs of mycelium from the margin of the *Pythium* species colony with a sterile plastic straw and cut into quarters with a sharp scalpel. Transfer four quarters to the petri dish with the blades of grass. Be sure the agar pieces are in contact with the grass. Incubate at 20°C in the light.

3. Cultures may be observed weekly with either a 40× dissecting scope or individual pieces of grass from the culture may be mounted in lactoglycerol (with or without acid fuschin) on microscope slides and viewed with a compound microscope. The slide may be made semipermanent by painting the clean, dry edges of the coverslip with nail polish. Draw and label all structures. To stimulate zoospore release from sporangia of *Pythium* species, chill (2°C to 4°C) the grass cultures for a few hours and then allow them to warm to room temperature. Observe sporangia, vesicle formation, and zoospores with the aid of a dissecting microscope.

4. Transfer *Phytophthora* species cultures to V8A and incubate in the dark at 20°C to 25°C for 48 to 72 h. *Note*: Colonies should be at least 2 cm in diameter. Aseptically remove agar plugs (2 mm) near the margin of the colony with a sterile drinking straw and place five of them into a sterile, empty 60-mm diameter petri dish.

5. Cover the agar plugs with 7 to 10 ml of NS-SES. Place dishes under continuous flourescent lights at 20°C to 25°C. Observe plugs with a dissecting microscope after 12 to 24 h for sporangia, or, if not present, after an additional 24 h. Draw and label all structures.

6. To initiate oospore formation in cultures of *Phytophthora* and *Pythium* species, transfer the organisms to 60-mm diameter petri dishes containing cV8A and incubate at 20°C to 25°C in the dark. Microscopically examine cultures weekly for up to six weeks. Draw and label all structures. This procedures works well for homothallic species of both *Pythium* and *Phytophthora*. Oospores are usually present after 2 weeks and are mature by 4 weeks.

Anticipated Results

Sporangia should form very quickly in all *Pythium* and *Phytophthora* cultures, and in 24 to 48 h for many *Pythium* and *Phytophthora* species in NS-SES. Gametangia and oospores usually take 2 to 4 weeks to form in the cultures.

Questions/Activities

- Why where homothallic species of *Phytophthora* and *Pythium* used?
- Are all species of *Phytophthora* and *Pythium* homothallic?
- What is the taxonomic significance of the origin (same hypha, below oogonium, etc.) of antheridia involved in sexual reproduction?
- What purpose do oospores serve in the life cycle of these organisms?
- Assign a ploidy level to each of the structures seen in the cultures.

Experiment 4. Coculture of *Peronospora Tabacina* and Tobacco Callus

Obligate pathogens, such as *P. tabacina* (syn. *P. hyoscyami* or blue mold), require a living host to grow and reproduce. This is a very specialized and complex relationship influenced strongly by environmental conditions. It is possible to establish the pathogen and host together in culture (see Chapter 38). In vitro cocultures allow active maintenance of isolates of pathogens; studies of physical, physiological, and molecular interactions between pathogen and host; and assessment of resistance or susceptibility of host cultivars to the pathogen. Coculture of *P. tabacina* and tobacco was first report by Izard et al. (1964) and then by Trigiano et al. (1984). In a very interesting recent report, Heist et al. (2001) devised a technique for culturing infected tobacco leaf tissue directly and maintaining the association of the organisms. Because we are more familiar and experienced with the 1984 study, this experiment will be based largely on modifications of the protocols found in Trigiano et al. (1984) and are intended for advanced undergraduate and graduate students. If you are located in a tobacco-growing area of the country, we suggest that this experiment only be attempted when tobacco is not being grown to prevent accidental escape of the pathogen.

General Considerations

Tobacco Cultures

Callus cultures can be established from almost any blue mold-susceptible tobacco cultivar. Remove the leaves from the stems of 7 to 9 leaf-stage plants, wash the stems in soapy water, and rinse well in tap water. Stem tissue can be surface sterilized in 20% commercial bleach for 10 min and then rinsed three times in sterile distilled water for 2 min each. Remove and discard the cut ends of the stem damaged by the bleach and split the remaining stem lengthwise. Excise the pith as small (about 1 cm³) pieces and place on Murashige and Skoog medium (1962) modified to contain 2 mg of indolacetic acid (IAA), 1 mg benzylaminopurine (BAP), 30 g sucrose, MS salts (Sigma, St. Louis, MO), 8 g of purified agar, in 1 l of distilled water. The pH of the medium should be adjusted to about 5.7 before sterilization by autoclaving. Medium should be poured into 10-cm diameter dishes and cultures incubated in a 16 h light/8 h dark photoperiod at 25°C. Callus should form on the excised pith pieces after two weeks and the entire culture may be subdivided and cultured on the same medium every three to four weeks.

Diseased Plants

Obtain sporangia of *P. tabacina* from the ATCC (permit required) or a friendly land grant university plant pathologist (permit required if out of your state). Establish the disease by spraying five to seven leaf-stage tobacco plants with a suspension of sporangia (2×10^4/ml works well). Place the plants in plastic bags overnight and then remove the next morning. Plants can be placed in a greenhouse or on a laboratory bench. Water plants by placing water directly on the soil or potting medium only; do not get the leaves wet. Chlorotic spots, which indicate the establishment of the pathogen, should be evident 7 to 10 days after inoculation. To induce sporulation, water soil generously and place pot and plant in a plastic bag overnight or for a minimum of 12 h. Sporangiophores and sporangia should be evident in the morning on the undersides of leaves (blue-white "fuzz" in chlorotic areas). Sporangia can be collected in distilled water and sprayed on new plants to restart the infection cycle.

There is an alternative procedure for producing sporangia to use for inoculation of callus cultures. At the first sign of leaf chlorosis (infection), lightly swab the lower surface of the area with 70% ethanol; do not soak the area. Repeat this on three successive days. Do not use alcohol the day before sporulation is induced as described previously. Swabbing with alcohol and maintaining dry leaves dramatically reduces contamination of the cocultures.

Materials

- Autoclaved glass microscope slides and filter paper (9-cm diameter, any grade will work)
- Sterile plastic 10-cm diameter petri dishes
- 100 ml of 1% (10 g agar in 1 l) water agar melted and maintained at 55°C in a water bath
- Flow hood and alcohol lamp
- Scalpel with #10 blade and long forceps

- Sterile 250-ml beaker containing sterile MS medium as described previously
- Tissue cultures of tobacco (see under general considerations)
- Tobacco plants infected with *P. tabacina* (see under general considerations)
- Dissecting microscope and compound microscope equipped with epifluoresence — 330- to 380-nm excitation and 420-nm absorption filter
- 0.05% (w/v) aniline blue in 0.03 *M* phosphate (K_3PO_4) buffer (Martin, 1959)
- 100 ml of 0.05 M potassium phosphate buffer, pH = 6.8 to 7.2
- Sterile distilled water

Follow the protocols listed in Procedure 21.4 to complete this experiment.

Procedure 21.4 Coculture of *Peronospora tabacina* and Tobacco Callus

1. Do this operation in a laminar flow hood and 3 to 5 days before sporangia are needed. Aseptically place two pieces of filter paper into a sterile 10-cm diameter plastic petri dish and moisten them with sterile, distilled water. Drain excess water from the dish. Dip a sterile glass microscope slide into sterile MS medium — about 75% of the slide should be immersed in the gel. Withdraw the slide and hold parallel to the surface of the flow hood and allow agar to solidify. Place slides on top of the moist filter papers in the petri dishes (Hock, 1974). Repeat the procedure.

2. Transfer some friable (breaks into small pieces or individual cells) callus to the agar on the microscope slides. A thin layer of cells or aggregates over the entire surface is best. Seal the dishes with parafilm and incubate as with petri dish cultures of callus. Transfer callus cultures to be used in experiments to fresh MS medium. Try to place the callus on the agar as nearly in the center of the dish as possible.

3. Do this operation on a clean laboratory bench. Melt 1% water agar and cool to about 55°C in a water bath. Place moist filter paper in the bottom portion of several petri dishes and close the dishes. Perform the next two steps quickly and carefully.

4. Excise several 4-cm² (~1 sq. inch) pieces of leaf tissue with sporangiophores and sporangia and place with the top of the leaf in contact with the filter paper. Close the lid to maintain high relative humidity.

5. Place a drop of 1% molten agar in the center of another dish top and quickly affix an infected leaf piece from Step 4 by slightly immersing the top side of the leaf into the agar. Remove the lid of a callus culture or a microscope slide culture and replace with a lid to which the infected leaf tissue is attached. Do not seal with parafilm and place the dish in the running flow hood. After 4 to 6 h, remove the lids with attached leaf pieces and replace with new, sterile lids. Incubate cultures in the dark at 25°C. An alternative method for inoculating the tobacco cultures is to transfer the leaf segments to the petri dish lids before sporangiophores and sporangia form. Seal the dishes with parafilm and incubate the cultures in the dark overnight. By the next morning sporulation will be evident. Remove the parafilm wrapping and place in a running flow hood. This method achieves higher inoculum levels, but generally results in more contaminated cultures.

6. Microscope slide cultures can be examined after 48 to 96 h directly with a compound microscope (10× and 20× objectives only). Ungerminated and germinated sporangia should be present as well as very long germ tubes and hyphae. Careful examination will reveal appressoria, haustoria, and intracellular hyphae.

7. Cells can be removed from the slides and suspended in either phosphate buffer (Figure 21.1) or aniline blue fluorescent stains. Higher magnification objectives with microscopes equipped with differential interference contrast (Figure 21.2) and epifluorescent (Figure 21.3) microscopy can be used to examine the host–pathogen interaction. *Caution*: Do not look directly at the UV light source. Cells can also be suspended in fixatives for light and electron microscopy (Trigiano et al., 1984).

8. Periodically observe the inoculated callus on agar culture for 2 weeks. Discard any obviously contaminated cultures. Blue mold hyphae are hyaline (glass-like, without color) and may be scanty or abundant on the surface of the calli. Sporulation will not occur in cultures incubated in the dark.

9. Incubate inoculated callus cultures at 25°C with a 14/10 h light/dark photoperiod. A few sporangiophores with sporangia should develop.

Anticipated Results

This procedure should result in coculture of *P. tabacina* and tobacco callus. Observe sporangia, germ tubes, inter-

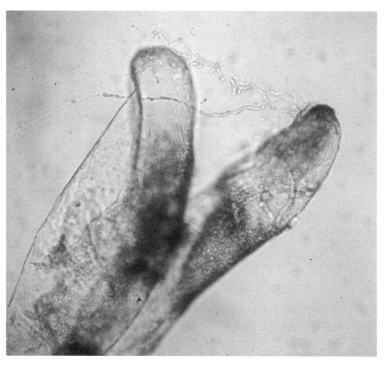

FIGURE 21.1 Whole mount of host cells (HC) with external hyphae. Note fungal growth in the host cells. (From Trigiano, R.N., C.G. van Dyke, H.W. Spurr, Jr. and D.J. Gray, 1984, *Phytopathology* 74: 280–285. Taken from American Phytopathological Society. With permission.)

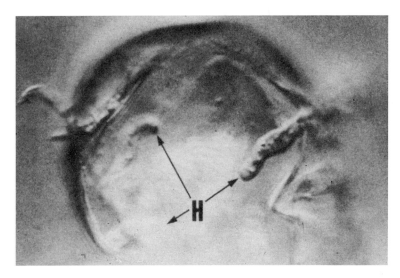

FIGURE 21.2 Differential interference contrast micrograph of a whole mount of three haustoria (H) in a host cell. (From Trigiano, R.N., C.G. van Dyke, H.W. Spurr, Jr. and D.J. Gray, 1984, *Phytopathology* 74: 280–285. Taken from American Phytopathological Society. With permission.)

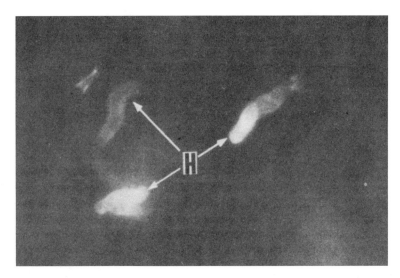

FIGURE 21.3 Fluorescence microscopy of the same haustoria (H) shown in Figure 21.2 stained with aniline blue and viewed with UV light. (From Trigiano, R.N., C.G. van Dyke, H.W. Spurr, Jr. and D.J. Gray, 1984, *Phytopathology* 74: 280–285. Taken from American Phytopathological Society. With permission.)

and intracellular hyphae, appressoria, and haustoria after 48 to 96 h. Details of the expected results are provided as comments in Procedure 21.4.

Questions

- Is the physical and physiological relationship between the host and pathogen similar in culture and intact leaf tissue?
- If *P. tabacina* sporangia typically produce very short germ tubes on leaf surfaces, why do germ tubes (hyphae) grow for long distances in callus culture?

- What are the benefits to establishing cocultures of pathogens with their hosts?
- Why do haustoria "glow" when stained with analine blue and viewed with UV light?

EXPERIMENT 5. OBSERVING MORNING GLORY LEAF TISSUE INFECTED WITH *ALBUGO CANDIDA*

White rust of morning glory is very common. Infected morning glory leaves have chlorotic halos on the upper surface and white pustules on the lower surfaces of the leaves (Figure 20.12). White pustules may appear pink and very crusty in the fall.

Materials

Each student will require the following items:

- Morning glory leaves infected with *A. candida*
- Compound microscope, glass microscope slides, and coverslips
- Water in a dropper bottle
- Razor blades (*Caution*: Very sharp)
- 0.1% calcoflor in water
- Epifluorescence microscope equipped with 395- to 420-nm excitation filter and 470-nm absorption filter

Follow the protocol in Procedure 21.5 to complete this experiment.

Procedure 21.5 Morphological Features of *Albugo candida* in Infected Morning Glory Plants

1. Observe pustules on morning glory leaves using a dissecting microscope (Figure 20.12).
2. With a razor blade, cut cross sections (as thin strips as possible) from morning glory leaves infected with *A. candida* and mount in water on a microscope slide. Try to include a portion of a pustule in the section. Draw sporangia and pustule morphology.
3. Cut other thin sections of leaf tissue and pustules and mount in calcoflor solution. View these sections by epifluorescence and draw all pathogen and host structures. *Caution*: Do not look directly at the UV light source.

Anticipated Results

Sectioned mounts of pustules should reveal many square- or rectangular-shaped sporangia. Knob-like haustoria may also be seen in host cells. Compare with a prepared slide if these structures cannot be observed in the fresh sections. If the tissue is observed with calcoflor, hyphae, sporangia, and haustoria should "glow" white. Plant cell walls will also "glow."

Questions

- Why is it relatively difficult to observe sporangiophores of *Albugo* compared to those of *Peronospora*?
- Why are *Albugo* sporangia box-shaped?
- Why does calcoflor stain *Albugo* structures? Would it stain structures of most oomycetes?

REFERENCES

Ferguson, A.J. and S.N. Jeffers. 1999. Detecting multiple species of *Phytophthora* in container mixes from ornamental crop nurseries. *Plant Dis.* 83: 1129–1136.

Heist, E.P., W.C. Nesmith and C.L. Schardl. 2001. Cocultures of *Peronospora tabacina* and *Nicotiana* species to study host-pathogen interactions. *Phytopathology* 91: 1224–1230.

Hock, H.S. 1974. Preparation of fungal hyphae grown on agar-coated microscope slides for electron microscopy. *Stain Technol.* 49: 318–320.

Izard, C., J. Lacharpagne and P. Schiltz. 1964. Comportement de *Peronospora tabacina* dans les cultures de tissus et le role de l'épiderme foliare. *SEITA (Serv. Exploit. Ind. Tab. Alumettes) Ann. Dir. Etud. Equip. Sect.* 2: 95–99.

Jeffers, S.N. and H.S. Aldwinckle. 1987. Enhancing detection of *Phytophthora cactorum* in naturally infested soil. *Phytopathology* 77: 1475–1482.

Jeffers, S.N. and S.B. Martin. 1986. Comparison of two media selective for *Phytophthora* and *Pythium* species. *Plant Dis.* 70: 1038–1043.

Martin, F.W. 1959. Staining and observing pollen tubes in the style by means of fluorescence. *Stain Technol.* 34: 125–128.

Murashige, T. and F. Skoog. 1962. A revised medium for rapid growth and bioassays with tobacco tissues. *Physiol. Plant.* 15: 473–497.

Trigiano, R.N., C.G. van Dyke, H.W. Spurr, Jr. and D.J. Gray. 1984. Infection and colonization of tobacco callus by *Peronospora tabacina*. *Phytopathology* 74: 280–285.

22 Parasitic Seed Plants, Protozoa, Algae, and Mosses

Mark T. Windham and Alan S. Windham

CHAPTER 22 CONCEPTS

- Parasitic seed plants can be chlorophyllous or achlorophyllous and can cause significant economic crop loss.

- Parasitic seed plants may range in size from being a few centimeters tall, such as Indian pipe, to several meters tall, such as buffalo nut.

- Dodder has a large host range, is entirely aerial, and may destroy entire ornamental plantings of annuals and perennials.

- Dwarf mistletoe is extremely destructive on black spruce in western North America.

- Witchweed is a extremely destructive pest of grass crops such as corn, millet, sorghum, and sugercane in Africa and Asia. The pest is established in North and South Carolina in the U.S.

- Flagellated protozoa can cause serious diseases of coffee and oil palm, cassava, and coffee.

- Algae can cause leaf spots on numerous plants, including southern magnolia and cultivated azaleas in coastal areas of the southern U.S.

Although most of the important plant pathogen groups have been covered in previous chapters, a few remaining plant pathogens or organisms that have been associated with disease symptoms are parasitic seed plants, protozoa, algae, and mosses. The most important pathogens among these groups are the parasitic seed plants. Parasitic seed plants have flowers and produce seed, but may be deficient in other typical characters associated with plants such as roots or chlorophyll. Some parasitic seed plants, such as buffalo nut (*Pyrularia pubera*), are shrubs that have green leaves and a root system whereas others, such as squawroot (*Conopholis americana*), Indian pipe (*Monotropa uniflora*), pine-sap (*M. hypopiths*), witchweed (*Striga lutea*), dodder (*Cuscuta* species), and dwarf mistletoe (*Arceuthobium* species) and leafy mistletoe (*Phoradendron* species), have a very small thallus and do not have roots or green leaves, or both. In this chapter several examples of parasitic seed plants, tropical protozoa that have been associated with several diseases, an alga whose colonies are associated with leaf spotting of magnolia and azalea, and ball moss that causes death of shrubs and trees in isolated areas of Louisiana, Texas, and Florida will be considered.

PARASITIC SEED PLANTS

DODDER

Dodder is also known by the names strangle weed and love vine and is a small plant with a thin yellow to orange stem. Infestations of dodder look like yellow-orange straw spread along roadsides. The stems of dodder are leafless, but contain abundant whitish flowers that form in early summer and are produced until frost (Figure 22.1). Seeds are tiny, brown to gray to black, and mature in a few weeks after the flowers bloom. Dodder has a very wide host range that includes alfalfa, potatoes, numerous herbaceous annuals and perennials, and young shrubs or seedlings of trees in nurseries. Dodder infections can kill young plants. Severe infestations can destroy entire plantings of wild flowers along a roadway or annuals or perennials in a flowerbed. Dodder can also be used as a bridge (vector) to transmit a number of plant viruses (Chapter 4), but the economic importance of virus transmission by dodder is unknown.

0-8493-1037-7/04/$0.00+$1.50
© 2004 by CRC Press LLC

FIGURE 22.1 Closeup of dodder on stem of lespedeza. Note knobby protuberances on dodder stems that are haustoria primordia.

Symptoms and Signs of Dodder Infection

Symptoms of infected plants include stunting, loss of vigor, poor reproduction (flowering), and death. Signs of dodder include yellow-orange stems of dodder entwined around plant stems, petioles, and foliage. Dodder does not have roots and is entirely aerial. Patches of dodder enlarge by growing from plant to plant and by new dodder plants emerging from seed produced during the growing season. In container nurseries, dodder can spread rapidly throughout the canopy of plants that are tightly packed together.

Disease Cycle

Dodder seed may remain dormant in soil for many years or may be introduced into a field or flowerbed at planting. Only a stem is produced when seeds germinate (Figure 22.2) and the young seedlings rotate until they come in contact with a host plant. If a host plant is not available, the seedling eventually dies. However, if a suitable host is found, the stem of the seedling wraps around the host stem and produces haustoria that penetrate the host stem. After successful infection(s) has taken place, the dodder stem begins to grow from one host plant to another if the host plants are close enough, and produce many small white flowers. Seeds mature in approximately 3 weeks and may contribute to the current epidemic or lie dormant until the next growing season. Seeds are spread by water, animals, tillage equipment, or in mixtures with host seed.

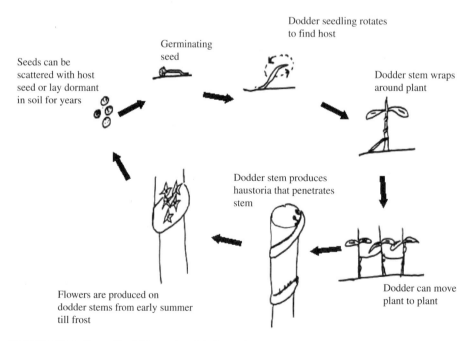

FIGURE 22.2 Generalized life cycle of dodder (*Crucuta* species).

TABLE 22.1
Economically Important Species of *Arceuthobium* (dwarf mistletoe) and Principal Hosts

Arceuthobium species	Host
A. abietinum f. species concoloris	White and grand firs
A. americanum	Jack, lodgepole and beach pine
A douglasii	Douglas fir
A. laricis	Western larch
A. pusillum	Alberta and black spruce
A. tsugense	Mountain and western hemlock
A. vaginatum subspecies vaginatum	Apache pine, rough bark Mexican pine, and ponderosa pine

Source: From Tainter, F.H. and F.A Baker, 1996, *Principles of Forest Pathology,* John Wiley & Sons, New York, 805 pp.

Control

Dodder infestations can be very difficult to control once the pathogen establishes in a field or flowerbed. Equipment should be cleaned before moving from an infested area to a noninfested area. Livestock in infested areas should be kept there and not moved to areas that are thought to be free of dodder. Once dodder infestations are established in a field, they are usually controlled with contact herbicides that destroy both the dodder and host plants before the parasite has a chance to flower. Cultivation or fire can be used to destroy dodder if done before flowering. Fumigation is possible in flowerbeds, but is impractical in fields due to the patchiness of dodder infestations.

DWARF MISTLETOES

Dwarf mistletoes are the single most economically important parasitic seed plant in North America and are the most important disease problem in conifer production in the western U.S. In Oregon and Washington, about 13% of the total wood production is lost to dwarf mistletoes annually. In Minnesota, infestations are primary in black spruce (*Picea mariana*) and about 11% is lost due to the disease. Losses in tree production are due to mortality, poor growth and wood quality, reduction in seed production, wind breakage, and predisposition of infected trees to other diseases and insects. There are nearly 40 species of dwarf mistletoe in the genus *Arceuthobium*. Economically important *Arceuthobium* species and their hosts are listed in Table 22.1.

Symptom and Signs

Infected portions of the tree become swollen and excessive branching at the infection site leads to **witches' broom** (Figure 22.3A and Figure 22.3B); multiple witches' brooms can occur on the same tree. Although the growth rate of infected tree parts increases, growth in the rest of the tree is retarded and growing points on the tree often die. Signs of dwarf mistletoe include small yellow to orange plants with sessile leaves and white flowers (Figure 22.4A). Developing berries are white and darken with age (Figure 22.4B).

Disease Cycle

Dwarf mistletoe is dioecious, and flowers are pollinated by insects. The berries contain one seed and are surrounded by **viscin**, a sticky mucilaginous pulp. The seeds are forcibly discharged for a distance of up to 15 m (Figure 22.5). The discharged seeds land directly on foliage of neighboring trees. Long-distance spread is by seed that become stuck to the feet of birds. During periods of rainy weather, the seeds slide down the needle (or off the feet of birds) and land on the twigs. Seed usually start to germinate within a few weeks, but germination may be delayed till spring. Once the seeds germinate and the radicle comes in contact with a rough area in the bark surface, a primary haustorium penetrates the limb. After the primary haustorium is successfully formed, the dwarf mistletoe plant develops an extensive absorption system that eventually includes the xylem of the host tree. After three or more years, an aerial shoot of the parasite emerges from the bark of the infected twig and flower production begins the following year. Dwarf mistletoe plants contain chloroplasts but fix little carbon and obtain practically all their nourishment from the host. Dwarf mistletoe is most severe in open stands along ridge tops (higher elevations) and is seldom a problem in bottomlands.

Control

Because dwarf mistletoes are obligate parasites, clear-cutting infected trees and prescribed burning to remove any ripe berries and limbs lying on the forest floor can be an effective control tactic. In many western areas, eradication of dwarf mistletoe is not practical and control strategies center around reducing the amount of dwarf mistle-

(A)

(B)

FIGURE 22.3 (A) Dwarf mistletoe infection of lodgepole pine. (B) Close up of infected limb.

toe after the current stand is harvested. After timber harvest, heavily infected trees are removed and are not used as sources of spruce seed. Plantings are accomplished in a manner to take advantage of natural barriers to dwarf mistletoe such as highways, large streams, nonhost species, etc. In recreation sites, the parasite may be controlled by pruning infected branches before seeds are produced. Resistance to this disease has not been successfully used except in the Rouge River valley in Oregon, where ponderosa pine (*Pinus ponderosa*) has drooping needles. The seeds slide down to the tip of needles and drop harmlessly

to the ground. However, seedlings taken from this valley revert to having erect needles. Thus, the ability of ponderosa pines in this area to escape the disease may be due to environmental conditions instead of a genetic trait.

LEAFY MISTLETOES

Leaf mistletoes occur throughout the world and usually attack hardwoods in forest and landscape areas. Heavily infected trees may begin to decline in vigor, but economic losses are small when compared to losses attributed to dwarf

(A)   (B)

FIGURE 22.4 (A) Dwarf mistletoe plant growing on conifer. (B) Closeup of dwarf mistletoe plant showing sessile leaves and ripening berries.

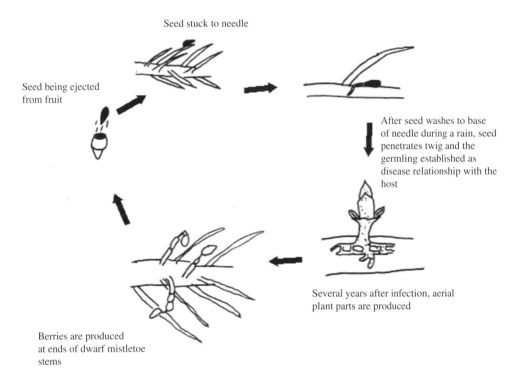

FIGURE 22.5 Generalized life cycle for dwarf mistletoe (*Arceuthobium* species).

mistletoe infections. In North America, infection by leafy mistletoes is usually due to *Phoradendron* species, but leafy mistletoes of *Viscum* species, common in Europe, are also found in California. Leafy mistletoes are evergreen plants and may be so numerous in a deciduous tree that they may make the tree appear to be an evergreen during the winter months (Figure 22.6). Symptoms of leafy mistletoe infections are similar to those of dwarf mistletoe in that infected branches become swollen and often form a witches' broom. Heavily infected trees may have reduced growth and limb death. Leafy mistletoes have white berries that are eaten by birds. The seeds are excreted by the birds and stick to the branches where the birds perch. Control of leafy mistletoes is usually not necessary, but infected branches can be pruned out. A comparison of dwarf and leaf mistletoes is given in Table 22.2.

FIGURE 22.6 Hardwood tree with multiple infections by a leafy mistletoe.

OTHER PARASITIC SEED PLANTS OF ECONOMIC IMPORTANCE

Witchweed

Witchweed is an economically serious parasite in Africa, Asia, Australia, and in limited areas in North and South Carolina. Witchweed can parasitize hosts such as corn, millet, sorghum, rice, sugarcane, tobacco, and cowpeas. Infected plants are stunted, usually wilted, chlorotic, and often die. Infected roots may contain numerous large haustoria from a single witchweed plant or haustoria from more than one witchweed plants. Flowers of witchweed are brightly colored with red or yellow petals that are showy. A single plant may produce nearly one-half million tiny seeds.

Witchweed is difficult to control and avoiding the introduction of seed into fields is paramount. **Quarantines** have been effective in limiting spread of the parasite in the Carolinas. Eradication of witchweed has reduced the infested area significantly since the disease was discovered in the mid-1950s. Witchweed can also be controlled by using trap or catch crops and by destroying the host plants and parasite with a herbicide before flowering and seedset.

Broomrapes

Broomrapes (*Orobanche minor* and *O. ramosa*) attack more than 200 species of plants and occur throughout much of the world. In India, broomrape infections in tobacco may destroy one-half of the crop, whereas yield losses on tobacco in the U.S. tobacco are rare. Plants attacked early in the season are stunted, whereas plants attacked later in the growing season suffer few effects from the parasite. Broomrape infections are usually clumped in a field. Broomrapes are whitish to yellowish to purplish plants that grow at the base of the host plant. Often, ten or more broomrape plants can be found attacking the same plant. Broomrape can produce more than one million seeds at the base of a single host plant.

Broomrapes, like witchweed, are hard to control in areas where infestations are severe. In India, some control may be obtained by weeding out broomrape plants before seeds are produced. Weeding must be done throughout the growing season because more broomrape plants will emerge after the initial weeding. Soil fumigation is effective in killing the seed.

FLAGELLATE PROTOZOA

Some protozoa in the family Trypanosomatidae are known to parasitize plants in tropical areas. Although thousands of protozoa can be found in the phloem of disease plants, formal proof of their pathogenicity has not been achieved because infections with pure cultures of the protozoa have not been accomplished (Koch's postulates have not been satisfied; Chapter 36). Diseases associated with flagellate protozoa include sudden wilt of oil palm, heartrot of coco-

---

TABLE 22.2
A Comparison of Dwarf and Leafy Mistletoes

Dwarf Mistletoes	Leafy Mistletoes
Found in western and northern parts of North America	Found in the southern half of North America
Attack conifers	Attack hardwoods
Propagated by forcibly ejected brown to gray seeds that stick to needles or to the feet of birds	Propagated by white seeds not forcibly discharged and are eaten and excreted by birds
Dioecious plants	Monoecious plants
Economically important in the U.S.	Seldom of economic importance in the U.S.
Yellow-orange to brown plants	Green plants

FIGURE 22.7 Colonies of *Cephaleuros virescens* on a leaf of southern magnolia.

nut palm, empty root of cassava, and phloem necrosis of coffee. Symptoms of these diseases include chlorosis (yellowing of fronds), stunting, and death. These diseases can spread very rapidly. For example, heartrot can spread to thousands of trees in one year. Protozoa that cause phloem necrosis in coffee and empty root of cassava can be transmitted by root grafts or grafting. Protozoa associated with sudden wilt of oil palm and heartrot of coconut palm are transmitted by pentatomid insects. Control is primarily by avoiding infected stock at transplanting. Control of potential vectors is of questionable value.

ALGAE

Although at least 15 species of three families of algae are known to parasitize plants, only three species of *Cephaleuros* are common worldwide and *C. virescens* is the only species common in the U.S. Parasitic algae affect more than 200 species of plants in the U.S. along coastal areas of North Carolina, extending south and westward to Texas. Economically important hosts that may suffer noticeable damage include southern magnolia (*Magnolia grandiflora*) and cultivated azaleas (*Rhododendron* species hybrids). Algal colonies are only successful if they form in small wounds and develop between the cuticle and epidermis. Superficial colonies on the leaf surface wash away during heavy rains. Host cells directly beneath the colony die, causing necrotic leaf spots. Many colonies can be found on a single leaf (Figure 22.7) and colonization is most severe during periods of warm, wet weather. Control is usually not recommended, but multiple sprays of copper are effective if needed.

MOSSES

Ball moss, *Tillandsia recurvata*, is a bromeliad that is closely related to Spanish moss. It is an epiphyte that is found in the southernmost U.S. (Florida, Louisiana, Texas, and Arizona) and southward to Argentina. Ball moss occurs on many deciduous and evergreen species. Large clusters of ball moss may completely encircle smaller branches and prevent buds from developing, which results in branch death. Other damage to the trees may be due to the weight of the epiphyte populations causing mechanical damage to host plants. Leaf abscission and branch death have also been attributed to abscission substances produced by the epiphyte. Control of ball moss is accomplished by using copper or sodium bicarbonate sprays.

Among pathogens covered in this chapter, the most economically important pathogens worldwide are dwarf mistletoes and witchweed. Yield losses from these two diseases can destroy plantings over a wide area and both diseases may cause pandemics. Other pathogens discussed in this chapter can cause high yield losses in specific locations. Flagellate protozoa may destroy plantings of palms, coffee, or cassava in particular locations. Dodder-infested flowerbeds can lead to severe limitations regarding the types of annuals or perennials that may be grown in those areas. Broomrapes may cause severe disease losses in tobacco in India. Control for most of these pathogens remains difficult and that is a uniting characteristic. Little resistance is known to any of these pathogens with the species of susceptible hosts. Chemical control any of the pathogens is almost impossible without also destroying the crop. Most of these pathogens have been the subject to lesser amounts of research than many fungal, bacterial, or viral pathogens. Until research scientists give pathogens in these groups more attention, many questions concerning their life cycles, infection processes, and disease control tactics will remain unanswered.

SUGGESTED READINGS

Agrios, G.N. 1997. *Plant Pathology*, 4th ed. Academic Press, New York, 635 pp.

Coyier, D.L. and M.K Roane (Eds.). 1986. *Compendium of Rhododendron and Azalea Diseases*. APS Press, St. Paul, MN, 65 pp.

Holcomb, G.E. 1995. Ball moss: An emerging pest on landscape trees in Baton Rogue. *Proc. Louisiana Acad. Sci.* 58:11–17.

Lucas, G.B. 1975. *Diseases of Tobacco*, 3rd ed. Biological Consulting Associates, Raleigh, NC, 621 pp.

Sinclair, W.A, H.H. Lyon and W.T. Johnson. 1987. *Diseases of Trees and Shrubs*. Comstock Publishing Associates, Ithaca, NY, 575 pp.

Tainter, F.H. and F.A. Baker. 1996. *Principles of Forest Pathology*. John Wiley & Sons, New York, 805 pp.

23 Abiotic Diseases

Alan S. Windham and Mark T. Windham

CHAPTER 23 CONCEPTS

- Abiotic diseases of plants are often caused by cultural practices or environmental factors.

- Abiotic diseases may be difficult to diagnose as the causal agent or factor may have dissipated prior to symptom development.

- Abiotic diseases may predispose plants to infection by plant pathogens.

- Sun scald occurs when leaves acclimated to low levels of light are exposed to full sun.

- Drought stress often predisposes woody plants to infection by canker-causing fungi.

- Bark splitting is often associated with winter injury or a sudden drop in temperature.

- The most common reasons for pesticide injury are the misuse or misapplication of pesticides, movement from the initial point of application due to vaporization, drift or movement in water, and injurious residue left from a previous crop.

- Symptoms associated with herbicide injury may be confused with symptoms produced by certain virus diseases.

Abiotic or noninfectious diseases are caused by cultural practices (Chapter 32) or environmental factors on plants in nature and also on cultivated crops (contrast with Chapter 36 for disease diagnosis). Although not true plant diseases (actually plant stress factors), damage by abiotic extremes such as light, water, and temperature can be quite severe under certain circumstances. Most plants grow best at optimum levels of environmental conditions. If, for example, temperature drops below or exceeds the optimum for growth, damage may occur. Also, although water is necessary for normal plant function, excessive amounts of water or insufficient amounts may cause injury.

Damage caused by abiotic diseases may be difficult to diagnose because symptoms may not appear until well after plants were exposed to suboptimal cultural conditions or environmental extremes. Another interesting note is that abiotic diseases may **predispose** plants to infection by plant pathogens. For instance, woody plants exposed to drought stress are more likely to be infected by canker-causing fungi such as *Botryosphaeria* species. Small grains and turfgrass grown in alkaline soils are at risk to infection by *Gaeumannomyces*, a soilborne fungus associated with take-all patch diseases. In many cases, damage from abiotic diseases is compounded by **biotic diseases** that follow.

LIGHT

Low light decreases plant vigor, slows growth, elongates internodes, and may reduce flowering and fruit. Suboptimum levels of light lead to a decreased carbohydrate production and damage to the plant's photosynthetic system through reduced chlorophyll production.

It is often difficult to separate the effects of high light levels and high temperature on leaves. Plants grown at low light intensities have leaves with little or no wax and cutin. **Sun scald** occurs when plant material is moved from low light conditions to high light. An example is moving a shade-loving plant such as rhododendron from a lathe house to a landscape bed in full sun. Also, leaves of woody shrubs may develop sun scald after pruning exposes leaves that are acclimated to low light intensities. Occasionally, plants grown at high elevations will develop sun scald when exposed to full sun after several days of cloudy weather.

LOW-TEMPERATURE INJURY

Freeze injury occurs when ice crystals form in intercellular and intracellular spaces. If the cell membrane is ruptured, the cell will die. Cold temperature injury to young trees

FIGURE 23.1 Winter injury to the foliage of southern magnolia (*Magnolia grandiflora*). Note marginal leaf desiccation.

FIGURE 23.2 Drought injury on Kousa dogwood (*Cornus kousa*). Note marginal leaf scorch.

often leads to bark splitting. Extremely low winter temperatures may cause severe injury to woody plants, killing all above-ground plant parts (see Plate 8A following page 80). However, woody shrubs may also be damaged by a sudden drop in temperature, which can lead to **bark splitting** on the lower stem. Fungi that cause **canker** and wood decay fungi may enter these wounds and cause branch dieback or death. Tender foliage or shoots may be injured by freezing temperature (Figure 23.1).

WATER STRESS

Drought stress and water stress are sometimes used interchangeably. However, water stress can also be used when discussing an excessive amount of water, such as flooding in poorly drained soils. Drought stress is normally used when discussing a shortage of natural rainfall.

DROUGHT STRESS

Wilting, chlorosis, shortened internodes, and poor flower and fruit production are all symptoms that have been associated with drought stress. Leaf scorch, which is a **marginal** or interveinal leaf necrosis, is often seen on deciduous plants exposed to drought stress (Figure 23.2). Evergreen plants such as conifers may shed needles in response to a shortage of water. Remember that any biotic disease that affects the root system, the vascular system, or the trunk and branches of a plant may produce symptoms that could be confused with those associated with drought stress.

One of the most common causes of losses of landscape plants is the lack of water. Annual flowering plants are especially sensitive to drought stress and most herbaceous perennial plants are less sensitive; however, this varies by

species and even cultivar. It is not uncommon to see native trees in forests die after several years of below-normal rainfall. Many woody plants are more susceptible to canker-causing fungi such as *Botryosphaeria* and *Seiridium* species if the plants have been exposed to significant levels of drought stress.

Sometimes drought stress may be localized. Soils with underlying rock or construction debris, sandy soils, or golf greens with hydrophobic areas may have localized areas of plants exhibiting wilting and chlorosis. Also, if pine bark media that is used in container nurseries is not stored properly, it may become infested with fungi, such as *Paecilomyces,* that cause the bark to become hydrophobic. Irrigation water forms channels and is not absorbed by the media in affected containers simulating drought stress. This sometimes leads to losses once plants are installed in landscape beds as the bark mix continues to repel water.

EXCESSIVE WATER/FLOODING

Flooding or prolonged periods of saturated soil conditions can lead to the decline or death of many cultivated plants. Chlorosis, wilting, and root necrosis are symptoms that may be exhibited by plants exposed to seasonal or periodic flooding. In waterlogged soils, low oxygen levels lead to root dysfunction and death. Water molds such as *Pythium* and *Phytophthora* are favored by the conditions found in saturated soils or growth media (Chapter 20).

During periods of cloudy weather, uptake of water may remain high while transpiration rates are slow. When this occurs, leaf tissue saturated with water ruptures, forming corky growth on the underside of leaves. This condition, edema, is fairly common on ornamental flowering plants such as ivy geranium. A similar condition, called intumescence, may be observed on sweet potato and ornamental sweet potato. The symptoms of intumescence occur on both the upper and lower leaf surface.

FIGURE 23.3 Hail injury to the trunk of flowering dogwood (*Cornus florida*).

LIGHTNING INJURY

The damage caused by lightning strikes is frequently observed in forest and shade trees. Large wounds are often created on the trunk of the tree as the charge moves down the cambium, blowing out strips of bark and sapwood. The wound may be nearly vertical or spiral down the trunk following the grain of the wood. Trees injured by lightning may soon die from the strike or decline after insects and wood decay fungi enter the wound.

Cultivated crops such as vegetables are sometimes injured by lightning strikes in fields. Circular areas of damaged plants are often visible within days of the strike. In succulent vegetables such as tomato, wilting may be visible within hours of a strike. Further damage often follows the initial strike in vegetables such as tomato, cabbage, and potato when necrotic pith tissue develops in the stem of affected plants. On golf greens, lightning sometimes strikes flag poles and the charge radiates out in a circular pattern across the green in an almost cobweb fashion.

HAIL

Cultivated crops may be ruined within a few minutes during a hailstorm. Hailstones quickly shred the leaves of many plants; the bark of young trees may be severely damaged (Figure 23.3). Bark damage on trees is often on one side only. Wounds from hail damage are often invaded by insect borers, canker-causing fungi, and wood decay fungi. In many cases, nursery stock severely damaged by hail have to be destroyed.

MINERAL DEFICIENCY AND TOXICITY

Normal plant growth is dependent on the availability of several mineral elements. Minerals such as nitrogen, phosphorus, potassium, magnesium, calcium, and sulfur are needed in larger amounts (macroelements) than elements such as iron, boron, zinc, copper, manganese, molybdenum, sodium, and chlorine (minor or trace elements). If these elements are not available in sufficient amounts for typical plant growth, a variety of symptoms may develop (Table 23.1). Chlorosis, necrosis, stunted growth, rosette, reddish-purple leaves, and leaf distortion are symptoms that have been associated with nutrient deficiencies. Also, if certain nutrients are available in excessive amounts, plant damage may occur.

Soil **pH** may have a significant effect on nutrient deficiency or toxicity (Figure 23.4). In alkaline soils, ericaceous plants such as blueberry, azalea, and rhododendron may show symptoms of iron deficiency (see Plate 8B following page 80). In contrast, manganese is sometimes toxic to crops such as tobacco that grow in acidic soils. In **saline** soils, sodium and chlorine ions may occur at damaging levels.

The diagnosis of deficiency or toxicity of mineral nutrients in cultivated crops is complex. Observe the crop showing symptoms of a nutrient problem. Are the symptoms on young leaves, older leaves, or both? Chlorosis of older leaves may indicate that a mobile nutrient such as nitrogen has moved from older foliage to newly developed leaves. Are affected leaves chlorotic or necrotic? Chlorosis is associated with nitrogen, magnesium, and sulfur deficiencies, whereas marginal necrosis of leaves is often linked to potassium deficiency. Keep in mind that visible foliar symptoms are more often associated with nutrient deficiencies than toxicities. The diagnosis of nutrient problems is often complicated by plant disease symptoms, insect damage, and pesticide injury.

Soil and plant tissue analyses are very helpful in diagnosing fertility problems. Besides revealing soil pH, soil analysis shows the potential availability of mineral nutrients. It is often helpful to collect a soil sample from a problem area and from an adjacent area where plants are growing normally for comparison. Plant tissue analysis gives a "snapshot" of the actual nutritional status of the plant. When collecting plant tissue for nutrient analysis, note the stage of growth. Collecting separate samples of old and young leaves can give information on nutrients that are mobile within the plant. Finally, be aware that fungicides containing copper, zinc, or manganese may affect results of plant tissue analysis.

TABLE 23.1
Symptoms of Nutrient Deficiencies in Plants

Nutrient	Symptoms
Nitrogen (N)	Chlorosis in older leaves, leaves smaller than normal, stunted plants
Phosphorus (P)	Purple to red leaves, smaller than normal leaves, limited root growth
Potassium (K)	Chlorosis in leaves, marginal chlorosis in older leaves
Magnesium (Mg)	Chlorosis in older leaves first, chlorosis may be interveinal
Calcium (Ca)	Leaf distortion such as cupping of leaves; fruit of some plants may rot on blossom end
Sulfur (S)	Chlorosis in young leaves
Iron (Fe)	New growth chlorotic, chlorosis often interveinal, major veins may be intensely green
Zinc (Zn)	Alternating bands of chlorosis in corn leaves; rosette or "little leaf"
Manganese (Mn)	Interveinal chlorosis, may progress to necrosis
Boron (B)	Stunting, distorted growth, meristem necrosis
Copper (Cu)	Tips of small grains and turfgrass are chlorotic, rosette of woody plants such as azalea

FIGURE 23.4 Rosette or witches' broom (on left) associated with copper deficiency of azalea (*Rhododendron* species).

PESTICIDE INJURY

The most common reasons for pesticide injury are the misuse or misapplication of pesticides, movement from the initial point of application due to vaporization, drift or movement in water, and injurious residue left from a previous crop. The inappropriate use of pesticides fre-quently leads to plant injury. For instance, a nonselective soil sterilant used to edge landscape beds may lead to disastrous results if the roots of desirable plants absorb the herbicide. Other common causes of pesticide injury are the drift of herbicides onto nontarget plants, use of the same sprayer for herbicides and other pesticide applica-tions, storing fertilizer and herbicides together, mislabeled pesticide containers, inappropriate tank mixtures, and planting crops into soil with harmful herbicide residues (Figure 23.5). The amount of damage caused by a partic-ular pesticide is often related to the type of pesticide and the concentration applied. Some plants may recover from the initial damage induced by some herbicides if the con-centration applied was low, such as in drift of spray drop-lets from an adjacent field.

There are numerous symptoms of pesticide injury, such as necrosis, chlorosis, witches' broom or rosette, strap-shaped leaves (see Plate 8C following page 80), cupping, and distinct leaf spots. It is important to remem-ber that the symptoms of many plant diseases caused by plant pathogens may be confused with the symptoms of pesticide injury, so a thorough investigation is important. The symptoms of some viral diseases may be mistaken for symptoms caused by **herbicide injury**.

The diagnosis of pesticide injury may be tedious and time consuming. Collect information about pesticide applications within the last two growing seasons as some

FIGURE 23.5 Phenoxy herbicide injury of flowering dogwood (*Cornus florida*). Note curled, strap-shaped leaves.

FIGURE 23.6 Sublethal dose of glyphosate herbicide on variegated periwinkle (*Vinca major*). Note dwarfed new growth (left).

damage may be the result of a preemergent herbicide applied the previous year. Ask if pesticides were applied to adjacent fields or utility right-of-ways. Look for diagnostic symptoms of injury. Check several plant species for similar symptoms. Question the applicator about the pesticide applied, the rate applied, and check the calibration of the sprayer. Collect plant and soil samples for analysis by a plant disease clinic, soil test laboratory (for soil mineral analysis and pH) and an analytical laboratory (for pesticide detection). It is helpful to collect samples as soon as possible after the damage occurs (Figure 23.6). Also, it is more economical to ask an analytical lab to assay the soil or plant material for a pesticide suspected of causing the problem rather than an open screen of many pesticides.

Bioassays may also be used to determine if damaging levels of pesticide residues remain in soils. Collect soil from problem areas and place in pots or flats in a green house and plant several plant species such as a small grain, radish, tomato, and cucumber into the suspect soil. Observe germination rates and look for symptoms development on young seedlings. Solving the cause of pesticide injury may be time consuming, but it is worth the effort in a crop of high value.

AIR POLLUTION INJURY

Air **pollution** that damages plants may originate from man-made sources or be produced naturally. Some of the more common pollutants that cause plant injury are ozone (O_3), sulfur dioxide (SO_2), and ethylene. Ozone is gener-

ated naturally during lightning strikes, but may also be produced when nitrogen dioxide from automobile exhaust combines with oxygen in the presence of ultraviolet light. Sulfur dioxide originates from several sources, including automobile exhaust and coal-fired steam plants. Ethylene may be produced by poorly vented furnaces (such as those used to heat greenhouses) and from plant material or fruit stored in poorly ventilated areas.

Ozone is one of the most damaging air pollutants. It may cause chlorotic stippling of the needles of conifers or chlorotic to purple discoloration of the leaves of deciduous plants such as shade trees. Sulfur dioxide may also cause chlorosis of foliage. However, in eastern white pine, the tips of the needles of affected trees turn bright red. Not all trees are affected equally. In a pine plantation, it is not unusual to see a low percentage of trees with symptoms associated with sulfur dioxide exposure. Ethylene is a plant hormone. However, plants that are exposed to abnormal levels of ethylene produce distorted foliage that is often confused with symptoms of virus diseases. Ethylene may also reduce fruit and flower production. Ultimately, the damage caused by air pollutants depends on the concentration of the pollutant, the time of exposure and the plant species. The diagnosis of air pollution injury is very difficult and is often tentative and based on the distribution of the damage, symptoms, and the plant species affected.

Abiotic plant diseases should not be discounted or overlooked as a cause of plant damage. Nearly half of all plant samples submitted to plant disease clinics exhibit symptoms associated with abiotic diseases such as drought stress, nutrient deficiencies, or pesticide injury. Diagnosis is not easy and often depends on the experience of the diagnostician (Chapter 36). Abiotic diseases related to cultural practices are easier to solve. Diseases associated with environmental or climatic factors are often harder to diagnose.

UNKNOWN CAUSES OF DISEASES

Galls (burls) and witches' brooms of many trees (see Plate 8D following page 80) and flattening of branches (fuscination) of herbaceous plants are the result of unknown causes. These symptoms maybe the result of pathogens, abiotic stresses, or genetic abnormalities.

SUGGESTED READINGS

Agrios, G.N. 1997. *Plant Pathology*, 4th ed., Academic Press, New York.

Fitter, A.H. and R.K.M. Hay. 1987. *Environmental Physiology of Plants*. Academic Press, New York.

Kramer, P.J. and T.T. Kozlowski. 1979. *Physiology of Woody Plants*. Academic Press, New York.

Levitt, J. 1980. *Responses of Plants to Environmental Stresses*, Vol. I. Academic Press, New York.

Levitt, J. 1980. *Responses of Plants to Environmental Stresses*, Vol. II. Academic Press, New York.

Marschner, H. 1986. *Mineral Nutrition of Higher Plants*. Academic Press, New York.

Part III

Molecular Tools for Studying Plant Pathogens

24 Studying the Ecology, Systematics, and Evolution of Plant Pathogens at the Molecular Level

Emnet Abesha and Gustavo Caetano-Anollés

CHAPTER 24 CONCEPTS

- The destructive role of plant pathogens is an important threat to agronomic and horticultural plants.

- The identity, spread, evolution, and impact of microbial pathogens and plant genes conferring specific resistance to disease can be characterized at the nucleic acid level.

- Nucleic acid markers reduce the extraordinarily complex information of a genome to discrete nucleic acid sequences, and these can be used effectively to analyze relatedness, phylogeny, and inheritance of genetic material.

- Molecular markers are generally based on the amplification and hybridization of nucleic acids.

- These and other molecular tools have advanced the study of plant–microbe interaction at the ecological and evolutionary levels.

Plants and pathogens interact in an endless race to evade and cause disease. This interplay dominates many important issues in plant pathology. From an evolutionary standpoint, the study of plant resistance genes involved in gene-for-gene interactions illustrates the plant–pathogen survival race (Bergelson et al., 2001). Leucine-repeat-encoding regions in these *R*-genes are fundamental to recognize pathogenicity determinants in the pathogen, the so-called avirulent (*Avr*) factors, and are known to evolve at unprecedented rates. These regions depict coevolutionary arm races between plants and pathogens. However, mechanisms of genetic exchange and adaptive evolution coexist with selective sweeps characteristic of the arm races, and are strongly influenced by demographic and ecological factors. This highlights the need to understand patterns of molecular diversity from perspectives centered on population and lineage. This is now possible with the ability to access genetic information at a wide scale in genomic research projects, its analysis with powerful bioinformatic tools, and the use of novel strategies capable of tracking genetic change at the molecular level. These advances involve harnessing the molecular biology of **nucleic acids**.

Nucleic acids are biological polymers that store and transmit genetic information in living organisms. These gene-encoding molecules are localized mainly in the nucleus of eukaryotic cells or in the nucleoid of prokaryotic organisms, but also in organelles such as mitochondria and chloroplasts. There are two kinds of nucleic acid molecules: deoxyribonucleic acid (DNA) and ribonucleic acid (RNA). With the exception of some viruses that use RNA as genetic material, DNA provides the genetic makeup of a genome, that is, the entire genetic complement of a living organism. In turn, RNA provides the expression vehicle of the entire complement of genes, that is, the transcriptome.

To understand the function and diversity of nucleic acids, one has to appreciate their structure. Structurally, nucleic acid molecules are linear chains of individual subunits called nucleotides. Nucleotides are made up of three essential components: a nitrogenous base, a sugar, and a phosphate moiety. Bases can be purines, which are double-ringed structures containing five- and six-member aromatic rings connected together [e.g., adenine (A) and guanine (G) base components in both DNA and RNA], or pyrimidines, which are single six-member aromatic ring structures [e.g., thymine (T; present in DNA), uracil (U; present in RNA) and cytosine (C)]. The sugar moiety is a pentose: ribose in RNA and 2′-deoxyribose in DNA. The pentose sugar is attached at its 5′ end to the phosphate moiety, and its 3′ end is used to establish phosphodiester bonds with other nucleotides to form polynucleotide chains. The ability to form phosphodiester bonds imparts a distinct directionality (from 5′ to 3′) to the molecule, which is both chemically and structurally important.

Both purine and pyrimidine bases are produced in multistep biochemical reactions in the cell and are capable of pairing with each other through hydrogen bonds. These bonds usually follow strict Watson–Crick base pairing rules in which A always pairs with T (or U in RNA) and G always does so with C, and, for example, are responsible of DNA being double-stranded and helical in structure. The two complementary DNA strands base pair with each other by twisting around a common axis in a helical pattern. The sugar and phosphate groups are water soluble and form a negatively charged backbone facing the outside of the double helix, whereas the bases tend to stack in the middle through hydrophobic and Van der Waals force interactions. These interactions stabilize the double-helical structure of DNA.

The great diversity of life that we see today is the result of differences in composition, degree of organization, and complexity found in the genetic material of living organisms. In other words, the entire genetic make-up of an organism dictates its identity by large. Accessing the total genomic sequence of an organism is therefore a powerful alternative to study evolutionary, cellular, and biochemical processes. Genome initiatives have therefore focused on pathogens in order to understand and combat disease. For example, bacteria establish pathogenic or symbiotic interactions by deploying common molecular mechanisms. These are generally a consequence of relatively large changes in genome repertoire (Ochman and Moran, 2001). When comparing pathogenic and non-pathogenic bacterial strains in comparative genomic exercises, genes required for virulence are generally confined to lineages and are the result of lateral transfer events of acquisition of genetic material. This occurs pervasively, but not indiscriminately, and involves large chromosomal regions known as pathogenicity islands, which are usually longer than 10^5 base pairs. In certain symbionts, large plasmids help encode symbiosis functions. For example, *Sinorhizobium meliloti* has two megaplasmids, pSymA and pSymB, (both totaling 3×10^6 base pairs and representing almost half of the overall genome). These megaplasmids appear as recent and separate acquisitions that contribute to the ability of the bacteria to associate with alfalfa and medics and fix atmospheric nitrogen (Galibert et al., 2001). Determining the complete sequence of pathogenic and nonpathogenic fungi will similarly help understand why some fungi are virulent and others harmless. Various projects are on the way to sequence the entire genome of a number of plant pathogens of fungal origin. However, sequence information for all of the most impor-

tant fungal pathogens will not be publicly available in the near future due to research funding limitations, and comparative genomics cannot be applied on a wide scale. Other strategies are therefore needed to evaluate diversity and help monitor and combat the diseases caused by emergent plant pathogens.

MOLECULAR MARKERS

Morphological and biochemical **characters** have in the past been widely used to investigate relatedness, phylogeny, and inheritance of genetic material. With the advent of genome analysis, it has become possible to accomplish the same task directly at the molecular level. **Molecular markers** are nucleic acid segments that behave as landmarks for genome analysis and are based on naturally occurring genetic variabilities (usually termed **polymorphisms**). Nucleic acid techniques reveal these markers by confining analysis of the complex information contained in a genome (typically 10^6 to 10^{10} base pairs) to one or more specific nucleic acid sequences (representing only 1 to 10^4 base pairs). To be useful, however, sequences must be informative when compared to each other and should therefore harbor detectable polymorphisms.

Molecular markers are powerful and reliable and offer advantages over morphological and biochemical alternatives. For example, they provide multiple characters, which are stable and detectable in all tissues regardless of growth and differentiation conditions, and are unaffected by the environment. Therefore, molecular markers have been widely applied in varied fields of biology, including forensic sciences, taxonomy, systematic and evolutionary biology, ecology, population genetics, epidemiology, breeding, and plant pathology.

Molecular markers can be categorized in several groups according to the way they are generated (c.f., Karp et al., 1997). Here, we group them in the following two categories:

- Hybridization-based techniques — These methods are based on the Watson–Crick complementary rules of base pairing and require the use of labeled nucleic acid molecules as hybridization probes. Probes can be short single-stranded nucleic acid segments (oligonucleotides) of synthetic origin or cloned DNA segments that bind to complementary nucleic acids, forming hybrid molecules. In conventional analyses, probes are hybridized to target DNA molecules that have been previously cut (digested) with restriction endonuclease enzymes and transferred (blotted) to a membrane support according to the Southern procedure (Southern, 1975).

When probes represent single copy genomic segments, the procedure generates **restriction fragment length polymorphic (RFLP)** markers. More advanced applications involve the use of oligonucleotide arrays ("chips") that confine individual oligonucleotides to defined physical addresses in solid supports such as glass or silicon (Southern, 1996). These hybridization arrays are very popular as they can be used successfully in both sequencing or genotyping applications.

- Amplification-based techniques — Nucleic acid amplification methods use oligonucleotides to drive the exponential accumulation of specific sequences from defined regions in a genome or transcriptome (Landegren, 1993). Thermostable DNA polymerase enzymes that copy (replicate) the accumulating nucleic acid segments with efficiency and high fidelity usually mediate these experimental strategies (Chapter 25).

A number of techniques based on the **polymerase chain reaction** (PCR) target unique genomic sequences. PCR uses two oligonucleotide primers about 20 nucleotides in length that hybridize specifically to opposite DNA strands flanking the region to be amplified. A series of temperature cycles separate (denature) the DNA strands, allow for the primers to bind (anneal) to the complementary sequences in each strand, and then permit the efficient extension of the annealed primers by the polymerase enzyme. This results in the specific amplification of the target region many million-fold to produce an amplified fragment with its termini defined by the 5' end of each primer. PCR-based techniques have been used to amplify single targets such as ribosomal RNA (rRNA) genes, mitochondrial and chloroplast DNA sequences, and repetitive DNA. For example, primers designed to target sequences flanking **simple sequence-repeat** (SSR) regions produce powerful nucleic acid markers, because these microsatellite sequences are highly variable and reveal many allele variants (Litt and Luty, 1989; Weber and May, 1989). However, one important limitation is the prerequisite for partial or total knowledge of the target sequence.

Techniques based on **arbitrarily amplified DNA (AAD)** target multiple sites in a genome or template molecules and are particularly useful because they do not require prior sequence information or cloned and characterized hybridization probes (Caetano-Anollés, 1996). Three AAD techniques have been the most popular: **randomly amplified polymorphic DNA** (RAPD; Williams et al., 1990), arbitrarily primed PCR (AP-PCR; Welsh and McClelland, 1990) and **DNA amplification fingerprint-**ing (DAF; Caetano-Anollés et al., 1991 — Chapter 25). They differ in length and sequence of the primers used, stringency of the amplification reaction, and procedure to separate and visualize the amplified fragments (Caetano-Anollés, 1996). For example, DAF amplifies discrete segments of DNA sequences by using short (5 to 8 nucleotide) oligonucleotide primers that generate arbitrary but characteristic fingerprint patterns. These patterns are rather complex and capable of uncovering sequence polymorphism and genome variability at different taxonomical levels. The AAD strategy relies on the natural existence of short repeat sequences scattered throughout the target nucleic acid (genome, DNA fragment, etc.) that are complementary, inverted, and close enough to each other. These repeats define primer annealing sites in opposite strands of the target DNA and permit amplification of multiple arbitrary sites. In DAF, the amplified fragments are then separated on polyacrylamide gels and detected by silver staining (Caetano-Anollés et al., 1991).

AAD techniques can also use semiarbitrary primers complementary to sequences that are interspersed in the genome, such as restriction sites, repetitive sequences, transposable elements, and microsatellites. Examples in this group include the AFLP technique (Vos et al., 1995), microsatellite-primed PCR (MP-PCR; e.g., Perring et al., 1993), and randomly amplified microsatellites (RAMS; Ender et al., 1996).

MOLECULAR MARKERS IN PLANT PATHOLOGY

The human population and its food demands are increasing at an annual rate of 1.4%. Although the population will probably stabilize by the year 2050, this will require an increase of about 25% in plant productivity. These demands can be partially met by diminishing the impact of pest and pathogens on crop yields. Global losses of agricultural crop production caused by fungal, bacterial, and viral pathogens represent 12% of plant productivity (James et al., 1990). These losses are about a third of total losses due to pests (e.g., insects, nematodes) and pathogens combined.

In spite of the many and different kinds of plant pathogens that come in contact with a plant, in most cases plants remain resistant to disease. This is because they possess resistance genes that specifically block pathogen invasion (Baker et al., 1997). Although *R*-genes provide disease resistance, only few are able to control the pathogen for an extended period of time, at least for many of the important plant diseases that affect agricultural production (e.g., wheat rust and rice blast; Crute, 1985). For this reason, many plant breeding initiatives have focused on the discovery and introgression of new *R*-genes into elite cultivars by traditional breeding systems (Chapter 31), despite it being time consuming and labor intensive. However, the use of molecular markers such as RFLP, AAD, and SSR, coupled with genetic engineering, can accelerate the process considerably. These techniques enable the screening of segregating populations at the molecular level rather than using disease phenotypes (Rommens and Kishore, 2000). Furthermore, advances in plant tissue culture and genetic engineering techniques have opened a whole new array of possibilities and methodologies for plant improvement involving the search for plant resistance to pathogen attack. These techniques allow the detection, isolation, modification, transfer, and expression of single genes, or group of related genes from one organism to another, enhancing the ability to produce key plant accessions resistant to disease.

The importance of molecular markers for the identification and analysis of genetic diversity among plant pathogens rivals that applied to the discovery and study of plant resistance genes. The genetics, ecology, and evolution of pathogenic bacteria, fungi, and viruses can be studied at the molecular level with the help of molecular markers and genomic tools. Questions on diversity both in hosts and pathogens can be addressed at the species, population, and within-population (individual) levels. At the species level, molecular markers not only help identify taxonomic units but also help determine the uniqueness of a given species. For example, DAF was used efficiently to investigate the introduction and spread of *Discula destructiva*, a recently introduced filamentous fungus that has devastated flowering dogwood populations in the North American deciduous forests (Trigiano et al., 1995). In this study, *D. destructiva* was found to be highly homogeneous at the genetic level and quite different from an undescribed *Discula* species, found also associated with the dogwood anthracnose and assumed to be indigenous. At the population level, molecular markers help assign isolates or strains to organismal groups sharing characteristics such as host range, virulence, ecological niche, and geographical distribution. The choice of the appropriate molecular markers for this purpose depends on the degree of diversity (polymorphism) that is anticipated. Markers producing multilocus profiles (e.g., RAPD and DAF) are good choices when analyzing closely related genotypes (Karp et al., 1997). Alternatively, targeting highly variable genomic loci can equally help uncover diversity in closely related organisms. For example, molecular marker strategies such as RAPD and the amplification and sequencing of specific RAPD loci have been used to identify and establish the **phylogenetic** relationship of strains of *Trichoderma atroviride*, a filamentous fungus used as biocontrol agent against soilborne fungal plant pathogens

(Hermosa et al., 2001). Similarly, DAF and sequencing of internal transcribed spacers (ITSs) of rRNA have been used to study patterns of evolution of *Discula* species (Caetano-Anollés et al., 2001). The steps required to analyze the ITS spacer region are shown in Figure 24.1. Individual loci harboring the region of analysis (in this case ITS2) were amplified by PCR from several species, including one used as outgroup for **phylogenetic tree** reconstruction. These fragments were sequenced and used to reconstruct phylogenetic relationships. In this example, *D. destructiva* was of recent origin and was closely related to the European pathogen *D. umbrinella* (c.f., Caetano-Anollés et al., 2001). Finally, assessing diversity at the individual level is important to establish the identity of individuals with respect to their parents. Multilocus profiling techniques are extremely valuable for this purpose. In our laboratory, we used DAF to study the fairy ring fungus *Marasmius oreades* (Abesha, et al., 2003). Fairy rings in turfgrasses are widespread and are very destructive to parks, golf courses, lawns, and pastures. The rings and arcs arise by suppression or stimulation of turfgrass growth by the mycelium of the basidiomycete fungus growing in the soil beneath the ring and the appearance of basidiocarps (mushrooms). These rings can be very old (up to 500 years of age) and are genetically highly homogeneous (Burnett and Evans, 1966). We collected basidiocarps from two different locales in the southern region of Norway where fairy rings have been well established and followed their growth in three consecutive years. We then analyzed genetic diversity as it relates to rings, ecological factors, and time. Fingerprinting patterns of individual isolates of *M. oreades* amplified with a single octamer primer are shown in Figure 24.2. Rings were genetically homogeneous whereas isolates collected from a same growing ring exhibited DNA polymorphisms, showing the existence of mutation or somatic recombination in hyphae of the perennial mycelium.

When studying populations at the genetic level, it is essential to understand patterns of organismal propagation, inheritance, and evolution. In fungi, genetic clones resulting from asexual reproduction represent elements defined by multilocus genotypes that usually do not involve sexual transfer (Anderson and Kohn, 1998). These clones are considered the units of fungal populations and the basis for studies that seek to understand sexual or asexual genetic recombination in fungi (Milgroom, 1996). Clones can be of recent origin or can be extremely ancient, can be physically connected and highly territorial (from centimeters to kilometers), or can separate from their origin for dispersal, and local populations can be almost as diverse as the entire metapopulation. Spatial connections can be broken by somatic incompatibility reactions

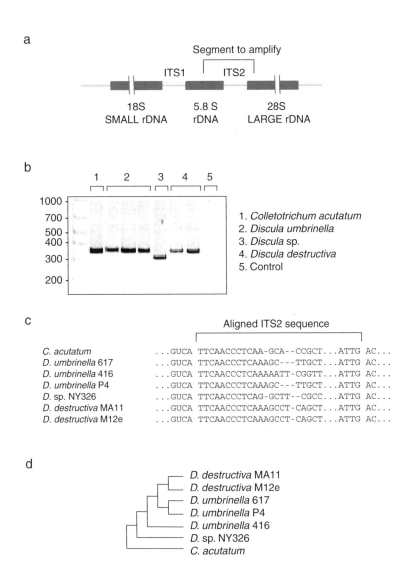

FIGUREURE 24.1 Use of the internal transcribed spacers (ITS) of rDNA to characterize fungi. (a) *Experimental design*: Primers complementary to regions flanking the ITS sequence to be amplified are selected and synthesized. Note that the two ITS spacers are variable in sequence and separate the different rRNA genes. (b) *PCR amplification*: The selected primers are used to drive the amplification reaction and amplified fragments are separated in polyacrylamide gels and stained with silver. Note that in some cases DNA fragments are of varying length and that a control without DNA was included in the analysis. Molecular markers are given in base pairs. (c) *Sequence analysis*: Fragments are purified and sequenced. Sequences are curated and aligned. d. *Phylogenetic reconstruction*: The aligned sequences are used as a data matrix to reconstruct phylogenetic relationships by the distance, parsimony, and maximum likelihood methods. In this example, the ITS2 sequence of *Colletotrichum acutatum* was used as outgroup to root the phylogenetic tree. (Data from Caetano-Anollés, G.R.N. Trigiano and M.Windham, 2001, *Curr. Genet.* 39: 346–354.)

between neighbors, creating clear patches of growth that favor the formation of many individual genotypes. The study of these spatially connected fungal clones offers a unique possibility to address important phenomena such as existence of barriers to gene flow, role of somatic incompatibility, evaluation of the cost and benefits of sex in populations of varying size, and role of deleterious mutation in evolution. These issues in fungal biology are also central to the dispersal and coevolutionary strategies of pathogenic organisms. Molecular markers provide tools to study the properties of the interactions between plant pathogens and their hosts at the population level and the evolutionary roles of sex and lineage. Further advancement in agriculture and sustainable management of natural

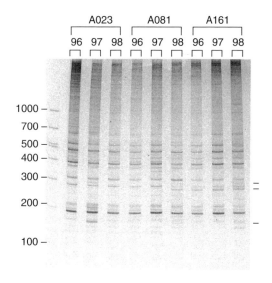

FIGURE 24.2 DAF analysis *of Marasmius oreades*. DNA extracted from the fruiting bodies of several isolates was amplified by using the oligonucleotide primer GTATCGCC, generated fingerprints were separated using polyacrylamide gels, and fragments were stained with silver. Isolates are labeled according to fairy ring and year that they were collected. Molecular markers are given in base pairs, and selected polymorphic bands are indicated.

resources will ultimately depend on an appropriate understanding of the ecology and evolution of plant disease, especially as it relates to how pathogens interact with the host, other organisms and the environment.

REFERENCES

Abesha, E.G., G. Caetano-Anollés and K. Hoiland. 2003. Patterns of propagation of the fairy ring fungus *Marasmuis oreades* in a Norwegain sand dune ecosystem revealed by DNA amplification fingerprinting. *Mycologia*, 95: in press.

Anderson, J.B. and L.M. Kohn. 1998. Genotyping, gene genealogies and genomics bring fungal population genetics above ground. *Trends Ecol. Evol.* 13: 444–449.

Baker, B., P. Zambrisky, B. Staskawicz and S.P. Dinesh-Kumar. 1997. Signaling in plant-microbe interactions. *Science* 276: 726–733.

Bergelson, J., M. Kreitman, E.A. Stahl and D. Tian. 2001. Evolutionary dynamics of plant *R*-genes. *Science* 292: 2281–2285.

Burnett, J.H. and E.J. Evans. 1966. Genetic homogeneity and the stability of the mating-type factors of "fairy rings" of *Marasmius oreades*. *Nature* 210: 1368–1369.

Caetano-Anollés, G. 1996. Scanning of nucleic acids by *in vitro* amplification: New developments and applications. *Nat. Biotechnol.* 14: 1668–1674.

Caetano-Anollés, G., B.J. Bassam and P.M. Gresshoff. 1991. DNA amplification fingerprinting using very short arbitrary primers. *Bio/technology* 9: 553–557.

Caetano-Anollés, G., R.N. Trigiano and M.Windham. 2001. Patterns of evolution in *Discula* fungi and the origin of dogwood anthracnose in North America, studied using arbitrarily amplified and ribosomal DNA. *Curr. Genet.* 39: 346–354.

Crute, I.R. 1985. The genetic basis of relationships between microbial parasites and their hosts, pp. 80–142, in *Mechanisms of Resistance to Plant Diseases*, Fraser R.S.S. (Ed.). Martinus Nijhoff/Dr. W Junk Publishers, Dordrecht.

Ender, A., K. Schwenk, T. Städler, B. Streit and B. Schierwater. 1996. RAPD identification of microsatellites in *Daphnia*. *Mol. Ecol.* 5: 437–441.

Galibert, F., T.M. Finan, S.R. Long, A. Pühler, P. Abola and F. Ampe. 2001. The composite genome of the legume symbiont *Sinorhizobium meliloti*. *Science* 293: 668–672.

Hermosa, M.R., I. Grondona, J.M. Diaz-Minguez, E.A. Iturriaga and E. Monte. 2001. Development of a strain-specific SCAR marker for the detection of *Trichoderma atroviride* 11, a biological control agent against soilborne fungal plant pathogens. *Curr. Genet.* 38: 343–350.

James, W.C., P.S. Teng and F.W. Nutter. 1990. Estimated losses of crops from plant pathogens, pp. 15–50, in *CRC Handbook of Pest Management*, Vol. 1, Pimental, D. (Ed.). CRC Press, Boca Raton. .

Karp, A., K.J. Edwards, M. Bruford, S. Funk, B. Vosman, M. Morgante, O. Seberg, A. Kremer, P. Boursot, P. Arctander, D. Tautz and G.M. Hewitt. 1997. Molecular technologies for biodiversity evaluation: opportunities and challenges. *Nat. Biotechnol.* 15: 625–628.

Landegren, U. 1993. Molecular mechanics of nucleic acid sequence amplification. *Trends Genet.* 9: 199–204.

Litt, M. and J.A. Luty. 1989. A hypervariable microsatellite revealed by *in vitro* amplification of a dinucleotide repeat within the cardiac muscle actin gene. *Am. J. Hum. Genet.* 44: 397–401.

Milgroom, M.G. 1996. Recombination and the multilocus structure of fungal populations. *Annu. Rev. Phytopathol.* 34: 457–477.

Ochman, H. and N.A. Moran. 2001. Genes lost and genes found: Evolution of bacterial pathogenesis and symbiosis. *Science* 292: 1096–1098.

Perring, T.M., A.D. Cooper, R.J. Rodriguez, C.A. Farrar and T.S. Bellows. 1993. Identification of whiteflies species by genomic and behavioral studies. *Science* 259: 74–77.

Rommens, C.M. and G.M. Kishore. 2000. Exploiting the full potential of disease-resistance genes for agricultural use. *Curr. Opin. Biotechnol.* 11: 120–125.

Southern, E.M. 1975. Detection of specific sequences among DNA fragments separated by gel electrophoresis. *J. Mol. Biol.* 98: 503–517.

Southern, E.M. 1996. DNA chips: Analyzing sequence by hybridization to oligonucleotides on a large scale. *Trends Genet.* 12: 110–115.

Trigiano, R.N., G. Caetano-Anollés, B.J. Bassam and M.T. Windham. 1995. DNA amplification fingerprinting evidence that *Discula destructiva*, the cause of dogwood anthracnose in North America, is an introduced pathogen. *Mycologia* 87: 490–500.

Vos, P., R. Hogers, M. Bleeker, M. Reijans, T. van de Lee, M. Hornes, A. Frijters, J. Pot, J. Peleman, M. Kuiper and M. Zabeau. 1995. AFLP: A new technique for DNA fingerprinting. *Nucl. Acids Res.* 23: 4407–4414.

Weber, J.L. and P.E. May. 1989. Abundant class of human polymorphisms which can be typed using the polymerase chain reaction. *Am. J. Hum. Genet.* 44: 388–396.

Welsh, J. and M. McClelland. 1990. Fingerprinting genomes using PCR with arbitrary primers. *Nucl. Acids Res.* 19: 861–866.

Williams, J.G.K., A.R. Kubelik, K.J. Livak, J.A. Rafalski and S.V. Tingey. 1990. DNA polymorphisms amplified by arbitrary primers are useful as genetic markers. *Nucl. Acids Res.* 18: 6531–6535.

25 Molecular Techniques Used to Study Systematics, Ecology, and Evolution of Plant Pathogens

Robert N. Trigiano, Malissa H. Ament, Ladare F. Habera, and Gustavo Caetano-Anollés

The primary objective of these laboratory exercises is to familiarize undergraduate and graduate students (and instructors) with two very powerful molecular techniques that are used either to characterize **DNA** of plant pathogens and other organisms or define relationships between organisms. Both techniques are similar in that they use the **polymerase chain reaction (PCR)** to amplify or increase copies of DNA, but differ primarily in the sequences of the **genomic DNA** that are targeted for amplification.

The first technique is **DNA amplification fingerprinting** or **DAF** (Caetano-Anollés et al., 1991). This technique is a DNA profiling protocol that employs relatively short **arbitrary primers** (5 to 10 base pairs) that target anonymous, but discrete, regions of genomic DNA. Amplification in the DAF technique produces a multitude of products of various sizes, which can be separated and visualized as bands on an acrylamide gel. The DAF procedure is partitioned into four independent laboratory exercises: DNA isolation, DNA amplification, gel electrophoresis and silver staining, and data collection and analysis. Although the DNA amplification and gel electrophoresis exercises are emphasized, very detailed, easy-to-follow instructions and protocols are provided for all aspects of the DNA fingerprinting process. The procedure is adapted largely from Trigiano and Caetano-Anollés (1998) with permission from the American Society for Horticultural Science, and we recommend that you obtain a copy for your reference.

The second technique involves the selective amplification of the internal transcribed spacer (ITS) regions that flank the 5.8S nuclear ribosomal unit (rRNA). The PCR reaction is completed with longer paired primers (18 to 22 base pairs) of known sequences and on amplification produces a single band or product. These exercises or similar ones have been successfully completed on the first attempt by several classes of novice undergraduate, graduate students, and other researchers.

DNA fingerprinting can be defined functionally as a sampling procedure capable of reducing the extraordinarily complex genetic information contained in DNA to a relatively simple and manageable series of bands or "bar codes," which represent only selected, but defined, portions of a genome. Comparison of **DNA profiles** or fingerprints from different, but closely related, organisms can reveal regions with unlike nucleotide sequences (**polymorphisms**) that uniquely identify individuals much like the distinctive patterns of a person's fingerprints. A note of caution about the limitations of arbitrary fingerprinting techniques is in order, especially in determining relationships among a group of organisms. One assumption of the techniques made by inexperienced researchers is that bands appearing at the same position for different samples (locus) in the gel are of the same base pair (sequence) composition. In reality this is not always the situation. An individual "band or locus" from a sample organism may actually either contain several different amplified products of similar weight that comigrate or the sequence of the product may be very different than products from other organisms in the comparison. One way to minimize these types of errors is to select very closely related organisms such as isolates of a fungus species or cultivars of plants. These limitations notwithstanding, DNA fingerprinting has been used in genetic and physical mapping, map-based cloning, ownership rights, molecular systematics, phylogenetic analysis, marker-assisted breeding, parentage testing, gene expression, and many other applications in the plant sciences, including plant pathology.

Prior to the 1990s, DNA characterization required molecular hybridization (Southern, 1975) or selective DNA amplification (Mullis et al., 1986; Erlich et al., 1991). These techniques demanded prior knowledge of DNA sequence information, clones, or characterized probes, and often required extensive experimentation (Caetano-Anollés, 1996). Since then, a multitude of techniques (see review Caetano-Anollés and Trigiano, 1997) has been developed that employ relatively short (5 to 20 nucleotides), arbitrary oligonucleotide primers to direct **DNA polymerase**-mediated amplification of discrete, but anonymous, segments of DNA. Among these techniques are random amplified polymorphic DNA (RAPD) analysis (Williams et al., 1990) and DNA amplification fingerprinting (DAF; Caetano-Anollés et al., 1991; Caetano-Anollés and Gresshoff, 1994). Both methods produce information that characterizes a genome somewhere between the level of the DNA sequence and chromosomes.

Arbitrary oligonucleotide primers amplify multiple genomic regions (amplicons), many of which are variant (polymorphic) and represent allelic differences that can be traced in inheritance studies or can be treated as characters that can be used in population or phylogenetic analyses. The amplification reaction occurs through the succession of temperature cycles. Under low stringency conditions (low annealing temperature and ionic environment), the primer with arbitrarily (user) defined sequence binds to many sites distributed in the genomic DNA template. DNA synthesis is initiated by a thermostable DNA polymerase, even in those cases where there is substantial mismatching between primer and template base sequences. Despite perfect or imperfect priming, the DNA polymerase continues the amplification process by the successive addition of template-complementary bases to the 3' terminus of the primer. Strand elongation is increased by raising the reaction temperature to an optimum level (usually about 72°C) and generally ends when the temperature is high enough to allow for the denaturation (disassociation) of the template DNA and the newly copied DNA strand. The separated strands now serve as template DNA when the reaction temperature is decreased to a point where primer annealing is permitted again (usually less than 60°C). Following this initial amplification cycle, successive changes in temperature result in the selective amplification of genomic regions bordered by primer annealing sites occurring in opposite strands and separated by no more than a few thousand nucleotides (bases). The outcome of the amplification reaction is primarily determined by a competition process in which **amplicons** that are the most stable (efficient) primer annealing sites adjoining the easily amplifiable sequences prevail over those that are inefficiently amplified. A model to explain the amplification of DNA with arbitrary primers was proposed (Caetano-Anollés et al., 1992) and later discussed in detail (Caetano-Anollés, 1993), and is based on the competitive effects of primer–template as well as other interactions established primarily in the first few cycles of the process. Essentially, the rare, but stable, primer–template duplexes are transformed into accumulating amplification products. The final outcome is the selection of only a small subset (5 to 100) of possible amplification products.

Within the sample's DNA sequences, polymorphisms arise from nucleotide substitutions that create, abolish, or modify particular primer annealing sites, which may alter the efficiency of amplification or priming. The resultant polymorphic RAPD or DAF fragments are useful DNA markers in general fingerprinting or mapping applications. These markers have been profusely applied in the study of many prokaryotic and eukaryotic organisms.

Although both RAPD and DAF analyses produce similar types of information, there are some differences between the two techniques: (1) DAF uses very short primers, usually 7 or 8 nucleotides in length, whereas RAPD typically uses 10 nucleotide primers. (2) DAF products are resolved by 5 to 10% polyacrylamide gel electrophoresis and silver staining (Bassam et al., 1991), whereas RAPD products are typically separated electrophoretically in agarose gels, stained with ethidium bromide, and visualized under UV light. (3) The **reaction mixture** or **cocktail** in DAF contains higher primer-to-template ratios than with RAPD and produces relatively complex banding profiles containing 30 to 40 products that are lesser than 700 base pairs in length. In turn, RAPD usually generates more simple patterns of 5 to 10 bands. (4) DAF polyacrylamide gels are backed by using polyester films and are amenable to permanent storage, whereas RAPD agarose gels are difficult to store and a photograph serves as the only permanent record. One could argue the relative merits of each fingerprinting technique, but from our experience, DAF is easier for students and instructors to learn and use and is very tolerant, almost forgiving, of some errors typically made by novices, such as inaccurate pipetting. etc. The data are permanently recorded in the form of a gel instead of a photograph, which is very gratifying to students and facilitates research by allowing repeated and close scrutiny of data. Some research laboratories also have the capability to scan and store data from gels as computer records.

The intention of these exercises is not to fully describe and explore the theoretical aspects of DNA fingerprinting, which may be otherwise obtained by reading the literature cited throughout this chapter. The educational objectives of the laboratory exercises are to acquaint students with the general concepts, techniques, and uses of DNA fingerprinting, and to remove some of the perceived mystique underlying molecular genetics.

A number of products are mentioned throughout the laboratory exercises. Complete information is provided in Table 25.1 should an instructor wish to order from any company mentioned in this chapter. These are simply what we normally use and do not constitute product endorsements by either the authors, the publisher, or the University of Tennessee, nor implied criticism of those products not mentioned. There are equally suitable, if not alternative, products and equipment that may be substituted for those listed herein.

Before beginning the exercises, a few essential generalities apply to all laboratories: (1) All pipette tips,

TABLE 25.1
Sources for Laboratory Equipment and Materials

Supplier	Product	Address	Phone, Fax, or URL
Applied Biosystems	DNA polymerase	850 Lincoln Center Drive, Foster City, CA 94404	800-327-3002; www.appliedbiosystems.com
Barnstead/Thermo-lyne Corp.	Nanopure water	2555 Kerpar Blvd., PO Box 797, Dubuque, IA 52004	800-553-0039; www.barnstead.com
Bio-Rad	Electrophoresis supplies; tips	2000 Alfred Nobel Dr., Hercules, CA 94547	800-424-6723; www.bio-rad.com
BioVentures, Inc.	Biomarkers	P.O. Box 2561, Murfreesboro, TN 37133	800-235-8938
BioWhittaker Molecular Applications	Gelbond PAG film	101 Thomaston St., Rockland, ME 04841	800-341-1574; www.bmaproducts.com
Electron Microscopy Sciences	16% Formaldehyde	P.O. Box 251, 321 Morris Rd., Fort Washington, PA 19034	800-523-5874
Ericomp, Inc.	Thermocycler	6044 Cornerstone Ct. W., Suite E, San Diego, CA 92121	800-541-8471
Exeter Software	NTSYS-pc, version 2.0	100 North Country Road, Sedtauket, NY 11733	516-689-7838
Fisher	Centrifuge tubes, acrylamide	P.O. Box 4829, Norcross, GA 30091	800-766-7000; www.fischersci.com
Gentra Systems, Inc.	DNA isolation kit	15200 25th Ave. N., Suite 104, Minneapolis, MN 55447	800-866-3039; www.gentra.com
Integrated DNA Technologies	Synthesis of primers	1710 Commercial Park, Coralville, IA 52241	800-328-2661; www.idtdna.com
Midwest Scientific	Flat loading tips	280 Vance Rd., Valley Park, MO 63088	800-227-9997; www.midsci.com
MJ Research	Thermocycler	149 Grove St., Watertown, MA 02172	800-729-2165
Pharmacia Biotech, Inc.	Mini-Fluorometer; plates	800 Contennial Ave., Piscataway, NJ 08855–1327	800-526-3593; www.apbiotech.com
Phenix	Pipette tips, centrifuge tubes	3540 Arden Road, Haywood, CA	800-767-0665; www.phenix1.com
Qiagen	QiaQuick PCR clean-up kits	9600 DeSoto Ave., Chatsworth, CA 91311	800-426-8157; www.quigen.com
Rainin Instrument Company	Pipettes and tips	Rainin Road, Box 4026, Woburn, MA 01888–4026	800-472-4646; www.rainin.com
Sierra-Lablogix, Inc.	Staining trays	1180-C Day Road, Gilroy, CA 95020	800-522-5624
US Biochemical	dNTPs	P.O. Box 22400, Cleveland, OH 44122	800-321-9322

eppendorf centrifuge tubes, water, and reagents used to assemble the amplification reactions either should be autoclaved or filter sterilized (0.22 μm) and made with sterile, high-quality water. (2) Participants should wear either latex or acetonitrile gloves to avoid hazardous materials (acrylamide and silver nitrate) and protect samples from **DNases** found on the skin (Dragon, 1993). (3) Where possible, use only ACS (American Chemical Society) certified pure chemicals and double-distilled or nanopure water (<16 MΩ/cm, Barnstead/Thermolyne Corp.), hereafter referred to as "pure." It is not necessary to use HPLC-grade water.

Characterization of genomes by DAF, as well as any other of the arbitrary primer-based techniques, always consists of at least four independent phases: isolation of DNA, amplification of DNA, **electrophoresis** and visualization of amplified products, and collection and analysis of data. Each of these steps requires between 4 and 6 h to complete. If class and laboratory time is limited, the instructor may opt to complete one or more of the laboratory sessions for the students. In fact, for large classes, the instructor may wish to complete the exercises as a demonstration. However, students will derive the most benefit by fully participating in each of the laboratory sessions. This chapter emphasizes DNA amplification and DAF product separation and visualization, and, to a lesser degree, DNA isolation and data analyses. We recommend that these laboratory experiments and procedures be completed by advanced undergraduate or graduate students working in teams of four or less.

EXERCISE 1: DNA AMPLIFICATION FINGERPRINTING

Following an examination of a few journal articles concerning some aspect of DNA analysis, it is evident that there is a multitude of methods to isolate genomic DNA, all of them more or less suitable for the purposes of this laboratory exercise. Fortunately, DAF reactions do not require the high quality or large quantity of DNA necessary, for example, in RFLP analysis. In our lab, we use a DNA isolation kit especially formulated for plants or fungi (Puregene®, Gentra Systems, Inc., Table 25.1) or a procedure developed specifically for fungi (Yoon et al., 1991). Unlike earlier methods for isolation of DNA, most commercially available kits avoid the use of highly toxic materials such as phenols. Regardless of the technique or kit used to isolate DNA, young, quickly growing cultures of Gram-negative bacteria or fungi should be used. Mycelia or bacterial cells should be stored at −70°C until needed. These laboratory exercises will be illustrated using *Fusarium oxysporum* isolates.

DNA ISOLATION

Materials

Each team of students will require the following materials:

- PureGene plant or fungus DNA isolation kit (one kit for the entire class will be sufficient) or follow Yoon et al. (1991)
- 100% Ethanol
- 70% Ethanol
- Sterile 1.5- and 0.65-ml eppendorf centrifuge tubes
- Sterile 100- and 1000-μl pipette tips and pipettors
- High-speed table top centrifuge
- Water bath at 60°C
- Liquid nitrogen (wear insulated gloves and eye protection) and Dewar vessel
- Sterile, chilled mortar and pestle for each isolate
- Mycelium (either fresh or frozen at −70°C).
- Sterile, pure water
- Other reagents required by DNA isolation kit (see manufacture's instructions)
- Insoluble polyvinylpolyprolidone (PPVP)

Follow the protocol outlined in Procedure 25.1 to complete DNA isolation.

Procedure 25.1 Isolation of DNA from Mycelium

1. For each isolate, place 25 mg or less of mycelium with about 25 mg polyvinylpolypyrrolidone (PPVP, which sequesters plant phenols) into a sterile mortar and pestle, add liquid nitrogen, and grind frozen mycelium to a powder. Add 500 μl of extraction buffer.

2. Continue to grind and freeze and thaw at least twice, adding additional extraction buffer if necessary. Slurry should be very thin and watery when melted.

3. Load about 400 μl of the slurry into sterile, 1.5-ml centrifuge tubes and float in a 60°C water bath for about 1 h. Centrifuge at maximum rotation (14,000 rpm) for 10 min to deposit (pellet) cellular debris and PPVP. Transfer supernatant to a new, sterile centrifuge tube and complete the kit's instructions except for the RNase step when using a fluorometer, and include when DNA concentration will be determined by a spectrophotometer (see Procedure 25.2).

4. At the end of the isolation procedure, do not redissolve the DNA in TE buffer; instead use 50 μl of sterile, pure water. *Note*: A large DNA pellet is unlikely, and, in fact, you may not see a distinctive pellet. A little faith is required now — there is DNA in the bottom of the tube. Heat the contents of the tubes in a 60°C water bath for about 2 min to help dissolve the DNA and refrigerate (4°C) overnight. The next morning, centrifuge for 2 min to pellet any undissolved particulate material, and then carefully pipette the supernatant containing the DNA into new, sterile, 0.65-ml centrifuge tubes labeled F1 to F7 and store at 4°C.

DETERMINING DNA CONCENTRATION

DAF is exceptionally tolerant of both the quality (purity) and quantity of DNA used in the reaction mixture. DNA concentration can be determined spectrophotometrically with a dedicated **fluorometer** (e.g., Mini-Fluorometer, Pharmacia Biotech, Table 25.1) invariably set at 365 nm. The fluorometer only reads DNA; RNA is not detected. Instructions for using the fluorometer are included with the instrument; dye and calf thymus DNA for standard concentrations of DNA may be purchased directly from Pharmacia Biotech (Table 25.1). Because most isolations will usually yield DNA concentrations between 10 and 75 ng/μl, we recommend preparing and calibrating the fluorometer with standards of similar concentrations.

If a dedicated fluorometer is not available, DNA content can be determined directly using a spectrophotometer

(Procedure 25.2). The 260/280 ratio of a pure double-stranded DNA preparation should be between 1.65 and 1.85. Although this ratio is dependent on the fractional GC content, higher ratios are often due to RNA contamination and lower values due to protein or phenol contamination. Thus, if determining DNA concentration using this method, it is imperative that the RNA in the sample be eliminated with the **RNase** step in the isolation procedure. Note that low concentrations of DNA are very difficult to read by using the spectrophotometer.

If you are using a spectrophotometer, follow the protocols outlined in Procedure 25.2 for this part of the exercise.

Procedure 25.2 Determining DNA Concentration by a Spectrophotometer

1. "Zero" spectrophotometer by pipetting 1 ml of distilled water into both sample and reference quartz cuvettes.
2. Pipette and mix by gentle inversion 2 μl of DNA into 1 ml of distilled water in the sample cuvette.
3. Read absorbance (optical density) at 260 nm (e.g., 0.012).
4. An OD of 1.0 = 50 μg/ml DNA; therefore, if the OD is 0.012, the entire sample contains 50 μg/ml $\times\, 0.012 = 0.6$ μg or 600 ng/ml. The amount of DNA in each μl = 600 ng/2 μl or 300 ng/μl.

ADJUSTING THE DNA CONCENTRATION

Typically, the concentration of DNA from most isolations is too high to be used directly as the template in the amplification stage of DAF. Optimum concentrations of DNA range from 0.02 to 2.0 ng/μl of the reaction mixture. Therefore, the DNA must be diluted with sterile, pure water to a more functional concentration, such as 5.0 ng/μl (Procedure 25.3). We have stored isolated DNA and dilutions at 4°C for more than 5 years without apparent degradation.

Follow the protocols outlined in Procedure 25.3 to complete this part of the exercise.

Procedure 25.3 Diluting DNA for Use in Reaction Mixtures

1. Determine concentration of DNA in isolation from fluorometer or Procedure 25.2. Assume that the concentration of isolated DNA is 79 ng/μl.
2. Make a 5 ng/μl solution by using the following formula:

$$C_1 \times V_1 = C_2 \times V_2$$

where C_1 is the concentration of the isolated DNA (ng/μl), V_1 the volume (μl) of the isolated DNA to dilute (arbitrarily used 20 μl), C_2 the concentration of diluted DNA (5 ng/μl), and V_2 the volume of diluted DNA (μl, unknown). Substitute in the equation and solve for V_2:

$$79 \times 20 = 5 \times V_2 = (79 \times 20)/5 = V_2$$

$V_2 = 316$ μl = total volume of diluted DNA.

3. As 20 μl of original DNA was used, the amount of sterile, pure water to add is:

$$316\ \mu l - 20\ \mu l = 296\ \mu l$$

Pipette 20 μl of the original DNA solution into a sterile 0.65-ml tube and add 296 μl of sterile, pure water. Mix thoroughly by vortexing and centrifuge (14,000 rpm for 5 sec) to remove air bubbles. This is the template concentration you will use for the amplification cocktail.

4. Store all DNA stocks at 4°C.

QUESTIONS

- Why is EDTA included in the extraction procedure?
- Why is the extraction solution buffered?
- What is genomic DNA?
- Why is it important to wear gloves and use sterile tubes?

EXERCISE 2: DNA AMPLIFICATION

This laboratory exercise requires careful experimental design and planning, and involves handling of many liquid reagents. By completing this exercise, students will gain experience and confidence with routine procedures in a molecular biology laboratory.

Let us begin with a few helpful hints. Plan ahead and write everything down! It is very easy to forget what has been and needs to be done; record keeping is an integral part of good laboratory practices. When in doubt, change sterile pipette tips. Do not risk cross contamination of solutions and templates or introduction of DNases to save a pipette tip. Mix, by vortexing, all stock solutions except DNA polymerase, and centrifuge tubes to remove large bubbles from the liquid before opening and to avoid aerosols. Look at the pipette tip to ensure that an appropriate amount of fluid has been taken into the lumen. Finally, always wear gloves to protect stock solutions and DNA from contamination with either bacteria or DNAases.

MATERIALS

Each team of students will require the following materials:

- AmpiTaq Stoffel Fragment DNA polymerase (Applied Biosystems, Table 25.1)
- 66 mM $MgCl_2$ for fungi. (Do not use the 10× $MgCl_2$ for plants included in the kit; use all other components.) Sterilize with a 0.22-μm filter.
- Primers (Integrated DNA Technologies, Inc., Table 25.1) with the sequences (5′ to 3′) GAGCCTGT (8.6A), CCTGTGAG (8.6B), CTAACGCC (8.6G), and CCGAGCTG (8.7A). The first number in the primer codes denotes oligonucleotide length and the second represents the approximate fractional GC content. Follow Procedure 25.4 to prepare the correct primer concentration.
- Deoxynucleoside triphosphates (dNTPs, US Biochemical, Table 25.1) are supplied as a set of four ampoules containing 25 μmol of each dNTP in 250 μl of water (100 mM). Simply combine the four ampoules (1 ml) in a sterile container and add 11.5 ml of sterile, pure water to produce a 2-mM solution containing all the necessary dNTPs. Dispense 250-μl aliquots into fifty 0.65-ml sterile centrifuge tubes and store at −20°C. This is the working dNTP concentration for the amplification reaction mixture.
- Heavy mineral oil (no need to sterilize; some **thermocyclers** do not require oil)
- Assorted sterile pipette tips
- Sterile 0.65-ml eppendorf tubes
- Sterile, pure water
- Thermocycler — PCR machine

Follow the protocols in Procedure 25.4 to prepare primers.

Procedure 25.4 Preparation of Primer Stocks

1. Prepare a 300-μM primer stock; for example, if 159 nmoles (on label) is provided by supplier:

$$159 \text{ nmoles primer}/x \text{ μl} = 300 \text{ μM}$$

$$x = 532 \text{ μl}$$

Add 532 μl of sterile, pure water to the manufacturer's tube containing an individual primer. Mix thoroughly by vortexing and centrifuge briefly.
2. Prepare 30 μM primer stock: Pipette 30 μl of 300 μM into a sterile 0.65-ml eppendorf tube and add 270 μl of sterile, pure water. Mix thoroughly by vortexing and centrifuge briefly. This is the stock to use in the amplification cocktail.
3. Store all stocks at −20°C in a nondefrosting refrigerator.

PLANNING THE EXPERIMENT

DNA from the seven isolates of your organism should be amplified with four primers. A total of twenty-eight sterile 0.65-ml centrifuge tubes are needed. Label the tubes 1 to 28 according to the scheme depicted in Figure 25.1 and record in a laboratory notebook.

AMPLIFICATION OF DNA

The assembly of the amplification cocktail is the heart of the DAF technique. Each group of students should make "master mixes" (one or more) containing all the ingredients common in each amplification reaction cocktail (Procedure 25.5). The only two variables in the *Fusarium* experiment are the DNA templates from the individual isolates and the primers. Master mixes should therefore contain sterile pure water, Stoffel buffer, dNTPs, magnesium chloride, Stoffel enzyme and a single primer; assembly should be in a sterile centrifuge tube labeled with the primer code (e.g., 8.6A). Add DNA template to individual reaction tubes later. Procedure 25.5 details how to make a master mix and provides a list of reagents and their final concentrations in 20 ml of the mixture. Start by removing the ingredients from the freezer, and after thawing, vortex and centrifuge briefly (except the enzyme). All persons involved in making the master mixes should wear gloves.

Follow the protocols in Procedure 25.5 to complete this portion of the experiment.

Procedure 25.5 Preparation and Assembly of Master Mixes for Each Primer

Note: Sufficient for eight reactions — always prepare more than you have samples to allow for pipetting errors.

1. Pipette 65.6 μl (8.2 × 8) of sterile, pure water into a sterile 0.65-ml tube.
2. Add 16 μl (2 × 8) of 10× Stoffel buffer provided by the manufacture (final conc. = 1×). (Final con-

Fungal DNA templates

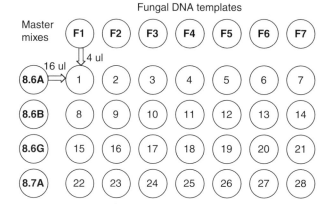

FIGURE 25.1 Scheme for dispensing master mixes and DNA templates. Pipette 16 μl of master mix into each row of seven sterile 0.65-ml reaction centrifuge tubes. Pipette 4 μl of DNA template into each reaction tube. Remember to change tips between tubes.

centrations of reagents are based on 20 μl reaction volumes after the addition of 4 μl of template DNA.)

3. Add 16 μl (2×8) of 2 mM dNTPs (final conc. = 200 μM).

4. Add 16 μl (2×8) of 30 μM primer stock (final conc. = 3 μM).

5. Add 9.6 μl (1.2×8) of 66 mM $MgCl_2$ (final conc. = 4.0 mM).

6. Add 4.8 μl (0.6×8) of DNA polymerase provided by manufacturer.

7. Mix thoroughly by vortexing and centrifuge briefly at 14,000 rpm to eliminate air bubbles.

8. Place the master mixes to the left of the four rows of seven reaction tubes (in a plastic flipper rack) labeled 1–7, 8–14, 15–21, and 22–28 as shown in Figure 25.1. Dispense 16 μl of the master mix into each of seven sterile 0.65-ml tubes. Master mix for an individual primer can be distributed to tubes without changing tips. However, a different tip should be used for each of the four master mixes since they contain different primers.

9. Remove the 5.0 ng/μl DNA stocks from the refrigerator, vortex, and briefly centrifuge at high speed. Place the DNA stocks at the top of the flipper rack above those reaction tubes 1, 2, 3, 4, 5, 6, and 7, which correspond to the isolates to be analyzed (Figure 25.1).

10. Pipette 4 μl from F1 into tube 1, then close the reaction tube and discard the tip. Repeat the sequence for tubes 8, 15, and 22, then close stock tube F1. Repeat the procedure for cultivars F2 through F7. Each tube now contains 20 μl.

11. Mix the contents of the tubes by vortexing and centrifuging briefly at high speed. Open the tubes and add a drop of heavy white mineral oil to each

to prevent evaporation and condensation during amplification. *Note*: Some thermocyclers do not require oil — see thermocycler manufacturer's instructions.

12. Place the tubes in a thermocycler for amplification of the DNA.

Thermocyclers are programmed to establish **annealing** (30°C to 55°C), **extension** (72°C), and **denaturing** (95°C) **temperatures** for prescribed times. This set of temperature regimens constitutes a cycle, which is repeated 30 to 40 times. However, because there may be significant differences in ramping times between thermocyclers, proceed with caution when searching for a suitable cycle. Ramping time can be thought of as the time necessary for the amplification mixture to go from one designated temperature to the next (e.g., from annealing to extension temperature). The Easy Twin Block System (Ericomp Inc., Table 25.1) has relatively slow ramping times compared to the DNA Engine PTC-200 (MJ Research, Table 25.1). Reproducible, clear profiles can be generated with the Ericomp machine by using a cycle of 10 sec at 95°C and 30°C, without an extension step. However, for the DNA Engine, which has faster ramping times, the cycle of 1 min at 95°C and 30°C with an extension step at 72°C for 30 sec works well. A general rule is to increase the annealing, extension, and denaturing times when the thermocycler has short ramping times. The entire amplification process takes between 2 and 6 h depending on the thermocycler.

RECOVERY AND STORAGE OF AMPLIFICATION PRODUCTS

You may skip this section if oil was not used in the amplification. Before removing the products, prepare several tubes containing 20 μl of blue water [i.e., a drop of xylene cyanol stock: 12 g urea and 2 ml of xylene cyanol stock solution (4 mg xylene cyanol in 8 ml of pure water) in 20 ml of water] under a drop of mineral oil. Students can easily see the blue color and practice pipetting the water without drawing any oil into the tip. Once the amplification program is completed, the reaction products are removed from beneath the oil. Label 28 sterile 0.65-ml centrifuge tubes by designating an experiment (e.g., 1) and the tube number (e.g., 1.1, 1.2, … 1.28). A P-100 (Rainin, Table 25.1) pipette or equivalent set on 22- or 23 μl with sterile Prot/Elec tip (Bio-Rad, Table 25.1) can be used. Perform the operation quickly and "squirt" the contents of the tip into a new sterile, prelabeled 0.65-ml centrifuge tube without touching the sides to avoid contamination with oil clinging to the tip. When working with the reaction tubes, be sure to change pipette tips between each sample. Store samples at 4°C as is or diluted 1:1 with sterile, pure water.

QUESTIONS

- Why is a "low" annealing temperature used in DAF?
- Why is magnesium ion necessary in the reaction mixture?
- When the DNA polymerase enzyme assembles the new DNA strand, are there copy errors?

EXERCISE 3: ELECTROPHORESIS AND STAINING OF DAF PRODUCTS

This laboratory exercise focuses on the electrophoretic separation of amplification products and should expose students to one of several electrophoretic techniques routinely used to characterize biological molecules. Experience with this technique should facilitate understanding and performance by students of similar procedures such as protein electrophoresis.

ASSEMBLING THE GEL APPARATUS

While wearing gloves, assemble two Protean II (or III) Electrophoresis Cells (Bio-Rad, Table 25.1) a day before the DAF products are to be separated electrophoretically. We recommend using 0.5-mm spacers that can be purchased separately for the Protean II apparatus and 0.75-mm spacer for the Protean III. GelBond flexible backing supports (sheets) can be purchased from BioWhittaker Molecular Applications (Table 25.1). Here are a few helpful hints in assembling the rigs. Meticulously clean the glass plates with running distilled water to remove any dust and **acrylamide** from previous experiments. Assemble the rig under running distilled water. Place the hydrophobic surface (the side that water beads on) of the backing film on and toward the large glass plate and rub the hydrophilic surface until all trapped air is evacuated. All gel rig components should be flush at the bottom. Lastly, do not overtighten the knobs (Protean II) — the glass plates will bow and produce a thickened center portion of the gel, which will not stain properly. The assembled apparatus should be examined carefully to ascertain that the glass plates, spacers, and backing film are flush with each other on the bottom. Run a fingernail across the bottom of the apparatus. If it does not feel smooth or if the fingernail gets "hung-up," the level of the glass plates, spacers, or support film needs to be adjusted. The gel rigs should be allowed to dry overnight in a place that is dark and dust-free.

CASTING THE GELS

Materials

Each team of students will require the following items:

- Acrylamide stock solution — toxic, wear gloves (Procedure 25.6)
- A 0.22-μm filter and 10-ml syringe
- TEMED — toxic, wear gloves
- 10% ammonium persulfate (100 mg/ml pure water) solution — Ammonium persulfate may be made in bulk, dispensed into 1.5-ml centrifuge tubes and frozen at −20°C
- Two assembled Protean II or III gel rigs, casting stand, and two 0.5-mm combs
- One 10-ml disposable pipette and pipette pump.
- One 25- or 50-ml beaker and stir bar
- Aluminum foil
- Acetonitrile gloves

Follow the protocols listed in Procedure 25.6 and Procedure 25.7 to make running buffer and acrylamide stock solution, and to assemble and pour gels. Always wear acetonitrile gloves when working with acrylamide and TEMED.

Procedure 25.6 Composition of 10× TBE Buffer and 10% Polyacrylamide Stock

- 10× TBE Buffer
  1. Dissolve 121.1 g Tris base, 51.4 g boric acid, and 3.7 g Na$_2$EDTA·2H$_2$O in 800 ml of pure water.
  2. Bring the final volume to 1 l with pure water; find pH of solution = 8.3. Store at room temperature.
  Note: If room is cool, salts may not remain dissolved. Try making a 5× buffer.
- 10% Polyacrylamide stock
  1. Dissolve 19.6 g acrylamide, 0.4 g PDA (piperazine diacrylamide), and 20.0 g urea in 130 ml of pure water. [*Caution:* Wear protective particle mask, gloves, eyeglasses and clothing — unpolymerized acrylamide is a potent neurotoxin; avoid skin contact and accidental inhalation of the compound.] Do not substitute BIS (*N,N′*-methylene bis acrylamide) for PDA as it adversely affects staining quality.
  2. Add 20.0 ml of 10× TBE buffer and 10.0 ml glycerol to the acrylamide solution.

3. Bring the final volume to 200 ml with pure water. Store at 4°C in a brown bottle and discard unused portion after 8 weeks.

Procedure 25.7 Casting Gels

1. Wear gloves! Mount the gel rigs (Protean II) onto the casting stand by using the gray rubber gaskets on the bottom. We usually place several equal length and width strips of parafilm wrap under the gasket to ensure a good seal with the glass plates. A very distinct snap should be heard as the rigs are set into place on the casting stand. Place the casting stand onto a large piece of aluminum foil on which a 10-ml syringe and 0.22-μm filter can be laid.

2. Pipette 10 ml of the 10% polyacrylamide stock (Procedure 25.6) into a 20-ml beaker containing a small magnetic stir bar. Place the pipette tips containing 15 μl of TEMED and 150 μl of 10% ammonium persulfate solutions into the stirring acrylamide solution and dispense. Dispose of the tips in a safe location. Stir for about 10 sec.

3. The following steps in casting the gel should be completed as quickly as possible (usually less than 2 min). Carefully draw the gel solution into a syringe, avoiding introduction of air into the barrel. If air bubbles are present, hold the syringe upside-down at 70° away from the body; the air should rise to the top. Slowly depress the plunger until the air is expelled. Mount a nonsterile 0.22-μm filter on the open end of the syringe. Slowly express a small amount of acrylamide to wet the filter and release any trapped air.

4. Place the tip filter in the middle of the ledge formed by the small (short) plate and quickly dispense the acrylamide solution into the space between the glass plates. Rotate the casting stand 180° and fill the second gel rig with acrylamide. If there are bubbles trapped in the gel, gently tap the inner (short) glass plate and with luck, they will rise to the top.

5. Position the 10- or 15-well combs about halfway (level) in each of the rigs and examine for small bubbles residing on the bottom surface of the teeth. If bubbles are present, remove and reposition the combs. For Protean III, completely insert combs.

6. Allow the acrylamide to polymerize for at least 20 min. If desired, the gels may be cast the day before the laboratory exercise and stored overnight lying flat on the bottom of a plastic container that is lined with wet paper towels. Be careful not to disturb the combs and store in the dark.

PREPARING SAMPLES FOR ELECTROPHORESIS

Materials

Each team of students will require the following items:

- A microtitre plate
- Loading buffer (0.25% bromophenol blue, 0.25% xylene cyanol, and 15% type 400 Ficoll in water)
- Molecular marker solution (1:10 or undiluted)
- P10 pipettor and tips
- Amplification products

Amplification products can be prepared for electrophoresis while the acrylamide is polymerizing. First, carefully pipette 3 μl of loading buffer into the number of wells in a 6 × 10 microtiter plate (Pharmacia Biotech, Inc.). In the case provided in Figure 25.2, each row in the plate would have eight wells filled, seven for the samples and one for the molecular weight marker. However, we encourage teams of students to use a separate plate for their DNA. Map the order of the samples in the gel in the laboratory notebook, for instance (from left to right), sample 1.1, 1.2, … 1.7, and M. Now, in the order that the samples will be placed in the gel, pipette 3 ml of each into their respective wells and mix by repipetting the solution several times. Change tips between samples. The last well is the molecular weight marker (M) consisting of a 50- to 1000-base-pair ladder (Biomarker Low, BioVen-

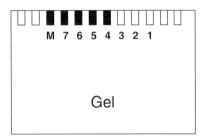

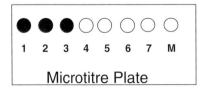

FIGURE 25.2 Loading DNA samples into the gel. Because the gel apparatus was rotated 180° when mounted on the central reservoir stand, the samples must be loaded in reverse order or from right to left in the gel. For example, load Sample 7 in the fourth well from the left.

tures, Inc.). Make the working solution by mixing 10 μl of biomarker with 90 μl of sterile, pure water. Biomarkers may also be purchased as ready to use — do not dilute. As with the other samples, 3 μl of biomarker is used per well. Be sure to replace the microtiter plate cover to prevent evaporation of the sample preparations.

PRERUNNING GELS AND PREPARING STAIN AND DEVELOPER SOLUTIONS

Materials

Each team of students will require the following items:

- 5× or 10× TBE running buffer
- 1-l graduated cylinder
- Protean II or III reservoir and central stand
- Tuberculin syringe with 25-gauge needle
- Power supply (two or three teams can share this item)

Make 1 l of 1× TBE buffer by mixing 100 ml of 10× TBE (Procedure 25.6) and 900 ml of water in a 1-l graduated cylinder and mix thoroughly. Dismount the two gel rigs from the casting stand and gently remove any polymerized acrylamide from the bottom of the plates with a laboratory tissue. Rotate the rigs 180° and snap into the central stand. Be careful not to touch or disturb the combs. When both rigs are mounted, the small plates of the rigs will be toward the interior, facing each other, and the outer plates will form the top buffer reservoir. Fill inner and outer reservoirs with a total of about 800 ml of 1× TBE. Carefully remove the combs from the gels by gently pulling straight up with equal pressure on both sides; do not damage the wells. Fill a 1-ml (cc) tuberculin syringe, equipped with a 1.5-inch long, 25-gauge needle with buffer from the central well. Gently insert the needle tip about one-quarter of the way into the top portion of a well and gently force the buffer into the well. This will flush the accumulated urea and errant bits of acrylamide from the wells. Repeat the process so that all wells of both gels are cleaned. Connect the apparatus to the power supply and set to a constant 180 to 200 V for 15 to 20 min.

While the gel is prerunning, there will be time to prepare both the silver stain and carbonate developer solutions. Both solutions may be prepared in bulk, including every constituent except formaldehyde and sodium thiosulfate. Silver nitrate solution is light sensitive and should be stored in a brown bottle. The sodium thiosulfate solution should be prepared weekly and stored in the refrigerator. If preparing developing and staining solutions for daily use, then plan on 75 ml for each gel.

Follow the protocols in Procedure 25.8 to prepare silver stain and developer solutions.

Procedure 25.8 Composition of Silver Stain and Developer Solutions

- Silver Stain
  1. For two gels, dissolve 0.15 g of ACS-certified silver nitrate in 150 ml of pure water.
  2. A few minutes before use, add either 750 μl of 16% or 325 μl of 37% formaldehyde.
- Developer
  1. For two gels, dissolve 4.5 g of ACS-certified sodium carbonate ($NaCO_3$) in 150 ml of pure water and chill to 2°C to 4°C.
  2. Add 75 μl of sodium thiosulfate solution (0.2 g/50ml).
  3. Before use, warm the solution to 8°C and add either 600 μl of 16% or 260 μl of 37% formaldehyde (open formaldehyde in a fume hood).

LOADING SAMPLES AND RUNNING THE GEL

After prerunning the gels, clean the wells in one gel as described previously. With a P10 (or equivalent) pipette adjusted to deliver 6.5 μl, load the samples into the wells by using flat tips (Midwest Scientific, Table 25.1). Because the gels were rotated 180°, load the gels in reverse order or from right to left as indicated in Figure 25.2. Draw the far right sample in the microtiter plate into a flat pipette tip. Keeping the flat tip parallel to the glass plate, guide it partially into the third well from the left side with the left index finger and gently dispense the sample into the well. Be careful not to damage the well. Eject the tip into a beaker of distilled water. Flat tips can be cleaned by first drawing water and then 95% ethanol through them with a vacuum. These tips do not need to be autoclaved. Load the next sample with a new flat tip. Load all the samples for this gel and repeat the procedure, including cleaning the wells, for the other gel. *Note:* Maintain the same sample loading order in each of the gels. Keeping the same order between gels will greatly facilitate data collection. Reconnect the power supply and run at a constant 180 to 200 V for about 1 to 1.5 h or until the blue tracking dye reaches the level of the bottom platinum electrode.

STAINING AND DEVELOPING GELS

Turn off the power and disconnect the gel apparatus from the power supply. Wearing gloves, disassemble the Protean II gels under distilled water by first loosening the four knobs and gently removing the glass plate sandwich from the apparatus. The gel RIGS simply release from the Protein III holder. Holding the "sandwich" with the large glass plate contacting the palm of the left hand and

TABLE 25.2
Troubleshooting Gels: Some Common Imperfections and Their Causes

- Bands in some lanes but not in others: DNA template missing or degraded in lanes with weak or no products.
- No amplification — all lanes blank: Missing ingredient in master mix, degraded primer, or, less likely, all DNA templates degraded.
- Dark streaks in lanes: Old loading buffer or dust particle on the bottom surface of the well.
- Lightly staining products in center of gel: Developer poured directly on gel; glass plates warped creating thickened gel in center; developer less than 8°C.
- Individual bands not straight but jagged: Bottom surface of well damaged.
- Lanes not straight, but deflected: Air bubble under support film.
- Bubbles in gel: Aspirated air from syringe or air bubble adhering to a dirty glass plate.
- Light bands or no bands in lanes: Not all primers work well with all organisms; try other primers.
- Black smudges in gel: Incomplete removal of silver stain solution before adding developer.

in a stream of or a pan of distilled water, insert the fingernail of the right index finger under the top corner of the small glass plate and gently pry it upward. Let the water do most of the work. The backing film and the gel may now be separated from the large plate and placed in a staining tray (Sierra-Lablogix, Inc., Table 25.1) or in lids from pipette boxes. Follow the staining and developing procedures outlined in Procedure 25.9. Remember to add formaldehyde to silver stain and developer solutions (Procedure 25.8) just prior to use. After silver staining is completed, quickly and thoroughly rinse the gels with pure water to remove all excess silver nitrate solution. Do not pour the cold developing solution directly on the gels; instead, introduce the solution onto the bottom of the staining dish and immediately place on a rotary shaker set at about 40 to 60 rpm. Continue shaking until the bands in the marker and sample lanes are dark and sharp or until the margins and the background of the gel start to discolor (overdeveloped). Stop with cold 7.5% acetic acid. The gels may be "hung to dry" in a dust-free environment after they are treated with anticracking solution (under a fume hood, add 100 ml of glacial acetic acid, 10 ml of glycerol, and 370 ml of 95% ethanol to 520 ml of pure water).

Procedure 25.9 Synopsis of Fixing, Staining, and Developing Gels

1. Fix gels in 7.5% acetic acid for 10 min on a rotary shaker (60 rpm).
2. Rinse gels with pure water three times, each for 2 min, on a rotary shaker (60 rpm).
3. Soak gels in silver stain for 20 to 30 min on a rotary shaker (40 rpm).
4. Rinse gels in pure water for 10 to 20 s.
5. Soak gels in 8°C developer for 5 to 8 min (or until bands are dark) on a rotary shaker (40 to 60 rpm).
6. Fix gels in cold (4°C) 7.5% acetic acid for 5 min on a rotary shaker (60 rpm).
7. Soak gels in pure water two times each for 5 min on a rotary shaker (60 rpm).
8. Soak gels in anticracking solution for 5 min on a rotary shaker (60 rpm).
9. Hang gels overnight to dry.

Label the gels with the experiment number and gel identification number (e.g., 1-A) with a permanent marker after the gels are dry in 12 to 24 h. Gels may be stored indefinitely in photo albums. Table 25.2 describes some of the more common gel imperfections, their causes, and remedies.

QUESTIONS

- What purpose does the loading buffer serve?
- How does the percentage of acrylamide affect the migration of amplified products?
- Do all amplified products appearing at the same level in the gel (loci) have the same sequence?
- Does a single band in the gel represent a single amplification product?

EXERCISE 4: DATA COLLECTION AND ANALYSES

DAF data will be analyzed by the Numerical Taxonomy and Multivariate Analysis System (NTSYSpc) program, version 2.0G (Exeter Software, Table 25.1). The analyses are easily understood and provide estimates of genetic distances and relationships between isolates.

DATA COLLECTION

Materials

- White light box (transilluminator)
- Computer with NTSYS-pc version 2.0

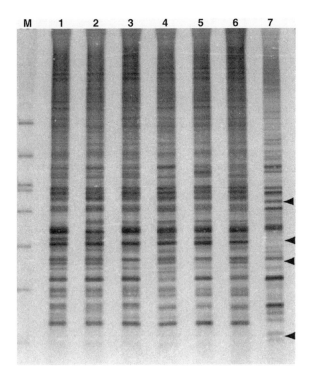

FIGURE 25.3 DNA profiles of *Fusarium oxysporum* isolates (F1 to F7: Lanes 1 to 7) using primer 8.6A. Note some of the many polymorphisms (arrowheads). M = molecular weight markers (from top: 1000, 700, 525, 500, 400, 300, and 200 bp).

- Clear plastic ruler
- Clear, 12 inch × 12 inch glass plate

View the dried gels on a light box and cover with a clear, glass plate. Beginning at about 700 bp, align common bands in the different sample lanes with a straight edge and enter the binary data: 1 = product present, 0 = product absent and 9 = missing data. For example, in Figure 25.3 (primer 8.6A) at about 675 bp, a prominent band appears in all sample lanes. The data for this character locus would be the following: 1 1 1 1 1 1 1. Continue to record data for the entire gel. Combine data from the primers and enter in the computer as shown in Table 25.3. The order in which the data from individual primers are entered into the data set is not important. Be careful to include a hard return after each line of the data set and eliminate any extraneous spaces within the lines. The first line of the data set should begin with the number "1" followed by the number of lines (character loci) in the data set, e.g. "154," then the number of samples (7L) and "1" then "9" if there were any missing values. If there were no missing values, enter "0" instead of "1." The next line contains the abbreviations for the samples — F1, F2 etc. Save the data set as an ASCII file. You may also use an Excel spreadsheet for data entry, which can be imported by NTSYSpc.

DATA ANALYSES

After your data is in the correct format, you are now ready to open NTSYSpc. At any time during your session, you can click on the "notebook" icon at the top of the screen to view a log of your analyses. This notebook can be saved as text, printed, etc. It can be a useful record of where to find output files for later reference, or to find out why the program will not run on your data.

Follow the instructions in Procedure 25.10 to analyze data.

Procedure 25.10 Cluster and Principal Component Analyses DAF Data by NTSYSpc

- Cluster Analysis
  1. Click the *Similarity* tab. Select the *SimQual* button. Double click the entry window to choose the file from the hard drive or diskette, or type the path and name of the data file. At the coefficient line, select **J** for Jaccard. Enter a path and filename for the output. For example, if the data file is *Fusarium*, call the output *Fusarsim*. Then click on *Compute* (Table 25.4).
  2. Click the *Clustering* tab. Select the *SAHN* button. Double click the entry window to choose the file from the hard drive or diskette, *Fusarsim* above, or type the path and filename of the similarity output. Type the path and filename for the output, for example, *Fusarclus*. Be sure that *UPGMA* is selected in the clustering method. Click *Compute*.
  3. To view the phenogram (Figure 25.4A), click the icon resembling a phenogram in the lower left of the window. To save the phenogram as a *.emf file, click *File/Save* and name the phenogram. To print, click *File/Print/OK*. To copy the phenogram into a word-processing document, click *Edit/Save bitmap*. Minimize NTSYS. Open the word-processing program. Right click in the body of the document to insert the phenogram. Click *Paste*. Save this word-processing file.
- Principal Component Analysis
  1. Click the *General* tab. Select the *Dcenter* button. Double click the entry window to choose the file from the hard drive or diskette, *Fusarsim* above, or enter the path and filename of the similarity output. Enter the path and filename for the output, for example, *FusarDC*. Leave the *Square Distances* box checked. Click *Compute*.

TABLE 25.3

An Example of Matrix Definitions and Binary Data for *Fusarium oxysporum* Isolates to be Analyzed by the NTSYSpc Program

This data set (rectangular = 1) includes 111 character loci from 7 isolates (initals) and contains missing data represented by a 9. Rows with all 1s are monomorphic; a row containing a single 1 identifies a unique marker for an isolate.

1	111	7L	9			
F1	F2	F3	F4	F5	F6	F7
0	1	0	0	0	1	1
1	1	1	1	1	1	1
1	1	1	1	1	9	1
0	0	1	0	0	0	0
.	.	.	.	.	.	.
1	1	1	1	1	1	1

TABLE 25.4

Example of a Similarity Matrix Generated Using Jaccard Coefficient for Seven Isolates of *Fusarium oxysporum*

Isolates	F1	F2	F3	F4	F5	F6	F7
F1	1.000						
F2	0.780	1.000					
F3	0.863	0.883	1.000				
F4	0.823	0.822	0.865	1.000			
F5	0.831	0.830	0.895	0.897	1.000		
F6	0.779	0.796	0.817	0.800	0.826	1.000	
F7	0.627	0.624	0.644	0.615	0.618	0.693	1.000

2. Click the *Ordination* tab. Select the *Eigen* button. Double click the entry window to choose the file from the hard drive or diskette, *FusarDC* above, or enter the path and filename of the Dcenter output. Enter the path and filename for the output, for example, *FusarEign*. Click *Compute*.

3. To view the three-dimensional PCA graph (Figure 25.4B), click the icon resembling a PCA graph in the lower left of the window. Be sure to *Click Options/Plot Options* and click the button by the word *Label* so that the graph is labeled. The view of the three-dimensional graph can adjusted by turning it or leveling it. To save the PCA graph as a *.emf file, click *File/Save* and name your graph. To print, click *File/Print/OK*. To copy your PCA graph into a word-processing document, click *Edit/Save bitmap*. Minimize NTSYS. Open the word-processing program. Right click in the body of the document, where you want to insert the PCA graph. Click *Paste*. Save this word-processing file.

QUESTIONS

- Are the isolates closely related to each other?
- How many character loci are needed to get an accurate representation of the relationships among isolates?
- Does the choice of coefficient in the similarity measure influence the calculated relationship between isolates?

EXERCISE 5: AMPLIFICATION OF INTERNAL TRANSCRIBED SPACERS (ITS) OF rRNA

Eukaryotic ribosomal genes are arranged in tandem repeats with the 5.8S coding region flanked by internal transcribed spacers (ITS; Figure 25.5). Although these regions are important for maturation of nuclear ribosomal RNA (rRNA), ITS regions are usually considered to be under low evolutionary pressure and therefore typically treated as nonfunctional sequences. The sequences of the ITS regions have been used in many phylogenetic studies and molecular systematics of fungi. This experiment will allow students to amplify one or both of the

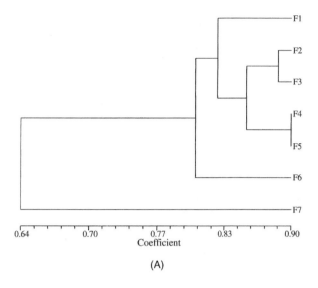

(A)

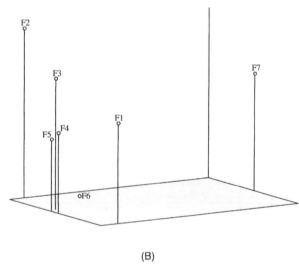

(B)

FIGURE 25.4 Analysis of *Fusarium* DAF data. (A) Diagrammatic representation of UPMGA or cluster analysis of the relationships between seven isolates of *Fusarium oxysporum*. (B) Diagrammatic representation of principal component analysis of the relationships between seven isolates of *F. oxysporum*.

ITS regions (1 and 2) and is largely based on Caetano-Anollés et al. (2001) and White et al. (1990). The experiment will not describe DNA sequencing (usually completed at a university center) or the analysis of sequence data. Instructors and students are encouraged to seek expert help on their campus.

MATERIALS

- AmpliTaq Gold DNA polymerase kit with buffer II (Applied Biosystems, Table 25.1)
- All materials under Experiment 2 (DNA amplification), except primers
- The following primers (5′-3′) in 3 μM concentrations:

For ITS1 region use: TCCGTAGGTGAACCT-GCGG (ITS1) and GCTGCGTTCTTCATC-GATGC (ITS2)

For ITS2 region use: GCATCGATGAA-GAACGCAGC (ITS3) and TCCTCCGCT-TATTGATATGC (ITS4)

- Stock fungal DNA from DAF experiment diluted 1:9 or 0.5 ng/μl
- QiaQuick PCR Purification Kit (Qiagen; Table 25.1)

Follow the protocol outlined in Procedure 25.11 to amplify the ITS region.

Procedure 25.11 Amplification of the ITS1 and ITS2 Regions

1. Pipette 46.4 μl (8 × 5.8) of pure water into a sterile, 0.65-ml eppendorf tube.
2. Pipette 16.0 μl (8 × 2) of 2 mM nucleotides into the tube.
3. Pipette 16.0 μl (8 × 2) of 10 × Buffer II (without magnesium chloride) into the tube.
4. Pipette 9.6 μl (8 × 1.2) of 25 mM magnesium chloride into the tube.
5. Pipette 16 μl (8 × 2) of each of either ITS1 and ITS2 or ITS3 or ITS4 primers (*Note*: As an alternative, use ITS1 and ITS4 primers and amplify both ITS1 and ITS2 regions plus the gene).
6. Pipette 8 μl (8 × 1) of AmpliTaq Gold DNA polymerase.
7. Vortex and centrifuge briefly. Dispense 16 μl of the reaction mixture into seven sterile 0.65-ml eppendorf tubes. Now add 4 μl of fungal DNA template (0.5 ng/μl) to each of the seven as indicated in Figure 25.1. Be sure to change pipette tips between different isolates. Vortex and centrifuge briefly.
8. Add drop of heavy, white mineral oil to each tube and place in thermocycler (some thermocyclers do not require oil – follow the manufacturer's instructions).

The thermocycler should be programmed as follows:

Step 1: 95°C for 9 min — 95°C activates the DNA polymerase enzyme that the Stoffel fragment polymerase did not require. This step is also termed a *hot start*.

Step 2: 96 C for 1 min. —Denatures DNA

Step 3: 56°C for 1 min — Primers anneal to sites

Step 4: 72°C for 1 min — Extension

Repeat Step 2 to Step 4 for 35 cycles

Step 5: 72°C for 7 min.

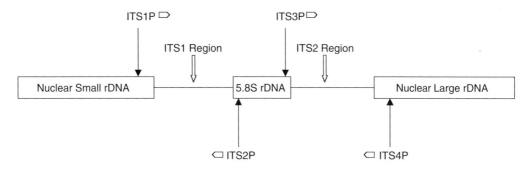

FIGURE 25.5 Diagrammatic representation of ITS and ribosomal regions. ITS1P and ITS2P are primers for ITS1 region and ITS3P and ITS4P are primers for ITS2 region. (Modified from White, T.J., T. Bruns, S. Lee and J. Taylor, 1990. pp. 315–321 in *PCR Protocols: A Guide to Methods and Applications.* Innis, M.A., J. Gelfand, J. Sninky and T.J .White (Eds.), Academic Press, San Diego.)

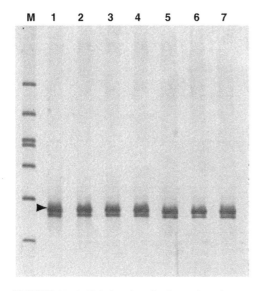

FIGURE 25.6 Gel showing single product (arrowhead at about 275 bp) of the ITS1 region from seven *Fusarium oxysporum* isolates (Lanes 1 to 7) produced by using primers ITS1 and ITS2. The lighter band below is probably an artifact. M = molecular weight markers (from top: 1000, 700, 525, 500, 400, 300, and 200 bp).

Recovery, electrophoresis, staining, and developing of the amplified product are the same as described under Experiment 1. Sometimes more than one product appears in the gel (Figure 25.6). The lighter product is an artifact and will not affect sequencing — the vast majority of product is the complete ITS region. If you were to run the product on an agarose gel and stain with ethidium bromide, the second lighter product would not be visible. You may or may not be able to detect visually differences in absolute number of base pairs in the ITS regions between *Fusarium* isolates.

Before the product can be sequenced, other remaining components of the reaction mixture must be removed. We use a QiaQuick PCR Purification Kit (Qiagen). The instructions in the kit are excellent and easy to follow.

Inquire at your sequencing center whether the DNA should be eluted in pure, sterile water or buffer. It is imperative to use sterile water that is between pH 7.0 and 8.5 for the QiaQuick kit. If you use distilled, pure water (pH = ~5.5), the DNA bound to the membrane will not be released. Most sequencing centers require the DNA concentration in the samples to be between 5 and 10 ng/μl and you must also supply some of the primers used in the amplification reaction.

QUESTIONS

- What are insertions and deletions?
- Why would you not send DNA in TE (Tris-EDTA) buffer to the sequencing center?
- Would any product be formed using only one primer in the amplification cocktail?

REFERENCES

Bassam, B.J., G. Caetano-Anollés and P.M. Gresshoff. 1991. Fast and sensitive silver staining of DNA in polyacrylamide gels. *Anal. Biochem.* 196: 80–83.

Caetano-Anollés, G. 1993. Amplifying DNA with arbitrary oligonucleotide primers. *PCR Meth. Appl.* 3: 85–94.

Caetano-Anollés, G. 1996. Scanning of nucleic acids by *in vitro* amplification: New developments and applications. *Nat. Biotechnol.* 14: 1668–1674.

Caetano-Anollés, G., B.J. Bassam and P.M. Gresshoff. 1991. DNA amplification fingerprinting using very short arbitrary oligonucleotide primers. *Bio/Technology* 9: 553–557.

Caetano-Anollés, G., B.J. Bassam and P.M. Gresshoff. 1992. Primer-template interactions during DNA amplification fingerprinting with single arbitrary oligonucleotides. *Mol. Gen. Genet.* 235: 157–165.

Caetano-Anollés, G. and P.M. Gresshoff. 1994. DNA amplification fingerprinting of plant genomes. *Meth. Mol. Cell. Biol.* 5: 62–70.

Caetano-Anollés, G. and R.N. Trigiano. 1997. Nucleic acid markers in agricultural biotechnology. *AgBiotech: News Inf.* 9: 235N-242N.

Caetano-Anollés, G., R.N. Trigiano and M.T. Windham. 2001. Patterns of evolution in *Discula* fungi and origin of dogwood anthracnose in North America, studied using arbitrarily amplified and ribosomal DNA. *Curr. Genet.* 39: 346–354.

Dragon, E.A. 1993. Handling reagents in the PCR laboratory. *PCR Meth. Appl.* 3: S8-S9.

Erlich, H.A., D. Gelfand and J.J. Sninsky. 1991. Recent advances in the polymerase chain reaction. *Science* 252: 1643–1651.

Mullis, K.B., F.A. Faloona, S. Scharf, R. Saiki, G. Horn and H. Erlich. 1986. Specific enzymes amplification of DNA *in vitro*: The polymerase chain reaction. *Cold Spring Harbor Symp. Quant. Biol.* 51: 263–273.

Southern, E.M. 1975. Detection of specific sequences among DNA fragments by gel electrophoresis. *J. Mol. Biol.* 98: 503–517.

Trigiano, R.N. and G. Caetano-Anollés. 1998. Laboratory exercises on DNA amplification fingerprinting for evaluating the molecular diversity of horticultural species. *HortTechnology* 8: 143–423.

White, T.J., T. Bruns, S. Lee and J. Taylor. 1990. Amplification and direct sequencing of fungal ribosomal DNA genes for phylogenetics, pp. 315–321 in *PCR Protocols: A Guide to Methods and Applications.* Innis, M.A., J. Gelfand, J. Sninky and T.J. White (Eds.), Academic Press, San Diego.

Williams, J.K., A.R. Kubelik, K.J. Livak, J.A. Rafalski and S.V. Tingey. 1990. DNA polymorphisms amplified by arbitrary primers are useful as genetic markers. *Nucl. Acid Res.* 18: 6531–6535.

Yoon, C.-K., D.A. Glawe and P.D. Shaw. 1991. A method for rapid small-scale preparation of fungal DNA. *Mycologia* 83: 835–838.

Part IV

Plant–Pathogen Interactions

26 Pathogen Attack Strategies

Larry E. Trevathan

CHAPTER 26 CONCEPTS

- Pathogenic microorganisms have certain characteristics that provide them with a competitive advantage in the interaction with hosts compared to the relationship of hosts with nonpathogenic microorganisms.

- Successful plant pathogens enter the host, overcome defenses, and cause disease.

- Saprophytic and parasitic microorganisms are (1) passively mobile and energy dependent, (2) actively mobile and energy dependent, or (3) passively mobile and energy independent.

- Passively mobile, energy-dependent fungal saprophytes produce abundant inoculum — a quantifiable element in the strategy for utilizing substrate.

- Facultative parasites and facultative saprophytes attack hosts, cause disease, and utilize dead organic matter as substrate.

- Endophytes are associated with plant hosts across a continuum of symbiosis variously expressed from antagonism to mutualism.

- When obligate parasites, which have lost the ability to effectively function as saprophytes, attack plants, an intimate relationship develops between the pathogen and the host.

- Mobility and energy-dependent movement of plant parasitic nematodes is facilitated by their shape and size.

- The passively mobile, energy-independent attack strategy of plant viruses results in the use of host components for virus production, conservation of viral nucleic acid sequences in hosts, and vectors, viral nucleic sequences in the genome of the plant host, and recombination during production of mature virus particles.

- Plant pathogenic organisms depend to some degree on the production of chemicals by the pathogen, by the host, or unique to the interaction between pathogen and host for successful colonization of susceptible tissues.

The coordinated charge of an organized army is a well-recognized strategy. Forces are chosen, equipped, armed, and trained to accomplish a specific objective. The execution of the objective is the result of a reasoned and planned strategy — an attack strategy. In the absence of the power to reason, do pathogens attack plants? Or is the expression of disease in plants simply the result of chance dissemination, impact, and infection of susceptible hosts by pathogenic organisms? If pathogens do attack plants, how do they do so?

Despite the anthropomorphic nature of the title of this chapter, both laypersons and professionals in the plant sciences recognize that some means of coordinated interaction exists between plants and pathogens. Plant diseases generally are not epidemic and do not endanger the survival of plant species within natural ecosystems. However, most profound and far-reaching plant disease epidemic imbalances can be directly attributed to human activities.

Only about 2% of all known fungal species cause disease in higher plants; the estimate for known bacterial species that cause plant disease is about 6% (Agrios, 1997). Therefore, the majority of both fungal and bacterial species are saprophytes that derive nutrition from dead organic matter. Certain characteristics provide a competitive advantage to some microorganisms in the interaction with hosts that other microorganisms do not possess. One example is the production of antibiotics (Figure 26.1). As a strategic element advantageous to pathogens that attack plants, the ability to produce antibiotics allows a particular microorganism to inhibit the growth of other microorganisms and thereby protect sites of colonization and sources of nutrition.

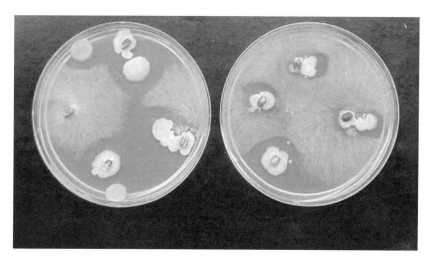

FIGURE 26.1 Antibiosis by bacteria inhibiting the growth of *Rhizoctonia solani*. Both organisms were initiated from wheat seed on potato dextrose agar.

FIGURE 26.2 Corn growing in no-till culture. (Courtesy of G. Triplett.)

SELECTION PRESSURE

Humans, as well as other forces of nature, act as agents of **selection pressure** to induce variability in both plants and plant pathogens. Plant hosts deploy genes in response to pathogens and their vectors, to changes in the vectors, and to changes in the pathogens resulting from mutation and genetic recombination. Some of the activities of humans that contribute to the process of genetic modification of host plants and pathogens include, but are not limited to, elimination or introduction of competing hosts, vectors, pathogens, or other organisms. This can result from the use of specific cultural practices or the adoption of practices not previously used. The modern-day practices of conservation and minimum-tillage have increased the severity of foliar diseases of maize (Figure 26.2). One

reason for this is the increased survivability of pathogens that cause these diseases in maize residue remaining on the soil surface. Whether through cultural practices or other means, human-made activities that result in the dissemination of host, vector, or pathogen germplasm into, or removal from, historical areas of adaptation contribute to genetic modification of both plants and pathogens.

ENERGY EXCHANGE AND INITIAL ATTACK

Adaptation by microorganisms that become plant pathogens is dependent on the ability to enter the plant, the ability to overcome resistance in the host, and the ability to cause disease (Oku, 1994). **Heterotrophic** plant pathogens (those living on organic food sources produced by other organisms) are dependent on autotrophs (organisms

FIGURE 26.3 *Ganoderma applanatum* growing on sugar maple. A single conk
is capable of producing 30 billion spores/day. (Courtesy of W. J. Stambaugh.)

that manufacture their own food) as a source of energy. In this relationship, a portion of the energy derived by heterotrophs from their respective autotrophic hosts is used to attack the autotrophic host or host tissues and complete the disease cycle or life cycle of the pathogen. Because pathogens have developed strategies for attacking host plants over a range of high to low energy requirements, host–pathogen interaction can be defined on the basis of energy exchange and the means by which an organism initiates attack of a host. At one limit of this range, the production and dispersal of large quantities of inocula by fungi or bacteria represent an attack strategy requiring a significant energy investment by the pathogen. By contrast, the transmission and establishment of other pathogens, such as viruses, do not require energy input on the part of the pathogen. On the basis of energy required and initial attack of the host, plant pathogens can be identified as (1) passively mobile and energy dependent, (2) actively mobile and energy dependent, or (3) passively mobile and energy independent. Saprophytes can also be categorized in this manner.

Competition among microorganisms for nutrients is a source of selection pressure that promotes the specialized relationship of parasitism on living plant material. Adaptation by hosts in this relationship has resulted in a range of reactions from resistant plants to susceptible plants. Microorganisms have evolved that function as saprophytes, as facultative parasites (primarily as saprophytes, but have the ability to parasitize plants), as **endophytes**, as facultative saprophytes (primarily as **parasites**, but have saprophytic capabilities), and as parasites. Those that utilize dead plant cells as an energy substrate are also referred to as necrotrophs. At the other extreme are biotrophs that

do not kill host cells, but derive nutrition from living tissue. **Hemibiotrophs** have developed relationships with plant hosts that include a period of dependence on living cells; host tissues are killed in latter stages of infection.

PASSIVELY MOBILE AND ENERGY DEPENDENT

Saprophytes

About 90% of known fungal species are strictly saprophytic. These microorganisms live on and decompose dead organic matter. This is accomplished primarily through processes that involve cellulose and **lignin** metabolism. Some of the most familiar of these microorganisms are the wood-rotting fungi (Chapter 19) that produce conspicuous fruiting bodies on the surface of decaying wood. They are also evident on living trees with symptoms of heart rot, the tissue composed of dead cells. Because these fungi do not attack and penetrate living wood, exposed heartwood must be available for fungal establishment within the host. The lack of aggressiveness by these fungi explains, in part, their strategy for attacking substrate, which is saprophytic in nature (Manion, 1981). They are examples of passively mobile, energy-dependent organisms that produce abundant propagules that are primarily disseminated by wind. A single conk of the wood-decay fungus, *Ganoderma applanatum*, is capable of producing 30 billion spores a day over a period of several months (Figure 26.3). This strategy of overwhelming inoculum production has served these organisms and the biosphere well; as essential degraders of organic material, they recycle elements that are bound and unavailable.

FIGURE 26.4 Wheat crown tissue infected by *Fusarium culmorum*; symptomatic secondary root. (From the *Compendium of Wheat Diseases*. American Phytopathological Society. With permission.)

FIGURE 26.5 Vascular discoloration in banana symptomatic for Panama disease caused by *Fusarium oxysporum* f. species *cubense*. (From APS Slide Set, *Diseases of Tropical Crops*. American Phytopathological Society. With permission.)

Facultative Parasites

Microorganisms that function primarily as saprophytes but have parasitic capabilities have developed strategies for attacking plants. Some that have been very successful as saprophytes have become pathogens. One of the more successful fungal species, and another example of a passively mobile, energy-dependent organism, is *Fusarium oxysporum*. This fungus is commonly isolated from asymptomatic roots of crop plants. It has the ability to persist without resorting to pathogenesis and produces many nonpathogenic strains that colonize a wide range of plant species. *Fusarium oxysporum* colonizes crop residue, quickly

reoccupies soil that has been fumigated to eliminate other organisms and nonpathogenic strains, and aggressively colonize the cortex of plant roots. Isolates of these strains are often not pathogenic to plants from which they were recovered.

Two basic types of plant disease are now recognized to be caused by *Fusarium* species: cortical rots and vascular wilts (Figure 26.4 and Figure 26.5). If pathogenic forms are derived from nonpathogenic forms, *F. oxysporum* could have developed the ability to infect vascular tissue by growing beyond the root cortex. The fungus elicits responses in plants that result in interference with water-conducting capacity and subsequently causes wilting.

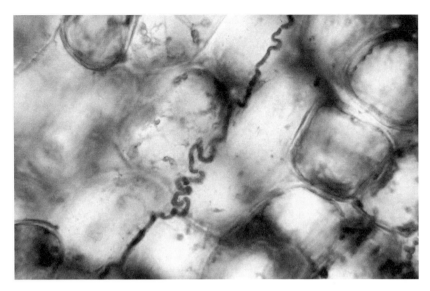

FIGURE 26.6 Stained mycelium of *Neotyphodium (Acremonium) coenophialum* within leaf sheath tissue of tall fescue.

Because many nonpathogenic strains of *F. oxysporum* colonize a wide range of cultivated and native plant species, there are many opportunities for pathogenic strains to develop. In fact, about 80 *formae speciales* (biotypes of a pathogenic species with differential ability to infect selected genera or species of susceptible plants) of *F. oxysporum* have been recognized.

Endophytes

An intermediate response between saprophytism and parasitism could result in limited pathogen growth leading to an endophytic association (Rayner, 1996). Such associations exist between plants and microorganisms and they are variously expressed from antagonism to **mutualism** across a continuum of symbiosis (dissimilar organisms living together in intimate association). In the specific relationship referred to as pleiotropic symbiosis, the fungal symbiont undergoes both sexual and asexual life cycles on separate flowering tillers of infected host plants (Schardl, 1996).

Perhaps the best-known example of this type of symbiosis is between grass hosts and clavicipitaceous endophytes. Approximately 10 million square miles, or 20% of the land surface area of the world, has been established in permanent pastures and meadows. Because nearly 10,000 species of grasses are distributed worldwide, it is not surprising that the relationship between these hosts and endophytes has been well studied. Obligately sexual *Epichloë* species are considered more antagonistic in these associations because of effects on reproduction of the host. The inflorescence of grasses infected by these fungi is aborted as a result of expression of the sexual cycle of the fungus (Chapter 15). Obligately asexual endophytes, such as Neotyphodium (*Acremonium*) species (Figure 26.6),

provide more mutualistic benefits in association with their grass hosts (Chapter 10). These fungi do not naturally survive outside of the hosts and are not transmitted in nature from one plant to another (Hill, 1994). Therefore, the host provides a specific growth niche for the fungus. Fungal mycelium is confined to the intercellular spaces of the leaf sheaths and stems of the plant. Stromata (compact masses of vegetative hyphae in or on which reproductive structures are formed) are not produced and the fungus passes clonally in seed from one generation of the host to the next. Asexual fungal endophytes enhance tillering and root growth and protect host plants from nematodes, disease-causing organisms, and herbivores by producing fungal **alkaloids** and **phenolic** fungitoxins. Ergot alkaloids and **indolediterpenes** (**lolitrems**) are neurotropic and toxic to vertebrates. Peramine and the saturated **aminopyrrolizidines** (**lolines**) deter insect feeding and are insecticidal, respectively.

Facultative Saprophytes

Facultative saprophytes live as disease-producing pathogens for most of their existence, but have the ability to complete their life cycle as saprophytes. Some examples of facultatively saprophytic fungi include *Venturia inaequalis*, which causes apple scab; *Phytophthora infestans*, the causal organism of the infamous late blight of potato; and *Colletotrichum lindemuthianum*, a common inhabitant of home gardens that causes anthracnose of bean. As relationships between organisms have become more specific over evolutionary time, the interaction between facultative saprophytes and hosts has become more refined. *Venturia inaequalis* (Chapter 15) is one of a group of fungi that is continually associated with host tissue regardless of whether the fungus is in a parasitic or

FIGURE 26.7 Scab on fruit of pear infected by *Venturia pirina*. (From the *Compendium of Apple and Pear Diseases*. American Phytopathological Society. With permission.)

saprophytic state. Other facultative saprophytes can parasitize a particular host and then grow saprophytically on host debris or any other available decaying plant material.

Venturia inaequalis is a very successful symbiont of apple (MacHardy et al., 2001). Because the host and fungus have coevolved, a genetic equilibrium has been reached. The apple host expresses vertical, horizontal, and ontogenic (age-related) resistance during the parasitic phase of the life cycle of the fungus. The resistance developed by the host in this association limits both colonization of and damage to apple by the fungus, is effective against different races of *V. inaequalis*, and is durable.

Because apple is a deciduous, temperate zone plant, the association between this host and *V. inaequalis* must account for periods of dormancy, as well as periods of active growth. The saprophytic and parasitic growth phases of the fungus contribute to biological fitness through adaptation to seasonal changes. These phases have evolved to correspond to sexual and asexual stages in the life cycle of *V. inaequalis*. The saprophytic sexual phase that occurs on infected, fallen apple leaves is critical to survival during the winter and to genetic recombination. These leaves are sites of production of pseudothecia or ascocarps with bitunicate asci in one to many unwalled locules produced in a stroma. The pseudothecia contain asci filled with ascospores. Pseudothecial development is dependent on fertilization between compatible mating types of *V. inaequalis*. The resultant ascospores are propagules of primary inoculum that mature early in the apple-growing season when susceptible tissues are present. Initial maturation of ascospores is synchronized with the initial emergence of host tissues from flower buds. Despite variations in the time of leaf fall, bud break, and the interval between these two events, *V. inaequalis* con-

sistently produces the first mature ascospores within a few days of apple bud break in the spring. During this time, parasitic infections are established in the canopy of apple trees as a result of local and distance transport of ascospore inoculum by air currents and wind. Both emerged and newly emerging tissues are susceptible to infection. Once infection is established, the parasitic asexual phase continues throughout the growing season as conidia develop in lesions resulting from initial ascosporic infections. Conidial inoculum is splash-dispersed downward in the tree canopy, and conidia germinate to initiate additional infections as long as free moisture is available (Figure 26.7). This increases the total amount of foliar tissue colonized by the fungus. When infected leaves abscise, fungal hyphae from subcuticular infections grow into the palisade and mesophyll tissues and form stromatic spheres that develop into pseudothecia. Abundant infection of abscised leaves increases the likelihood of pairing of compatible mating types of the fungus and the formation of pseudothecia. This return to the saprophytic sexual phase permits the fungus to survive during periods of host dormancy.

Most plant pathogenic bacteria are facultative saprophytes that use local and long distance means to reach plants and rapid generation times as attack strategies. Some examples include splashing rain, contaminated irrigation water, contaminated planting stock, mechanical transmission, and insect vectors. Long-distance effects can include these factors and are associated with significant meteorological events. One means of distance transport is in the form of bacterial aerosols that can be moved along weather front gradients. Movement of insect vectors, contaminated with bacteria, along weather front gradients is also effective in distance movement.

Because bacteria are short-lived microorganisms, any attack strategy must include movement in a protected state that results in arrival on host substrate in a viable condition. This is effectively accomplished in aerosols that provide a moist environment for large numbers of bacteria that are readily transmitted. The impact of simulated raindrops 5 mm in diameter on potato stems with 2% infection by the bacterium, *Erwinia carotovora*, will result in the production of 800 million colony forming bacteria per hectare. Based on measurements made on dry leaves, rain is not absolutely necessary for the generation of bacterial aerosols. Because a single bacterium is capable of initiating the disease process, the number of colony-forming bacteria and the percentage of infected plants in the field are predicted to be directly proportional.

Facultatively saprophytic plant pathogenic bacteria are cell or tissue specific during the parasitic phase. Those that are specific to vascular tissue and are xylem-limited (Figure 26.8) have developed mechanisms to concentrate and absorb nutrients within the xylem of plants where the liquid volume flow is relatively high and, as a result, the concentration of organic compounds is low (Purcell and Hopkins, 1996). Extracellular strands, produced at the ends of rod-shaped bacteria, bind them to xylem vessel walls, concentrate digestive enzymes, serve as a food reservoir, and bind nutrients, which makes them available to bacteria in a dilute nutritional environment.

Obligate Parasites

Obligate parasites have lost the ability to effectively function as saprophytes. Because of this, these organisms have strict dependence on living cells and have evolved intimate relationships with their hosts that include increasingly effective means of attacking those hosts. Characteristically, they have restricted host ranges that reflect the specificity of the relationship and have elaborate mechanisms to enter plants. In the case of some obligately parasitic fungi, this is accomplished primarily through the formation of specialized infection structures known as **appressoria** (Dean, 1997). Adhesion to plant surfaces prior to infection is a critical step in the attack strategy and subsequent establishment of these fungi. Spores of these organisms may be coated with a mucilaginous material or produce adhesive material on contact with a suitable surface. These spores germinate to produce germ tubes that grow toward particular targets. On reaching the target, appressoria are commonly formed at the tips of germ tubes.

Because obligately parasitic fungi are highly evolved, they are able to respond to environmental, physical, and chemical stimuli. Environmental factors such as light, temperature, moisture, and humidity stimulate the germination of spores of these fungi, other fungi, and other organisms on plant surfaces. Resultant germ tubes growing over

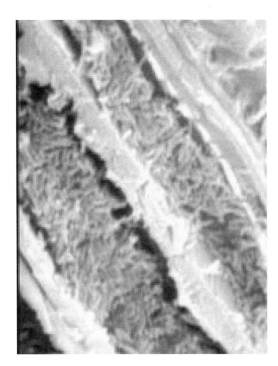

FIGURE 26.8 Pierce's disease bacterium, *Xylella fastidiosa*, inhabiting xylem vessels of perwinkle. (From R. Brlansky. With permission.)

the plant surface respond to physical and chemical factors derived from the host. Although the growth of germ tubes of some fungi appears to be indiscriminate, others are influenced by leaf surface topography. This response is referred to as **thigmotropism** and results in germ tube growth perpendicular to the parallel arrangement of anticlinal cell walls. In the interaction between wheat and the stem rust fungus, *Puccinia graminis* f. species *tritici* (Chapter 18), this directional growth has the advantage of increasing contact with junctions of anticlinal cell walls and stomata, which are the two primary sites of penetration of the leaf surface by the fungus. Appressoria formation at these sites is an adaptive response resulting from chemical interactions. The interaction may be generated by environmental, physical, or chemical stimuli and the mechanism is referred to as signal transduction.

Once an appressorium forms on the leaf surface, infection hyphae are produced that penetrate the leaf and initiate the infection process. If the appressorium forms over a stoma, infection hyphae swell and a vesicle develops in the substomatal cavity. The vesicle forms a thick-walled haustorial mother cell that penetrates the mesophyll cell wall and produces a **haustorium**, a specialized feeding cell. Regardless of the manner of penetration of the host and initiation of infection, the appressoria also provide distinct advantages. Because this structure is essentially sealed to the plant surface, other organisms are excluded and both the site of colonization and the source of nutrition are protected.

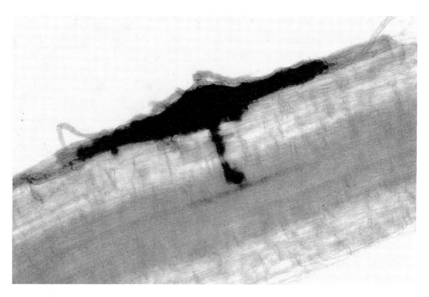

FIGURE 26.9 Dome-shaped infection cushion of *Rhizoctonia solani* on a corn root producing constricted infection peg extending into the stele.

If dependence on infection structures such as appressoria is evidence of greater specialization in more advanced fungi, variation in the expression of such characteristics should be found throughout fungal phyla. In fact, appressoria are found in fungi from different fungal phyla, and there is also variability in appressorium production within fungal species. For example, *Rhizoctonia solani* produces appressoria, penetrates directly in the absence of appressoria, produces an infection cushion, or penetrates through stomata (Figure 26.9). Appressoria produced by different fungi differentiate germ tubes of differing length. If the appressoria develop soon after spore germination, germ tube length is restricted. In some species, such as *V. inaequalis*, germ tube length is so short that the appressorium appears to be directly connected to the spore.

ACTIVELY MOBILE AND ENERGY DEPENDENT

Nematodes

Of those organisms that move actively in search of susceptible hosts and expend energy in the process, plant pathogenic nematodes (Chapter 8) are the most obvious. It is not difficult to find organisms with tiny, elongate, eel-shaped bodies moving through the aqueous phase of the soil matrix. The total number of nematode species is estimated to exceed the total diversity of insect species. Most nematodes are free living and feed on microorganisms and organic matter. Those with a flexible stylet, or mouth spear, account for about 20 known genera and can penetrate plant cell walls for food uptake.

Morphological adaptations, such as stylet development, often accompanied by physiological adaptations,

result in the development of specific pathogen attack strategies. Nowhere is this more obvious than in nematodes that parasitize plants by the stylet (Hussey and Williamson, 1998). To use the stylet to feed in the cells of a plant root, plant parasitic nematodes must be able to locate a host. The strategy of host location by active movement has evolved nematodes that range from those that are primarily migratory ectoparasites to those that are primarily sedentary endoparasites. The morphological and physiological adaptations in plant parasitic nematodes that make them efficient pathogens result in corresponding morphological and physiological changes in the plant host. This is most obvious in the association between plants and nematodes that are primarily sedentary endoparasites. A reorganization of host root morphology and physiology occurs at the site of interaction. This results in complex feeding structures that provide permanent sources of nutrition for nematode development and reproduction.

Host location by plant pathogenic nematodes is accomplished through the use of chemoreceptors known as **sensilla** or sense organs. These organs are concentrated at the anterior (head) end of the nematode. They are complemented by **proprioreceptors** (pressure, position, or movement) and mechanoreceptors. Once a nematode has found a host, the strategy for attacking that host becomes more localized and specific through the use of chemoreceptors to detect chemotactic signals from plant roots. The nematode may spend a limited or an extended period of time at a specific site. Migratory ectoparasitic nematodes only use the stylet to penetrate root tissues at selected sites for a limited period of time. At those sites, these nematodes may attack epidermal cells or subsurface tissues (Figure 26.10). Those nematode species that feed on epi-

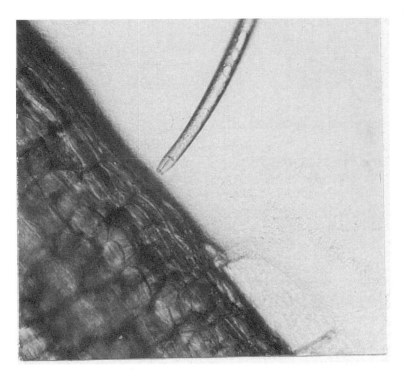

FIGURE 26.10 The migratory ectoparasitic nematode, *Quinisulcius acutus*, probing feeding sites on a sorghum root.

dermal cells have evolved short stylets, which inject secretions that predigest cytoplasm prior to ingestion. Other species with long, needle-like stylets liquefy the cytoplasm and nucleoplasm of cells of subsurface tissues. Sedentary ectoparasites, which remain at one location for an extended period, feed from a single cortical cell in a similar manner. These nematodes may remain vermiform, or wormlike, and change feeding sites or attach permanently to the root surface and become rotund in shape.

Migratory ectoendoparasitic nematodes exhibit advanced adaptation by feeding both ectoparasitically and endoparasitically on roots. These nematodes remain vermiform and the shape facilitates the migratory feeding habit. As endoparasites, these nematodes invade roots partially and feed on cortical or stelar cells. Nematodes that are entirely migratory endoparasites feed only in cortical cells. However, any developmental stage of these nematodes is capable of invading or leaving root tissues. Intracellular movement results in death of invaded cells and of adjacent, uninvaded cells.

The morphological changes referred to earlier in plants attacked by sedentary endoparasites are the result of specific physiological responses at discrete feeding sites. Giant cells and **syncytia**, which form at these sites, are strong nutrient sinks (Williamson and Hussey, 1996). Giant cells are the result of abnormal cell growth following nuclear division. Syncytia form when cells fuse as a result of the dissolution of cell walls. These morphological responses are readily observed in plant host tissues. Main-

tenance of these structures is dependent on continuous stimulation by the nematode and results in a prolonged feeding association. As a result, nematodes that feed in this manner have evolved the sedentary, endoparasitic habit. This very specific symbiotic relationship has reduced the necessity for mobility by the nematode and morphological and physiological changes in these pathogens mirror those in the host. When feeding begins, the body volume of the nematode increases. This increased body volume renders the nematode incapable of leaving the roots and locomotory musculature is lost.

PASSIVELY MOBILE AND ENERGY INDEPENDENT

Viruses

Some of the most specific pathogen attack strategies in plants involve viruses and the relationship of viruses (Chapter 4) to their vectors. Plant pathogenic viruses are composed primarily of nucleic acid surrounded by a protein coat. As a result of the nature of the replication process of plant viruses, there are opportunities for recombination in the production of mature virus particles. The release of viral nucleic acid into the host system allows the virus to modulate the synthesis of mature virions produced by the host. A significant advantage to the virus is conservation of nucleic acid sequences in the host genome or in a vector genome. A host or a vector can function as a source of precursors for virus replication and the host and vector both can serve as a reservoir of viral genomic material. The evolutionary history of such a relationship should contain landmarks of association between the virus and the host species. Indeed, DNA sequences of geminiviruses have been found in the nuclear genome of tobacco and other *Nicotiana* species. The distribution of viral DNA sequences provides evidence for the presence of geminiviruses in *Nicotiana* species predating the cultivation of tobacco.

In addition to acting as refugia of genomic sequences of plant pathogenic viruses, vectors also serve as agents of physical transfer. Organisms not normally considered to be vectors of plant viruses might provide one means for viruses to attack plants. Complete replication of a geminivirus can take place in the bacterium *Agrobacterium tumefaciens* (Rigden et al., 1996). If geminiviruses were derived from prokaryotic replicons, an integration event facilitated by bacteria such as *A. tumefaciens* may be responsible for the initial association of these viruses with plants.

In vectored relationships involving plant viruses and hosts, there are advantages to the virus that may not accrue to other pathogens and the hosts that they attack. The virus passes in what could be described as a protected state from host plants to vectors and back. The virus also becomes a benefactor of the feeding preference of the vector for plants that are sources of virus. Virus transmission may be nonspecific and accomplished by several species of vectors within a single taxonomic group. Specificity in virus/vector relationships means the virus is successfully transmitted by only a limited number of vector species. This type of vector-specific transmission is actually mediated by the virus and not by the vector. Because a number of plant viruses, including reoviruses, marifiviruses, rhabdoviruses, tospoviruses and tenuiviruses, replicate in both plants and insects, arthropod viruses may be ancestral to plant viruses or vice versa.

In certain virus/host/vector relationships, the tolerance or susceptibility of the host to the virus has limited influence on the suitability of that host as a source of virus for the vector. One of the best examples is plum pox virus that causes Sharka disease in plums, apricots, nectarines, almonds, cherries, and peaches. Previously reported as a serious disease in Europe, North Africa, the Middle East, and South America, the disease was not known in North America until 1999, when the diagnosis was confirmed in trees from Adams County, Pennsylvania. *Plum pox virus* is transmitted by more than 20 different species of aphids. However, the ability to transmit the virus is not directly related to the ability of aphid species to colonize host trees. Transient aphids can spread this virus just as efficiently as aphids that colonize *Prunus* species.

CHEMICAL ATTACK

Pathogenic organisms that are obligately parasitic on plant hosts depend more on mechanical pressure for penetration of surface barriers than do organisms that have evolved nonobligate, necrotrophic relationships. However, all plant pathogenic organisms depend to some degree on the production of chemicals for successful colonization of host plants. These may be produced by the pathogen, by the host, or may be unique to the interaction between the pathogen and the host. Enzymes that degrade the cell wall (Chapter 27) and toxins are common in these associations, but other substances, such as growth regulators and polysaccharides, are also involved in the production of disease. Substances produced by plant pathogens as normal metabolic constituents, regardless of concentration, are referred to as constitutive; those produced only in response to the appropriate host substrate are adaptive or inductive.

EXTRACELLULAR ENZYMES

Plant pathogens produce a variety of enzymes, which, regardless of whether they are constitutive or inductive, catalyze reactions within host plants that contribute to disease development. Mobilization of different enzymes is directed at barriers that pathogens confront when attempting to enter plant cells. Some of the more useful enzymes (Chapter 10 and Chapter 27) include cutinases, pectinases, cellulases, hemicellulases, ligninases, proteases, lipases, and amylases that direct a regulated order of penetration of the cell cuticle and cell wall. These enzymes facilitate an attack strategy for penetrating the middle lamella, composed primarily of pectin; the primary and secondary cell walls, composed primarily of pectin, cellulose, and hemicellulose; and the cell membrane, composed primarily of protein and lipid. Ligninases are active in degrading lignin found in the middle lamella, secondary cell wall, the epidermis, and the hypodermal cell wall. Finally, amylases break down starch within plant cells to glucose, which is used as an energy source by pathogens. Enzymes that degrade the cell wall are most useful in attack strategies of necrotrophic pathogens with wide host ranges. For example, *Botrytis* species attack more than 1000 plant species. Bacterial plant pathogens that cause tissue maceration and extensive cell death are also characterized by having wide host ranges. Bacterial species such as *Erwinia* species produce pectinases that cause tissue disorganization (Figure 26.11). Although specific genes have been identified that are absolute requirements for pathogenesis, the exact role of specific enzymes remains elusive because of the multigenic nature of gene expression in enzyme production by bacterial plant pathogens. Gene sequences have been identified from plant pathogenic bacteria that express homology to sequences

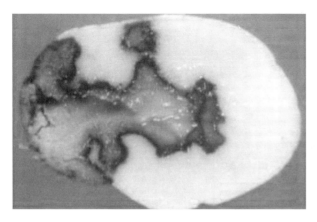

FIGURE 26.11 Bacterial soft rot of potato tuber caused by *Erwinia carotovora* pv. *carotovora*. (American Phytopathological Society. With permission.)

in genes from bacteria that attack animals. During the evolution of plants, animals, and bacteria, certain mechanisms of attack used by bacteria have been conserved and are common to the colonization of both animals and plants.

HORMONES

Constitutive and inductive hormones are distinctly and specifically advantageous to pathogens that attack plants. Certain genes in plants that are involved in defense responses are downregulated by hormones and susceptibility to some pathogens is actually induced by hormones. The hypersensitive response in plants to pathogen attack, a specific defense reaction, is also inhibited by hormones.

The association between plants and microorganisms that produce or induce hormones has resulted in the development of diseases with or without hypertrophic symptomology. Hypertrophy, or abnormal cell enlargement, develops in response to high concentrations of auxin or cytokinin, or both, that alter the regulated hormone balance in host plants. The production of the naturally occurring auxin, indole-3-acetic acid (IAA), is widespread in fungi and bacteria. The involvement of IAA in disease symptomatology, characterized by abnormal cell development, is consistent with the activity of this hormone in plants because it is required for both cell elongation and cell differentiation. Both IAA and cytokinins are produced by the *Pseudomonas* and *Agrobacterium* species, that cause neoplastic or tumor-like diseases. **Hyperauxiny** is also common in diseases caused by downy mildew fungi that are characterized by accelerated plant growth (Figure 26.12). These hormones provide evidence of the interaction of different chemical substances in disease development. IAA increases the plasticity of cell walls, making pectin, cellulose, and protein components more accessible to enzymatic degradation.

TOXINS

Host-specific toxins and nonspecific toxins are important in plant/pathogen interactions. With few exceptions, toxigenic substances injurious to plants are secondary metabolites. By contrast, mammalian toxins are products of primary metabolism and the most common are proteins produced by bacteria. Long-term associations between plants and pathogens have resulted in the production of secondary metabolites and means of neutralizing them, which has been demonstrated in both plants and pathogens.

Toxins selective for plant hosts (HSTs) have very specific activity, are low-molecular-weight secondary metabolites and are known only from fungal pathogens, such as *Alternaria* and *Cochliobolus* species. These toxins are agents of virulence or pathogenicity because fungi producing HSTs cause more disease than those without HST capability. The HST victorin, produced by *C. victoriae*, is a highly phytotoxic and selective compound. This HST is active against sensitive oats at 10 pg/ml, but does not affect resistant oats or other plant species at concentrations one millionfold higher (Walton and Panaccione, 1993).

EXTRACELLULAR POLYSACCHARIDES

The requirement for a stable environment for plant pathogenic bacteria has resulted in the evolution of protective mechanisms at the interface with host plants. One such mechanism is the production of extracellular polysaccharides (EPS) (Chapter 6). Some of what is known about the role of EPS in attack strategies by plant pathogenic bacteria leads to the conclusion that these materials can have a primary role in disease causation. Polysaccharides promote colonization and disease development at the interface of bacteria and host plants. The function of polysaccharides in pathogenesis is most evident in plant diseases caused by bacteria that express symptoms of water soaking and wilting. Cell-free extracts of EPS of *Ralstonia solanacearum* cause wilts in tomato in the absence of the bacterium.

Other lines of evidence support a less direct role of EPS in disease development. These polysaccharides encapsulate bacterial cells, which enhances survival. Prolonged water soaking of plant tissue associated with EPS production may provide an environment conducive to maximum multiplication of bacteria and optimum production of inoculum. In the association of *R. solanacearum* and tomato, the wild-type bacterium is able to multiply within the plant, but mutants that are deficient in EPS production do not multiply as well *in planta*. The bacterium that causes fire blight of apple and pear, *Erwinia amylovora*, enters the host through flowers. The EPS produced by this bacterium, levan, is important for moisture retention to protect bacteria, reduced osmolarity of extra-

FIGURE 26.12 Hyperauxiny of plants symptomatic for downy mildew.

cellular fluids in the nectaries and release of readily available substrates such as glucose. All these factors are important in the infection process of flowers.

CONCLUSION

Both plants and plant pathogens continue to evolve, as does the relationship between these organisms. The coevolution of attack and counterattack, defense and counterdefense between plant hosts and invading pathogens will continue. Plants will refine cellular defense mechanisms, for example, RNA silencing, to counteract pathogens that depend on the insertion or synthesis of RNA, such as plant pathogenic viruses. In response, plant viruses will encode proteins that suppress RNA silencing (Vance and Vaucheret, 2001). Perhaps, this constant response and counterresponse between host and pathogen will provide the basis for describing pathogen attack strategies in the next introductory plant pathology text.

REFERENCES

Agrios, G.N. 1997. *Plant Pathology.* Academic Press, New York.

Dean, R.A. 1997. Signal pathways and appressorium morphogenesis. *Annu. Rev. Phytopathol.* 35: 211–234.

Hill, N.S. 1994. Ecological relationships of Balansiae-infected graminoids, pp. 59–71 in *Biotechnology of Endophytic Fungi of Grasses,* Bacon, C.W. and J.F. White, Jr. (Eds), CRC Press, Boca Raton, FL .

Hussey, R.S. and V.M. Williamson. 1998. Physiological and molecular aspects of nematode parasitism, pp. 87–108 in *Plant and Nematode Interactions, Agronomy Monograph 36,* American Society of Agronomy, Barker, K.R., G.A. Pederson and G.L. Windham. (Eds.), Crop Science Society of America, Soil Science Society of America, Madison, WI.

MacHardy, W.E., D.M. Gadoury and C. Gessler. 2001. Parasitic and biological fitness of *Venturia inaequalis*: Relationship to disease management strategies. *Plant Dis.* 85: 1036–1051.

Manion, P.D. 1981. *Tree Disease Concepts.* Prentice-Hall, Englewood Cliffs, NJ.

Oku, H. 1994. *Plant Pathogenesis and Disease Control.* CRC Press, Boca Raton, FL.

Purcell, A.H. and D.L. Hopkins. 1996. Fastidious xylem-limited bacterial plant pathogens. *Annu. Rev. Phytopathol.* 34: 131–151.

Rayner, A.D.M. 1996. Antagonism and synergism in the plant surface colonization strategies of fungi, pp. 139–154 in *Aerial Plant Surface Microbiology,* Morris, C.E., P.C. Nicot and C. Nguyen-the (Eds.), Plenum Press, New York.

Rigden, J.E., I.B. Dry, L.R. Krake and M.A. Rezaian. 1996. Plant virus DNA replication processes in *Agrobacterium*: Insight into the origins of geminiviruses? *Proc. Natl. Acad. Sci. USA* 93: 10280–10284.

Schardl, C.L. 1996. Epichloë species: fungal symboints of grasses. *Annu. Rev. Phytopathol.* 34: 109–130.

Vance, V. and H. Vaucheret. 2001. RNA silencing in plants — defense and counterdefense. *Science* 292: 2277–2280.

Walton, J.D. and D.G. Panaccione. 1993. Host-selective toxins and disease specificity: Perspectives and progress. *Annu. Rev. Phytopathol.* 31: 275–303.

Williamson, V.M. and R.S. Hussey. 1996. Nematode pathogenesis and resistance in plants. *Plant Cell* 8: 1735–1745.

27 Detecting and Measuring Extracellular Enzymes of Fungi and Bacteria

Robert N. Trigiano and Malissa H. Ament

Fungi, fungi-like organisms, and bacteria do not "eat" or find nutrition in the ways that many animals do — they lack organized digestive systems in the traditional sense. Instead, they absorb all of the essentials for life directly from the environment. However, many of the simple molecules, such as sugars that are easily transported into the organism, are present only as complex polymers such as cellulose or starch in the environment. Complex carbohydrates and other classes of polymers cannot be used unmodified by the microorganisms. Therefore, these organisms must have a means to degrade the carbohydrate polymers, proteins, etc., into their constituent smaller molecules.

Regardless of their relationship to substratum (pathogenic or saprophytic), these classes of organisms produce **extracellular enzymes**, which interact with the environment outside of the cell or hypha. These enzymes are complex proteins that are manufactured in the bacterial cell or fungal hypha, transported across the plasmalemma (cell membrane) and cell wall to the outside environment. These are typically hydrolytic or oxidative and systematically degrade or break down very complex plant (and animal) polymers, such as cellulose, lipids, starch, lignin, pectin, and protein, to simple molecules. In turn, the simple molecules such as sugars, amino acids, and fatty acids resulting from enzymatic actions are absorbed through the cell wall, transported across the plasmalemma, and used for growth, energy, reproduction and other life processes. Some enzymes that degrade plant cell walls may also be involved in plant pathogenesis (Chapter 6, Chapter 10, and Chapter 26). For example, the fungus *Rhizopus stolonifera* and the bacterium *Pectobacterium (Erwinia) carotovora* produce a battery of **pectinolytic** enzymes (pectinases) that degrade the middle lamella, which is composed chiefly of pectin and lies between plant cells. These enzymes cause soft rots. Many obligate parasites, such as *Peronospora tabacina* (blue mold of tobacco), use cell wall degrading enzymes (e.g., cellulases) to establish contact between haustoria and plant cell membranes.

The laboratory exercises in this chapter will be concerned with detection of the activities of extracellular enzymes produced by microorganisms. There are three basic methods used to measure enzyme activities: detection of products of the enzymatic reaction (reducing sugars, acids, change in pH and viscosity, etc.); depletion of enzyme substrate (phenolic compounds, etc.); and exhaustion of a cofactor (ATP and others) in the reaction. The enzymes discussed in the exercises generally fall into the following two broad classes: constitutive and inducible. Constitutive enzymes are produced by the organisms at all times, albeit at a low level perhaps. Inducible enzymes are produced only when the microorganism is grown in the presence of the enzyme substrate. The time for the enzyme to be induced and produced to levels of detection will depend on the organism and the specific enzyme.

In the first exercise designed for undergraduate students, **amylase**, **lipase**, and **polyphenol oxidase** activities will be qualitatively determined by using solid agar media in which the enzyme substrate has been incorporated. These three enzymes are usually considered to be constitutive, but the quantity of enzyme produced may be influenced by the presence of the substrate. The second exercise devised for graduate students or special projects for advanced undergraduates will quantitatively determine **cellulolytic** (cellulase) activity of fungi grown in liquid medium. Cellulases are usually considered inducible and the test organisms will require contact with a cellulose or modified cellulose substrate before any appreciable enzyme activity can be detected.

EXERCISES

EXPERIMENT 1: QUALITATIVE DETERMINATION OF SOME ENZYME ACTIVITIES

Often the only information desired from an experiment is whether or not the fungus produces a specific enzyme. For example, the fungi that cause dogwood anthracnose, *Discula destructiva*, and an undescribed species of *Discula* can be distinguished from each other by their ability to produce polyphenol oxidases, a presumptive test for the ability to degrade lignin. Qualitative techniques are ideally suited to achieve this goal. Typically, evaluation for the ability to produce an enzyme can be accomplished by

0-8493-1037-7/04/$0.00+$1.50

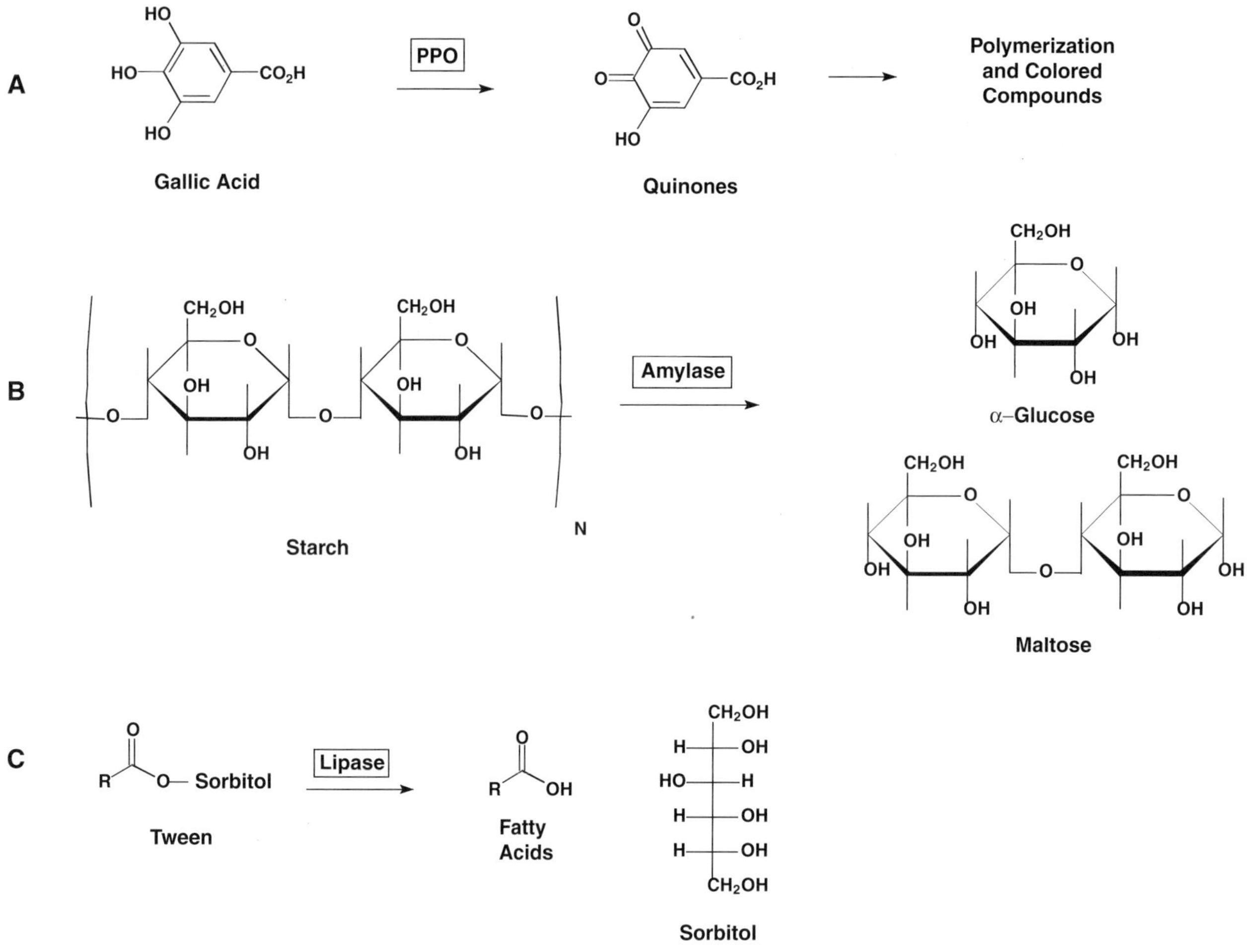

FIGURE 27.1 Substrates and products of polyphenol oxidase, amylase, and lipase. (A) Polyphenol oxidase (PPO) oxidizes adjacent hydroxyl groups on gallic acid (3,4,5-trihydroxybenzoic acid) and creates quinones, which are unstable and spontaneously polymerize to form pigmented (colored) products. (B) Starch is a α1–4 polymer of D-glucose residues. α-amylase hydrolyzes the bond between adjacent glucose units to form a random mixture of D-glucose and maltose (two D-glucose molecules linked α1–4) residues. (C) The Tweens are synthetic fats with a sorbitol (a sugar alcohol with five carbons instead of glycerol, which has three carbons) backbone esterified to various fatty acids (R, e.g., lauric or oleic acid). Note only carbon 1 is shown. Lipase hydrolyzes the ester bond between the carbon in sorbitol and the carbonyl carbon of the fatty acid to form sorbitol and a free fatty acid. Changes in pH and calcium bonding with the free fatty acids combine to produce the white, flocculent precipitate in the medium. (Courtesy of Dr. J. Green, University of Tennessee.)

using an agar medium in which the substrate for the enzyme has been incorporated. The fungus is allowed to grow on the agar and positive enzyme activity is indicated by a color change or precipitation of a product. For some other qualitative methods, staining for the original substrate can be used. In this experiment, we will qualitatively evaluate different fungi for the ability to produce amylase, lipase, and polyphenol oxidase. The substrates and enzymatic products are shown in Figure 27.1.

Materials

Each student or team of students will require the following items:

- Duplicates of 10-cm diameter petri dishes of five to eight cultures of different fungi grown on the appropriate inoculum medium (Table 27.1). We suggest not using slow-growing fungi such as *Geotrichum* or *Neotyphodium* or prolific spore- or sporangia-producing fungi such as *Penicillium* or *Rhizopus*, respectively.
- One dish of uninoculated (without fungi) NA and YEME to use as controls (Table 27.1)
- Three 60-mm diameter petri dishes containing enzyme assay medium for each fungus
- A supply of autoclaved plastic drinking straws of 5- to 7-mm diameter

TABLE 27.1
Medium for Growing Inoculum

Amylases and Lipases	Polyphenol Oxidase
Nutrient agar (NA)	Malt extract – yeast extract (YEME)
Nutrient broth (Difco Lab, Detroit, MI) 8 g	Malt extract (Difco Lab, Detroit, MI) 20 g
Agar 20 g	Yeast extract (Difco Lab, Detroit, MI) 1 g
Distilled water 1 L	Agar 20 g
	Distilled water 1 L

Note: Combine all ingredients and autoclave at 121°C for 20 min. Dispense media into sterile petri dishes (10-cm diameter) when cooled, but not hardened.

- Laminar flow hood (optional)
- Alcohol burner
- Scalpel and number 11 blade or stainless steel spatula
- Growth room or incubator
- Iodine reagent consisting of 15 g KI and 3 g I_2 per liter of distilled water
- Parafilm
- Large forceps
- Aluminum foil

Follow the protocols outlined in Procedure 27.1 to complete the experiment.

Procedure 27.1 Detection of Extracellular Amylase (AMY), Lipase (LIP), and Polyphenol Oxidase (PPO) Activity

1. Choose any number of fungi (we suggest five to eight species or include some isolates of the same species) and grow cultures for inoculum in 10-cm diameter petri dishes, using the appropriate medium for each of the enzymes (see Table 27.1). Two cultures of each fungus will be adequate for each student or team of students. Also provide a blank petri dish with medium (without fungus growth) for each enzyme.

2. Cut 10 to 12 equal-diameter plugs of inoculum from the perimeter of the mycelium with a sterile plastic straw for each fungus (Figure 27.2). If the plugs get stuck in the straw, squeeze them out into the dish with sterile forceps. Repeat the process for each fungus and the blank agar, using different straws.

3. With a sterile scalpel or spatula, transfer plugs with mycelium to the center of three dishes containing AMY enzyme assay medium. The mycelium side of the plug should be in contact with the medium.

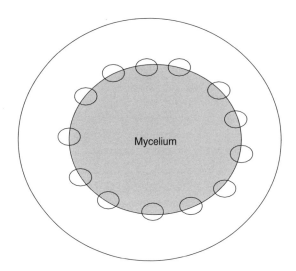

FIGURE 27.2 Harvesting mycelial plugs from cultures. Push the open end of a sterile plastic straw through the periphery of the colony as designated by the circles. Transfer the plugs to the center of the assay medium with the mycelium side of the plug in contact with the medium.

In a fourth dish, place a blank agar plug. Be sure to flame-sterilize the scalpel between transferring isolates of fungi. Repeat the process for LIP and PPO assay media. Seal each dish with parafilm and place the cultures in an incubator at 25°C. Wrap the PPO cultures in aluminum foil to prevent exposure to light.

4. Examine the PPO cultures for darkened plugs or assay medium, or both, after 24 and again after 48 h. Be sure to compare to the blanks or control cultures to inoculated cultures. Dark coloration indicates PPO activity and polymerization of the oxidized gallic acid. Measure the diameter of the discolored area and record observations.

5. Examine the AMY cultures periodically for mycelial growth. After the mycelium has covered 50% to 75% of the agar surface, flood one of the cultures with iodine reagent; there is no need to maintain

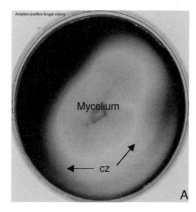
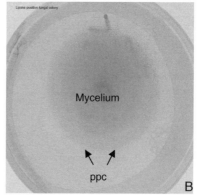

FIGURE 27.3 Assay media that are positive for enzymatic activity. (A) Amylase-positive dish. Cleared zone (cz) indicates the lack of starch (amylase positive), whereas the dark (blue) zone has intact starch polymers. (B) Lipase-positive dish. Note the flocculent (white) precipitation (ppc) in the medium that indicates lipase activity.

axenic cultures at this time. With a scalpel, make numerous cuts through the mycelium in one dish and observe any color changes in the assay medium. Cleared zones underneath or in advance of the mycelium indicate amylase activity, whereas blue coloration denotes intact starch polymers and lack of amylase activity (Figure 27.3A). If the test is negative for amylase at this time, allow the fungus in the other dishes to grow for an additional week and retest with the iodine reagent. Record observations.

6. Observe the LIP cultures for white flocculent inclusions either beneath or in advance of the mycelial mat (Figure 27.3B). Observation of the medium with a dissecting microscope may help to see the inclusions. Precipitation is a positive indication of lipase activity. As with the amylase assay, if the cultures are negative for lipase at this time, allow another week for growth and look for precipitation. Note where the precipitation occurs in the medium and record observations.

Anticipated Results

Polyphenol oxidase activity should be expressed as either darkened inoculum plugs or darkened assay medium within 24 to 48 h. Generally, the darker and more widespread the discoloration of the medium, the greater the amount of PPO that has been produced. In some cultures, the plugs may be very slightly discolored and it may be difficult to determine whether the fungal isolate produces PPO. Many fungi produce PPO, and generally isolates of the same species are all capable of producing the enzyme. However, there may be great variance within species of the same genus and this information may be useful in taxonomic considerations. Some fungi grow on PPO assay medium whereas others do not. Growth or lack of growth of individual species is probably influenced by low pH of the medium and the toxicity of gallic acid.

Most species of fungi produce both amylase and lipase and the enzymes are easily detectable after a short period (5 to 10 days) in culture. A positive test for LIP is a white flocculent precipitate in the medium. Amylase-positive cultures are indicated by clear zones in the agar after staining with iodine solution. As with PPO, it is possible that not all species classified in the same genus will produce the enzymes.

Questions

- What is an enzyme?
- If a fungus does not grow on PPO assay medium (Table 27.2) then how can the assay medium turn dark?
- Besides participating in lignin degradation, what other advantage might PPO confer on a plant pathogenic fungus?
- How do enzymes move through the assay medium and what does it mean if enzyme activity is apparent well beyond the perimeter of the fungal colony?
- What role does calcium chloride play in the LIP assay medium? (*Hint*: What is precipitated in the medium?)

TABLE 27.2
Enzyme Assay Media

Amylase (Amy)	Lipase (Lip)[a]	Polyphenol Oxidase (PPO)[b]
Soluble starch 2 g	Peptone 8 g	Solution A
Nutrient broth 8 g	$CaCl_2 \cdot H_2O$ 0.1 g	Gallic acid 5 g
Agar 20 g	Agar 20 g	Distilled water 250 ml
Distilled water 1 l	Distilled water 990 ml	Solution B
		Malt extract 5 g
	Tween 20 or 80 10 ml	Agar 20 g
		Distilled water 750 ml

[a] Autoclave Tween [polyoxyethylene sorbitan monolaurate (20) or monooleate (80)] and base medium separately at 121°C; then add to base medium after cooled, but not hardened. Swirl to mix and dispense into sterile, plastic petri dishes (60-mm diameter).

[b] Autoclave Solutions A and B separately at 121°C; then combine after cooled, but not hardened. Swirl to mix and dispense into sterile, plastic petri dishes (60-mm diameter). This medium should be stored in the dark or wrapped in aluminum foil to exclude light.

Sources: Enzyme assay media for amylase from Society of American Bacteriologists, 1957. *Manual of Microbiological Methods*. McGraw-Hill, New York, 315 pp.; for lipase from Sierra, G., 1957. *Antonie Van Leeuwenhoek Ned. Tijdschr. Hyg.* 23: 15–22; and for polyphenol oxidase from Davidson, R.W., W.A. Campbell, and D.J. Blaisdell, 1938. *J. Agric. Res.* 57: 682–695.

EXPERIMENT 2: QUANTITATIVE MEASUREMENT OF CELLULOLYTIC ACTIVITY

Quantitative measurement allows more accurate assessment of enzyme activities of individual isolates and comparison of activities between organisms. The procedures described in this experiment are designed to detect products of cellulolytic activity colorimetrically. However, this experiment may be adapted to any enzyme system, such as pectinases and xylanases, where **reducing sugars** (or other compounds with **anomeric** carbons) are produced.

Crystalline cellulose, 1–4β linkages of D-glucose residues, is very insoluble in water at neutral or slightly acid pH typically found in growth media. An individual polymer consists of 500 to 15,000 D-glucose units and has a molecular weight in the range of about 5×10^4 to 2.5×10^6 Da. Therefore, carboxymethylcelluloses (CMCs), which are far more soluble in water, are used as a substrate in most media. CMC has methyl groups esterified through the carboxyl group on carbon 6. Solubility is achieved through the partial positive charge imparted by the methyl group(s) to the molecule. Not all available carboxyl groups are methylated, and CMC is available in a wide range of percentage substitutions.

Cellulases are inducible enzymes and in order for organisms to produce them, cellulose or substituted cellulose compounds must be present in the growth medium. Cellulases are hydrolytic (adding a water molecule between individual units) and actually consist of a number or battery of enzymes. The C_1 enzyme or 1,4-β-glucan cellobiohydrolase enzymatically hydrolyzes crystalline cellulose to form cellibiose, a disaccharide. The C_{x1} enzyme (1,4-β-exoglucanase) cleaves off individual glucose residues from the end of the chain, whereas the C_{x2} enzyme (1,4-β-endoglucanase) randomly hydrolyzes the cellulose polymer into a mixture of cellulodextrans of various lengths. The C_{x3} enzyme is a 1,4-β-glucosidase (cellibiosase) that hydrolyzes cellobiose into two D-glucose residues (Figure 27.4).

Experiment 2 is divided into three parts: fungal growth and protein isolation; developing standard curves and measuring enzyme activity; and calculating cellulolytic activity. Fungal cultures take about 2 to 3 weeks to grow. Two 4- to 6-h laboratory periods should be scheduled to complete the remaining tasks. If this experiment is used as a class activity, we suggest that the instructor initiate the cultures, and, if time is a limiting factor, isolate total protein. Teams of students may also be assigned portions of the experiment. For example, a team of four students can help isolate the protein. Two of the students can work on determining protein concentration while the other two measure reducing sugars.

FUNGAL GROWTH AND PROTEIN ISOLATION

Materials

Each student or team of students will require the following materials and cultures:

- Five to eight fungal isolates grown on nutrient medium described in Table 27.1
- Liquid cellulolytic induction medium (Table 27.3; Reese and Mandels, 1963)
- Three 125-ml Erlenmyer flasks of the above medium for each fungal isolate
- Alcohol lamp
- Sterile straws (6- to 8-mm diameter)

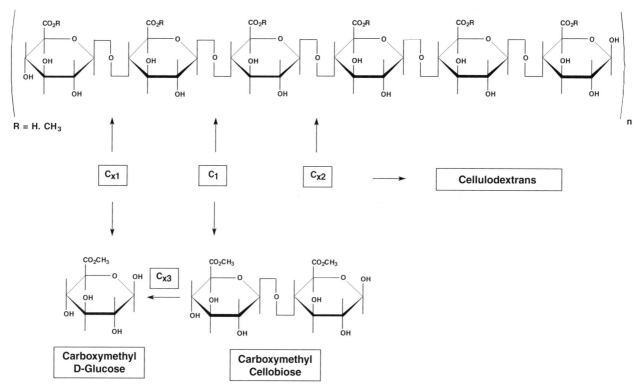

FIGURE 27.4 Diagrammatic representation of the action of cellulolytic enzymes on the substrate carboxymethyl cellulose (CMC). The C_1 enzyme (1,4-β-glucan cellobiohydrolase) hydrolyzes the 1–4β bond on the nonreducing end of the polymer to yield carboxymethyl cellobiose molecules. C_{x3} or cellobiase (1,4-β-glucosidase) hydrolyzes cellobiose to yield carboxymethyl D-glucose residues, which is a reducing sugar. C_{x1} is exocellulase (1,4-β-exoglucanase) that cleaves one carboxylmethyl D-glucose residue from the nonreducing end of the CMC molecule, whereas C_{x2} is an endocellulase (1,4-β-endoglucanase) that internally hydrolyzes bonds to produce various-length cellulodextrans. All of the cellulodextrans can be substrates for C_1 or C_{x1} enzymes, and thus C_{x2} may increase the apparent activity of the other enzymes by providing more substrate (if substrate is a limiting factor in the reaction mix). (Courtesy of Dr. J. Green, University of Tennessee.)

- Scalpels fitted with number 11 blades
- Incubator at 25°C
- Büchner funnel, side-arm flask, and vacuum source
- Whatman #1 filter papers and a 250-ml graduated cylinder
- 750 to 1000 g ammonium sulfate
- 250-ml beakers, magnetic stir bars, and stir plates
- Dialysis tubing (Spectrum Laboratories, Inc., Rancho Dominguez, CA) 3500 exclusion and clamps (suture string will also work)
- Ice bath
- Table top centrifuge
- 50-ml centrifuge tubes (do not need to be sterile)
- Top-loading balance
- Squeeze bottle of distilled water
- Distilled water
- Refrigerator

Follow the protocols outlined in Procedure 27.2 to complete this part of the exercise.

Procedure 27.2 Growth of Fungi and Isolation of Protein

1. Grow five to eight isolates of fungi (can be different isolates of the same species) in petri dishes containing nutrient agar as described in Table 27.1.
2. Prepare sufficient liquid cellulase induction medium (about 150 ml per isolate) to dispense 35 ml into each of four capped 125-ml flasks for all isolates included in the experiment. After autoclaving the medium in the flask, some of the CMC may be precipitated as fine white threads at the bottom of the flask, but this will not affect the induction of cellulase. Prepare inoculum plugs by using sterile straws as shown in Figure 27.2 and dissect the plugs into quarters with a scalpel and number 11 blade. Transfer four quarters to each of three flasks

TABLE 27.3
Composition of Liquid Cellulase Induction and Assay Medium

Cellulase Induction Medium	Cellulase Assay Medium
KH_2PO_4 2.0 g	Citric acid·H_2O 10.5 g
$(NH_4)_2SO_4$ 2.5 g	KH_2PO_4 17.4 g
Urea 0.3 g	Water to 1 l
$MgSO_4$·7 H_2O 0.3 g	CMC^a 5.5 g
$CaCl_2$·2 H_2O 0.5 g	pH = 5.0–5.2
Peptone 1.0 g	Do not autoclave, store at 4°C
Water to 1 l	
CMC^a 10.0 g	
pH = 5.5^b	
Autoclave 121°C for 15 min	

[a] pH medium before adding CMC. May require extensive stirring to dissolve CMC.
[b] CMC (carboxylmethyl cellulose) may precipitate after autoclaving.

per isolate. Prepare a control (cut from an uninoc- ulated agar dish — no mycelium) to inoculate the fourth flask. Label the flask with the species name, isolate, and date. Repeat the inoculation step for each species, using a different straw each time. Place all flasks in a 25°C incubator for 14 to 21 days, depending on the growth rate(s) of the fungi selected for the experiment.

3. Examine all flasks for contamination, especially for bacteria. The medium should be clear; cloudiness usually indicates bacterial contamination. Place a piece of Whatman #1 filter paper in the Büchner funnel and attach to the side-arm flask. Moisten the paper with distilled water and apply a gentle (not greater than 15 psi) vacuum. Pour the contents of a flask into the funnel and express the medium through the paper into the flask. Repeat this proce- dure for the remaining two inoculated flasks for the same species or isolate. The paper may need to be changed between flasks depending on the nature of the mycelium. Pour the contents of the side-arm flask into a graduated cylinder, record the volume, and transfer the liquid to a 250-ml beaker. Rinse the Büchner funnel, side-arm flask, and graduated cylinder with distilled water between species or isolates. Repeat entire procedure for the remaining fungal isolates and the uninoculated flasks. Place used filter papers in a biohazard bag and autoclave before disposal.

4. Add 66 g of ammonium sulfate for every 100 ml of induction medium in the 250-ml beakers. For example, if 110 ml of medium was collected, then add 66 g × (110/100) = 66 g × 1.1 or 72.6 g of ammonium sulfate. Stir until all the ammonium sulfate is dissolved and place beakers on ice for at least 30 min.

5. Stir briefly and dispense the liquid into 50-ml cen- trifuge tubes (do not add more than about 45 ml to each tube) and label. Be sure to balance opposite tubes by weighing on a balance (within 1 g is okay) and centrifuge for 10 min at about 3000 to 5000 g. Decant and discard the supernatant and invert the tube on a paper towel. A pellet may or may not be present in the bottom of the centrifuge tubes. After the excess liquid has drained, dissolve the total protein by adding 0.5 ml of distilled water per tube and shake or vortex. Allow to stand for 10 min. Combine the contents of the tubes for each species or isolate.

6. Cut 10- to 15-cm lengths of dialysis tubing and soak in distilled water or autoclave for three min. Tubing should be rinsed several times in distilled water before use. Express all of the water from the tube and clamp one end. Load the total protein from one species into each of the tubes. Clamp the other end of the tube, taking care not to introduce air into the lumen. Immerse the loaded dialysis tubing into chilled, distilled water and place in the refrigerator at 2°C to 4°C overnight.

7. The next morning, aliquot 0.5 ml from the dialysis tubes to several 1.5-ml labeled eppendorf tubes. The number of tubes needed per isolate will vary. The eppendorf tubes may be stored at −70°C for future use, or placed on ice for use later this day. Note that the samples can only be unfrozen once from −70°C without permanently denaturing the enzymes or significantly decreasing the activity.

Anticipated Results

Most fungi will grow well in cellulase-inducing medium and the enzyme(s) should be present in the medium

TABLE 27.4
Dinitrosalicylic Acid Reagent for Detecting Reducing Sugars

3,5-Dinitrosalicylic acid	8.0 g
Crystalline phenol	6.9 g
Sodium bisulfite	6.9 g
Sodium-potassium tartrate	2.55 g
Water to	1 l

Prepare at least 24 h before use and store in brown bottle at 4°C.

Caution: Wear gloves and eye protection. Prepare solution under fume hood.

between 1 and 3 weeks after inoculation. Ammonium sulfate yields three osmotically active particles when dissolved in water and will salt out most proteins (e.g., cellulases) at 90% saturation. It is most important that no undissolved salt crystals remain when starting the centrifugation step (salt crystals will cause excess water to enter the dialysis tube and dilute the protein). A mat of other material, mostly carbohydrates, will be present at the top of the centrifuge tubes after centrifugation. This is normal and should be discarded. After overnight dialysis, the contents of the tubes should be increased — some tubes may be very swollen and turgid. The quantity of water in the tube is dependent on the amount of salt that was carried over after centrifugation. In many experiments, each tube will yield about 4 ml of an aqueous solution of protein. Salting out and dialysis concentrates the protein 15 to 25 times than that found in the inoculated medium.

DEVELOPING STANDARD CURVES AND MEASURING ENZYME ACTIVITY

To determine enzyme activity, it is necessary to have solutions of known concentration of total protein and enzyme product (in this case reducing sugar) to compare to unknowns. Standard solutions are prepared in several concentrations and equations calculated to predict intermediate values of unknowns. The equations are formulated by regressing absorbance on concentration and usually follow Beer's law. That is, the relationship between known concentrations and absorbance is linear except at very low or high concentration of product, in this case protein or reducing sugar. In this portion of the experiment, the students will develop standard curves for total protein by the method developed by Lowry et al. (1951) and for reducing sugar with dinitrosalicylic acid (DNSA) reagent (Miller, 1959).

Materials

Each student or team of students will require the following materials:

- D-glucose and bovine serum albumin (BSA)
- Several 1-, 5-, and 10-ml pipettes
- Visible light spectrophotometer
- Fume hood
- Glass test tubes (greater than 10-ml capacity) and racks
- 50-ml centrifuge tubes (not necessarily sterile)
- Disposable, plastic cuvettes
- Folin-Ciocalteau reagent (Fisher Scientific Company). *Caution*: This reagent is caustic: wear hand and eye protection
- 2% (w/v) sodium carbonate in 0.1 N sodium hydroxide solution
- 2% (w/v) aqueous sodium tartrate solution
- 1% (w/v) aqueous cupric sulfate
- Nitrile gloves and safety glasses
- DNSA reagent (Table 27.4). *Caution*: This reagent is caustic; wear hand and eye protection and heat under fume hood only
- 100°C water bath or hot, freshly autoclaved water
- 40°C water bath
- Vortex
- Computer with programming capable of completing linear regression

Follow the protocols provided in Procedure 27.3 to Procedure 27.5 to complete this section of the experiment.

Procedure 27.3 Developing a Standard Curve for Protein

1. Dissolve 100 mg of bovine serum albumin (BSA, protein) in 100 ml of distilled water. This makes a

1 mg BSA/ml solution. Make the following standard solutions by dilutions: 0.1, 0.2, 0.3, and 0.5 mg BSA/ml. Use the following formula:

$$C_1 \times V_1 = C_2 \times V_2$$

where C is the concentration of protein and V the volume. For example, to make a 0.5 mg BSA/ml, start with 10 ml of 1.0 mg BSA/ml.

$$1.0 \text{ mg BSA/ml} \times 10 \text{ ml} = 0.5 \text{ mg BSA/ml} \times V_2$$

Solve the equation for V_2, which is the total volume of the desired solution.

$$10 \text{ mg} / 0.5 \text{ mg/ml, or } 20 \text{ ml} = V_2$$

Because 10 ml volume was initially used:

$$V_2 - V_1 = 10 \text{ ml}$$

Add 10 ml of distilled water to the original 10 ml of 1.0 mg BSA/ml. There should be 20 ml of 0.5 mg BSA/ml. The following amounts of distilled water should be added to 10 ml of 1.0 mg BSA/ml to make the appropriate standards: 90 ml water for 0.1 mg BSA/ml; 40 ml of water for 0.2 mg BSA/ml; and 23.4 ml of water for 0.3 mg BSA/ml. Mix all solutions well, transfer to 50-ml centrifuge labeled tubes, and if not used within the day, store at −20°C.

2. Prepare 200 ml of a 0.1 N NaOH solution by dissolving 0.8 g of NaOH in 200 ml of distilled water. NaOH can be difficult to weigh accurately. Alternatively, dissolve 4 g NaOH in 100 ml (1 N) and dilute 20 ml of this solution with 180 ml of distilled water to make a 0.1 N solution of NaOH. To 200 ml of 0.1 N NaOH add 4 g of sodium carbonate. Stir until the sodium carbonate is completely dissolved and pipette 1.0 ml each of 1.0% (w/v) cupric sulfate and 2.0% (w/v) sodium tartrate into the solution and mix thoroughly. At this time, dilute 20 ml of Folin-Ciocalteau reagent with an equal amount of distilled water. Wear gloves and eye protection when making this solution.

3. Carefully pipette 1 ml of each of the BSA standard concentrations, including a water control (0 mg BSA/ml) control, into glass test tubes. Prepare three replications of each BSA concentration. For each fungal species or isolate, pipette 1 ml of the total protein isolated from induction medium into a glass test tube. Add 5 ml of the cupric sulfate solution (described in Step 2) to each of the test

tubes and mix well by using a vortex mixer at low speed. After 10 min at room temperature, quickly add 0.5 ml of the diluted Folin-Ciocalteau reagent (Step 2) to each of the tubes and mix. Set aside at room temperature for 30 min.

4. After the 30 min, pipette the standards and unknown samples into disposable plastic cuvettes. Determine the absorbance values in a spectrophotometer set at 500 nm and record. Establish the baseline by using the 0 BSA samples as a reference. After reading, dispose of the liquid (and cuvettes) in an approved, labeled, hazardous waste container.

5. Calculating standard curves by linear regression and determining protein in unknowns are discussed in Procedure 27.6.

Procedure 27.4 Developing a Standard Curve for Reducing Sugars

1. Follow the instructions for preparation of standard BSA solutions in Step 1 of Procedure 27.3, but instead use D-glucose. Prepare the following standard concentrations: 0.1, 0.2, 0.3, 0.4, and 0.6 mg D-glucose/ml.

2. Pipette 1 ml of each standard D-glucose solution (including 0 mg D-glucose/ml) into each of the three glass test tubes. Reducing sugars (RS) in the cellulolytic assay medium samples can be assayed at this time if Procedure 27.5 has been completed. Wearing gloves and eye protection, pipette 3 ml of DNSA reagent (Table 27.4) into each of the tubes.

3. In a fume hood, place the test tube rack with tubes into a 100°C water bath for 15 min. If a water bath is unavailable, the rack can be placed in a plastic dishwashing tub and freshly autoclaved water added.

4. Remove the test tubes and rack from the heated water and allow to cool to room temperature (about 30 min).

5. Transfer D-glucose standards and unknowns to disposable plastic cuvettes, read on a spectrophotometer set at 550 nm, and record the absorbance values. If a dual beam spectrophotometer is available, establish the baseline by using the 0 D-glucose samples as a reference. After reading, dispose of the liquid and cuvettes in an approved, labeled, hazardous-waste container.

6. Calculation of standard curves by linear regression and determination of unknowns are discussed in Procedure 27.7. However, for illustrative purposes,

the equation $Y = 2.34 X + (-0.27)$ for reducing sugars was calculated from actual data (not shown).

Procedure 27.5 Assaying Isolated Proteins for Cellulase Activity

1. Before starting the procedure, set a water bath to 40°C. Select two 1.5-ml eppendorf tubes containing 0.5 ml of protein for each fungal species or isolate. If the samples were frozen at −70°C, allow to come to room temperature — do not heat to defrost. Be sure the tube is labeled and securely closed. Place one of the tubes into boiling water for 15 min to irreversibly denature any enzymes. This will serve as a control.

2. Prepare cellulase liquid assay medium the day before as described in Table 27.3 and store in the refrigerator (4°C). Allow the assay medium to come to room temperature and pipette 4.5 ml into test tubes. Depending on the volume of the protein isolation, additional replicates can be prepared for unknown samples. Dispense 0.5 ml of nondenatured isolated protein (not boiled) into test tubes for each of the unknowns. Dispense 0.5 ml of boiled preparations from Step 1 into separate tubes. Also, prepare at least one tube to which 0.5 ml of sterile water has been added to reaction mixture. Mix the contents of all the tubes thoroughly and cap with aluminum foil to prevent water loss.

3. Incubate the test tubes in a 40°C water bath for 1 h. *Note*: If after the reducing sugar assay some samples are negative or very low, incubate the samples for an additional hour.

4. Complete the assay for reducing sugars by using the DNSA procedure described in Procedure 27.4. Note that the unknowns may be assayed for reducing sugar at the same time that the standard concentrations of D-glucose are determined.

Anticipated Results

Measurement of protein concentration is quick, easy, and accurate by Lowry's method. The blank or 0 protein mixture will develop a very slight blue tint and with increasing protein concentrations, the mixture will be progressively darker. The highest standard concentration of protein will be almost opaque. Absorbance measurements of the unknown concentrations should lie within the range of absorbances of the standards. A general rule is not to interpolate beyond the range of standards. Most unknown preparations will fall between 0.1 and 0.3 mg protein/ml.

However, if any unknown is not within the range, additional standard concentrations, especially above 0.5 mg/ml, may be prepared and the regression line recalculated. Alternatively, the unknown can be diluted 1:1 or 1:2 with water and reassessed. If the concentration is low, the sample can be concentrated by evaporation of water and then remeasured. Two reminders: always use the equation of the line to compute the protein concentration of the unknown and dispose of all reagents properly.

The color of DNSA becomes increasingly red to dark red with higher concentrations of reducing sugars. The reducing sugar content from unknown assays will usually fall between 0.3 and 0.5 mg D-glucose/ml. The same considerations for determining proteins apply to reducing sugars. If the absorbance of the unknown sample is beyond that of the most concentrated standard, dilute the sample 1:1 or 1:2 with assay mixture and reassess the concentration with DNSA. Once again, use the regression equation to compute reducing sugars in the unknown samples. Also, heat DNSA and samples under the fume hood and dispose of all reagents properly.

Questions

- Why are three replications of each standard concentration used to estimate the regression line?
- Is the total protein isolated composed entirely of cellulases?
- The activities of which cellulases are measured with this assay procedure? How can the activity of C_{x2} be measured?
- Why is it necessary to include a boiled enzyme preparation in the assay?

CALCULATION OF ENZYME ACTIVITY

All the essential information for calculating enzyme activity is now available. The hypothetical example will use the standard curves for the data as calculated by Excel. The final expression of data will be in μmoles of reducing sugars (as D-glucose equivalents)/mg of crude protein/hour. Follow the steps outlined in Procedure 27.6 to calculate the regression equations for protein and reducing sugars and Procedure 27.7 to determine enzyme activity of the unknown samples.

Procedure 27.6 Construction of Standard Curves and Determination of Unknowns

1. To calculate protein and reducing sugars standard curves, data should be entered in columns, as shown in Table 27.5. Column A contains the inde-

TABLE 27.5
Sample Data for Protein Standard Curve Entered in Excel Spreadsheet

Protein Concentration (mg/ml)	Absorbance (a.u.)
0.0	0.000
0.1	0.115
0.1	0.116
0.1	0.125
0.2	0.239
0.2	0.233
0.2	0.235
0.3	0.340
0.3	0.335
0.3	0.334
0.5	0.518
0.5	0.521
0.5	0.519

TABLE 27.6
Regression Output for Protein Data Using Excel and Quattro Pro

	Excel	Quattro Pro
Constant (Y intercept or b)	0.029	0.022
Standard error of Y estimate	0.006	0.012
R-squared	0.996	0.995
No. of observations	13	13
Degrees of freedom	11	11
X coefficient (slope or m)	0.992	1.015
Standard error of coefficient	0.019	0.021
Estimated equation	$Y = 0.99X + 0.029$	$Y = 1.02X + 0.022$

pendent variable (concentration of the standards) and Column B the dependent variable (absorbance values recorded from the spectrophotometer). To build a data set that can be properly analyzed, you should have multiple samples and corresponding readings for each independent variable. The first measurement taken was to zero the spectrometer. As there were not many samples to read, the zero measurement was not repeated. For each of the protein concentrations (0, 0.1, 0.2, 0.3, and 0.5 mg BSA/ml for example), three samples were analyzed and their absorbance measurements were recorded. Note that data generated by the class may be similar, but not exactly the same as sample data presented in Table 27.5. The data may be graphed, but it is not necessary in order to calculate either protein or reducing sugar concentration. There are other programs available that furnish more control over the appearance of the graph than either Microsoft Excel or Corel Quattro Pro provide. To use the Regression tool in Excel, the version must have the Analysis Tool Pack installed.

2. For linear regression, enter your data using Microsoft Excel, click Tools/Data Analy-sis/Regression/OK. In the appropriate box, select the range of Y values. These values are the dependent variables (absorbance values) in the data set. Select the range of X values in the next box. These are independent variables (concentration). Select an area in the spreadsheet in which Excel can place the output of the regression calculations. Click OK. In the results displayed in Table 27.6, the X coefficient (or slope of the line, m) is labeled "0" under the "coefficients" heading, and the Y-axis intercept (or b) is labeled "intercept" under the "coefficients" heading. The R^2 value is found at "R Square" under "Regression Statistics" and is a measure of how well the data agrees with the estimated line. From these calculated values, you can construct the equation of the regression line, $y = mx + b$, where m is the slope of the line and b is the Y intercept. The equation from this analysis is $Y = 0.99X + 0.03$. Use this equation to determine the unknown concentrations. Remember that Y is the absorbance value for the unknown; solve for X, the concentration. Microsoft Excel provides a tool to graph the data and apply a trendline to complete regression analysis, but that method will not be addressed

here. The regression equation is the most accurate way to determine concentration. Trying to determine it using the graph provides only a good estimate of the value.

3. Corel Quattro Pro can also be used to calculate a linear regression. Data should be typed in the format shown in Table 27.5. Click Tools/Numeric Tools/Regression. In the appropriate boxes, select the independent (concentration) and dependent (absorbance) values, and an area on the spreadsheet in which the output of the regression analysis can appear. Be sure that the Y-intercept radio button is set to "Compute" and click OK. The results from analysis of the data listed in Table 27.5 are shown in Table 27.6 and calculated regression equation is $Y = 1.01X + 0.02$. There are minor differences in the values of m and b depending on which program (Excel or Quattro Pro) was used to estimate the regression line. Discrepancies are due to variations in the algorithms used in the programs. Very slight differences in the regression equations will not cause major differences in the calculations in either protein or reducing sugar concentrations. Choose one spreadsheet program and consistently use it to avoid introducing minor errors into the statistical analysis.

Procedure 27.7 Calculation of Enzyme Activity

1. Calculate total protein in unknown samples. The estimated equation of the regression line given in Step 3 of Procedure 27.6:

$$Y = 0.99X + 0.03$$

where Y is the absorbance measurement [arbitrary units (a.u.)] and X the concentration of protein (mg/ml). If the absorbance of an unknown sample (e.g., #1) is 0.33 a.u., make the following substitution into the previous equation:

$$0.33 = 0.99X + 0.03$$

Solving for X, the concentration yields:

$$(0.33 - 0.03)/0.99 = X \text{ or } 0.30 \text{ mg total protein/ml}$$

Note: Not all of the protein in the sample is cellulases.

2. Calculate reducing sugar (RS)/ml in the reaction mixture. The example of estimated equation for the regression line given in Step 6 of Procedure 27.4 is:

$$Y = 2.34X + (-0.27)$$

where Y is the absorbance measurement (a.u.) and X the concentration of RS (mg/ml).

If the absorbance of an unknown RS sample (e.g., #1) is 0.52 a.u., make the following substitution into the previous equation:

$$0.52 = 2.34X + (-0.27)$$

Solving for X, the concentration yields:

$$(0.52 + 0.27)/2.34 = X \text{ or } 0.34 \text{ mg RS/ml}$$
of reaction assay medium

3. Calculate total RS in the 5 ml of reaction assay medium by the following equation:

$$\text{Total RS} = \text{mg RS/ml} \times 5 \text{ ml}$$
(total volume of the reaction mixture)

Substituting in the equation:

$$0.34 \text{ mg RS/ml} \times 5 \text{ ml reaction mixture} = 1.7 \text{ mg RS}$$

Remember to incorporate any dilutions of the assay volume (e.g., if the assay was diluted 1:1 with assay medium, then the total volume would equal 10 ml).

4. Calculate mg RS/mg crude protein in the assay. Remember that the assay protocol used only 0.5 ml of protein extract (divide concentration by two to determine the amount of protein incubated with the assay mixture). From Step 1, the sample contained 0.30 mg total protein/ml.

$$0.30 \text{ mg total protein/ml} \times 0.5 \text{ ml} = 0.15 \text{ mg}$$
of total protein

From Step 3, the total RS in the reaction mixture was 1.7 mg. Divide the RS by the mg of total protein used:

$$1.7 \text{ mg RS/0.15 mg total protein} = 11.33 \text{ RS/mg}$$
total protein

This calculation provides a measure of the amount of RS each mg of crude protein can produce.

5. Calculate mg RS/mg crude protein/h that the assay medium was incubated at 40°C.

$$11.33 \text{ mg RS/mg total protein/h} = 11.33 \text{ RS/mg}$$
total protein/h

Most incubation times are 1 h, but occasionally 2 h are used. This calculation provides a rate for formation of the product.

6. Calculate µmoles of RS by using the molecular weight of D-glucose (180.2 g). First, convert g to mg.

$$180.2 \text{ g D-glucose/mole} \times 1000 \text{ mg/g} = 180,200 \text{ mg D-glucose/mole}$$

Now, calculate the number of moles of RS produce in the reaction:

$$11.33 \text{ mg RS}/180,200 \text{ mg/mole} = 6.29 \times 10^{-5} \text{ moles RS}$$

Lastly, convert moles to µmoles:

$$0.0000629 \text{ moles} \times 1,000,000 \text{ µmoles/mole} = 62.9 \text{ µmoles}$$

The final expression of enzyme activity becomes:

$$62.9 \text{ µmoles RS/mg total protein/h}$$

Questions

- Why is it important to calculate a rate of enzyme activity for each fungus? Why not compare only the reducing sugar produced in each enzyme assay mixture?
- Will the enzyme activity rate remain constant over time (i.e., after 1, 2, 3, or 4 weeks of culture)?
- How would the enzyme activity change if the incubation temperature were decreased to 30°C or 20°C or increased to 50°C or 60°C?

REFERENCES

Davidson, R.W., W.A. Campbell and D.J. Blaisdell. 1938. Differentiation of wood decaying fungi by their reaction on gallic or tannic acid medium. *J. Agric. Res.* 57: 682–695.

Lowry, O.H., N.J. Rosenbrough., A.L. Farr and R.J. Randell. 1951. Protein measurement with the Folin phenol reagent. *J. Biol. Chem.* 193: 265–275.

Miller, G.L. 1959. Use of dinitrosalicylic acid reagent for determination of reducing sugar. *Anal. Chem.* 31: 426–428.

Reese, E.T. and M. Mandels. 1963. Enzymic hydrolysis of cellulose and its derivatives, pp. 139–143 in *Methods in Carbohydrate Chemistry*, Whisler, D.L. (Ed.), Academic Press, New York.

Sierra, G. 1957. A simple method for the detection of lipolytic activity of microorganisms and some observations on the influence of the contact between cells and fatty substrates. *Antonie Van Leeuwenhoek Ned. Tijdschr. Hyg.* 23: 15–22.

Society of American Bacteriologists. 1951. *Manual of Microbiological Methods.* McGraw-Hill, New York, 315 pp.

28 Host Defenses

Kimberly D. Gwinn, Sharon E. Greene, James F. Green, and David J. Trently

CHAPTER 28 CONCEPTS

- Most plants are resistant to most pathogens.

- Resistant plants produce passive barriers that the pathogen cannot overcome or they must be able to activate successful defense(s) that arrest pathogen development.

- Although both active and passive defenses can be important at any stage of the disease cycle, passive defense is usually more important during prepenetration and penetration, and active defense is usually most important in the infection stage.

- Active disease responses require signal recognition/transduction, followed by gene activation.

- Examples of passive defenses are inhibitory plant surface chemicals, thick cuticle, lignified tissues, and phytoanticipins. Examples of active defenses are antimicrobial proteins, phytoalexins, the hypersensitive response, and systemic resistance.

Healthy plants grow in an atmosphere crowded with fungal spores, bacterial cells, and viruses; in soil, healthy roots predominate despite the high numbers of fungal spores, bacterial cells, and nematodes that thrive in the rhizosphere (soil immediately around the roots of the plant). In the face of this onslaught of potential pathogens, plants defend themselves with an arsenal of weapons, and, as a result, most plants are resistant to most pathogens. Plants have developed defense strategies that successful pathogens must overcome. Although plants defend themselves against potential pathogens in different ways, scientists have categorized strategies into the following two basic categories: **passive** (present before pathogen recognition) or **active** (induced after pathogen recognition by the host). Knowledge and exploitation of host defenses can lead to new pathogen control strategies (e.g., fungicides that turn on resistance, transgenic plants that can silence viruses, plants that overproduce bioactive natural products, or plants that produce antimicrobial proteins).

Successful host defenses disrupt the disease cycle (Chapter 2), primarily in the prepenetration, penetration, or infection phases. In general, passive defenses against pathogen attack are more prevalent in the prepenetration and penetration phases, and active defenses are more important in the infection phase.

PREPENETRATION

The barriers most important during prepenetration are passive. Plant surfaces are colonized by numerous microorganisms, both pathogenic and nonpathogenic. The extent to which an organism can colonize the plant surface is related to the chemical nature and topography of the surface. If the host does not have the necessary physical or chemical attributes, then the pathogen cannot recognize, attach, or colonize the plant surface. The nature and quantity of available nutrients may increase the fitness of the pathogen. Pathogens must attach to the surface and outcompete existing nonpathogenic organisms. All of these organisms compete for nutrients (carbon, nitrogen, and essential inorganic molecules) that are present on the surface as well as for physical sites (e.g., depressions on the leaf surfaces; Mercier and Lindow, 2000). When some pathogens contact host chemicals, they form structures for attachment and penetration. For some fungi, formation of penetration structures is induced by physical contact with the ridges and valleys of the host surface. Changes in host surface chemistry or topography can limit the pathogen.

Most pathogens change or multiply when they contact the plant surface. Fungal spores germinate, nematode eggs hatch, and bacteria multiply. The chemical and physical attributes of the plant surface can affect these processes. Lipopolysaccharides (LPS), which are cell surface components of Gram-negative bacteria (Chapter 6), allow the bacterium to exclude antimicrobial substances, including those on the plant surface and those induced by active defense by the plant. Bacteria that lack LPS are less effective pathogens.

The thickness of physical barriers plays an important role in host defense during prepenetration, penetration, and infection. In prepenetration, waxes on the surface of many plant parts limit the availability of the free water that many pathogens need to multiply (bacteria) or germinate (fungi). The roles of cuticle and cell wall thickness are discussed in the section on penetration.

PENETRATION

Passive

Physical barriers can limit pathogen penetration. The thickness of the cuticle plays an essential role in host defense against some pathogens: a thin cuticle layer is more easily penetrated by some fungi. The cuticle is composed primarily of cutin, a lipid. Suberin, a similar compound associated with cork cells, can be a passive defense or formed in an active response. Cell wall thickness plays a role both in the epidermal cells and in the cells being invaded by a pathogen. Many fungi require enzymes to penetrate the cuticle and cell walls (Chapter 10 and Chapter 27), but some plants contain compounds that inhibit these enzymes (Petrini and Ouellette, 1994).

Active

Plants have developed a number of active mechanisms to limit the entry of pathogens that evade preestablished barriers. The key to active defense is host recognition of the pathogen. Plants react to the presence of a pathogen by signal recognition, followed by signal transduction, and finally gene activation. Gene activation results in end products that contain, inhibit, or kill the invader. The response to invasive organisms, whether they are pathogens or mutualists, is fundamentally similar (Andrews and Harris, 2000). The following three categories of active defense have been described: primary, secondary, and systemic. Primary responses are those limited to the cell in contact with the pathogen. Secondary responses are induced in cells adjacent to the cell in contact with the pathogen. Systemic responses are induced throughout the plant.

An understanding of **signal recognition/transduction** is essential to understanding active defense mechanisms. In response to pathogen penetration, a signal recognition/transduction pathway is activated. This involves ion fluxes, oxidative bursts, protein phosphorylation, and signaling molecules. An ion flux is an efflux of Cl^- and K^+ ions coupled with an influx of H^+ and Ca^{++}; Ca^{++} is believed to be the central mediating component of early plant defense responses, although the mechanism by which it controls signaling is not known. An oxidative burst is the rapid production of hydrogen peroxide (H_2O_2) and is believed to signal cell death, overwhelm antioxidant cell protection, lead to rapid cell wall reinforcement, and induce gene expression. Protein phosphorylation is a primary means to regulate transcription. Phosphorylations

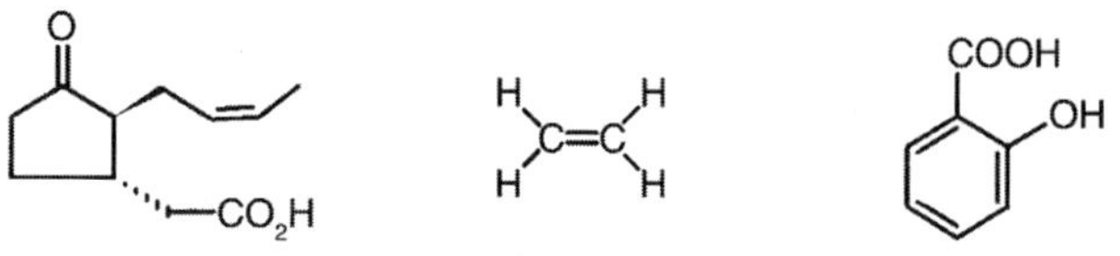

FIGURE 28.1 Some examples of signal compounds. The active defense systems of all plants studied thus far are induced by these molecules.

can increase binding to DNA or signal the protein to move to the nucleus. Signal molecules are the active defense systems of all plants studied thus far and are induced by the same molecules — salicylic acid, jasmonic acid, and ethylene (Stuiver and Custers, 2001; Figure 28.1). Jasmonic acid and its methyl ester, methyl jasmonate, increase in response to pathogen attack both locally and systemically and are both preexisting and induced compounds. These molecules are produced at penetration (for some fungal and nematode diseases) and in early infection for bacterial, viral, and some fungal and nematode diseases. An *Arabidopsis* mutant impaired in the jasmonic acid response has increased susceptibility to necrotrophic fungi, but not to biotrophic fungi. Ethylene, a gaseous plant growth regulator, is synthesized in both compatible and incompatible reactions. Ethylene appears to mediate resistance against necrotrophic fungal pathogens and also nonhost resistance. Treatment of plants with salicylic acid, the active component in aspirin, induces systemic resistance in many plants. Salicylic acid or its methyl ester, methyl salicylate, were once believed to be the long-distance signals in systemic acquired resistance; however, several recent studies have shown that this is unlikely.

In response to these signals, genes are activated, resulting in increased translation (reading the DNA code) and transcription (producing the protein sequences coded by the DNA). Enzymes that are directly important in defense or in the biosynthesis of defense compounds are produced.

Plants can limit penetration of many pathogens to the first cell that is attacked or to cells in the immediate vicinity (i.e., primary and secondary responses). Papillae are cell wall appositions deposited between the plasma membrane and the cell wall. Papilla formation is generally preceded by the aggregation of the cytoplasm in the cell under attack. They consist primarily of callose, a polymer of β1,3 D-glucose residues, but also may contain cellulose, phenolic compounds, and lignin. The penetration peg of the fungal pathogen is either stopped by the papilla or grows through it. Halos, locally modified regions of the host cell wall around a penetration site, are rich in substances such as phenolics, silicon, lipids, proteins (peroxidases), and lignin. Halos are often larger at sites of unsuc-

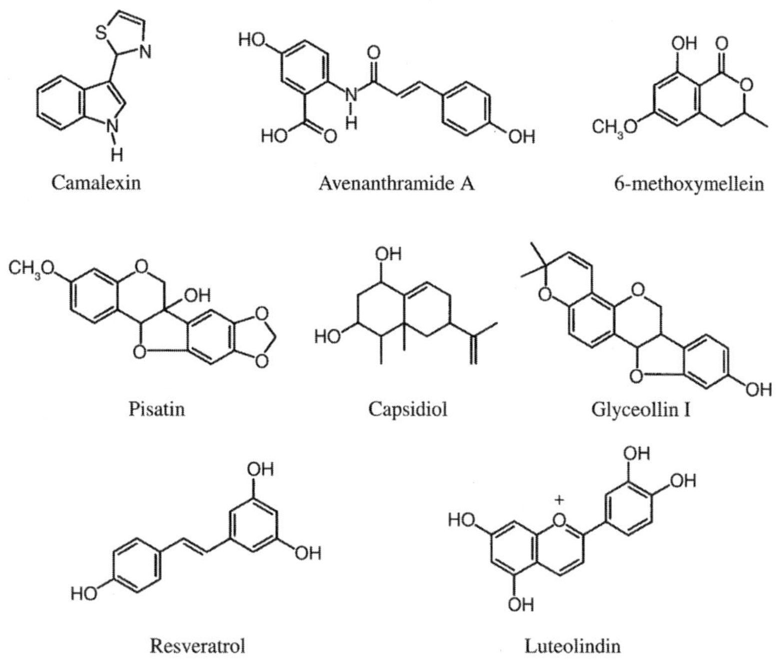

FIGURE 28.2 Structures of selected phytoalexins.

cessful penetration than at sites of successful penetration. Epidermal cells adjacent to the papilla or adjacent to an appressorium are often lignified. This induced lignin may be chemically different from the lignin in nonchallenged tissue. Suberization, depositing of insoluble polymers at the cell wall, also increases resistance to some fungal pathogens.

INFECTION

PASSIVE

Plants are capable of synthesizing several thousand different low-molecular-weight compounds (secondary metabolites) and many of these products are present at concentrations that affect pathogens. Antimicrobial natural products produced by plants are grouped into the following two classes: the **phytoanticipins** and the **phytoalexins**. Phytoanticipins are preestablished low-molecular-weight compounds that are stored in the plant cell or are released from a glucoside. Phytoalexins are formed in response to the pathogen and so are discussed in the section on active defense responses. Some compounds are phytoalexins in one species and phytoanticipins in another. Most antimicrobial natural products are broad spectrum; specificity, if it exists, is determined by the pathogen's ability to break down the compounds. In a few studies, plants with lower amount of antimicrobial compounds have been shown to be more susceptible to disease. Crops may be susceptible to many pathogens because selective breeding for other characteristics has decreased fitness by

reducing numbers and concentrations of natural products (Dixon, 2001). Many phytoanticipins are stored within vesicles, whereas others are stored in the glycosylated form until cells are damaged. Glycosylation (chemically bonding to a glucose molecule) converts a reactive and toxic phytoanticipin to a stable, nonreactive storage form that is more likely to be water soluble. Transgenic plants have been engineered that make glycosidic phytoanticipins and demonstrate increased disease resistance.

ACTIVE

Phytoalexins

Phytoalexins are antimicrobial natural products that are produced after infection or elicitation by abiotic agents. Plants produce natural products by a number of metabolic pathways; phytoalexins (Figure 28.2) are chemically diverse because they are manufactured by a number (or combination) of biosynthetic pathways. However, species in the same plant families tend to make phytoalexins that are derived from the same pathway and therefore are very similar in chemical structure. Although it is generally accepted that phytoalexins play a role in host defense, they are likely only one part of an overall defense strategy. However, phytoalexin production may simply be a response that is correlated with the expression of defense. Several lines of evidence support a role for phytoalexins in host defense and include the following: (1) in several gene-for-gene systems, resistance is associated with phytoalexin production; (2) phytoalexins accumulate rapidly to inhibitory concentrations at the site of pathogen development; (3) pathogens can overcome host resistance by phytoalexin detoxification; and (4) plants genetically transformed to overproduce phytoalexins are more resistant to disease (Hammerschmidt, 1999). Phytoalexins are produced by the cell under attack (primary response) as well as in the adjacent cells (secondary response).

Antimicrobial Proteins

Antimicrobial proteins have been detected in many plant species and tissues. The widespread localization of antimicrobial proteins (e.g., chitinases and β-glucanases) in plants coupled with their activity against pathogens in vitro suggests that these enzymes may serve a protective role. Plants that are transformed with genes that code for these antimicrobial proteins can be more resistant to pathogens.

Hypersensitive Response

The active responses described above are nonspecific; they occur in response to pathogens and other organisms. The hypersensitive response, however, is highly specific and occurs only when the product of a pathogen avirulence gene interacts with the product of a plant resistance gene. Activation of this gene in gene-for-gene resistance results in a cascade of reactions within the cell. The **hypersensitive response (HR)**, a rapid death of a few host cells that limits the progression of the infection, is a manifestation of recognition of the pathogen avirulence gene product. Typically, an HR includes signal transduction, programmed cell death, increased activation of defense-related genes (e.g., synthesis of phytoalexins, salicylic acid, and antimicrobial proteins), and a distant induction of general defense mechanisms that serve to protect the plant [i.e., systemic acquired resistance (SAR)].

Elicitation of the primary responses results from the recognition of a pathogen effector protein by the host receptor protein. Most effector proteins have no apparent enzymatic activity, but are capable of binding to the receptor protein. When the two proteins interact, a signal transduction pathway is activated and the primary response is initiated. A rapid burst of oxidative metabolism leads to the production of superoxide and subsequent production of hydrogen peroxide that precedes the development of visual symptoms of an HR.

Plants appear to have adapted **programmed cell death**, a general process commonly associated with reproductive and xylem tissue development, as a host defense response. The attacked cell and several cells around it die in response to chemical signals. The sacrifice of these cells isolates the pathogen and is a particularly good resistance mechanism for biotrophic pathogens. The hypersensitive response is not limited to biotrophic pathogens. Pathogens may be killed by the reactive oxygen species generated by the oxidative burst or by the antimicrobial compounds and enzymes formed in response to gene activation.

Systemic Resistance

Although plant systems do not truly mimic the immune system of mammals, they do have the ability to better resist pathogens after exposure to other organisms. Infection or colonization of the plant by one organism can induce host resistance to other pathogens. There are two major types of systemic resistance: **systemical acquired resistance (SAR)** and **induced systemic resistance (ISR)**. In SAR, the attacking organism is a necrotrophic pathogen. Some bacteria that colonize the roots, but do not cause disease, may induce ISR. Although both result in systemic host resistance, methods and mechanisms differ significantly. In SAR, salicylic acid or methyl salicylate is produced as a primary and secondary response. An unknown compound then moves into the phloem and transports the signal to distal portions of the plant. Salicylic acid then accumulates in the distal portions of the plant and is converted to methyl salicylate, which is volatile and may serve as a signal to neighboring plants. Antimicrobial proteins appear in the distal portions of the plant. In ISR, both jasmonic acid and ethylene signal distal portions of the plant; antimicrobial proteins are not produced.

The case studies presented in Table 28.1 to Table 28.3 provide summaries of resistance mechanisms in well-known pathogens, illustrating the complexity of host resistance. Whenever possible, examples of active and passive defenses are used.

EXERCISES

The following laboratory exercises are designed to demonstrate the effects of essential oils (commonly used as candy flavorings) on the growth of plant pathogenic fungi and to demonstrate the impact of wound healing on disease.

EXPERIMENT 1. EFFECT OF VOLATILE COMPOUNDS FROM CANDY FLAVORINGS ON THE GROWTH OF PLANT PATHOGENIC FUNGI

Essential oils are highly volatile substances isolated from an odiferous plant; the term essential was used because these oils were thought to contain the essence of odor and flavor (Linskens and Jackson, 1991). The oil bears the name of the genus or common name of the plant from which it is derived and can be somewhat misleading because chemistry can be highly variable within a genus or species. Since antiquity, essential oils have been used as perfumes, medicines, and flavorings. Essential oils are well known for their antibacterial, antifungal, antiherbi-

TABLE 28.1
Case Study: *Tobacco Mosaic Virus*

- Viral pathogen — *Tobacco mosaic virus*. In some cultivars of solanaceous crops, symptoms of TMV are mild to severe mottling, chlorosis, and dwarfing of leaves. Tobacco cultivars with the N gene produce an HR. All viruses are obligate biotrophic organisms.
- Methyl salicylate, released from necrotic lesions on tobacco caused by TMV, increases the resistance of uninfected neighboring plants. The PR1 gene is also induced in the neighboring plants.
- Concentration of a protein located in the plastid can control host reaction to TMV. Overexpression of this protein increases the rapidity of the HR; low concentrations lead to a suppression of the HR.
- Rapid cell death occurs in plants with an N gene except when the plants are tranformed with viral genes that slow down the HR and cause the plant to become susceptible.
- At least three protein products have been demonstrated to function as avirulence determinants: the replicase protein (in lines carrying the *N* gene), the coat protein (in tobacco lines carrying the N′ gene, and the movement protein (in tomato lines carrying the Tm-2 and the Tm-2[2]. (Hutcheson, 1998).

TABLE 28.2
Case Study: Bacterial Wilt

- Bacterial pathogen — *Ralstonia solanacearum*. This bacterium causes a bacterial wilt of many hosts. The vascular tissue of stems, roots, and tubers turns brown and when cut oozes a stream of bacteria. *R. solanacearum* colonizes both the soil, which is nutrient-poor, and the inside of a plant, which is nutrient-rich but well defended.
- Bacterial cells attach to the plant roots and form microcolonies at sites on the root that are vulnerable in two major passive defense systems: the sites of lateral root emergence and the root elongation zone.
- While entering the plant cortex and the vascular system, the pathogen causes minimal tissue/cell damage and either avoids or suppresses host recognition. The exopolysaccharide (EPS1) may be responsible for masking bacterial structures that are targets of host recognition because strains of the bacterium that do not produce EPS1 simply agglutinate and degenerate in the cortex, perhaps due to defense responses.
- Bacteria sense that they have arrived in the plant cell and induce genes for host infection, the so-called *hrp* genes. The pathogen has more than 20 *hrp* genes. Inactivation of the *hrp* genes causes nearly complete loss of ability to cause disease and incites a hypersensitive response on resistant plants. The *hrp* genes encode proteins that produce a secretion apparatus that must allow translocation of avirulence factors across the bacterial membrane for delivery to the host cell. One protein (PopA) encoded by the *hrp* genes causes an HR-like response on plant tissues; another protein (PopC) is similar in structure to host receptor proteins and may interfere with host defenses.
- In plants that have the corresponding resistance genes, the plant recognizes the pathogen proteins and responds with cell wall thickening, separation of the cell membrane from the wall surface, and formation of wall appositions (Schell, 2000).
- The lipopolysaccharide produced by *R. solanacearum* prevents the HR induction in many dicots. Similar to ISR, salicylic acid does not act as a signal in this localized induced response and it does not require necrosis. Addition of the lipopolysaccharide to plants can induce production of defense-related proteins (e.g., peroxidase in tobacco and synthesis of phytoalexins) and changes in the plant cell surface. The lipopolysaccharide does not induce the oxidative burst typically associated with an HR (Dow et al., 2000).

TABLE 28.3
Case Study: Powdery Mildew

- Fungal pathogen — *Blumeria graminis* causes powdery mildew fungus of wheat and barley.
- Although host cuticle plays an important role in resistance to many pathogens, *B. graminis* effectively degrades the barley cuticle using esterase enzymes that break the ester bonds that hold together the cutin molecule. Esterases are produced within two hours of the conidia landing on the host
- Host genes govern resistance at different stages of the interaction (Vanacker et al., 2001) . In plants containing the *mlo* gene, penetration of the fungus is stopped by the formation of effective papillae; this type of resistance is effective against all races of the pathogen. Occasionally a few mildew colonies appear in *mlo*-resistant barley. Silicon-rich halos are observed in wheat and barley infected with *B. graminis*. Silicon levels are higher in regions of failed penetration attempts than in sites of successful penetration. Peroxidases and hydrolytic enzymes are also found in the halos. In plants containing the *Mlg* gene, fungal growth was arrested within papillae of cells that subsequently undergo an HR. In barley plants that have single gene controlled, race-specific resistance to *B. graminis*, failure of the papillae response leads to successful penetration of the cell; when the pathogen race is incompatible with the host, then an HR is initiated (Hückelhoven and Kogel, 1998).

vore, and antioxidant activities, and have been proposed as natural, safe pesticides. Essential oils of many plants (e.g., sage, oregano, citrus, and various mints) contain antifungal phytochemicals (compounds produced by plants; Kurita et al., 1981). Effects of essential oils on postharvest and grain spoilage pathogens (or similar species) have been studied most often.

Candy flavorings are easily manipulated, nontoxic sources of phytochemicals. Many flavorings (clove, lemon, orange, peppermint, and spearmint) are extracted without solvents as pure oils. These contain several antifungal phytochemicals, most notably various monoterpenes.

General Considerations

Candy flavorings that are essential oils can be purchased at local grocery or specialty shops, but care should be taken to avoid flavorings that contain alcohols or other solvents because these will confound the results. Clove, lemon, orange, peppermint, and spearmint oils produced by Lorann Gourmet (Lorann Oils, Inc., MI) can be used with good results.

Any pathogen that is easily cultured on solid media can be used in these experiments. For example, *Alternaria*, *Fusarium*, and *Sclerotinia* species are inhibited by compounds found in candy flavorings.

These experiments are designed for teams of students. Each student should have an opportunity to transfer the mycelial plug to the culture medium, pipette oils onto the filter paper, and measure fungal colony diameter.

Materials

Each team of students requires the following items:

- For each pathogen to be tested, 14 10-cm diameter petri dishes containing a fungal growth medium are needed. This is based on three oils at two concentrations run in duplicate plus dishes for nonamended controls. The choice of medium should be dictated by the choice of pathogen. Potato dextrose agar works well for most pathogens.
- Two actively growing cultures of each pathogen
- Cork borers (5- to 7-mm diameter), sterile plastic straws, or other tools for cutting similar-size mycelial plugs
- Bunsen burner and 70% ethanol
- Dissecting probe
- 10- to 100-μl pipette and tips
- Parafilm
- Filter paper
- Plastic ruler

Follow the protocols listed in Procedure 28.1 to complete this experiment.

Procedure 28.1 Effect of Essential Oils on Fungal Growth

1. Label petri dishes with your name, genus of pathogen, amount and type of flavoring, and date.
2. Cut several disks (ca. 5 mm) from the edge of an actively growing culture with a cork borer or sterile plastic straw.
3. Place one agar disk, mycelium side down, in the center of the culture medium, or if available, place a single sclerotium in the center of the dish. Inoculate all dishes in the same manner.
4. Layer two pieces of filter paper inside the lid of each dish. They should fit snugly and the bottom layer should be in contact with the inside of the lid.
5. Pipette a drop (0, 5 or 50 μl) of candy flavoring in the center of the filter paper. Invert the lid and fit the bottom of the dish into the lid. The petri dish should be upside-down with the mycelium plug above the filter paper. Each treatment should be repeated in a separate dish. Wrap dishes with Parafilm®.
6. Incubate the cultures at room temperature for one week. Dishes containing the different flavorings should be stored in separate areas. Observe growth of the fungi daily. Control treatments (0 μl) should not be allowed to grow to the edge of the petri dish. Data should be recorded when the mycelial growth of the controls nears the edge of the dishes or after 1 week.
7. Measure and record the diameter of each fungal colony. Determine the mean diameter for each treatment. Calculate comparative growth index (CGI) from the following equation:

$$CGI = 100\left(\frac{D_c - D_t}{D_c}\right) - 100$$

where D_c is the mean colony diameter for controls and D_t the mean colony diameter for each treatment. Values will be greater than zero if mean growth in the treatment exceeds mean growth of control. Values will be less than zero if mean growth in the treatment is less than the control.

Anticipated Results

Treatment with essential oils should reduce the growth of all fungi; that is, comparative growth index should be less than zero for all compounds. Growth should be inhibited more by the higher concentration (50 μL) of the oil than the lower concentration (5 μL). Some oils will reduce growth more than others. For example, fungal growth in an atmosphere of lemon oil should be less than growth in an atmosphere of orange oil; both should be less than controls.

Questions

- What role might antimicrobial phytochemicals play in plant defense?
- Would a pathogen of the orange rind (e.g., *Penicillium*) be more or less sensitive or have the same sensitivity to volatile compounds produced by orange than would *Fusarium*, which is usually considered a nonpathogen of orange? Why or why not?

EXPERIMENT 2. PHYSICAL DEFENSES OF POTATO AGAINST SOFT ROTTING BACTERIA

Plants can defend themselves in many ways when they are wounded or when a pathogen tries to attack. These methods include induced structural and biochemical defenses. In this experiment, potato pieces that have had or have not had time to induce defensive mechanisms will be challenged with the soft rot bacterial pathogen *Pectobacterium* (*Erwinia*) *carotovora*. Host tissues will be examined for induced defensive mechanisms and how these mechanisms affect infection by the pathogen (M. Windham, personal communication).

Materials

Each student, student team, or class will require the following supplies:

- *Pectobacterium (Erwinia) carotovora* isolate
- Several 250-ml flasks
- Nutrient broth (Difco)
- Rotary shaker
- Flats of soil and labels
- Potatoes
- Compound microscope
- Microscope glass slides and coverslips
- Water
- Eyedropper
- Knife to cut potatoes
- Scalpel to make thin sections
- Greenhouse or incubator

Follow the protocols listed in Procedure 28.2 to complete the experiment.

Procedure 28.2 Physical Defenses of Potatoes Against Soft Rotting Bacteria

1. One week before lab, start cultures of *P. carotovora* in 250-ml flasks containing 75 ml of nutrient broth. Incubate at room temperature on a rotary shaker set for 100 rpm.
2. Two days before lab, pour the contents of two flasks into the soil of each flat and mix well.
3. One day before lab, take half the potatoes and cut them into pieces that are about 5 cm^2.
4. The day of the experiment, cut the remaining potatoes into pieces that are about 5 cm^2 in size.
5. With the scalpel, make thin sections from a potato cut the day before and from a freshly cut piece of potato. Place thin sections of both on a microscope slide with a drop of water and place a cover glass on top of the specimens. Examine the thin sections microscopically and record any differences in the sections.
6. Take pieces of freshly cut potato and pieces that were cut the previous day and lay them down (wounded side down) on the surface of the soil. Place flats in the greenhouse for one week. Keep flats moist, but not flooded.
7. After a 1 week incubation, remove the potato pieces and examine them for soft rot symptoms. Record your results.

Anticipated results

The response of the potatoes to wounding can be observed microscopically. In the sections from the potato cut a day prior to the experiment, a white, milky film should be evident. The film should be missing from the freshly sliced potatoes. There should also be differences in disease incidence and disease severity between the 1-day-old pieces of potato and the freshly wounded pieces of potato after they have incubated for 1 week on the surface of the soil infested with *P. carotovora*.

Questions

- What type of defense mechanisms had developed when you examined the potato pieces microscopically?

- What is suberin?
- What would you recommend to potato growers after conducting this experiment?

REFERENCES

Andrews, J.H. and R.F. Harris. 2000. The ecology and biogeography to microorganism on plant surface. *Annu. Rev. Phytopathol.* 38: 145–80

Dixon, R.A. 2001. Natural products and plant disease resistance. *Nature* 411: 643–647.

Dow, M., M.-A. Newman and E. von Roepenack. 2000. The induction and modulation of plant defense responses by bacterial lipopolysaccharides. *Annu. Rev. Phytopathol.* 38: 241–261.

Hammerschmidt, R. 1999. Phytoalexins: What have we learned after 60 years? *Annu. Rev. Phytopathol.* 37: 285–306.

Hückelhoven, R. and K-H. Kogel. 1998. Tissue-specific generation at interaction sites in resistant and susceptible near isogenic barley lines attacked by the powdery mildew fungus (*Erysiphe graminis* f. sp. *hordei*). *Mol. Plant Microbe Interact.* 11: 292–300.

Hutcheson, S.W. 1998. Current concepts of active defense in plants. *Annu. Rev. Phytopathol.* 36: 59–90.

Kurita, N., M. Miyaji, R. Kurane and Y. Takahara. 1981. Antifungal activity of components of essential oils. *Agric. Biol. Chem.* 45: 945–952.

Linskens, H.F. and J.F. Jackson. 1991. *Modern Methods of Plant Analysis, Vol. 12: Essential Oils and Waxes.* Springer Verlag, New York.

Mercier, J. and S.E. Lindow, 2000. Role of leaf sugars in colonization of plants by bacterial epiphytes. *Appl. Environ. Microbiol.* 66: 369–374.

Petrini, O. and G.B. Ouellette (Eds.). 1994. *Host Wall Alterations by Parasitic Fungi.* APS Press, St. Paul, MN.

Schell, M.A. 2000. Control of virulence and pathogenicity genes of *Ralstonia solanacearu-m* by an elaborate sensory network. *Annu. Rev. Phytopathol.* 38: 263–292.

Stuiver, M.H. and J.H.H.V. Custers. 2001. Engineering disease resistance in plants *Nature*, 411: 865–868.

Vanacker, H., T.L.W. Carver and C.H. Foyer. 2000. Early H_2O_2 accumulation in mesophyll cells leads to induction of glutathione during the hypersensitive response in the barley-powdery mildew interaction. *Plant Physiol.* 123: 1289–1300.

29 Disruption of Plant Function

Melissa B. Riley

CHAPTER 29 CONCEPTS

- Plant disease symptoms often result from plant pathogens affecting the normal physiological activities of the plant.

- Effects of a pathogen on the normal physiology of the plant can be at the macroscopic level (i.e., production of galls) or on the microscopic level (i.e., alteration of cell membrane permeability).

- Major plant physiological activities affected by plant pathogens include photosynthesis, respiration, production of plant growth hormones, absorption/translocation of water and nutrients, and transcription and translation.

- Study of plant pathogens led to the discovery of important compounds associated with normal plant activity — gibberellins, which are plant growth hormones, are produced by some fungal plant pathogens.

- Reduced photosynthesis and increased respiration induced by plant pathogens ultimately result in reduced plant growth and yield.

- Disruptions of absorption/translocation within a plant can result in mineral deficiency symptoms.

- One of first responses of a plant to pathogen attack is alteration in cell membrane permeability.

Since the beginnings of time, humans have observed the effects of plant pathogens on plants, although they did not know about plant pathogens or how they caused the observed effects. In many cases, the processes that result in symptom development are not completely understood. Plant disease has been defined as alteration in the normal **physiology** of a plant, but this definition requires knowledge of normal plant physiology. Normal plant physiology includes many processes such as **photosynthesis**; **respiration**; **absorption** and **translocation** of water; transport of photosynthetic products; production of compounds such as plant growth **hormones, enzymes**, proteins, carbohydrates, lipids, and nucleic acids by the plant; and movement of materials between individual cells. Changes in any of these processes may have an effect on the overall appearance of the plant, which is observed as symptoms of disease. How do plant pathogens interrupt these processes? Responses may be something as small as the alteration of the metabolism within a cell or the movement of materials across a cell membrane to an overall plant response such as reduction in crop yield. We have learned much about the effects of plant pathogens on the normal physiology of plants, but still many questions remain, especially related to the exact sequence of events that results in the expression of symptoms.

Some common symptoms observed in response to plant pathogens and their possible relationship to normal plant physiology, and examples of plant pathogens are provided in Table 29.1. This chapter will examine how plant pathogens can disrupt the normal functions of a plant. Specifically, the effects of these pathogens on the major activities of the plant, including photosynthesis; respiration; production of plant growth hormones; absorption and translocation of water and nutrients; protein, carbohydrate, and lipid production; and cell **permeability** will be considered.

PHOTOSYNTHESIS

Essentially all organisms on earth depend on the process of photosynthesis in which plants absorb solar energy and convert it into carbohydrates that can be further utilized as energy sources. Photosynthesis occurs in the **chloroplasts** of plant cells where carbon dioxide and water in the presence of solar energy and chlorophyll are ultimately converted to carbohydrates. This process is generally expressed by the following formula:

$$6\ CO_2 + 6\ H_2O + Light + Chlorophyll \rightarrow C_6H_{12}O_6 + 6\ O_2$$

TABLE 29.1
Common Plant Disease Symptoms, the Physiological Processes Affected, and Examples of a Disease and the Associated Plant Pathogen Affecting the Process

Symptom	Physiological Function	Example Disease/Pathogen
Chlorosis	Photosynthesis	Tobacco mosaic virus
Wilting	Xylem transport	Bacterial wilt/tomato and tobacco — *Ralstonia solanacearum*
Hyperplasia — cell division	Growth hormone regulation	Crown gall — *Agrobacterium tumefaciens*
		Black knot/plum — *Dibotryon morbosum*
Necrosis	Many different functions	Fire blight/apple — *Erwinia amylovora*
Hypertrophy — cell enlargement	Growth hormone regulation	Root knots — *Meloidogyne incognita* (root-knot nematode)
Leaf Abscission	Growth hormone regulation	Coffee rust — *Hemileia vastatrix*
Etiolation	Growth hormone regulation	Bakanae "foolish seedling" disease of rice — *Gibberella fujikuroi*
Stunting	Many different functions	Many different viral diseases
Abnormal leaf formation	Growth hormone regulation, respiration	Cucumber mosaic virus – ornamentals

The process can be divided into a light and a dark portion. During the light phase, solar energy produces reduced chemical compounds in the form of NADPH (reduced nicotinamide adenine dinucleotide phosphate) and ATP (adenosine 5′ triphosphate). During the dark phase, energy captured in the NADPH and ATP is utilized to convert carbon dioxide into carbohydrates. These reactions can occur through two major pathways: (1) the Calvin cycle in C_3 plants and (2) the C_4 cycle in C_4 plants. Alteration or inhibition in the overall photosynthetic activity of the plant can result in obvious gross symptoms in the plant, such as chlorosis or reduced growth and yield. Reduction in the photosynthetic activity of the plant can be accomplished in many different ways. One way is from a reduction in total leaf area due to leaf necrosis and thereby destruction of photosynthetic tissue. A study of the effects of rust and anthracnose diseases on the photosynthetic competence of diseased bean leaves revealed that rust infections reduced photosynthetic rates in direct proportion to visible lesion area, whereas the reduction in photosynthetic activity in anthracnose-infected leaves was much greater than could be attributed to the visibly infected area (Lopes and Berger, 2001). Photosynthesis in the green areas beyond the necrotic symptoms was severely impaired in anthracnose-infected leaves, indicating that the pathogen was affecting photosynthesis beyond the development of necrotic areas. Chlorosis is a common symptom observed in response to plant pathogens and can result from inhibition of chlorophyll synthesis (Almási et al., 2000), increases in the rate of chlorophyll degradation (Strelkov et al., 1998), and reduction in the chloroplast size and number (Kema et al., 1996).

Blumeriella jaapii, the causal agent of cherry leaf spot, seems to mainly interfere with the enzymatic process of the Calvin cycle and these effects were observed prior to the occurrence of visible symptoms (Niederleitner and Knoppik, 1997). The activity of three specific enzymes of the Calvin cycle, including ribulose-1,5-bisphosphate carboxylase/oxygenase (RUBISCO), decreased in powdery mildew-infected wheat leaves (Wright et al., 1995). RUBISCO activity was also decreased in *Arabidopsis thaliana* in response to *Albugo candida* (white blister rust; Chapter 20; Tang et al., 1996) and in sugar beet infected with beet curly top virus (Swiech et al., 2001). An interesting report to note is that carbon fixation associated with the dark phase of photosynthesis in uninfected bean leaves on plants infected with *Uromyces phaseoli* (bean rust; Chapter 18) increased compared to bean leaves of uninfected controls (Murray and Walters, 1992). These results indicated that the pathogen can affect the photosynthetic activity in an area of the plant that is not infected. This can be a result of the plant trying to compensate for the loss of activity in infected leaves. In studies of the wheat cultivar Miriam, which is susceptible but tolerant to *Septoria tritici* blotch (STB), the rate of carbon fixation per unit chlorophyll and per green leaf area was higher than that observed in healthy plants. Some plants exhibit enhanced photosynthesis in remaining green tissue to compensate for loss of photosynthetic tissue due to the pathogen (Zuckerman et al., 1997).

The photosynthetic system of plants is affected in many ways by pathogens. Currently, studies often look at the effects on only a few enzymes. In the near future, application of functional genomics techniques to assess disease situations will permit more complete descriptions of disrupted photosynthetic systems. Tests will be developed wherein it will be possible to appraise the effects of a pathogen on each of the enzymes in the photosynthetic process. Regardless of the details of disease process, the overall effect of plant pathogens that interfere with the photosynthetic potential is that less carbohydrates are produced. Ultimately, disease is manifested in reduced growth and yield. In a study on the effects of widespread viral infections of orchids, specifically *Oncidium*, healthy plants following virus elimination had a 17% increase in plant height, 65% increase in the inflorescence size, and 21% increase in photosynthetic capacity (Chia and He, 1999).

RESPIRATION

Respiration in plants involves oxidative processes in which complex molecules, such as the carbohydrates produced during photosynthesis, are broken down into carbon dioxide, water, and energy. Energy is transferred and available to life processes when ADP is converted into ATP. The conversion reaction of ADP to ATP is referred to as **oxidative phosphorylation,** and ATP serves as the energy component needed for almost all the operations in the plant cell. There are numerous metabolic pathways that serve to generate the substrates needed for oxidative phosphorylation and include the following: **glycolysis** (often referred to as Embden–Meyerhof pathway), tricarboxylic acid cycle (TCA), and oxidation of lipids. Another pathway that may be important, but that is not directly connected to oxidative phosphorylation, is the oxidative pentose phosphate pathway in which NADPH is generated along with pentose phosphates that can enter the glycolysis pathway. Respiration in plant pathogens (except viruses) has metabolic pathways similar to those of the plant host. Because of this it is difficult to differentiate between increases in respiration in plants in response to a plant pathogen or to determine if increases in respiration are due to the activity of the pathogen. Increases in respiration have generally been measured by determining increases in oxygen consumption. Because viral pathogens do not have respiratory pathways, they have been used to determine the effect of plant pathogens on respiration. All oxygen consumption in a viral infection can be directly correlated with plant host respiration.

Respiration usually increases in response to an invasion of a plant by a pathogen. Respiration increases more rapidly in plants exhibiting a resistance reaction due to the requirement for energy to rapidly produce and mobilize defense mechanisms. The rapid rise in respiration is followed by a decrease back to normal rates. In plants exhibiting a susceptible response to the pathogen, respiration levels increase more slowly, but continue to rise. Elevated respiration rates can be the result of increased activity of many different enzymes, faster breakdown of carbohydrates, including starch, and an uncoupling of oxidative phosphorylation. Uncoupling (ADP is not converted to ATP) results in increased concentration of ADP, which stimulates respiration. The activity of six enzymes associated with glycolysis and mitochondrial respiration of *Cucurbita pepo* infected with *Cucumber mosaic virus* (CMV) increased within lesions, whereas levels of RUBISCO involved in photosynthesis decreased only slightly (Técsi et al., 1996). Instead of using viruses as a

nonrespiring pathogen, heat-killed bacteria have been employed as the elicitor of host responses. Oxygen uptake in tobacco cells treated with dead bacteria increased within 4 min and lasted for 10 min; thereafter, respiratory rates returned to a steady state that was approximately twice the initial rate (Baker et al., 2000).

ABSORPTION AND TRANSLOCATION OF WATER AND NUTRIENTS

Many pathogens, including viruses, fungi, bacteria, and nematodes, disrupt the absorption and translocation of nutrients in plants. In some cases this may be on a small scale, such as within a single leaf, but in other cases the pathogen can affect the absorption and translocation of water and nutrients throughout the entire plant. This disruption can result from destruction of the root system, blockage of movement of water and nutrients in the xylem or phloem, or the redirection of host nutrients.

Some of the most obvious symptoms noted in response to plant pathogens that attack the plant roots are wilting, chlorosis, and general decline of the host plant. Root systems of plants infected with *Phytophthora*, *Pythium*, or *Thielaviopsis* species are often nonfunctional. Rotting and destruction of roots make it difficult, if not impossible, for the plant to absorb and transport water and nutrients normally obtained by the uninfected root system. Many nematodes cause disruption of absorption and translocation of water and nutrients in the host plant through the destruction of the small feeder roots. Ectoparasitic nematodes, such as ring nematodes (*Mesocriconema*) in turf grasses and peach trees, can lead to a significant reduction of feeder roots that can lead to chlorosis as well as plant mineral deficiency symptoms (Chapter 28).

Other plant pathogens are able to affect the absorption and translocation of water and nutrients within the plant because of their presence in the xylem and phloem tissues. *Ralstonia (Pseudomonas) solanacearum*, the causal agent of bacterial wilt in tobacco, tomato, eggplant, and potato, and Moko disease in bananas, invades the xylem through wounds (Chapter 6). The bacteria spreads through the plant, rapidly multiplies within the xylem, and produce copious amounts of exopolysaccharides (EPSs) that essentially dam or block water flow. Loss of water translocation results in rapid wilting of young plants, wilting of a portion of more mature plants, chlorosis, and general decline. Bacterial streaming can be observed when the stem of a young infected tobacco or tomato plant exhibiting wilt symptoms due to *R. solanacearum* is cut and immediately placed in water. Fungal agents such as *Fusarium oxysporum* f. species *lycopersici* (wilt in tomato) and *Ophiostoma novo-ulmi* (Dutch elm disease) decrease water flow through the xylem. This decrease in water flow can be attributed to physical obstruction of vessels with hyphae, secretion of polysaccharides and pectolytic enzymes, production of gums/mucilages, and formation of **tyloses** in the xylem by the host plant. The most obvious result in blockage of the vascular elements is the reduction in the water movement, but it also diminishes the transport of essential minerals throughout the plant.

The presence of various plant pathogens can also result in the redirection of nutrients and resources within the plant. Pathogens divert resources such as carbohydrates and amino acids of the host for their own growth and reproduction. Leaves of *Arabidopsis thaliana* infected with *Albugo candida* showed increased levels of both soluble carbohydrates and starch when compared to uninfected leaves. Activities of both wall-bound and soluble invertases, enzymes involved in sucrose hydrolysis, were higher in infected leaves compared to those in control leaves. Additionally, an invertase isozyme was present in infected leaves and not found in healthy tissues (Tang et al., 1996). In studies of CMV infection in cotyledons of marrow plants, virus replication and synthesis of viral protein created a strong sink within lesions, resulting in increased photosynthesis and starch accumulation. There was more than twice as much starch hydrolase activity within lesions when compared to that in areas outside the lesions and to healthy leaves (Técsi et al., 1996).

Healthy melon plants have stachyose as a major component of sugars in phloem. Absolute levels of stachyose in phloem sap were similar in CMV-infected leaves, but sucrose levels dramatically increased and the sucrose to stachyose ratio was 15- to 40-fold higher compared to that of healthy plants. Elevated sucrose concentrations did not appear to result from stachyose hydrolysis because there was no corresponding increase in galactose. Therefore, it is possible that CMV infection affected the movement of sucrose and was not due to metabolism (Shalitin and Wolf, 2000).

PLANT GROWTH HORMONES

Normal plant growth is controlled by five major plant growth hormones, which have numerous effects and interactions within the plant (Table 29.2). Alterations in the levels of these hormones can have significant effects on the plant's growth. Many different plant pathogens cause the host to change the hormone concentration in the plant or in some cases they actually produce these growth-controlling substances themselves. Symptoms observed in response to plant pathogens such as galls or tumor growth, excessive branching, leaf **abscission** or **epinasty**, abnormal leaf shape, and abnormal fruit ripening can be attributed to alterations in the levels of various hormones.

The pathogen *Gibberella fujikuroi*, causal agent of Bakanae or "foolish seedling" disease in rice, produces gibberellins in the host plant and causes elongation of the internodes. The disease results in spindly plants that often **lodge**. Prior to the identification of gibberellins as an actual plant growth hormone a Japanese scientist identified gibberellins from cultural filtrate of *G. fujikuroi*. This is an example of how the study of plant pathogens and their activities has been important to other areas of plant research. Conversely, *Fusarium* wilt of oil palm, which results in a stunting, may be caused by an inhibition of gibberellin synthesis, because similar symptoms were observed following the application of an inhibitor of gibberellin synthesis. Application of gibberellin to infected palms resulted in a partial, but not complete, elimination of symptoms (Mepsted et al., 1995).

Some of the best and most complete evidence to demonstrate production of plant growth hormones by a plant pathogen is with *Agrobacterium tumefaciens*, the causal agent of crown gall. This bacterium has a relatively small, self-replicating circular piece of DNA referred to as the Ti (tumor inducing) plasmid. The plasmid has the ability to transfer a section of its DNA, referred to as the T-DNA, into the host cell, where it becomes incorporated into a host cell chromosome within the nucleus. T-DNA contains genes for the production of opines, cytokinin, and auxin. Opines (amino acids) are an unusual nutrient source that cannot be utilized by the plant cell or by many other organisms except *Agrobacterium*, which has genes for opine metabolism. The production of cytokinin and auxin resulting from the T-DNA incorporated into the plant cell is not under the normal control mechanisms of the plant. The overproduction of these hormones results in gall development due to uncontrolled cell division (**hyperplasia**) and enlargement (**hypertrophy**). The ultimate result is the observed symptom, which gives the disease its common name, crown gall.

Pseudomonas savastanoi, the causal agent of olive knot disease, contains genes for auxin production on a plasmid as well as on its bacterial chromosome. In this case, the bacteria rather than the plant host produce the auxin that induces gall formation. In studies of *Erwinia herbicola* pv. *gypsophilae*, cytokinin and auxin genes are present in a pathogenicity-associated plasmid, which induces the production of a gall. The size of root galls observed in response to *Plasmodiophora brassicae*, the causal agent of clubroot of cabbage (Chapter 11), was correlated with the free indole acetic acid content in the clubs, indicating the role of auxins in symptom development (Grsic-Rausch et al., 2000).

Ethylene is a gaseous plant growth hormone. It has many different effects in the plant as shown in Table 29.2. Early in the 1900s, it was noted that oranges shipped with bananas promoted banana ripening. Ethylene, which caused fruit ripening, was not produced by the oranges, but by the fungus *Penicillium digitatum*, the cause of green mold on oranges. A nonethylene producing mutant showed that ethylene production has no significant role in the pathogenicity of *P. digitatum* even though low levels

TABLE 29.2
Major Groups of Plant Growth Hormones and Their Associated Physiological Activities within the Plant

Hormone Class	Physiological Activity
Auxin	Promote cell elongation
(indole-3-acetic acid or IAA)	Induce high levels of ethylene formation resulting in growth inhibition
	Induce cambial cell division
	Initiate root formation
Gibberellins	Promote stem elongation by cell elongation
(GA: many forms)	Stimulate α-amylase production in seeds
	Stimulate flower production
	Retard leaf and fruit senescence
Cytokinins	Promote cell division and differentiation
(Zeatin (Z), Z riboside (ZR), and ZR phosphate)	Inhibit senescence of plant organs
	Induce stomatal opening
	Suppress auxin-induced apical dominance
Ethylene	Stimulate fruit ripening
	Stimulate leaf and flower abscission
Abscisic acid	Induce maturation of seeds including development of desiccation tolerance
(ABA)	Induces water stress associated stomatal closure

of ethylene predispose fruit to postharvest diseases. Many plant pathogens are known to induce ethylene production in the host as a response to infection. Ethylene then induces many other responses within the plant, such as the production of various pathogenesis-related (PR) proteins (Chapter 28) and defense-related compounds. Ethylene production by the host or pathogen can result in premature ripening of fruit, leaf epinasty (downward bending), leaf abscission, and chlorosis.

Abscisic acid (ABA) in general is viewed as a growth inhibitor, and increases in its production have been noted in some plant diseases. ABA production associated with these plant–pathogen interactions resulted from a plant host response to the plant pathogen rather than being produced by the plant pathogen. ABA-producing strains of plant pathogenic fungi have been identified, but the involvement of ABA in pathogenesis has not been thoroughly investigated.

ALTERATION OF CELL PERMEABILITY

The plant cell membrane consists of a bilayer of phospholipids with embedded proteins. Membranes exist within the cell wall and surrounding plant organelles, such as the chloroplasts and mitochondria. Proteins within the cell membrane are important in the regulation and movement of materials across the membrane and are extremely important to the integrity of the cell. When this integrity is altered, the membrane is unable to control the movement of materials across the membrane and substances may be lost from the cell. When plant pathogens attack plants, one of the first responses that can generally be detected is the alteration of membrane **permeability**. Membrane integrity is measured by determining electro-

lyte (charged ions, such as K^+ or Cl^-) leakage from cells. Several toxins associated with plant pathogens have been shown to affect cell membrane permeability or integrity. Some examples of these toxins include T-toxin (*Cochliobolus heterostrophus*: southern corn leaf blight), HC-toxin (*Cochliobolus carbonum*: leaf spot disease in maize), AM-toxin (*Alternaria alternata*: alternaria leaf blotch of apple), and syringomycein E, syringotoxin, and syringopeptin (*Pseudomonas syringae* pv. *syringae*: necrotic lesions in a broad range of monocot and dicot species, resulting in tip dieback, bud and flower blast, spots and blisters on fruit, stem canker, and leaf blight). Many of these toxins are cytotoxic in plant cells at nanomole (10^{-9} M) concentrations and cause necrosis by forming ion channels, which freely allow the passage of divalent cations (e.g., calcium and magnesium). In a study of the cultivar-specific necrosis toxin produced by *Pyrenophora tritici-repentis* (tan spot, a widespread foliar disease of wheat), electrolyte leakage was observed in a toxin-sensitive cultivar after 4 h, but was not observed in a toxin-insensitive cultivar (Kwon et al., 1996).

TRANSCRIPTION AND TRANSLATION

The processes of **transcription** and **translation** are often affected by plant pathogens, but are most clearly observed with plant viruses. Transcription is the process whereby information associated with a DNA sequence or gene is copied into a complementary piece of RNA referred to as messenger RNA (mRNA) (Figure 29.1). RNA polymerase recognizes the start of a gene on the DNA, combines with the DNA separating the two nucleotide strands and then progresses down the DNA section making the complementary strand of RNA until a stop code is encountered.

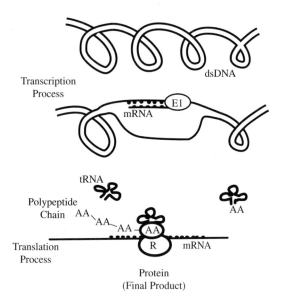

FIGURE 29.1 Transcription and translation process for the formation of proteins based on information located on dsDNA. dsDNA = double stranded DNA, mRNA = messenger RNA, tRNA = transfer RNA, AA = amino acid, E1 = RNA polymerase, R = ribosome.

Translation is the process by which this piece of mRNA is copied into a protein whose amino acid sequence is determined by the sequence of nucleotides in the mRNA. This process involves a ribosome binding to the mRNA and essentially reading three nucleotides at a time that correspond to specific amino acids, which are brought to the ribosome active site by a transfer RNA (tRNA). The amino acids are connected by peptide bonds forming a protein. Some processing of the protein may be required after the amino acid sequence is complete. Included in the proteins produced by this process are the enzymes that are responsible for the production of other materials within the plant cell such as the polysaccharides and lipids.

Virus particles that infect plants consist of genetic information in the form of ssDNA, dsRNA, or ssRNA surrounded by a protein coat (Chapter 4). Viruses do not have the ability to self-replicate, but instead use the cellular machinery of the host. The genetic material moves into the host cell, where it undergoes translation and transcription, allowing the genetic material to be produced along with the protein coat. The viral genome serves as the template for transcription for the production of the nucleic acid component of the virus and translation serves for the production of the components of the viral protein coat.

Much of the previous work concerning the effect of plant pathogens on translation and transcription has involved the detection of changes in the production of specific enzymes, such as the enzymes involved in glycolysis or chlorophyll biosynthesis and degradation to determine the effect of a plant pathogen on photosynthesis. The use of mRNA differential display and other molecular techniques for gene discovery and other molecular tools will supply much more information and will help us understand the exact process of how a plant pathogen affects the physiology of the host.

LABORATORY EXERCISES

"Seeing is believing" and "Prove it to me" are common phases that are used in all kinds of situations. These ideas are particularly important in research — we always want to know what the effect of the plant pathogen is on the plant. The overall objective of the following exercises is to illustrate some of the effects that plant pathogens have on the physiological processes of the plant. We see the results when we observe symptoms, but in these exercises we will look at the effects of plant pathogens on specific physiological activities of the plant. The first exercise can be conducted with many different plants. The only requirement is that the plant pathogen causes symptoms resulting in a chlorotic or mosaic appearance on leaves. Healthy plants with uninfected leaves are used for comparison to the concentration of chlorophyll present in the infected leaf. The second exercise was developed to be used with the bacterial pathogen *Ralstonia solanacearum* and tomato plants. This exercise illustrates the blockage of the xylem system by bacteria and how these bacteria can be isolated from the xylem stream on nutrient agar dishes. The final exercise was developed to illustrate how a plant pathogen, *Agrobacterium tumefaciens*, can alter the normal growth patterns of plant tissue.

EXPERIMENT 1. EFFECTS OF A PLANT PATHOGEN ON CHLOROPHYLL

Chlorophyll is probably one of the most important molecules on earth. In combination with other compounds and enzymes in the plant, chlorophyll is able to harvest light energy and transfer it into a usable form of chemical energy that the plant utilizes for growth and reproduction. This stored chemical energy also supplies energy for many other forms of life on earth, including humans and plant pathogens. Any time we eat food, we are utilizing this energy either directly or indirectly. The amount of chlorophyll associated with cells has a direct effect on the amount of stored energy that a plant ultimately produces. A reduction in the chlorophyll can lead to effects such as stunting due to the loss of the energy conversion potential. Chlorophyll extractions such as the one outlined here (Moran, 1982) can be easily done and provide a quick measurement of the effects of plant pathogens on chlorophyll content in plants.

Materials

Each student or team of students will require the following items:

- Leaves from a plant exhibiting chlorotic or mosaic symptoms. The specific plant is not that important, but it is vital to have a healthy plant that can be used for comparison. Examples of plants that can be used include tobacco plants infected with *Tobacco mosaic virus* or cucumber plants infected with *Cucumber mosaic virus*.
- Balance to determine the weight of individual leaves
- #5 cork borer
- Ruler
- Test tubes
- Dimethylformamide (DMF). *Caution*: Use in a chemical hood only and wear rubber gloves to prevent absorption through skin. Obtain information on material safety data sheet prior to use.
- Disposable rubber gloves
- Spectrophotometer to measure
- Disposable or quartz cuvettes and spectrophotometer
- Pipettes for transferring DMF
- Incubator set at 10°C without lighting
- Calculator

Follow the instructions in Procedure 29.1 to complete the experiment.

Procedure 29.1 Determination of Total Chlorophyll Present in Leaf Area to Compare Healthy and Infected Leaf Areas and to Compare Effects of Different Pathogens

1. Choose leaves from a plant that is exhibiting symptoms such as chlorosis or mosaic patterns. Nonsymptomatic leaves must also be obtained from the same plant or from the same general area of a healthy plant. Leaves associated with different diseased plants can be compared in order to compare the effects of different plant pathogens.
2. With a # 5 cork borer (1-cm diameter) cut 10 discs from each type of leaf area (normal and symptomatic). Calculate the area of the leaf disks by the following formula:

$$\text{Area} = \pi r^2 \times 10$$

where $\pi = 3.14$ and $r = 0.5 \times$ diameter, or, in this case, 0.5 cm.

3. Weigh the leaf disks and place disks in 5 ml cold dimethylformamide (DMF). Incubate in the dark at 10°C for a minimum of 48 h.
4. Determine absorbance of the DMF at 664 nm and 647 nm and calculate the total chlorophyll by the following formula:

$$\text{Total chlorophyll} = 7.04 \, \text{Abs}_{664} + 20.27$$
$$\text{Abs}_{647} = \mu g \text{ chlorophyll/ml}$$

5. Convert the µg/ml to µg/cm^2 by multiplying the total chlorophyll by 5 ml (DMF solution volume) and dividing by the total leaf area associated with the discs as calculated in Step 2. Convert the µg/ml to µg/gm fresh weight by multiplying the chlorophyll total by 5 ml and dividing by the weight in grams obtained in Step 2.

Anticipated Results

The results of this experiment provide a basis for suggesting that one pathogen or strain of the pathogen causes more severe reactions in the plant than another. Reductions in chlorphyll have a direct effect on plant growth. If the chlorophyll is reduced by 10% by one pathogen and 40% by a second pathogen, it is clear that this second pathogen is more severely affecting the plant based on the effect on chlorophyll. In many cases, however, reductions in chlorophyll are not the only effect a plant pathogen has, so other tests may be needed to quantify the total effect that the plant pathogen has on the plant. This experiment should have definite results. An additional component to the experiment may be to determine the variability of results for a specific group illustrating the purpose of having replicates in experiments.

Questions

- What is the percent reduction in the levels of chlorophyll associated with the infected leaves compared to the healthy leaves?
- What was the variation in chlorophyll values within the laboratory for the infected leaves and healthy leaves? Minimum value? Maximum value? Were there similar levels of variation for the healthy and infected groups or was there more variation associated with the infected leaves? How can you explain the variation within these groups?
- Why would it be important to know the variation in chlorophyll levels of plants prior to set-

ting up a research project investigating the effects of a plant pathogen on chlorophyll levels of a plant?

- How can plant pathogens affect the total level of chlorophyll in a plant?

EXPERIMENT 2. DISRUPTION OF WATER TRANSLOCATION

Disruption of the absorption and translocation of water within the plant can occur when plant pathogens cause a destruction of the root system, block the movement of water and nutrients in the xylem and phloem, or redirect the movement of host nutrients. The blockage of the water flow within plants is a common mechanism for plant pathogens that results in wilting of the plant and ultimately may kill the plant. Many of the pathogens that cause these symptoms are very good saprophytes in the soil and survive for many years even in the absence of a host. *Ralstonia solanacearum* used in this exercise is common in tropical and subtropical areas and causes severe diseases in tobacco, tomato, potato, and eggplant in warm areas outside the tropics. Several races of this organism have various host ranges, and a race that attacks tomato plants is used in this exercise.

Materials

Each student or team of students will require the following items:

- 2- to 4-week-old tomato plants (Marion or Rutgers varieties that are susceptible and various other varieties may be included to compare susceptibility of different varieties)
- 24- to 48-h-old culture of *Ralstonia solanacearum* (race pathogenic to tomato plants) grown in nutrient broth or trypicase soy broth (Difco, Lansing, MI)
- 1-ml sterile syringes with 0.5 inch needles
- Trypicase soy broth or nutrient broth
- 30°C growth chamber for tomato plants
- Glass microscope slides and cover slips
- Sterile distilled water in dropper bottle
- Nutrient agar or trypticase soy agar in 10-cm diameter petri dishes
- Shaker
- 28°C to 30°C bacterial culture incubator
- Bacterial transfer loop that can be sterilized by flaming
- Alcohol burner
- Safranin O (0.25% solution in 10% ethanol, 90% water; should be made up initially in ethanol)
- Disposable nitrile gloves
- Paper towels
- Sink with water
- Compound microscope with oil immersion lens

Follow the instructions in Procedure 29.2 to complete the experiment.

Procedure 29.2 Disruption of Vascular Flow in Tomato Infected with *Ralstonia solanacearum*

1. Obtain tomato plants approximately 2 to 4 weeks old. (Marion and Rutgers varieties are extremely susceptible, but other varieties can be included for comparison of resistance to *Ralstonia solanacearum*.)
2. Grow a virulent strain of *R. solanacearum* in nutrient broth or trypicase soy broth on a shaker for 24 h or until medium is turbid.
3. Using a 1-ml (cc) sterile syringe with 0.5-inch needle (smallest diameter possible), draw up 0.2 ml of bacterial suspension.
4. Inject the bacterial suspension just below the surface and into the lower stem of the tomato plant. Incubate the inoculated plants at approximately 30°C until wilt symptoms develop. This usually occurs within 2 to 7 days. Generally, the older the tomato plant, the longer it will take for the plant to wilt.
5. After the plants wilt, place a couple of drops of sterile distilled water on a glass slide. Cut the tomato stem across the area above the site of inoculation and immediately touch the cut stem to the water on the slide and observe bacteria streaming from the cut surface (water on slide should become turbid due to the presence of bacteria).
6. Take a sterile loop and streak the water–bacteria suspension from the slide onto either nutrient agar or trypticase soy agar in petri dishes and seal with parafilm. Incubate the sealed dishes at 28°C to 30°C for 24 h and observe bacterial growth.
7. Touch loop to the bacteria–water mix on the glass slide prepared in Step 5 and transfer a drop to a clean slide. Spread the suspension and air dry.
8. Heat-fix the bacteria to the slide by passing the slide quickly through the flame of a alcohol burner. Be careful not to get the slide too hot.
9. Add a couple of drops of safranin stain to the slide for 30 to 60 sec; then rinse with water (wear gloves).
10. Blot, do not rub, the slide dry with a paper towel. Rubbing will remove the bacteria.
11. Observe the stained bacteria by using an oil immersion lens on a compound microscope.

Anticipated Results

Tomato plants infected with *R. solanacearum* will generally wilt within 2 to 4 days, depending on the strain and the age of the tomato plant. Older plants will take longer to wilt. The temperature of the growth chamber is vital. If the temperature is less than 30°C, the wilting of the plant may be delayed or may not occur. Bacterial colonies should appear within 48 h on the nutrient agar dishes on which the bacteria were transferred from the slide. Bacteria should be obvious on the stained slide made from the xylem exudates–water mix.

Questions

- Is the wilting associated with *R. solanacearum* infection of tomato strictly due to the presence of bacteria in the xylem stream blocking the passage of water?
- What compounds do some bacteria produce that may also be important in the blockage of water flow in a plant?
- How are bacteria such as *R. solanacearum* disseminated in nature? What can be done to manage diseases caused by bacteria such as *R. solanacearum*? What are the limitations?
- Were there any observable differences between varieties of tomato used in the experiment?
- How many different types of bacterial colonies were isolated from the xylem exudate? Are these other organisms involved in the wilting observed? What might you need to do to determine their involvement?

EXPERIMENT 3. EFFECT OF *AGROBACTERIUM* SPECIES ON NORMAL PLANT GROWTH

Plant pathogens are known to either produce normal plant growth hormones or have the ability to increase or alter the production of plant growth hormones by the plant. Virulent strains of *Agrobacterium* contain plasmids that contain genes for the production of various plant growth hormones. The genes associated with the plant growth hormones are transferred to the plant during the infection process, and as a result more plant hormones are produced that interfere with normal plant function and development. The presence of these extra levels of hormones causes various abnormalities in the normal growth of the plant. Tomato plants serve as experimental subjects and can rapidly produce callus or root growth following inoculation with *Agrobacterium*.

Materials

Each student or team of students requires the following items::

- Tomato plants 4 to 6 weeks old (various varieties can be used for comparison to "Marion" and "Rutgers" varieties, which are susceptible)
- 24- to 48-h-old culture of virulent strain of *A. tumefaciens* (induce gall formation) or *A. rhizogenes* (induce root formation) growing on nutrient agar or trypticase soy agar
- Sterile scalpel with #10 blade
- Shaker
- Parafilm
- Balance
- Dissecting microscope
- Growth chamber, greenhouse, or window where tomato plants can be maintained for several weeks

Follow the instructions in Procedure 29.3 to complete the experiment.

Procedure 29.3 Induction of Gall or Root Formation by Inoculation of Tomato Plants with *Agrobacterium tumefaciens* or *A. rhizogenes*

1. Obtain tomato plants approximately 4 to 6 weeks old. (Marion and Rutgers varieties are susceptible, but other varieties can be included for comparison of gall or root formation following inoculation.)
2. Grow a virulent strain of *A. tumefaciens* (for induction of gall formation) or *A. rhizogenes* (for induction of root formation) on nutrient agar or trypticase soy agar for 24 h or until growth is observable on the dish.
3. Take a sterile scalpel and touch it to the bacterial growth in the petri dish.
4. Cut the surface of the tomato stem with the scalpel coated with the bacteria and then wrap a piece of parafilm around the cut. Be careful not to cut completely through the tomato stem.
5. For the control treatment take a sterile scalpel and touch it to the surface of a sterile agar and then cut the surface of a tomato stem with the scalpel followed by wrapping the parafilm.
6. Remove the parafilm after one day. Observe the tomato plants weekly for the production of callus tissue (galls) or roots. Abnormal growth should be observed after approximately 2 to 3 weeks. Plants can be incubated in a window, greenhouse, or growth chamber.

7. Compare callus/root growth in different tomato varieties by measuring the size/number of abnormal growths and then cutting the abnormal growth associated with the plant and weighing the tumor/root growth.

Anticipated Results

The tomato plants should be observed on a weekly basis. The plants should begin to show symptoms of gall or root formation within approximately 2 to 3 weeks. The first indication of abnormal growth will be a roughness on the surface of the stem where the plant was inoculated. Plant height should be measured for all plants prior to inoculation and at the end of the experiment. Control plants that are cut with the scalpel but are not inoculated with *Agrobacterium* should be included. The total weight of abnormal callus or root growth at the end of the experiment (after approximately 6 weeks) should be determined and should be related to the resistance of the various tomato varieties to *Agrobacterium*.

Questions

- What was the average callus or abnormal root growth for each of the varieties of tomatoes used in the experiment? Does the average weight of callus or root growth correspond to the resistance to *Agrobacterium*? Why or why not?
- Was there any response on the control plants? If so, what was that response? Why did it occur?
- Did the infection with *Agrobacterium* have any effect on the overall height increase of the infected plant compared to that of the control? Would you expect it to have an effect? Why or why not?
- Why is parafilm placed around the plant after it is inoculated?
- What are the methods used for the management of *Agrobacterium* in nature?

REFERENCES

Almási, A., D. Apatini, K. Bóka, B. Böddi and R. Gáborjányi. 2000. BSMV infection inhibits chlorophyll biosynthesis in barley plants. *Physiol. Mol. Plant Pathol.* 56: 227–233.

Baker, C.J., E.W. Orlandi and K.L. Deahl. 2000. Oxygen metabolism in plant/bacteria interactions: characterization of the oxygen uptake response of plant suspension cells. *Physiol. Mol. Plant Pathol.* 57: 159–167

Chia, T.-F. and J. He. 1999. Photosynthetic capacity in *Oncidium* (Orchidaceae) plants after virus eradication. *Environ. Exp. Bot.* 42: 11–16.

Grsic-Rausch, S., P. Kobelt, J.M. Siemens, M. Bischoff and J. Ludwig-Müller. 2000. Expression and localization of nitrilase during symptom development of the clubroot disease in *Arabidopsis. Plant Physiol.* 122: 369–378.

Kema, G.H.J., D. Yu, F.H.J. Rijkenberg, M.W. Shaw and R.P. Baayen. 1996. Histology of the pathogenesis of *Mycosphaerella graminicola* in wheat. *Phytopathology* 86: 777–786.

Kwon, C.Y., J.B. Rasmussen, L.J. Fancl and S.W. Meinhardt. 1996. A quantitative bioassay for necrosis toxin from *Pyrenophora tritici-repentis* based on electrolyte leakage. *Phytopathology* 86: 1360–1363.

Lopes, D.B. and R.D. Berger. 2001. The effects of rust and anthracnose on the phytosynthetic competence of diseased bean leaves. *Phytopathology* 91: 212–220.

Mepsted, R., J. Flood and R.M. Cooper. 1995. Fusarium wilt of oil palm I. possible causes of stunting. *Physiol. Mol. Plant Pathol.* 46: 361–372.

Moran, R. 1982. Formulae for determination of chlophyllous pigments extracted with N,N-dimethylformamide. *Plant Physiol.* 69: 1376–1381.

Murray, D.C. and D.R. Walters. 1992. Increased photosynthesis and resistance to rust infection in upper, uninfected leaves of rusted broad bean (*Vicia faba* L.). *New Phytol.* 120: 235–242.

Niederleitner, S. and D. Knoppik. 1997. Effects of the cherry leaf spot pathogen *Blumeriella jaapii* on gas exchange before and after expression of symptoms on cherry leaves. *Physiol. Mol. Plant Pathol.* 51: 145–153.

Shalitin, D. and S. Wolf. 2000. Cucumber mosaic virus infection affects sugar transport in melon plants. *Plant Physiol.* 123: 597–604.

Strelkov, S.E., L. Lamari and G.M. Ballance. 1998. Induced chlorophyll degradation by a chlorosis toxin from *Pyrenophora tritici-repentis. Can. J. Plant Pathol.* 20: 428–435.

Swiech, R., S. Browning, D. Molsen, D.C. Stenger and G.P. Holbrook. 2001. Photosynthetic responses of sugar beet and *Nicotiana benthamiana* Domin, infected with beet curly top virus. *Physiol. Mol. Plant Pathol.* 58: 43–52.

Tang, X., S.A. Rolfe and J.D. Scholes. 1996. The effect of *Albugo candida* (white blister rust) on the photosynthetic and carbohydrate metabolism of leaves of *Arabidopsis thaliana. Plant Cell Environ.* 19: 967–975.

Técsi, L.I., A.M. Smith, A.J. Maule and R.C. Leegood. 1996. A spacial analysis of physiological changes associated with infection of cotyledons of marrow plants with cucumber mosaic virus. *Plant Physiol.* 111: 975–985.

Wright, D.P., B.C. Baldwin, M.C. Shephard and J.D. Scholes. 1995. Source-sink relationships in wheat leaves infected with powdery mildew. II. changes in the regulation of the Calvin cycle. *Physiol. Mol. Plant Pathol.* 47: 255–267.

Zuckerman, E., A.E. Eshel and Z. Eyal. 1997. Physiological aspects related to tolerance of spring wheat cultivars to *Septoria tritici* blotch. *Phytopathology* 87: 60–65.

Part V

Epidemiology and Disease Control

30 Plant Disease Epidemiology

Kira L. Bowen

CHAPTER 30 CONCEPTS

- Epidemiology is the study of properties of pathogens, hosts, and the environment that lead to an increase in disease in a population.

- Polycylic pathogens that disperse readily usually cause the most damaging epidemics.

- Temperature, moisture, wind, soil properties, radiation, and other parameters may comprise a conducive environment.

- Host developmental stage and population uniformity can affect epidemic development.

- Disease spreads in space as it increases in incidence (numbers of affected plants).

- Mathematical models that allow disease prediction also contribute to disease management.

Epidemiology is the study of factors that lead to an increase in disease in a population. Understanding the reasons why diseases increase in populations of plants and what can influence those increases can contribute to decisions concerning plant disease management. The ability of humans to produce food and fiber is, in part, limited by their ability to manage plant diseases. Thus, plant disease epidemiology has contributed to the highly technological culture in which we currently live. Concepts related to plant disease epidemiology will be discussed in this chapter.

COMPONENTS OF AN EPIDEMIC

Disease occurs on a plant when a virulent pathogen and a susceptible host interact in a conducive environment. For disease to increase, these three components — pathogen, host, and environment — must continue to interact over time. In a population of plants, disease becomes important when the damage caused by disease increases to the extent that there are social and economic impacts. There are numerous examples of plant disease epidemics that have had a profound impact on human history and these illustrate the interaction needed over time among the pathogen, host, and environment. Well-known examples include the potato late blight epidemic of 1845, chestnut blight of the early 20th century, and the southern corn leaf blight epidemic of 1970.

POTATO LATE BLIGHT

The potato, *Solanum tuberosum*, native to the American continent, did well as a crop in the cool damp climate of Ireland. Prior to the 1840s, farmers of Ireland had become dependent on the potato for food and the crop was grown throughout the country. Therefore, two of the three necessary components for a disease epidemic were present — the susceptible crop and an environment that was conducive to disease. *Phytophthora infestans*, the fungal-like oomycete that causes potato late blight (Chapter 20), was introduced into Ireland by early 1844, providing the third element for an epidemic. It is probable that potato late blight had become widely established by the end of the 1844 growing season (Andrivon, 1996), but had remained at low levels. In 1845, excessively cool and wet weather prevailed throughout Europe. This weather, highly conducive to the development of *P. infestans*, allowed rapid increase in the severity of late blight and resulted in the loss of the entire potato crop by the end of the 1845 growing season. Because farmers were almost entirely reliant on the potato for food, this loss devastated the Irish population, caused the Irish potato famine, and began the discipline of plant pathology.

CHESTNUT BLIGHT

The American chestnut, *Castanea dentata*, was an important and predominant tree species in native forests of the eastern U.S. through the 18th century. The rot-resistant wood was important in the timber industry and the bark was a source of tannins for the leather industry. American

0-8493-1037-7/04/$0.00+$1.50
© 2004 by CRC Press LLC

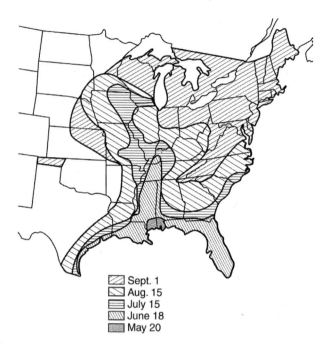

FIGURE 30.1 Disease progress of southern corn leaf blight during the 1970 growing season. (Adapted from Moore, W.F., 1970. *Plant Dis. Rep.* 54: 1104–1108).

chestnut also provided food in the form of nuts. In 1904, American chestnut trees in the New York Zoological Park began wilting and suddenly dying, apparently from a canker disease. During the next year, the disease, chestnut blight, was noticed on American chestnuts in other areas of New York. By 1911, chestnut blight had spread through New Jersey, south to Virginia, and north into Massachusetts. Communities in the Appalachians depended on the American chestnut tree for their livelihood. Attempts to stop the spread of chestnut blight were unsuccessful and mature chestnuts trees died throughout the eastern U.S.

Chestnut blight is caused by the pathogen *Cryphonectria parasitica* (previously *Endothia parasitica*) and is spread when sticky conidia exuding from cankers adhere to birds and insects. This pathogen was brought to the U.S. with botanical specimens from China. In China, *C. parasitica* is **endemic**, where it affects native Chinese chestnuts, causing low levels of disease. American chestnuts had no resistance to *C. parasitica* and there were no elements of the environment in the U.S. that effectively checked its spread. *Cryphonectria parasitica* continued to spread and kill mature American chestnuts throughout the eastern U.S. The chestnut blight epidemic changed the makeup of American forests in less than 50 years, but only after introduction of the virulent pathogen into the U.S.

SOUTHERN CORN LEAF BLIGHT

Southern corn leaf blight, caused by *Cochliobolus heterostrophus* (formerly *Bipolaris maydis*), is endemic to

the U.S. Through the 1960s, hybrid field corn cultivars (*Zea mays*) were generally resistant to this disease. However, this reaction changed due to a change in the way hybrid seed was produced. The change in seed production involved the use of a type of cytoplasmic male sterility, *Tcms*, which allowed the production of hybrid seed without the labor-intensive removal of pollen-producing tassels. Hybrid seed is produced by planting several rows of a desired "female parent" corn inbred (A) between one or two rows of the "male parent" corn inbred (B). Pollen-producing tassels are removed from the female parents in these hybrid seed production fields. Because all pollen in the field is from the male parent, all seed produced on the female parent is a result of the cross (A × B) and is hybrid seed. Sterile pollen is produced by corn plants with *Tcms* and this sterility was incorporated, through selection, into inbred lines used as female parents. *Tcms* was incorporated into numerous corn breeding lines used as female parents in hybrid seed production, and the hybrid plants produced from these crosses still contained traces of *Tcms* cytoplasm.

By 1970, 80% of hybrid corn in the U.S. was produced by using *Tcms*, creating a genetic uniformity throughout corn-producing regions. About that time, corn cultivars that had previously been resistant to southern corn leaf blight were becoming diseased. As *Tcms* had been increasingly incorporated into corn germplasm, a new race of *C. heterostrophus* had developed. This race, dubbed Race T, was more aggressive than the former dominant Race O of the pathogen and was highly virulent to *Tcms* corn. *Cochliobolus heterostrophus* Race T developed quickly on the susceptible host populations of *Tcms* plants in the hot moist conditions prevailing in the southern corn-producing regions of the U.S. (Figure 30.1). Losses due to this disease were 100% in some fields in the South where the epidemic apparently initiated. Although the moist weather of the South contributed to disease development, the southern corn leaf blight epidemic of 1970 was a result of human creation of the susceptible host population.

These three examples of historically important plant disease epidemics illustrate the need for the interaction of each of the components of the disease triangle (Chapter 2) over time. If the weather had not been highly conducive to the rapid development of *P. infestans* in Ireland in 1845, the Irish potato famine may never have happened. American chestnut trees were an important part of eastern U.S. forests until introduction of the virulent pathogen *C. parasitica* to the U.S. Southern corn leaf blight had been adequately managed with plant resistance until the host was altered to a susceptible population. These three plant disease epidemics also illustrate additional concepts relating to plant disease development and epidemiology.

PATHOGEN

Chestnut blight arose in the U.S. after the importation of infected botanical specimens for a collection prior to 1907. Similarly, potato late blight probably traces back to a shipment of tubers into Belgium from the Americas. These infected specimens and tubers provided the **initial inoculum** of virulent pathogens for these two plant disease epidemics. Given that the favorable environment and susceptible hosts were present upon introduction of the initial inoculum, the pathogen increased, causing more disease. In our current global economy, plant disease epidemics are rarely initiated by the introduction of an exotic pathogen to a new location. More commonly, initial inoculum for an epidemic is the quantity of the pathogen that survives a period without a host, as through a winter (also called primary inoculum), or the quantity that arrives at a location after a dispersal event. The initial inoculum for a plant disease epidemic may come from infected seed or may be the viable propagules of a soilborne pathogen at planting time. Initial inoculum can be from a single infected plant brought into a greenhouse or even from a vector for many viral and bacterial diseases of plants. Of course, the greater the quantity of initial inoculum, the greater is a disease level at the start of an epidemic.

Potato late blight and southern corn leaf blight disease levels increased in a matter of days or weeks, whereas the increase in chestnut blight was observed over years. Chestnut blight increased over years, not because the tree has a long life span, but because of inherent characteristics of the pathogen, *C. parasitica*. *Cryptonectria parasitica* is considered **monocyclic**, that is, this pathogen reproduces only once in a growing season. There are pathogens of annual plants that are considered monocyclic, such as species of *Sclerotium* that cause diseases on numerous plants including southern stem rot of peanut, white rot of onion, and southern stem blight of tomato. *Sclerotium rolfsii* survives as sclerotia and these germinate and cause plant infections during a growing season. At the end of the growing season, new sclerotia are produced on infected plants to serve as inoculum for the following growing season. Like chestnut blight, many canker diseases are monocyclic. Azalea leaf gall, caused by *Exobasidium vaccinii*, common in the spring in southern landscapes, is also a monocyclic disease (Chapter 19). Certain microcyclic rusts that have no repeating urediniospore stage, such as cedar-apple rust caused by *Gymnosporagnium juniperi-virginianae*, are also monocyclic diseases (Chapter 18). Plant disease epidemics that are due to monocyclic diseases will only increase over years, not a few months.

Southern corn leaf blight and potato late blight are both **polycyclic** diseases. Pathogens that cause polycyclic diseases multiply several times in a growing season. It is likely that only one or a few potato tubers with a viable infection of *P. infestans* survived the trip from the Americas to Europe, then to Ireland by early 1844. That infected tuber produced a plant with foliage infections from which sporangia or zoospores were produced. Sporangia are easily spread on wind currents, infecting more plants in moist conditions and producing new sporangia in as little as 5 days when weather conditions are optimum. Thus, in a 90-day growing season, this disease cycle could have repeated 17 times, increasing disease intensity with each cycle. Most plant disease epidemics that occur unexpectedly are polycyclic because of the rapidity with which they can develop.

Polycyclic diseases occur on perennial plants, just as they do on annual crops such as potatoes or corn. Apple scab, caused by *Venturia inaequalis*, for example, is a polycyclic disease of apple trees (Chapter 15). *Erwinia amylovora*, the bacterial cause of fire blight, will reproduce and cause increasing numbers of infections as long as moist conditions prevail. Many rust diseases that have the repeating urediniospore stage, including zoysia rust and leaf and stem rusts of cereals, are polycyclic, as are powdery mildew diseases (Chapter 14). Conidia of *Sphaerotheca macularis* and *Oidium* species, causing powdery mildew of hop and the ornamental banana shrub, respectively, can infect their hosts and produce more conidia in 5 to 8 days under favorable conditions. Diseases of plants caused by viruses are also often polycyclic, especially when caused by viruses that are vectored by insects. *Cucumber mosaic virus* and *Tomato spotted wilt virus* are examples of polycyclic diseases caused by viruses. Diseases caused by nematodes can also be polycyclic. Rootknot nematodes (*Meloidogyne* species) complete a life cycle in 3 to 4 weeks when soil temperatures are 25°C to 30°C. In the southern U.S., these nematodes go through three to four reproductive cycles in a growing season.

There are also pathogens that cause diseases that are **polyetic**. Polyetic diseases are those that take several years to develop and reproduce from the time infection is initiated. Because the pathogen *Cronartium ribicola* can take several years to grow into the main stem of the tree before sporulating, white pine blister rust is a polyetic disease. Mistletoe parasites are also polyetic because they reproduce only after a few years of plant growth.

Another aspect of the pathogen that affects epidemic development is the diversity within its population. Different individuals or isolates of the same pathogen may vary in the host cultivar that they can infect and these are called **races** of a pathogen. Races have been identified in a number of plant pathogens, including *Phytophthora sojae*, cause of soybean root rot, and *Xanthomonas campestris* pv. *vesicatoria*, cause of bacterial leaf spot of pepper. Differences in a pathogen population may also affect the pathogen aggressiveness, as seen with Race T of *C. heterostrophus*. When a new race of a pathogen develops, plant epidemics can become severe in crops previously thought to be resistant to that pathogen. This is what happened in wheat with *Puccinia graminis* f. species *tritici* through the early part of the 20th century. Wheat cultivars were developed and released to growers with improved resistance to stem rust, then within a few years the disease once again became severe on the crop.

ENVIRONMENT

Many of the most destructive diseases of plants are polycyclic diseases and each of them differ in some way. Disease cycles for southern corn leaf blight and potato leaf blight in optimum conditions can be very short, whereas many disease cycles are somewhat longer. Disease cycles, from infection by inoculum through production of more inoculum, lengthen in time with less than optimum conditions. Temperature and moisture, and the interaction of these two components of weather, are often the most important factors affecting the length of a disease cycle. For example, the southern corn leaf blight disease cycle can be completed in three days when moist conditions prevail and temperatures are between 20°C and 30°C. If drying occurs, especially if relative humidity remains below 90% for 18 or more hours in a day, the disease cycle lengthens. Urediniospores of *P. graminis* f. species *tritici*, stem rust of wheat, cause infection and produce more urediniospores in as little as 7 days in ideal conditions at 30°C. However, under normal field conditions, this cycle repeats every 14 to 21 days (Roelfs, 1985).

In the field, temperature and moisture fluctuations detract from ideal conditions for pathogen development. Such fluctuations may be second by second as wind currents vary, but we are usually most aware of those that occur diurnally, in a 24-h cycle, or due to moving weather systems. Often, unfavorable weather conditions limit disease development substantially, as can be seen with decreasing levels of powdery mildew in winter wheat as daytime temperatures warm or minimal black spot on roses during dry seasons. Diseases that seem to disappear during the summer, such as brown patch of warm-season turfgrasses caused by *Rhizoctonia* species, are actually limited by high temperatures and reinitiate in late summer as temperatures decrease.

Temperature and moisture are readily understood as influences on pathogen development. As suggested previously, other aspects of the environment can affect pathogen development. Wind, for example, plays several important roles. Wind currents are one of the most important means by which plant disease inoculum is **dispersed** or becomes spread over a geographical area. It is easy to imagine how a fungal spore can be carried on the wind from one plant to another in a field, but it is also possible for inoculum to move even greater distances. Generally, faster wind currents carry inoculum propagules further and smaller propagules are carried the furthest. Uredin-

iospores of *P. graminis* f.species *tritici*, which are relatively small spores, have been found to be carried by wind up to 680 km (>420 miles) from their source. Wind currents also have direct effects on temperature and moisture because they can cool and dry surfaces, thus affecting the length of the disease cycle.

Solar radiation, the energy from the sun, is another aspect of the environment that influences disease development. Radiation can inhibit or stimulate germination of a number of fungal spores. Most bacterial cells quickly lose viability with exposure to radiation. Dogwood anthracnose lesions develop more slowly and have reduced sporulation on foliage in full sun than in full shade. Conversely, spores of *Botrytis* species, causing gray mold of flowering plants, have been found to germinate only with exposure to particular UV wavelengths. Thus, the prevalence of overcast vs. cloudless skies can contribute to the development of an epidemic.

Other aspects of the environment can affect disease development. Soil, for example, can have physical and chemical characteristics that can profoundly affect disease development, particularly diseases by soilborne pathogens. Nematodes develop and move most readily in moist, but not waterlogged, soil, and near the soil surface where there is access to oxygen. Similarly, species of *Sclerotium* have a high oxygen requirement, so root rots by *Sclerotium* tend to begin near the soil surface. Soil acidity is one of the chemical properties that can affect disease development as seen with take-all of wheat and turfgrasses by *Gaeumannomyces graminis*, which is favored in more neutral (higher pH) soils. Soil nutrients can also affect the development of both soilborne and airborne foliar diseases. For example, nitrogen fertilization has been associated with increased potato late blight severity. High N is also conducive to brown patch development in turf caused by *Rhizoctonia* species and excessive fertilization can initiate a severe outbreak of the disease.

HOST

In addition to the influences from the environment, the genetic makeup of a host population can affect an epidemic. A higher degree of genetic uniformity among plants is likely to allow more rapid and more severe disease development. In any population of plants, there will be some differences between individual plants that can affect the amount of disease resulting from an infection. However, plants that are self-pollinated, such as wheat, will have less variability between individuals in a population than those that are cross-pollinated, such as corn. Also, plants that are vegetatively propagated, such as potatoes and many ornamentals, have lower variability between individuals than even self-pollinated plants. Thus, epidemics would be expected to develop most rapidly across a population of clonally or vegetatively propagated plants, less rapidly in self-pollinating plants and more slowly in cross-pollinated populations. In wild or natural plant populations, where plant species are intermingled and different ages and origins of plants exist together, disease epidemics are considered rare. This rarity of plant disease epidemics in mixed species populations is primarily due to the probability of pathogen propagules landing on an appropriate host.

Humans, of course, are responsible for creating massive populations of identical plants that can readily succumb to disease problems. Humans have also discovered that certain plants have resistance or the inherent ability to withstand pathogen infection. Two types of resistance are recognized. One is easy to select for or breed into plants because it is usually encoded for by a single gene. In addition, this **single-gene resistance** frequently prevents the development of any symptoms of disease (Chapter 31). The disadvantage of this type of resistance is that a change in the pathogen can negate the resistance and allow as much disease as if the plant were susceptible. Thus, adaptation of the pathogen to single-gene resistance can result in the development of a severe plant disease epidemic. In wheat, for example, numerous genes have been identified that encode for resistance to stem rust and these genes can be found in different cultivars of wheat. Because races of *P. graminis* f. species *tritici* shift from year to year, different cultivars need to be planted from one year to the next to avoid a devastating disease epidemic. The other type of resistance is **multigenic resistance**, and this is encoded by several to many genes. Multigenic resistance is difficult to incorporate into a single cultivar, but is not as readily overcome as single-gene resistance. However, multigene resistance usually does allow some disease development, although this resistance is effective with all races of a pathogen. Epidemics do develop in crops with multigenic resistance, but disease severity generally stays low (Vanderplank, 1963).

Tissue age or age of host plants can also influence the rate of disease development. Many diseases, such as apple scab, are more severe on young tissue because pathogens infect this tissue more readily. Plants are often more susceptible to diseases when they are young or when the plant or particular plant parts are actively growing. Young tissue can be more susceptible to disease because natural barriers (e.g., cuticular coats) have yet to develop. There are also plant diseases, such as Sclerotinia blight of peanut caused by *Sclerotinia minor*, charcoal rot of soybean caused by *Macrophomina phaseolina* and chestnut blight, that do not develop until plants have entered their reproductive stages.

Plant growth provides more tissue that has the potential to become diseased. If disease is not increasing as rapidly as the plant is growing, the proportion of diseased tissue on that plant will appear to decrease. Similarly, if defoliation occurs due to disease, as happens with peanut leaf spot diseases and black spot of rose, the proportion of the plant that is diseased may appear to decrease. Many plants have limited or finite growth during a season so that disease can eventually affect the entire plant if conditions remain appropriate. Thus, cultivars of plants that have longer growing seasons may allow an epidemic to develop longer and suffer heavier damage due to a particular disease than other, shorter-season cultivars.

As plants grow, they also go through various developmental stages that involve physiological changes and these changes can influence disease development. Although it might be tedious work, it is easy to understand that various aspects of plant size can be measured. Plant height, numbers of leaves, total area of foliage, root depth, root volume, and root mass, if monitored over time, will reflect plant growth. Host development, too, can be monitored and for many plants distinct **host growth stage** keys or diagrams are available (Campbell and Madden, 1990). In corn, for example, distinct stages were delimited by Hanway (1963) and include emergence, tassel formation, silk development, and several stages in kernel maturation. Each of these developmental stages is separated in time, but not necessarily by equal time increments. Growth stage keys have been developed for most agronomic crops, including cereals, soybeans, and peanuts.

DISEASE

Prior to the complete devastation of their crops in 1845, Irish farmers did not notice late blight on their potatoes even though it had likely occurred in the preceding year. Late blight that may have affected potato plants in 1844 would have been easy to overlook, in part because the concept of plant disease was unknown. However, even today disease is difficult to find in any plant population when damage levels are low. Most people who work regularly with plants will not notice symptoms of a disease until it has increased to about 1% intensity. This initial observation of disease is called **onset** and is often preceded by several to many reproductive cycles by the pathogen following introduction of the initial inoculum. Yet the difference between onset at ~1% disease level and severely diseased plants at 80% damage might only be another one or few reproductive cycles. Thus, it appears that disease development is slow at the beginning of an epidemic. This can be illustrated with a disease that may occur, for example, in an acre of 20,000 plants. If a single lesion occupies 0.1% of the foliage of a single plant, then it would account for about 0.000005% of all foliage in the acre of plants. If that single lesion produces inoculum that results in a twenty-fold increase in numbers of lesions, then the second "generation" of disease lesions would occupy 0.0001% of foliage in the acre, and so on. It would take more than four disease cycles before disease at ~1% is readily noticed, and only 1.6 more cycles before 100% of foliage is diseased if disease increase continued at this rate (Figure 30.2).

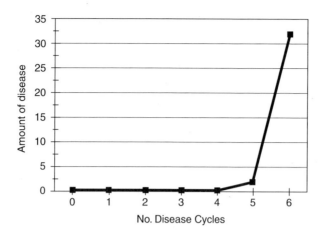

FIGURE 30.2 Graph showing increase in disease that starts from a single lesion affecting 0.000005% of foliage on one plant among 20,000 plants, when each disease cycle allows the development of 20 new lesions from each older lesion.

Epidemics are not always rapid developments of severe disease; epidemics occur when there is any increase in disease over time. Thus, even the increase of one plant to two plants being diseased reflects an epidemic. An increase in **disease incidence**, the number or proportion of plants in a population that are affected by disease, is due to spread or dispersal of disease-causing inoculum. When formerly healthy plants become diseased due to inoculum spread from another plant, this is termed **alloinfection**. Some inoculum spreads no further than to leaves of the originally diseased plant. However, even this **auto-infection** of the same plant leads to a greater disease level through an increase in severity. **Disease severity** represents the proportion of tissue of a single plant affected by disease. It is probable that potato late blight was widely established through Ireland at the end of the 1844 growing season and that disease incidence was high, with more than 50% of individual plants affected. However, the level of disease on individual plants was probably fairly low, perhaps less than 5% disease severity.

Disease severity and disease incidence are usually positively related to one another; that is, as disease severity increases so does disease incidence. However, a number of diseases increase only in incidence because a single infection affects the entire plant, frequently causing plant death. Southern wilt on tomato by *Ralstonia solanacearum* and Verticillium wilts of many plants including redbud, snapdragon, and most vegetables are just some of the diseases that only increase in incidence over time in a plant population. Although there may be differences in the stage of disease progress at some point in time among infected plants in a population, these differences are more likely due to when infection occurred, genetic differences between the infected plants, or even due to minute differences in the environment of the soil.

DISEASE RATING SCALES

Disease severity, or the proportion of the plant that is affected by disease, can range from 0 to 100%. However, the entire range of disease severity is difficult to distinguish with precision, especially when disease levels are moderate or between 20 and 80%. The Weber–Fechner law of visual discrimination relates to this ability to distinguish moderate levels of plant disease severity. This law states that the human ability to see differences decreases by the logarithm of the intensity of the stimulus. In other words, when the stimulus is one or few disease lesions on a leaf amidst healthy green tissue, it is easier to distinguish 1 from 3 or 5% disease severity than to tell 20 from 40% severity. Similarly, when a leaf is nearly completely diseased, as happens with southern corn leaf blight, it is relatively easy to distinguish 95 from 99% disease severity, but more difficult to determine 65 from 85% disease. It is easier to distinguish between high levels of disease as severity approaches 100% because the visual stimulus becomes the healthy tissue amidst diseased tissue.

DISEASE DIAGRAMS

As an aid in assessing disease severity on plants, disease diagrams have been developed for a number of diseases on several important crops. Such diagrams provide pictorial aids for determining particular disease levels (Figure 30.3). While frequently used in training, such pictorial aids improve the precision of disease ratings, even by experienced personnel. In addition, these aids provide consistency in disease evaluations made at different times and among different evaluators. Current technology allows us to easily develop customized diagrams for specific diseases on any host.

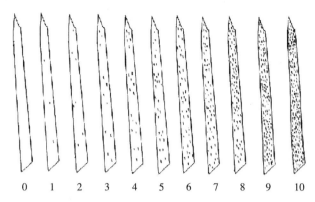

FIGURE 30.3 Disease severity diagram of rating scale used for leaf rust of wheat. (From Burch and Bowen, unpublished.)

COMPLICATIONS

Disease assessment, especially with the help of rating scales and disease diagrams, is not always straightforward. Often, disease symptoms vary in color and these colors may not be distinct from one another. Similarly, more than one disease may be affecting a single leaf or plant at any point in time.

DISEASE IN SPACE

As illustrated by the southern corn leaf blight and chestnut blight epidemics, increases in disease incidence over time also lead to an increase in the size of the area in which a disease can be found (Figure 30.1). At a single point in time, severity on individual plants is expected to decrease as distance increases from the original infection. This is called a **disease gradient**. Gradients can be due to gradual changes in the microclimate or soil type, but are most often seen with increasing distances from the first infection and are due to dispersal. Those pathogens that are most easily dispersed, such as powdery mildews and *P. infestans*, will infect plants further and more quickly from an original infection site than pathogens with larger propagules that are dispersed with more difficulty. *Phytophthora infestans* is said to have a "flat" dispersal gradient because of its ease of spread (Figure 30.4). Conversely, nematodes and *G. graminis* are considered to have "steep" dispersal gradients because their spread is due to their own movement and growth through soil, which is limited. Actually, dispersal gradients are not usually applicable to diseases caused by nematodes. Rather, because the affected area is very limited, diseases caused by nematodes and other soilborne pathogens are considered "focal diseases."

A number of diseases not only cause tissue damage, but also cause defoliation or premature loss of foliage. In addition to defoliation, disease development can cause lodging or the falling of plants, stunting or a decrease in size, wilting, and diminished quality (sugar, protein con-

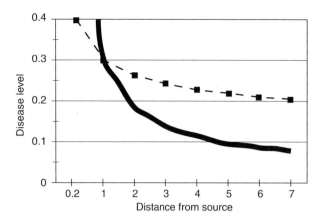

FIGURE 30.4 Graphical representation of disease gradients or the decrease of disease with increasing distance from source. Solid line represents "steeper" disease gradient.

tent, or toxins). Any of these disease characteristics can increase in intensity over time.

DISEASE ANALYSIS OR MODELS FOR EPIDEMICS

In trying to understand plant disease epidemics, mathematical models have been used as a means of simplification. These mathematical models provide a quantitative way to represent the sometimes complicated processes of an epidemic. Mathematical models have also allowed comparisons of epidemics from different times or places and provided a framework for forecasting future disease levels.

Exponential Model

One model that has been applied to plant disease development over time is the exponential model (sometimes called the logarithmic model, Figure 30.5):

$$y_t = y_0 e^{rt} \qquad (30.1)$$

where y_t is the amount of affected tissue after t time intervals, given a starting disease level y_0 and a rate parameter (r), where e is the mathematical constant 2.718. Note that the graph of this model looks similar to one presented previously (Figure 30.2) in that disease stays low during the first few cycles, then rapidly increases. The linear form of this model is frequently easier to understand:

$$\ln y_t = \ln y_0 + rt \qquad (30.2)$$

Terminology

In applying this and similar models to plant disease, several things should be noted. For example, the variable t can represent different time intervals, depending on the pathosystem. As noted earlier in this chapter, the time interval applicable to late blight of potato would be a day,

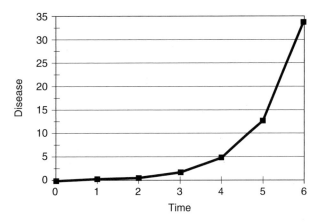

FIGURE 30.5 Exponential model of disease development over time.

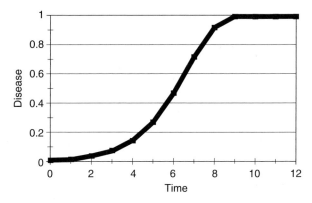

FIGURE 30.6 Graph of increase of disease in time as described by the logistic model.

whereas for chestnut blight a year would be more applicable. Similarly, the infection rate r differs among various pathosystems. Calculations done on a number of plant disease epidemics indicate that $0.10 < r < 0.56$ for diseases of annual plants. The starting disease level is y_0 and represents the initial inoculum level.

The infection rate of this model is where most of the simplification comes in relative to understanding epidemics. Numerous factors influence the rate of epidemic development and these are incorporated into the value of r. In real life, an increase in plant disease over time appears to be somewhat erratic — sometimes increasing rapidly, sometimes seeming to decrease. Quantitative models smooth out these apparent stops and starts of disease increase, as seen in the difference between Figure 30.2 and Figure 30.5. In addition, these models for plant disease increase in time do not account for any latent period, but assume that new infections are immediately visible and instantaneously infectious. All of these fluctuations and stops and starts in disease development are incorporated into the infection rate r.

Logistic Model

The exponential model allows for infinite increase in the value of y, but usually there is some finite limit to the amount of plant tissue that can become diseased. This is especially true when dealing with disease severity, when the proportion of disease cannot become greater than 100%, or when considering the proportion of plants in a finite population that may be affected by disease. Therefore, the exponential model is not often applicable to disease increase in plant populations. Disease proportions in a model for plant disease increase must be limited to 100%, as allowed with the logistic model (Figure 30.6):

$$y_t = 1/(1 + e^{[(\ln(y_0/(1-y_0))) + rt]}) \qquad (30.3)$$

The linearized form of the logistic model is:

$$\ln(y_t/(1 - y_t)) = \ln(y_0/(1 - y_0)) + rt \qquad (30.4)$$

When disease levels are low, less than 10%, the exponential model is very similar to the logistic model. Thus, either the exponential model or the logistic will apply during early stages of an epidemic.

Simple Assumptions

These models idealize disease progress because they assume constant and uniform values for each of the variables. Both the exponential and logistic models also simplify the processes of disease development. One or both of these models have been fitted to numerous plant disease epidemics, from potato late blight development to the development of dollar spot by *Rustroemia floccosum* (*Sclerotinia homeocarpa*) on creeping bentgrass. Vanderplank, in his 1963 book *Plant Diseases: Epidemics and Control*, was the first to quantitatively analyze plant disease epidemics. These initial analyses were criticized and built on and ultimately led to the plant disease epidemiology as an autonomous area of study. Models other than the two presented have been applied to plant disease progress curves. Information on these additional models, such as Gompertz or Bertalanffy-Richards, can be found in more advanced texts (Campbell and Madden, 1990).

CONTROL APPLICATIONS

By using the linearized forms (Equation 30.2 and Equation 30.4) of these models of disease increase, it is easy to see how changes in parameter values on the right side of the equation affect the parameter values on the left side. For example, a decrease in the value of y_0 will decrease y_t if the values of other parameters stay the same. Thus, these equations help in understanding the effects of many control measures.

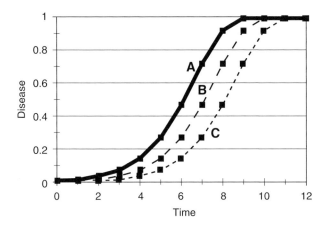

FIGURE 30.7 Graphs of disease in time according to the logistic model with three levels of initial inoculum. (A) $y_0 = 0.01$; (B) $y_0 = 0.005$ (50% of A); and (C) $y_0 = 0.0025$ (50% of B).

Effects on y_0

Crop rotation, for example, decreases the amount of viable inoculum of a soilborne pathogen, thus decreasing y_0. Initial inoculum is also decreased through sanitation, which is the removal of inoculum, and through the use of protectant fungicides, which are fatal to inoculum. Single-gene resistance is also considered to affect y_0. A decrease in initial inoculum, if substantial enough, can result in lower disease at the end of a finite time period, such as a growing season, compared to no decrease in y_0 (Figure 30.7). The figure shows that with $y_0 = 0.01$ (original), by $t = 9$, disease is 100%. However, when initial inoculum is decreased such that $y_0 = 0.005$, disease at $t = 9$ is 92%. A further decrease to $y_0 = 0.0025$ results in disease at 72% at $t = 9$. Lower disease levels are desirable because lower losses are associated with lower disease.

Effects on t

Control practices can also affect the t variable of models for disease increase. For example, early harvest or a short-season cultivar would end an epidemic earlier, for example at $t = 8$ rather than $t = 9$. In Figure 30.7 line C, an earlier harvest at $t = 8$ rather than at $t = 9$ could mean disease severity of 47% instead of 72%.

Effects on r

Control measures also can affect the infection rate, r, of these models. For example, multigene resistance in a plant is believed to reduce r compared to susceptibility. The use of fungicides that inhibit sterol biosynthesis (Chapter 33), which delay fungal development and reduce inoculum production, reduces the rate of disease increase. Cultural methods such as pruning a fruit tree to increase airflow, thereby decreasing duration of moist periods, and use of drip irrigation instead of sprinkler irrigation, also reduces the value of r in these models (Figure 30.8).

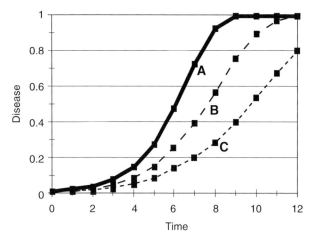

FIGURE 30.8 Graphs of disease in time according to the logistic model with three levels for the rate parameter. (A) $r = 1.0$; (B) $r = 0.75$ (25% less than for A); and (C) $r = 0.56$ (25% less than B).

Each of the components of these models acts together and distinctively for each pathosystem. In general, polycyclic diseases have low y_0 but high r values. Because polycyclic diseases can develop very quickly, the level of y_0 has less influence on the epidemic than does r. Thus, control strategies for diseases with high r values should be aimed at decreasing r. Black spot disease of rose is a polycyclic disease with a high r value in the southern U.S. Reduction of y_0 through sanitation by removal of fallen infected foliage is recommended for minimizing black spot. However, on many rose plants, fungicide applications are still needed in order to manage this disease. Similarly, peanut leaf spot diseases, caused by *Cercospora arachidicola* and *Cercosporidium personatum*, are polycyclic diseases with high values for r. Initial inoculum for these leaf spot diseases is believed to originate from plant debris remaining in the field. Because these fungal pathogens only infect peanut, the only source of inoculum is from previously infected tissue. However, if peanuts are grown in a field not previously planted to this crop, leaf spot diseases develop rapidly from the small quantities of inoculum that are carried to these new fields by wind. Powdery mildew diseases of many plants are additional examples of polycyclic diseases with high r values.

Conversely, most monocyclic diseases have low r values, but relatively high y_0 values. Most effective control strategies for monocyclic diseases would be to reduce y_0. As presented earlier, southern stem rot of peanut is a monocyclic disease. If peanuts are planted in a field not previously cropped to peanuts, the incidence of southern stem rot will be very low through the growing season. Thus, crop rotation is a recommended and effective strategy for minimizing southern stem rot of peanut, a monocyclic disease.

While increase of disease severity over time may be fit to the logistic model, a simpler means of comparing

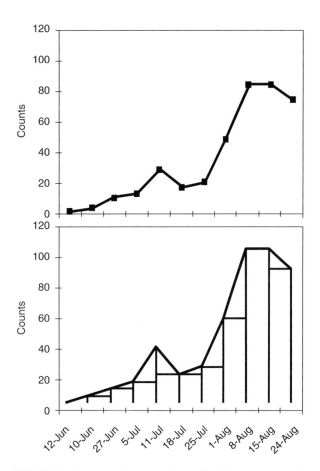

FIGURE 30.9 Geometric approach to calculating area under the disease progress curve (AUDPC) by summing all areas of rectangles and triangles that fit under disease curve. Area of rectangle = (width × height) or (difference between adjacent time intervals × difference between adjacent disease levels). Area of triangle = $\frac{1}{2}$ (width × height) or $\frac{1}{2}$ (difference between adjacent time intervals × difference between adjacent disease levels).

different epidemics is frequently desired. In these situations, the total area under the graphed curve of the epidemic might be used. The area under the curve or area under disease progress curve (AUDPC) is relatively easy to calculate and provides a single numerical value for quantifying an epidemic. This area might be quantified by calculus, or more simply with geometry, where the curve of the epidemic is broken into rectangles and triangles and the areas of these figures are summed (Figure 30.9). Several epidemics can be compared using AUDPC values.

SPATIAL MODELING

As has been mentioned, when there is disease increase in a population, the increase occurs over time as well as over geographical area. Spatial aspects of epidemics can also be described by mathematical models. One such model is

an exponential model with different parameters than the logistic model:

$$y = ae^{-bx} \qquad (30.5)$$

The linear form being:

$$\log y = \log a - b(\log x)$$

where y is disease at some distance x from an infection source, a represents the amount of viable inoculum, and b reflects the difficulty with which the pathogen is dispersed. As the value of b approaches zero, there is decreasing difficulty in dispersal. Thus, diseases caused by pathogens that disperse readily, such as *Botrytis* species with small conidia, have small values for b. Values for b for soilborne diseases are high, approaching 1, because the pathogens causing many of these diseases spread primarily by means of their own growth or movement.

Spatial analyses of plant diseases can contribute to improved management of those problems. For example, spatial analyses have been used with geostatistics and provided documentation for the role of weeds as alternative hosts in tomato virus management in the Del Fuerte Valley of Mexico (Nelson et al., 1999).

CROP LOSSES

In each of the three examples starting this chapter, plants and crops were killed or yields were substantially reduced due to plant disease epidemics. Potato late blight killed potato plants, and any tubers that did develop and eventually harvested were likely to rot in storage due to infection by *P. infestans*. Similarly, mature American chestnut trees died from infection with *C. parasitica* throughout eastern U.S. forests. Although not all plant pathogens kill their hosts, any disease will detrimentally affect plant growth and development. In the vast majority of diseases, leaf spots or even a few rotten roots will not negatively impact human lives. However, when disease incidence or severity increases to the point that most of a plant's foliage becomes necrotic or limbs of trees do not produce fruit, this is cause for concern.

One of the primary goals in gaining knowledge about plant diseases is to reduce crop losses to an acceptable level. An acceptable level of loss may be that which cannot be controlled with affordable management strategies. As the value of the yield of a crop or an ornamental plant increases, more can be done to manage any disease that is detrimental to crop yield or the aesthetics of the ornamental. However, not all levels of disease have noticeably detrimental effects on yield or plant development. So, first the relationship between disease and yield must be determined and understood.

The relationship between disease and yield (or losses in yield) has been developed for numerous pathosystems. With some plants, such as wheat, specific growth stages affect different components of the productivity of that plant and any disease occurring during that growth stage can subsequently affect yield. For example, the flag leaf, or the leaf just below the grain head of wheat, has been documented to contribute most to grain fill. For this reason, yield loss models for diseases of cereals often use the percent disease severity on the flag leaf (e.g., Burleigh et al., 1972; Large and Doling, 1962) to estimate losses due to disease. Of course, to use such estimates of loss, disease severity must be determined with some accuracy, which justifies the use of standardized rating scales as previously presented in this chapter. Another type of loss relationship can be illustrated with southern stem rot in peanuts where each locus of a *Sclerotium rolfsii* infection at plant maturity, up to 30 cm in length of 30.5-m row, causes 0.9 to 2.9% loss in yield quantity (Bowen et al., 1992). Knowledge of this relationship can aid a grower in determining how yields could change if disease incidence increased or decreased, and may provide justification for implementation of control measures in the following year. More useful, perhaps, would be a means of estimating crop loss due to disease before the occurrence of disease. Such knowledge would allow the use of control measures before damage and losses have developed.

DISEASE FORECASTING AND MANAGEMENT

One of the uses of the mathematical models that describe plant disease development in time is to predict or forecast future disease levels. Knowledge about a future disease level can assist in making disease management decisions.

With monocyclic diseases, disease levels might be forecast by a direct assessment of the amount of initial inoculum at the beginning of the season. Indeed, this has been done with white rot of onion, caused by *Sclerotium cepivorum*, in New Jersey (Adams, 1981) and for sclerotium rot of sugar beets, caused by *S. rolfsii* in sugar beets in Uruguay (Backman et al., 1981). With both of these diseases, sclerotia in the soil provide the initial inoculum, which is enumerated for a forecast. Similarly, nematode populations are determined prior to planting to estimate the potential damage in a crop. When initial inoculum levels are high, substantial disease is predicted and the recommendation is made to plant the crop elsewhere or to take protective measures. Another disease for which disease is predicted from initial inoculum is Stewart's wilt of corn, caused by *Pantoea stewartii* (previously *Erwinia stewartii*). The assessment of the initial inoculum, however, is indirect because the corn flea beetle vector of *P. stewartii* is assessed rather than the pathogen itself. The corn flea beetle population is adversely affected by cold temperatures during winter months. Thus, if the average winter temperature is at or below freezing, few flea beetles survive and the risk of high disease is substantially reduced. Knowledge of flea beetle survival helps growers determine which cultivar they should plant, especially when cultivars differ in their resistance to Stewart's wilt (Pataky et al., 1995).

Initial inoculum also plays a role in the forecast for apple scab. However, the actual amount of initial inoculum is not quantified; rather, the weather that is conducive for release of spores and infection by initial inoculum is analyzed. When infection is favored, the forecast recommends the initiation of a fungicidal spray program. Thus, regular fungicidal sprays are used to manage this polycyclic disease, but the forecast is largely based on weather that is favorable for initial inoculum.

Most polycyclic diseases are best controlled by reducing the infection rate. Weather is one of the primary influences on the rate of development of an epidemic. Thus, some of the best forecasts of polycyclic diseases are based on weather conditions that are favorable for pathogen development. In Alabama, for example, the AU-Pnut leaf spot advisory provides a recommendation for fungicide applications on peanuts when weather is "predicted" as favorable for infection. This advisory is based on the accumulation of rain events beginning at planting or following a period of protection after a fungicide spray. Further north, in Virginia and North Carolina, temperature can be a limiting factor for the development of peanut leaf spot diseases, so the Virginia peanut leaf spot advisory system includes temperature as well as moisture in predicting favorable conditions for disease increase. These two forecast systems, based on reducing the infection rate of an epidemic, have been shown to save growers one to two fungicide applications per year compared to applications made every 14 days through a growing season.

Moisture and temperature are also used to predict potato late blight, using BLITECAST. This system computes daily disease severity values that are based on numbers of hours with RH > 90% and the average temperature during moist periods. Greater disease severity values are predicted with longer periods of high RH and higher temperatures allow some shortening of high RH periods. Disease severity values for BLITECAST are accumulated over a 7-day interval, and along with numbers of rain days, are used in recommending whether to spray. In addition to the recommendation to spray or not spray, BLITECAST also suggests spray intervals depending on the accumulated disease severity values. A similar forecast system that uses moisture and temperature for determining disease severity values is TOM-CAST, a management aid for control of early blight (caused by *Alternaria solani*), Septoria leaf spot, and fruit anthracnose (caused

TABLE 30.1
Two Rating Scales to Assess Disease — the Horsfall–Barratt Scale Based on the Weber–Fechner Law of Visual Acuity and an Alternative Scale Based on 1/5th of the Plant

Horsfall—Barratt		Alternative Rating Scale	
Grade	Disease Ranges (% Severity or Incidence)	Grade	Disease Severity Range (%)
0	0	0	0
1	0–3	1	0–20
2	3–6	2	20–40
3	6–12	3	40–60
4	12–25	4	60–80
5	25–50	5	80–100
6	50–75		
7	75–88		
8	88–94		
9	94–97		
10	97–100		
11	100		

by *Colletotrichum coccodes*) in tomatoes. Disease severity values in TOM-CAST are highest with wet weather and temperatures between 70°F and 80°F. Fungicide applications are recommended by TOM-CAST when disease severity values have accumulated to particular levels. Each of these forecasting systems provides recommendations to decrease the number of fungicide applications when the environment is not conducive for disease development.

In addition to the four plant disease forecasting systems presented here, numerous additional systems have been developed. Many of these systems have been implemented with some success. The reasons that forecasting systems for plant diseases have not been more widely used are varied. One reason that growers do not adapt to using a plant disease forecast is that they consider the possible reduction in pesticide applications as too risky. Adding to the perception of risk is that pesticide labels do not address the possible use of forecasting systems. Another reason for a lack of implementation has been the limited availability of appropriate data for a forecast. However, this is changing with the more widespread application of the Internet and ease of obtaining customized and local weather information.

Plant disease forecasts and advisory systems are tools that have been developed through epidemiological studies. Plant disease epidemiology, or the study of the factors that allow disease increase, provides information that can improve the efficiency of plant and crop production. Numerous opportunities exist for continued application of knowledge gained through epidemiology for improving our ability to manage plant diseases.

EXERCISES

EXERCISE 1. DISEASE ASSESSMENT

Horsfall and Barratt developed a disease rating scale based on the Weber–Fechner law and the difficulty of distinguishing among moderate levels of disease severity. This rating scale consists of 12 levels or grades, each of which includes disease severity ranges that decrease as disease approaches 0 or 100% (Table 30.1). In addition, disease severity ranges are symmetrical around 50%. The Horsfall–Barratt disease rating scale is only one example of many that can be used to rate disease severity in order to monitor increase in disease over time. For example, in evaluating disease on a plant, it is relatively easy to think of whether the severity encompasses 1/5, 2/5, 3/5, etc., of the plant and this consideration can be implemented as a rating scale with 0 for no disease and 5 additional levels (Table 30.1). One advantage of the Horsfall–Barratt scale over one based on fifths of a plant is that there are more grades or levels in the scale. The more restricted ranges of disease at low and high levels allow greater accuracy in keeping track of an increase in disease over time. Rating systems with fewer grades may be more useful to compare the effectiveness of fungicidal products or the susceptibility of a number of cultivars.

Materials

Each student or team of students will require the following items:

- Two sets of diseased leaves or plants, at least 10 leaves per student per leaf set. Sets of diseased leaves collected one or two weeks apart.
- Copies of disease diagrams can be used and this exercise can be completed in one laboratory session.

Follow the experimental outline provided in Procedure 30.1 to complete this part of the exercise.

Procedure 30.1 Disease Assessment

1. By using Table 30.1, assign a grade (or rating value) to each leaf of the first set of leaves.
2. Calculate an average disease rating from the first set of diseased leaves.
3. Assign a grade (or rating value) to each leaf of the second set of leaves.
4. Calculate an average disease rating from the second set of diseased leaves.
5. Compare average disease ratings from the two sets of leaves.

Anticipated Results

Students will gain an appreciation for differential disease levels, and for the "art" of assessing disease severity. Ideally, using in vivo materials, students can see how disease severity might develop from one disease rating to the second. If used, copies of disease diagrams can be organized by the instructor to simulate disease development over time or to compare student improvement with increased numbers of assessments.

Questions

- What levels of disease were the most difficult to rate?
- Were there complicating factors affecting the symptoms you were rating?
- What was taken into consideration in deciding a value to assign to each leaf?

Exercise 2. Disease Progress

In trying to understand plant disease epidemics, mathematical models have been used as a means of simplification. These mathematical models provide a quantitative way to represent the sometimes complicated processes that comprise an epidemic. The logistic model is often used to depict plant disease development:

$$y_t = 1/(1 + e^{-[(\ln(y_0/(1-y_0))) + rt]})$$

In this model, y_t is the amount of affected tissue after t time intervals, given a starting disease level y_0 and a rate parameter (r), where e is the mathematical constant 2.718.

Materials

Each student or team of students will require the following items:

- Calculator
- Graph paper

Follow the experimental outline provided in Procedure 30.2 to complete this part of the exercise.

Procedure 30.2 Disease Progress

1. By using Equation 30.3 or Equation 30.4, and $y_0 = 0.01$, $r = 0.48$, calculate y_t values for $t = 0, 1, 2, ..., 20$.
2. Graph the data, calculated in Step 1, on graph paper. Connect data points to create Line 1.
3. Repeat Step 1 and Step 2, using $y_0 = 0.01$, $r = 0.24$ for Line 2.
4. Repeat Steps 1 and 2, using $y_0 = 0.005$, $r = 0.48$ for Line 3.
5. Compare plotted lines.

Anticipated Results

Students will generate three disease progress curves from the models. From the plots of these progress curves, effects of changes in initial inoculum or rate of infection, or both, can be seen.

Questions

- How did the plotted lines differ?
- What plant disease management method might have been taken to reduce y_0, as shown in the differences between Lines 1 and 3?
- What approach might be taken to reduce r, as demonstrated by Lines 1 and 2?
- What management strategies might be more effective to reduce the theoretical disease that is modeled?

REFERENCES

Adams, P.B. 1981. Forecasting onion white rot disease. *Phytopathology* 71: 1178–1181.

Andrivon, D. 1996. The origin of *Phytophthora infestans* populations present in Europe in the 1840s: A critical review of historical and scientific evidence. *Plant Pathol.* 45: 1027–1035.

Backman, P.A., R. Rodriguez-Kabana, M.C. Caulin, E. Beltramini and N. Ziliani. 1981. Using the soil-tray technique to predict the incidence of *Sclerotium* rot in sugarbeets. *Plant Dis.* 65: 419–421.

Bowen, K.L., A.K. Hagan, and R. Weeks. 1992. Seven years of *Sclerotium rolfsii* in peanut fields: Yield losses and means of minimization. *Plant Dis.* 76: 982–985.

Burleigh, J.R., M.G. Eversmeyer and A.E. Roelfs. 1972. Linear equations for predicting wheat leaf rust. *Phytopathology* 62: 947–953.

Campbell, C.L. and L. V. Madden. 1990. *Introduction to Plant Disease Epidemiology.* John Wiley & Sons, New York.

Hanway, J.J. 1963. Growth stages of corn (*Zea mays* L.). *Agron. J.* 55: 487–492.

Large, E.C. and D.A. Doling. 1962. The measurement of cereal mildew and its effect on yield. *Plant Pathol.* 11: 47–57.

Moore, W.F. 1970. Origin and spread of southern corn leaf blight in 1970. *Plant Dis. Rep.* 54: 1104–1108.

Nelson, M.R., T.V. Orum and R. Jaime-Garcia. 1999. Applications of geographic information systems and geostatistics in plant disease epidemiology and management. *Plant Dis.* 83: 308–319.

Pataky, J.K., J.A. Hawk, T. Weldekidan and P. Fallah Moghaddam. 1995. Incidence and severity of Stewart's bacterial wilt on sequential plants of resistant and susceptible sweet corn hybrids. *Plant Dis.* 79: 1202–1207.

Roelfs, A.P. 1985. Wheat and Rye Stem Rust, pp. 3–37 in *The Cereal Rusts,* Vol. II., Roelfs, A.P. and W.R. Bushnell (Eds.), Academic Press, New York.

Vanderplank, J.E. 1963. *Plant Diseases: Epidemics and Control.* Academic Press, New York.

31 Host Resistance

Jerald K. Pataky and Martin L. Carson

CHAPTER 31 CONCEPTS

- A resistant phenotype occurs when growth, reproduction, or disease-producing activities of the pathogen are reduced, and fewer symptoms of disease are observed.

- Because plant disease results from the interaction of a complex series of events and pathways in both the host and pathogen, it is easy to envision multiple ways the process can be interrupted or altered giving rise to a resistant reaction.

- Among a plant species, there often is a range of responses to pathogens that varies continuously from a small amount of disease (resistant) to a substantial amount of disease (susceptible). Discontinuous categories of resistant reactions also occur.

- Populations of pathogens vary genetically both for their ability to infect hosts with specific resistance genes (virulence), and for the degree of damage they cause (aggressiveness).

- Host resistance places selection pressure on pathogen populations for genotypes with virulence against that resistance (i.e., races).

- Although all isolates within a race have common virulence, they may be genetically diverse for other traits.

- To develop a disease resistant cultivar, sources of resistance must be identified, and resistance must be combined in cultivars that have desirable traits.

- The greatest genetic diversity for resistance usually occurs where the host and pathogen have coevolved, which often is the center of origin of host species.

- The benefit of genetic uniformity for most traits is considerably greater than the risk of potentially severe epidemics; however, genetic uniformity for disease resistance carries a substantial risk even though it may be greatly beneficial when it works.

- Breeders and pathologists have tried to cope with variable pathogens by constantly searching for new sources of resistance and by selecting combinations of resistance genes that provide resistance to the spectrum of races present in a growing region.

Planting resistant cultivars is one of the most efficient and effective methods of disease control. Host resistance eliminates or minimizes losses due to diseases and reduces the need for and cost of other controls. Resistance is compatible with other methods of disease and pest control and it can be integrated easily in pest management programs (Chapter 35).

Disease resistance is one of the most important factors contributing to the long-term stability of crop production. For example, wheat stem rust (*Puccinia graminis* f. species *tritici*) periodically devastated the U.S. wheat crop during the first half of the 20th century, but there has not been a major epidemic since the early 1950s largely due

to the use of durable combinations of resistance genes in modern wheat varieties. Resistant varieties produce higher and more consistent yields and have improved the quality of the world's major food crops. Cultivars that are resistant to many diseases are among the major achievements of plant breeders and pathologists in the past century. Resistance will continue to play an integral role in the success of agriculture well into the 21st century.

TYPES OF RESISTANCE

Resistance encompasses a wide variety of host–pathogen interactions. Resistant reactions vary in both degree and kind. A resistant phenotype reduces the growth, reproduction, or disease-producing activities of the pathogen. Disease symptoms are less severe on resistant hosts than on susceptible hosts. Susceptible phenotypes are unable to restrict the growth, reproduction, or disease-producing activities of the pathogen, and symptoms are severe. Among a plant species, reactions to a specific disease often vary continuously from highly resistant to highly susceptible, much like height of people varies continuously from tall to short. When variation is continuous, host reactions differ only in degree of symptom severity and resistance and **susceptibility** represent the two extremes. In other cases, host reactions can be placed into discrete categories as a result of hypersensitive reactions (Chapter 28) or resistant reactions that result in distinctly different kinds of phenotypes, much like humans may be either "normal" height or genetically dwarf.

Placing host reactions into broad categories (e.g., highly resistant, resistant, moderately resistant, moderately susceptible, susceptible) can be very useful because this type of classification gives an indication of how a cultivar will respond when grown in an area that is favorable for disease development. Nevertheless, categorizing host reactions can be a source of confusion and disagreement, especially when trying to separate continuous reactions. This problem is somewhat analogous to trying to classify height among people. For example, some people would say that a 6-ft man is tall, whereas others (maybe a basketball coach) would consider 6 feet to be average height or even short. In this example, the relative height of a 6-ft man depends on the population that is being sampled. Compared to professional basketball players, 6 ft is relatively short, but compared to all men in the U.S., 6 ft is above average height. The same problem exists when trying to compare host reactions to disease. For example, if the reaction of a corn hybrid to Stewart's bacterial wilt is given a score of 4 on a 1 to 9 scale (i.e., *Pantoea stewartii* spreads nonsystemically more than 30 cm from the site of infection), the hybrid may be classified differently among three different groups of corn germplasm (Pataky et al., 2000). Among all dent corn hybrids that are grown widely in the corn belt of the U.S., a hybrid with a Stewart's wilt rating of 4 is worse than average (e.g., moderately susceptible). Among a representative sample of corn lines collected from throughout the world, a variety with a Stewart's wilt rating of 4 is close to average (e.g., moderate). Among early maturing sweet corn, a hybrid with a Stewart's wilt rating of 4 is better than average (e.g., moderately resistant). Thus, resistance is relative to the population that is being sampled.

The arbitrary nature of boundaries is another problem that occurs when trying to classify levels of resistance among populations with continuous distributions of disease reactions. For example, if cultivars with <5% disease severity are classified as resistant and those with 5 to 15% disease severity are classified as moderately resistant, a cultivar with 7% leaf area infected would be classified as moderately resistant even though it is more similar to a resistant cultivar than it is to a moderately resistant cultivar with 14% severity.

Plant breeders and pathologists use different approaches and statistical procedures to deal with these types of classification problems, and, consequently, disease reactions are reported in a variety of ways. Numerical scales (such as those from 1 to 9 that correspond to resistant to susceptible) and ordinal classes (such as HR — highly resistant, R — resistant, MR — moderately resistant, M — moderate, MS — moderately susceptible, and S — susceptible) are just a few of the many approaches used to describe host reactions to diseases. Although a uniform method of reporting host reactions does not exist, resistance ratings should correspond to different degrees of protection from crop damage.

The expression of resistance genes requires that plants interact with the pathogen in an environment that is suitable for disease development. To separate host reactions to diseases, host lines must be compared under equivalent conditions. For many diseases, reliable inoculation methods can be used to screen for resistance. Inoculation helps ensure that all lines being evaluated are treated similarly. For some diseases, host reactions are evaluated in nurseries that are infected naturally. If inoculation procedures or disease nurseries are not uniform, extra care must be taken to ensure that all lines in an experiment are subjected to the same "disease pressure." For example, when disease nurseries are used to evaluate reactions to soilborne diseases, differences in inoculum density frequently occur as a result of the clustered spatial arrangement of soilborne pathogens. In such cases, moderately susceptible lines evaluated at low inoculum densities can have phenotypes (e.g., amount of disease symptoms) similar to moderately resistant lines evaluated at high inoculum densities (Figure 31.1). Separation of reactions can also be affected by environmental conditions. For example, a group of resistant maize lines that could not be separated for reactions to northern corn leaf blight (NCLB) during a dry season in Wooster, OH, displayed a range of reactions with intermediate classes when evaluated at two locations in Uganda where the environment was more conducive to the development of NCLB (Table 31.1). The aggressiveness of pathogen isolates or strains used to inoculate plants in a nursery also can affect the ability to separate disease reactions. In general, more aggressive isolates separate cultivar reactions more effectively than less aggressive isolates. To ensure that the results of disease nurseries are accurate, a set of standard cultivars with known reactions to diseases usually is included.

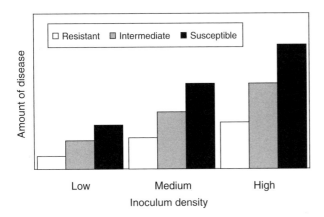

FIGURE 31.1 Amount of disease on cultivars with resistant, intermediate, or susceptible reactions when evaluated in field plots with low, medium, or high inoculum density. The phenotype (e.g., amount of disease) may be similar when a resistant cultivar evaluated at a high inoculum density is compared to a susceptible cultivar evaluated at a low inoculum density.

ADJECTIVES THAT DESCRIBE RESISTANCE

Many adjectives are used to describe various types of host resistance. Frequently, pairs or groups of adjectives are used for specific characteristics. Misuse or haphazard use of some adjectives have resulted in their meaning becoming confounded or confused. Confusion occurs sometimes because different people have different meanings for terms. Confusion also occurs when an adjective used to describe one characteristic is thought to imply something about another characteristic. Although these implied associations frequently are correct, there are several examples of resistance where the implications are not correct. For example, monogenic is an adjective used to describe resistance that is inherited as a single gene. Frequently, mono-genic resistance is "race-specific" (i.e., virulent and avirulent races of the pathogen occur), but there are several examples of monogenic resistance for which virulent races of a pathogen are not known (i.e., nonrace specific). Many of the most common characteristics of resistance and their corresponding adjectives are discussed below.

Because resistance is a phenotype, there is no reason that a resistant phenotype cannot be conditioned genetically by any mechanism that conditions the expression of any other plant trait. **Monogenic**, **oligogenic**, and **polygenic** are used to describe the number of nuclear genes involved in the inheritance of resistance. Monogenic resistance is conditioned by a single gene, and is sometimes called **single-gene** resistance. Oligogenic is used to describe resistance that is conditioned by a few genes. If resistance is inherited in a monogenic or an oligogenic manner, resistant and susceptible phenotypes occur in genetically segregating populations, and these phenotypes can be placed in discrete categories that can be fit to Mendelian ratios. Polygenic refers to resistance that is inherited from several genes. Although the inheritance of polygenic resistance is more complex than that of oligogenic or monogenic resistance, polygenic resistance usually involves fewer genes than other traits that are inherited polygenically, such as yield. Most experimental estimates of the numbers of loci controlling polygenic resistance are less than ten and commonly only two to four. In populations segregating for polygenic resistance, disease reactions exhibit continuous variation and cannot be classified into discrete Mendelian classes. In a few instances, resistance or susceptibility is conditioned by cytoplasmic genes. Because cytoplasmic genes are transmitted from the female parent of a cross, these traits are said to be inherited maternally.

TABLE 31.1
Severity (%) of Northern Corn Leaf Blight (NCLB) on Maize Lines Evaluated in Uganda and Ohio

Maize Line	Kabanyolo, Uganda	Nmuloye, Uganda	Wooster, Ohio
B73	75[a]	59	28
Mo17	40	20	1
EV8428-SR	21	20	3
EV8429-SR	16	17	2
Gysau TZB-SR	14	16	1
Jos	8	11	1
KWCA-SR	6	7	1
EV8349-SR	5	10	1
Population 42	2	4	0
Babungo 3	1	2	0
LSD (0.05)	6.8	6.1	2.1

Note: Resistant lines were differentiated more easily in Uganda where the environment was extremely conducive for development of NCLB.

[a] Percent leaf area infected with NCLB (Adipala et al., 1993).

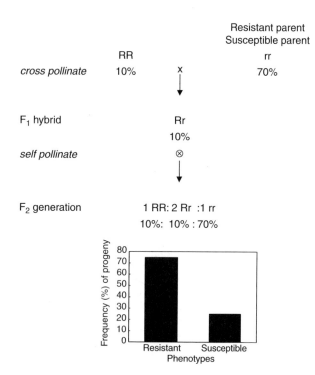

FIGURE 31.2 Qualitative resistance. Progeny in F_1 and F_2 populations of a cross of disease resistant and susceptible parents when resistance is inherited as a completely dominant, single gene. The resistant parent has a phenotype with 10% disease severity and the susceptible parent has 70% severity. The F_2 segregates 3:1 for resistant and susceptible reactions.

Resistance is dominant (*R*) if heterozygotes (*Rr*) express the resistant reaction completely. Resistance is recessive (*r*) if heterozygotes (*Rr*) have a completely susceptible reaction. If there is a complete lack of dominance (i.e., additive gene action), heterozygotes have a reaction that is intermediate between the resistant and susceptible parents. Partial dominance occurs when heterozygotes have a reaction that is more resistant than the midparent value between the resistant and susceptible parents.

Qualitative and **quantitative** are adjectives used to differentiate kinds of disease resistance or to describe expression of traits. Qualitative resistance refers to host reactions that can be placed in distinct categories that form the basis of Mendelian ratios. One or a few genes condition qualitative traits; hence, qualitative resistance is monogenic or oligogenic. Quantitative resistance refers to host reactions that have no distinct classes because reactions vary continuously from resistant to susceptible. A few or many genes are involved in the expression of a quantitative trait and the effects of individual genes are usually small and hard to detect. The difference between qualitative and quantitative resistance is obvious if we compare progeny in the F_1 and F_2 populations from a cross of a resistant and a susceptible inbred parent. In this example, we assume a phenotype with 10% disease severity for the resistant parent and 70% disease severity for the sus-

ceptible parent; gene expression is completely dominant for qualitative resistance and additive for quantitative resistance, and one gene conditions qualitative resistance whereas three independent genes are involved in quantitative resistance. For qualitative resistance, progeny in the F_1 generation have the same value as the resistant parent due to dominance. In the F_2, there is a 3:1 ratio of resistant to susceptible progeny (Figure 31.2). For quantitative resistance, reactions of F_1 progeny are intermediate (40%) between the resistant and susceptible parents due to the lack of dominance, but progeny in the F_2 display a continuous range of phenotypes between the reactions of the resistant and susceptible parents (Figure 31.3; Table 31.2). If environmental conditions create slight variation in reactions, the phenotypes in the F_2 population cannot be separated into distinct classes. Thus, a continuous range of host reactions that appears to be associated with quantitative resistance could be conditioned by as few as three genes with additive gene action. Other factors, such as interactions among genes and partial dominance, may

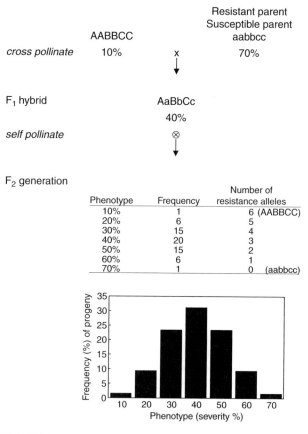

FIGURE 31.3 Quantitative resistance. Progeny in F_1 and F_2 populations of a cross of disease resistant and susceptible parents when resistance is inherited from three genes with additive gene action. The resistant parent has a phenotype with 10% disease severity and the susceptible parent has 70% severity. Each allele for resistance decreases severity by 10%. See Table 31.2 for genotypes in each phenotypic class.

TABLE 31.2
Phenotypes (Based on % Disease Severity), Frequency of Genotypes, and Number of Resistance Alleles in an F₂ Population Segregating for Disease Resistance at Three Loci with Strictly Additive Gene Action

F₂ Phenotype (%)	Genotype(s) and frequency	Number of Resistance Alleles
10	*AABBCC* (1/64)	6
20	*AABBCc, AABbCC, AaBBCC* (6/64)	5
30	*AABBcc, AABbCc, AAbbCC, AaBBCc, AaBbCC, aaBBCC* (15/64)	4
40	*AABbcc, AAbbCc, AaBBcc, AaBbCc, AabbCC, aaBBCc, aaBbCC* (20/64)	3
50	*AAbbcc, AaBbcc, AabbCc, aaBBcc, aaBbCc, aabbCC* (15/64)	2
60	*Aabbcc, aaBbcc, aabbCc* (6/64)	1
70	*aabbcc* (1/64)	0

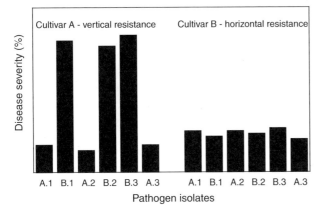

FIGURE 31.4 Vertical and horizontal resistance. Race-specific resistance (Cultivar A) is depicted by "vertical" differences in the height of histograms that illustrate disease reactions in response to different pathogen isolates. Race-nonspecific resistance (Cultivar B) is depicted by a "horizontal" decrease in the height of all histograms that illustrate disease reactions in response to different pathogen isolates.

affect the expression of resistant or susceptible reactions. Resistance can be expressed and controlled genetically in as many ways as any other phenotype.

Resistance frequently is divided into two types, **general** and **specific,** to differentiate host–pathogen interactions. General resistance is effective against all biotypes of the pathogen and is sometimes called race-nonspecific resistance. Durable resistance is defined as resistance that remains effective during its prolonged and widespread use in an environment favorable to the disease. General resistance is sometimes called durable resistance because changes in frequency of virulence genes in the pathogen population do not affect this resistance. Specific or race-specific resistance is effective against some races of a pathogen (i.e., avirulent races), but ineffective against virulent races. The adjectives vertical and horizontal also are used to describe race-specific and race-nonspecific resistance. Vertical and horizontal refer to the appearance of the histograms that depict reaction of hosts to different races of a pathogen (Figure 31.4). Vertical and horizontal

also have additional meanings. Vertical resistance is considered to be inherited monogenically, qualitatively expressed, and effective against initial inoculum of a polycyclic disease. Horizontal resistance is thought to be polygenic in inheritance, quantitative, and effective at reducing the rate at which polycyclic diseases develop. Since all race-specific resistance does not share the other characteristics of vertical resistance, and all race-nonspecific resistance does not share the other characteristics of horizontal resistance, the terms vertical and horizontal should not be used synonymously for race specific and race nonspecific even though they frequently are used this way. Similarly, all qualitative and quantitative resistance does not share all of the characteristics of vertical and horizontal resistance so these adjectives also are not synonymous. Caution should be exercised in describing resistance as vertical or horizontal because of the confounded meanings of these adjectives.

Several other adjectives are used to describe specific aspects of certain types of resistance. Complete resistance describes reactions in which some aspect of disease development (usually symptom development or pathogen reproduction) is stopped entirely. In some instances, complete resistance is called immunity, even though plants do not have immune systems. Some types of complete resistance are due to **hypersensitive responses** in which rapid biochemical and histological changes elicited by the pathogen result in necrosis or death of the invaded host cells thereby localizing infection. Because single, dominant genes frequently condition complete resistance or hypersensitive reactions, the gene responsible for the reaction can be used as the adjective to describe this type of resistance. For example, resistance in soybean to Phytophthora root rot is conditioned by several different single genes designated *Rps* (resistant to *Phytophthora sojae*). *Phytophthora* resistance in some soybean cultivars is identified by the specific resistance allele; for example, 1k refers to resistance conveyed by the *k* allele at the *Rp1* locus, and, similarly, 1c refers to resistance conveyed by the *c* allele at the *Rp1* locus. Several variations of this system are used to identify simply inherited resistance. For example, in corn, Rp resistance refers to resistance to the common rust fungus, *Puccinia sorghi*. Rp resistance is conveyed by one of nearly 25 different single *Rp* genes. Resistance to the northern corn leaf blight fungus, *Exserohilum turcicum* (syn. *Helminthosporim turcicum*), is conveyed by one of at least four different *Ht* genes (Ht = resistance to *H. turcicum*); therefore, this resistance is often referred to as Ht resistance. A variation of using host genes to describe resistance is to identify resistance based on races of the pathogen that are controlled by the resistance. For example, resistance to Fusarium wilt in tomato is designated by its effectiveness against race 1 or race 2.

The term **partial** resistance is used by some pathologists to describe resistant reactions that are not complete. Some disease symptoms or pathogen reproduction occurs on partially resistant plants. Partial resistance does not imply anything about the genetics, inheritance, or race specificity of the resistance. Partial resistance is similar to quantitative resistance except that there is no direct implication that the resistance is inherited as a quantitative trait. Field resistance is sometimes used to describe host lines that have less disease than other lines under field conditions, but may or may not exhibit resistance when grown in a greenhouse or growth chamber. Slow-rusting or rate-reducing are used to describe resistance in which rust or other diseases develop at a slower rate on resistant lines than on susceptible lines. Various components of the infection cycle (e.g., reduced infection frequency, longer incubation and latent periods, shorter infectious period, and reduced lesion numbers, lesion size, or sporulation) have been shown to be associated with rate-reducing resistance. Adult-plant resistance refers to resistant reactions that do not occur in seedlings. In some cases, there are documented biological explanations for differences in host responses at different growth stages. For example, juvenile corn leaves are more susceptible to infection by the common rust fungus, *P. sorghi,* than adult leaves are. As leaves of a corn plant change from juvenile to adult, at about the fourth to seventh internode, leaves differ in various traits such as types and amounts of cuticular waxes, thickness of cuticle, cell shape, and presence of trichomes. Severity of rust infection is probably affected by these morphological differences. All corn displays some adult-plant resistance to common rust regardless of the level of general resistance or susceptibility. In other cases, differences in seedling and adult-plant reactions to disease are the result of disease development. Because adult plants are exposed to several cycles of infection by polycyclic pathogens, slight differences in partial resistance that might be difficult to detect from a single cycle of infection of seedlings are sometimes quite apparent among adult plants.

The diversity in types of resistance and the adjectives used to describe these characteristics can be confusing even for plant pathologists and breeders who work on disease resistance. Nevertheless, it is important to have a general understanding of the adjectives used to describe types of resistance in order to know what kind of resistance you are dealing with. More importantly, recognize that all resistance is not alike even for a single pathogen of one host species. Many different types of resistance reactions may occur within a host species because resistant phenotypes may arise in many ways. Because plant disease results from the interaction of a complex series of events and pathways in both the host and pathogen, it is easy to envision multiple ways the process can be interrupted or altered.

PATHOGEN VARIATION

It would be neglectful to discuss host resistance without briefly mentioning genetic variation among pathogens. Just as hosts vary genetically for resistant and susceptible reactions to diseases, pathogens display a wide array of genetic variability. Terminology associated with pathogen variation and ability to infect hosts can be confusing. Shaner et al. (1992) presented an overview of uses of various terms in a review of the nomenclature and concepts of virulence and pathogenicity. For our purposes, we will focus on the following three characteristics of pathogens: (1) the ability to infect a host species, (2) host–pathogen interactions, and (3) the relative amount of damage on a host.

Pathogenicity is the ability of the pathogen to cause disease. For example, *Ustilago maydis* is pathogenic on corn because it causes the disease known as common smut. This fungus in nonpathogenic on oak trees because the host–pathogen interaction is incompatible and disease does not develop. Another way of thinking about pathogenicity is that corn is a host for *U. maydis* and oak is a nonhost.

Virulence is the ability of a pathogen to cause a compatible (susceptible) reaction on a host cultivar with genetic resistance. **Avirulence** is the inability of a pathogen to cause a compatible (susceptible) reaction on a host cultivar with genetic resistance. For a pathogen to be virulent or avirulent, it must first be pathogenic on a host species. Nonpathogens cannot be virulent or avirulent. A **race** of a pathogen is a population of isolates that have the same virulence. Although all isolates within a race have common virulence, they may be genetically diverse for other traits.

Races of pathogens are typically identified by their virulence on a set of differential varieties with known resistance genes. For example, races of *E. turcicum*, which causes northern corn leaf blight of maize, are determined by their reaction on a differential set of inbred lines that carry Ht resistance genes (Table 31.3). Each inbred line contains one of the four different known Ht genes for resistance.

TABLE 31.3
Race Identification[a] of Isolates of *Exserohilum turcicum* Based on Reactions of a Set of Differential Host Lines Carrying Ht Resistance Genes

Old Race[a]	Ht Resistance Genes					Virulence formula Effective/Ineffective Genes	New race
	none	Ht1	Ht2	Ht3	HtN		
1	S	R	R	R	R	123N/0	0
2	S	S	R	R	R	23N/1	1
3	S	R	S	S	R	1N/23	23
4	S	R	S	S	S	1/23N	23N

[a]Sixteen races could exist based on all possible combinations of resistant and susceptible reactions on four independent host resistance genes. *Note:* Old racial designation based on chronological order in which isolates were identified. New racial designation based on ineffective host genes.

Just as height varies in a population of people, populations of virulent pathogens may have continuous variation in their ability to cause different amounts of disease on a specific host. **Aggressiveness** is the amount of disease caused by an isolate of the pathogen. Aggressiveness is sometimes used synonymously with parasitic fitness, because the amount of reproduction of the pathogen on the host usually is directly proportional to the amount of disease. The same components of the infection cycle that may be affected by partial resistance in the host may also be affected by the aggressiveness of a pathogen strain. For example, when compared to weakly aggressive isolates, extremely aggressive isolates may have increased infection frequency; shorter incubation and latent periods; a longer infectious period; and increased lesion numbers, lesion size, or sporulation.

By these definitions, virulence and aggressiveness in the pathogen are complementary to race-specific resistance and general resistance in the host. Therefore, some people prefer the terms **specific virulence** and **general virulence** in place of virulence and aggressiveness.

Although a wide range of host–pathogen interactions occur, "resistance" is used most frequently to refer to host–pathogen interactions that involve specific resistance and virulence. In this case, resistant reactions are expressed as qualitative traits that occur when hosts with a specific resistance gene are attacked by isolates of the pathogen that have the corresponding avirulence gene. Susceptible reactions occur when either the host lacks the resistance gene or the pathogen lacks the corresponding gene for avirulence. We now know that resistance genes and avirulence genes are functional genes in the host and pathogen, respectively. Susceptibility and virulence arise when these genes are either absent or rendered nonfunctional by mutations. These interactions have been described as the "quadratic check" that illustrates four possible combinations of host–pathogen interactions and two races of the pathogen (Figure 31.5). The gene-for-gene concept refers to genes for avirulence in the pathogen that correspond to genes for specific resistance in the host.

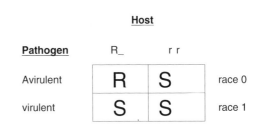

FIGURE 31.5 Quadratic check. Four possible combinations of host–pathogen interactions among an avirulent and a virulent race of a pathogen and hosts with and without a dominant, race-specific resistance gene (R).

If the host has more than one gene for specific resistance, the number of potential races of the pathogen increases exponentially, that is, the number of races = 2^N where N is the number of host resistance genes. If the host has two different independent genes for specific resistance, there is a potential for four pathogen races. Three resistance genes could result in eight races, and so on.

When several genes or sources of resistance occur, the number of pathogen races can be extremely large. To avoid confusion, races of some pathogens are designated by the ineffective host genes. The **virulence formula** for an isolate of a pathogen lists the effective host genes and ineffective host genes. For example, in corn there are at least four single dominant genes (Ht_1, Ht_2, Ht_3, and Ht_N) that independently convey resistance against the northern corn leaf blight fungus, *E. turcicum*. Because four different host resistance genes exist, there are 16 potential races of the pathogen (i.e., $2^4 = 16$). The old racial designations (races 1, 2, 3, 4, etc.) have been replaced by racial designations based on ineffective host genes from the virulence formula (Table 31.3). Old race 1 is now race 0 because these isolates do not have any specific virulence (i.e., the virulence formula is 123N/0 = effective host genes/ineffective host genes). Old race 3 is now race 23 because these isolates are virulent against the resistance genes Ht_2 and Ht_3 (i.e., the virulence formula is 1N/23).

TABLE 31.4
Indicator Lines (Host Differentials) Used to Identify HG Types for Populations of Soybean Cyst Nematode

HG Type	Indicator Line (Host Differential)
1	PI 54840 (Peking)
2	PI 88788
3	PI 90763
4	PI 437654
5	PI 209332
6	PI 89772
7	PI 548316 (Cloud)

TABLE 31.5
Examples of HG Types: Five Populations of SCN and Their Corresponding HG Type Based on an Index of Reproduction above 10[a]

		Index of Reproduction						
HG type[b]	Lee 74	(1) PI 54840	(2) PI 88788	(3) PI 90763	(4) PI 437654	(5) PI 209332	(6) PI 89772	(7) PI 548316
0	100	0	4	1	0	0	1	1
2	100	1	17	3	0	1	7	3
3	100	2	6	19	0	0	0	1
23	100	2	42	35	0	5	3	2
235	100	1	45	31	0	14	3	3

[a] Reproduction above 10% of that on Lee 74.

[b] HG types designate indicator lines on which the reproduction of a population is at least 10% of that on Lee 74.

A similar system has been proposed to describe populations of soybean cyst nematode, SCN (*Heterodera glycines*). HG-types are used to identify populations of SCN based on their ability of efficiently reproduce on certain host differentials (Tables 31.4). If reproduction of a population of SCN on any one of seven soybean indicator lines (i.e., a set of host differentials) is at least 10% of the reproduction on the standard susceptible, Lee 74, then that indicator line is included in the HG-type for that population of SCN just like ineffective host genes from a virulence formula are included in the racial designation of a fungal isolate. For example, a SCN population for which reproduction on PI 88788 is at least than 10% of that on Lee 74 is given an HG type of 2 (Table 31.5); a population for which reproduction on PI 88788 and PI 90763 are greater than 10% of that on Lee 74 is given an HG type of 23; a population for which reproduction on PI 88788, PI 90763, and PI 209332 are greater than 10% of that on Lee 74 is given an HG type of 235; and so on (Table 31.5).

Quantitative or partial resistance usually appears to be race nonspecific, so virulence and avirulence are not applicable; however, a few researchers have reported small, race-specific effects of isolates against cultivars with quantitative resistance. For example, isolates of barley rust that were allowed to reproduce for several generations on a host cultivar with partial resistance caused more disease on the partially resistant cultivar than on other cultivars (Clifford and Clothier, 1974). In some cases, both specific virulence and general virulence (aggressiveness) were increased for the rust isolates that reproduced on the partially resistant cultivar.

From this brief overview of pathogen variation, it should be obvious that the amount of variation in the pathogen for disease-producing traits is similar to variation in the host for resistant or susceptible reactions. Consequently, plant pathologists and breeders must continually identify new sources of resistance and incorporate new resistance into cultivars to compete with constant changes and adaptation in the genetic composition of pathogen populations.

BREEDING FOR DISEASE RESISTANCE

Two steps are required to develop a disease-resistant cultivar: sources of resistance must be identified and resistance must be combined in cultivars that have desirable traits necessary for production.

SOURCES OF DISEASE RESISTANCE

When a new disease or a new race of a common pathogen occurs, when an old disease increases in prevalence, or when the disease reaction of a particular variety needs to be improved, plant breeders and pathologists look for sources of resistance. Resistance often occurs in germplasm that has been maintained for a long period of time in an area where the host and pathogen have coevolved. Frequently, this is near the center of origin of the host. For example, resistance to many diseases of peanut is found in germplasm that originated from South America where peanut is native (Wynne et al. , 1991). On the other hand, resistance sometimes is found in germplasm that was introduced and selected in an area where the pathogen is endemic. For example, the best sources of resistance in maize to Stewart's bacterial wilt are found in germplasm originating from the Ohio River Valley of the U.S. (Pataky et al., 2000). Germplasm from Central America, the center of origin of maize, is not highly resistant to Stewart's wilt probably because the disease rarely occurs there.

Breeders and pathologists use whatever genetic variation for resistance is available, but resistance can be incorporated more easily from sources that are closely related to adapted, elite cultivars than from unrelated sources. Hence, there is a relatively straightforward strategy of where to begin looking for resistance.

- Adapted cultivars — resistance found in adapted cultivars is extremely useful because the cultivars already possess other traits necessary for production. In some cases, an existing resistant cultivar may simply replace a susceptible cultivar in commercial production. More important for resistance breeding, an adapted cultivar probably does not have excessively deleterious traits that must be improved when incorporating resistance into new, adapted cultivars. Hence, most types of resistance that are found in adapted cultivars can be incorporated into new cultivars with relative ease. This is particularly useful when trying to improve levels of quantitative resistance in elite cultivars.
- Breeders' stock — breeders' stock are lines that are relatively well adapted, but have not been released as cultivars because they are deficient in one or a few important traits. Resistance that is found in breeders' stock is used in breeding programs nearly as easily as resistance found in adapted cultivars.
- Cultivated germplasm — accessions maintained at plant introduction stations and other collections of germplasm usually represent the greatest amount of easily accessible genetic diversity within a species. Many sources of resistance that are in use today were initially discovered among these collections. For example, soybean cyst nematode resistance originally incorporated into many commercially available soybean cultivars was derived from one of three sources of cultivated germplasm: PI88788, PI90763, or the cultivars Pickett and Peking. Subsequently, other sources of resistance were found among cultivated germplasm (Table 31.4). The ease with which resistance from germplasm is incorporated into adapted cultivars depends on whether inheritance of resistance is simple or complex and whether or not the source of resistance is relatively well adapted. Simply inherited resistance (e.g., monogenic resistance) can be incorporated from relatively unadapted germplasm, but resistance that is more complex in inheritance (e.g., quantitative resistance) is more difficult to acquire from exotic or unadapted germplasm. The use of molecular markers may improve our ability to incorporate quantitative resistance into adapted germplasm from exotic sources.
- Wild or related species — it is much more difficult to incorporate resistance from wild or related species than from germplasm of the same species; nevertheless, wild or related species can be extremely valuable sources of resistance. The use of wild species for improving disease resistance in food crops has had great success in wheat, potato, and tomato (Jones et al., 1995; Lenne and Wood, 1991). For example, stem rust resistance in wheat is one of the most important instances in the past century where resistance was transferred from related species. Usually, the ability to transfer resistance from related species into an adapted cultivar is limited to simply inherited resistance. This limitation may change in the future if breeders and pathologists are able to use molecular techniques to identify multiple genes for resistance in related species and to transfer those genes through marker-assisted selection.
- Artificially induced variation within a plant species — when genetic variation for resistance

does not exist within a species or related species, breeders and pathologists may try to find new genetic variation through alternative approaches such as mutation breeding, somaclonal variation, and genetic engineering. Mutation breeding and somaclonal variation have not been highly successful techniques from which new sources of resistance have been incorporated into cultivars (Daub, 1986). Some new sources of resistance have been produced through genetic engineering, such as viral coat protein-mediated resistance to viruses. Similar to resistance from wild species, artificially induced variation must be relatively simply inherited to be incorporated easily into adapted cultivars.

At first, it may seem that there is little need to continue to search for new sources of resistance once reliable sources of resistance are identified and incorporated into adapted cultivars. However, the need for additional sources of resistance is never-ending because genetic variation in the pathogen allows for the occurrence of and selection for virulent isolates (i.e., new races). Therefore, in each crop species, it should be the primary objective of at least one breeder/pathologist to identify diverse sources and types of resistance to each disease of importance and to transfer those resistances into elite, adapted lines that can be used to develop commercial cultivars.

INCORPORATING RESISTANCE

Methods to breed for disease resistance are no different than those used to breed for any other trait of importance. Backcross methods, pedigree methods, population improvement, or any other breeding technique that might be used for another trait can be used for resistance, depending on the host and the type of resistance. Adequate disease pressure usually is required to select disease resistant phenotypes. Typically, this requires environmental conditions that are adequate for disease development and a reliable inoculation method to ensure that the pathogen is present. As pathogens can vary in virulence and aggressiveness, it is important to use a collection of pathogen isolates or biotypes that are representative of the area in which the cultivar is likely to be grown. Similarly, it is important to develop cultivars with resistance to each of the diseases that are likely to be important in an area. For pathogens that are difficult to manipulate (e.g., insect-vectored viruses) or diseases that occur sporadically, marker-assisted selection may play an increasingly important role in resistance breeding. Even so, reliable phenotypic data is absolutely necessary to correctly identify markers associated with resistance genes.

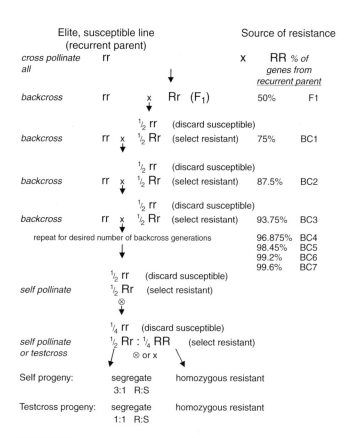

FIGURE 31.6 A general scheme for backcrossing a single dominant resistance gene (R) into a susceptible recurrent parent.

Simply inherited resistance usually is incorporated into elite, susceptible lines through backcrossing (Figure 31.6). An elite, susceptible line is crossed with a source of resistance. The resulting F_1 is then crossed "back" to the elite susceptible line (recurrent parent) to produce the first backcross generation. In each backcross generation, progeny are screened for resistant and susceptible phentoypes (1:1 segregation if resistance is conveyed by a single, dominant gene), and resistant progeny are backcrossed with the recurrent parent into which the resistance is being transferred. The exact methods and number of generations of backcrossing vary depending on the specific objective of the program. Following the last generation of backcrossing, the resistant progeny must be self-pollinated and progeny from those pollinations must be screened to identify lines that are homozygous for resistance. Backcross breeding is an easy way to transfer simply inherited forms of resistance, but it is inherently conservative. Often, the recurrent parent may be considered obsolete in its agronomic performance by the time the backcross-derived cultivar is released. Although used most commonly to transfer single dominant genes, backcrossing can also be used successfully with appropriate modifications to transfer recessive genes, multiple dominant genes, and even some types of partial resistance.

Usually, breeding for disease resistance is part of an overall program of cultivar development. Disease resistance often is considered just one of many traits, such as yield, quality, lodging resistance, maturity, or insect resistance, that must be considered when developing breeding or source populations that are subjected to selection in the breeding process. When synthesizing a breeding population from which to initiate selection, the breeder usually tries to select parents with complementary traits. The breeder hopes to select individuals that have the proper combination of favorable traits from the parents. In crops where the desired end product consists of a pure line cultivar or inbred line for making F_1 hybrids, the source population may consist simply of the F_2 population from a cross of two parent lines. Conversely, the source population may be a much more complicated three, four, or multiway cross of selected parents. Usually, the less agronomically desirable or poorly adapted the source of disease resistance, the lower the percentage of parentage it represents in the source population. Selection for disease resistance often is initiated early in the inbreeding process to eliminate the most susceptible individuals. Selection may be simply on an individual plant basis if the resistance is oligogenically inherited or heritability is high. Alternatively, selection may be on a family (e.g., F_3 or F_4) basis if heritabilities are lower. A reasonably large population should be maintained in order to have a sufficient number of resistant families remaining for agronomic testing in later generations.

The two basic methods used to develop pure line cultivars or inbred lines through the process of continued self-pollination are the bulk method and the pedigree method. In the pedigree method, the identity (i.e., pedigree) of individual families is maintained throughout the inbreeding process, so that any family (e.g., F_5) may be traced back to an individual plant in the original source population. The pedigree method allows selection of individual families at each stage of the inbreeding process and can be an effective means of selecting disease resistance of lower heritability. It requires substantial record keeping and may use more field space than the bulk method. In the bulk method, seeds from individual plants in each generation are bulked or combined with each generation such that it is impossible to trace the pedigree of any plant. Bulk populations are usually self-pollinated for several generations until they consist of a mixture of homozygous pure lines. Seed from individual plants are then planted in progeny rows for replicated testing and selection. The bulk method allows selection on an individual plant basis during early stages of the inbreeding process that can be an effective means of eliminating extremely susceptible individuals if the heritability of disease resistance is high. The bulk method eliminates the need for extensive record keeping and allows a breeder to work with a large number

of breeding populations at once because each population takes a minimal amount of space in the nursery.

In crops where the final product is an F_1 hybrid between two inbred lines, the parental lines are commonly developed by the pedigree method that is also used frequently in self-pollinated crops such as small grains and beans. In most crops where hybrids are grown, there are known heterotic patterns or breeding groups that exhibit hybrid vigor when crossed. For example, in maize hybrids grown in the U.S., almost all hybrids can trace the pedigree of their parental inbred lines to either the Iowa Stiff Stalk synthetic (typically used as female parents of the hybrid) or the open-pollinated variety Lancaster Surecrop (typically used as the male parent). Disease resistance and other traits of medium-to-high heritability are usually selected during the inbreeding process. Yield can only be assessed in hybrid combination with an appropriate tester, which may be an inbred line, a set of inbred lines, or a cross of two lines from the opposite heterotic group. Although the stage of inbreeding at which testcross performance is assessed varies greatly among hybrid crops and individual breeders, the value or worth of an inbred line is judged by its performance in hybrid combination and not by its performance per se. Thus, in developing disease-resistant hybrids, it may be sufficient for only one parent of an F_1 hybrid to carry resistance if that parent transfers a sufficient level of resistance to the hybrid. Similarly, a hybrid with resistance to multiple diseases may be developed more easily or efficiently if the resistances of the two parental inbred lines are complementary.

Lines with high levels of quantitative resistance can be derived if the source population first undergoes some method of cyclical population improvement. A source population may be created by crossing elite lines with sources of quantitative resistance. After an adequate amount of random mating, the population is screened for disease reactions and other important traits. A certain proportion of the most resistant progeny is selected and intermated to advance the population to the next cycle of selection. The exact selection intensity and methods of population improvement vary depending on several factors including heritability of resistance. During any cycle of selection, breeders may try to extract resistant lines from the population. The frequency of lines with high levels of resistance should be greater in the advanced cycles of selection than in the early cycles of selection because the distribution and mean of the population is continuously moving toward a more resistant phenotype (Figure 31.7).

A phenomenon that is sometimes referred to as the Vertifolia effect can occur when qualitative resistance prevents selection for quantitative resistance. Vertifolia was a cultivar of potato that had a single, dominant gene for qualitative resistance to late blight. This resistance was

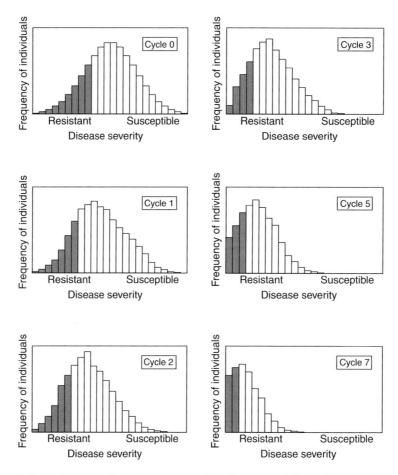

FIGURE 31.7 Population improvement. The frequency of lines with high levels of resistance is greater in advanced cycles of selection because the distribution and mean of the population is continuously moving toward a more resistant phentoype. Shaded area represents plants selected to form the next cycle.

effective until a virulent race of *Phytophthora infestans* occurred. When the virulent race became prominent, Vertifolia was more susceptible than potato cultivars that had been developed without the R gene because the R gene prevented potato breeders from selecting quantitative resistance while developing Vertifolia.

GENETIC DIVERSITY, SELECTION PRESSURE, AND MONITORING

People often mistakenly believe that genetic diversity is necessary for all of the important traits in a crop in order to avoid significant risks of severe epidemics of diseases or pests. Although any genetic uniformity poses a miniscule risk because of unknown pleiotrophic effects of uniform genes, most crops are uniform for many traits. For example, all sweet corn grown in the U.S. is uniform for the *su1* gene or the *sh2* gene that increase levels of kernel sugars. The benefit of uniformity for most traits is considerably greater than the risk of potentially severe epi-

demics. Host resistance to diseases is an exception to this rule. Genetic uniformity for disease resistance carries a substantial risk even though it is greatly beneficial when it works.

Host resistance places selection pressure on populations of pathogens. If genetic variation within the pathogen population allows some biotypes to reproduce on resistant hosts more rapidly than other biotypes, the frequency of biotypes that are more fit will increase in the population and a new, virulent race will occur. This interaction is similar to the response of other populations to selection. For example, antibiotics place selection pressure on populations of bacterial pathogens that can result in an increased frequency of antibiotic-resistant strains of bacteria by elimination of susceptible strains. Similarly, synthetic and naturally produced insecticides place selection pressure on populations of insects that can result in an increased frequency of insecticide-resistant insects. Fungicides place selection pressure on populations of fungi that can result in an increased frequency of fungicide-resistant isolates of fungi. Therefore, it should not be surprising that host resistance places selection pressure on populations of plant pathogens that can result in an increased frequency of virulent (resistance-resistant) pathogens (i.e., races).

In each example described in the previous paragraph, the occurrence of resistant populations (e.g., virulent races) can be avoided or reduced by altering selection pressure. For example, in human medicine, doctors prescribe a variety of antibiotics for patients who are seriously ill with bacterial infections in order to prevent the development of an antibiotic-resistant strain. Similarly, reduced number of applications of insecticides and fungicides and alternating applications of compounds with different modes of action or mixtures of these compounds are recommended to prevent insecticide-resistant pests and fungicide-resistant pathogens. A plan to delay the occurrence of *Bt*-resistant insects is based on the maintenance of *Bt*-susceptible, wild type populations in refugia where selection pressure has been removed completely. To prevent the occurrence of virulent pathogens in response to selection pressure created by host resistance to disease, disease resistance must be genetically diverse.

The durability associated with quantitative resistance is related to genetic diversity. Because most quantitative resistance is inherited polygenically, or at least oligogenically, it is unlikely that all quantitatively resistant cultivars carry all of the same genes for resistance. Therefore, each quantitatively resistant cultivar exerts slightly different selection pressure on pathogen populations. Also, because most forms of quantitative resistance allow at least some reproduction of the pathogen, quantitative resistance exerts less selection pressure on the pathogen population than qualitative resistance that often totally suppresses pathogen reproduction. Genetic diversity inherently associated with quantitative resistance may also be due to a different mechanism of resistance or modes of actions of resistance genes. A quantitatively resistant phenotype may be the result of several different factors such as reduced infection frequency, longer incubation and latent periods, reduced lesion size, reduced sporulation, or shorter infectious periods. Each of these factors may be affected by more than one physiological or physical mechanism in the host. Pathogen populations must become adapted to each of these factors in order to become highly virulent. In this case, general virulence (aggressiveness) of the pathogen population as well as specific virulence would probably change.

Monogenic resistance is more likely to result in genetic uniformity in crops because the same gene may be used widely in many or most resistant cultivars. In spite of the potential risk associated with this uniformity, monogenic resistance sometimes is used for a prolonged period of time before virulence becomes frequent. For example, the Ht_1 gene that conveyed chlorotic-lesion resistance to

northern corn leaf blight was used widely in dent corn grown in the U.S. from the mid-1960s through the late 1970s before virulence (i.e., race 1) became frequent in the North American population of *E. turcicum*. Similarly, in the mid-1980s, the *Rp1-D* gene that conveys resistance to common rust was incorporated into most sweet corn hybrids grown for processing in the midwestern U.S. Virulence against this gene was not observed in midwestern populations of *P. sorghi* until 1999. During the 15 years that the *Rp1-D* gene was used effectively without the occurrence of virulence, *Rp*-resistant hybrids increased the on-farm value of processing sweet corn in the Midwest by at least $60 million based on conservative estimates of yield losses that did not occur and costs of fungicides that were not applied. Clearly, the economic benefits of monogenic resistance can be substantial in certain situations even though experience proves that this type of resistance frequently has selected for new, virulent races of pathogens.

From the very outset of modern efforts at breeding for disease resistance, humans have had to cope with the problem of "shifty pathogens." With some diseases such as the Fusarium wilts of various crops, resistance has remained durable for many years. For others such as the cereal rusts, resistance genes often are ineffective before they are deployed in a cultivar. Even when effective cereal rust resistance gene is deployed, it is not uncommon for it to remain effective for only five years or less. Breeders and pathologists have tried to cope with variable pathogens by constantly searching for new sources of resistance and by selecting combinations of resistance genes that provide resistance to the spectrum of races present in a growing region. These efforts depend on knowledge of the relative frequency, distribution, and virulence of races present in areas where the crop is being grown.

Two basic approaches of monitoring pathogen populations are race surveys and monitor plots. Many examples of race surveys are being used to document diversity in pathogen populations. Usually, scientists in a region cooperate to collect isolates that are then assayed for virulence characteristics (i.e., races) using a set of differential varieties. For example, the USDA Cereal Disease Laboratory in St. Paul, MN, annually collects samples of leaf and stem rust from wheat grown throughout North America and race-types these collections on differential varieties. Wheat breeders use this information to determine which combinations of resistance genes will be effective against prevalent races. The rust survey also serves as an early warning system to alert breeders of the presence of new races virulent on existing cultivars, thus allowing breeders to respond before significant losses have occurred. Data from the rust survey also have been useful as a historical resource from which to examine the influence of resistance gene deployment on pathogen populations.

The second approach, monitor plots, may be used alone or in conjunction with race surveys. Monitor plots consist of field plots of differential varieties planted at various strategically selected locations in a region. The occurrence and severity of disease is monitored in these plots throughout the growing season. By itself, information on disease development on differential varieties may give a rough estimate of the frequency of virulence to specific resistance genes, but it may not provide information on specific virulence combinations present in races. However, samples of isolates are usually collected from monitor plots and they may be used later to identify races as well as for other purposes. Monitor plots also can serve as an effective early warning system for new diseases, new races of a pathogen, or an increase in prevalence of a previously less-important disease. The Dekalb Plant Genetics Maize Pathogen Monitoring Project is an excellent example of how a private plant breeding company used monitor plots to gather useful information on variation and distribution of maize pathogens throughout North America (Smith, 1977, 1984).

DEPLOYMENT OF RESISTANCE GENES

Several approaches are either being used or have been proposed to increase the durability of disease resistance. Probably the most common means employed to prolong the useful life of a resistant cultivar is to combine more than one effective specific resistance gene into a single cultivar. The rationale behind this approach is that if virulence toward a specific resistance gene occurs as a very rare random mutational event, then the probability of multiple mutations conferring virulence to multiple resistance genes is so low that the resistance gene combination will have prolonged durability. This probability explanation for the apparent durability of certain combinations of stem rust resistance genes in wheat has been criticized. Often it is only certain combinations of resistance genes that exhibit durability. For example, certain combinations of particular seedling *Sr* genes with *Sr2* (an adult plant resistance gene) appear to confer durable resistance to stem rust of wheat. As a general rule, the combination of multiple resistance genes can be effective, but experience will dictate which combinations appear to be durable and which are not.

A related approach to achieve durability is gene pyramiding, that is, putting as many resistance genes as possible into a cultivar. Proponents of this approach often do not make the distinction between race-specific single genes and polygenes for partial resistance. Some proponents of gene pyramiding consider polygenes for partial resistance merely to be defeated race-specific genes that now have a small, incomplete effect on resistance. This alleged residual effect of defeated race-specific genes is

sometimes referred to as ghost resistance. Evidence for ghost genes is contradictory at best, and does not appear to be of much significance in most pathosystems. The value of arbitrarily pyramiding resistance genes without regard to their specificity or record of durability is questionable. Perhaps a better and more efficient approach would be to pyramid combinations of genes for partial resistance with race-specific genes that appear to be durable in combination.

Multiline cultivars or cultivar mixtures also have been promoted as a means to achieve durable resistance to highly variable pathogens such as the cereal rusts. A multiline cultivar consists of a set of near-isogenic, backcross-derived lines, each of which contains a different race-specific single gene for disease resistance. Because only a fraction of the pathogen population is virulent on a single component line of a multiline variety, the net effect is a reduction in disease development on the multiline as a whole. It is assumed that stabilizing selection (selection against unnecessary genes for virulence) will prevent selection for complex races with virulence on several components of the multiline. Multiline varieties have been developed and appear to have resistance in field tests, but their durability when grown over a large area for a prolonged period of time remains to be demonstrated. Theoretical simulations have shown that stabilizing selection must be relatively strong to prevent the buildup of complex races virulent on multilines. Experimental evidence for strong stabilizing selection operating in agroecosystems is inconsistent, so it is not clearly evident that multilines will be as durable as proposed. High-yielding, uniform, and easily maintained multilines have been difficult to develop. These problems have limited the wide-scale acceptance of this approach.

Regional deployment of race-specific resistance genes in a given epidemiological area is another strategy that has been proposed to prolong the durability of specific resistance. Specifically, it has been proposed that oat varieties grown in different regions of the central North American "Puccinia path" carry different genes for crown rust resistance. The crown rust fungus, *P. coronata*, and other cereal rust fungi overwinter in the southern U.S. and in northern Mexico. These pathogens spread northward each summer with the developing crop. Proponents of regional deployment of resistance believe that this annual northward spread of pathogens could be prevented or delayed by using different resistance genes in the southern states from those used in the Midwest and Canada. In theory, races selected for virulence to resistance genes deployed in the southern states would be avirulent on resistance genes used in the northern states. Although this approach is appealing, it requires cooperation among breeders from several states and three countries. Breeders and growers in southern states and

Mexico must agree not to use certain resistance genes. That decision could be detrimental to producers in the South if virulent races develop on southern varieties, but not on the varieties used in the North.

Temporal deployment of resistance genes is a variation of the strategy of regional deployment. In this case, cultivars with different resistance genes are planted each growing season in order to alter selection pressure on pathogen populations with each crop. Deploying different resistance genes in time has been proposed as a useful way to prolong the durability of resistance to many soil-borne pathogens that are not easily disseminated, such as soybean cyst nematode and Phytophthora root rot.

Polygenically controlled, partial disease resistance has been assumed to be more durable than monogenic forms of resistance. Presumably, this durability is the result of the polygenic nature of the resistance and the incomplete nature of resistance that allows some pathogen reproduction, and thus reduces selection pressure on the pathogen population. Although polygenic resistance appears to be controlled by several genes, each of which contributes a small portion to the total resistance, most experimental estimates of the numbers of loci controlling polygenic resistance are less than ten and commonly only two to four. Polygenic resistance appears to be found in a wide array of crops to diverse pathogens, but its use depends on the crop and the ease with which it can be manipulated in a breeding program. Perhaps the best example of the successful use of polygenic forms of resistance is in maize where hybrids rely almost exclusively on polygenic, partial resistance for control of various leaf blights. This situation reflects the ease with which maize can be manipulated by an array of breeding methods and the widespread availability of resistant germplasm. In other pathosystems such as the cereal rusts, polygenic resistance has not been widely exploited due to difficulties in effectively selecting for it in traditional pedigree breeding programs and the availability of easily scored and transferred monogenic forms of resistance. Once useable levels of polygenic resistance have been transferred into adapted, elite germplasm, there are fewer barriers to its use in many breeding programs.

Although polygenic resistance has been effectively used for many years in crops such as maize without any evidence that resistance has significantly eroded, it would be wrong to assume that pathogens cannot eventually adapt to polygenic resistance. Some experimental evidence suggests that pathogens have the capacity to at least partially overcome polygenic forms of resistance. However, this adaptation probably would be a slow, gradual process. If breeders continually locate and incorporate new sources of polygenic resistance (and presumably new polygenes) into new cultivars, adaptation by the pathogen

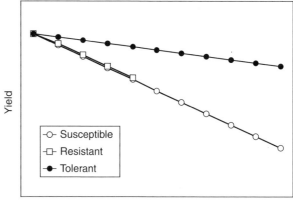

FIGURE 31.8 Tolerance. Based on the academic definition of true tolerance, the slope of the relationship between yield and disease severity is less negative for a tolerant cultivar than for a susceptible or resistant cultivar because a tolerant cultivar sustains less damage under the same level of infection. Note that the slope of the yield–disease relationship is the same for the resistant and susceptible cultivars, but disease is less severe on the resistant cultivar.

to a given set of polygenes would be to little avail because new genes would continue to be deployed in the host.

TOLERANCE

Disease **tolerance** is a term that often evokes different responses from individuals because this term has different meanings depending on who is using it. Plant pathologists, particularly those in academia, often define tolerance or "true tolerance" as the ability of a cultivar or variety to sustain less damage when the amount of infection (i.e., disease severity) is the same as on a susceptible cultivar. In other words, the slope of the relationship between yield and disease severity is less negative for a tolerant cultivar than for a susceptible or resistant cultivar (Figure 31.8). Although this is not a difficult concept, it is impractical to select for "true tolerance" in a breeding program because tolerance is difficult to measure experimentally. Due to variability inherent in measuring traits such as yield, yield loss, and disease severity, it is difficult to show that two cultivars have equivalent amounts of disease, but suffer significantly different amounts of yield losses. In many cases, tolerance can be demonstrated in one environment, but not in another. The practical value of disease tolerance also has been criticized because tolerance often is confounded with low yield potential. That is, low-yielding cultivars are affected less by disease than high-yielding cultivars simply because low-yielding cultivars do not need as much healthy tissue to reach their yield potential as do high-yielding cultivars. Nevertheless, there are examples of susceptible cultivars that sustain less damage from a particular disease than other susceptible cultivars, and these cultivars have true tolerance.

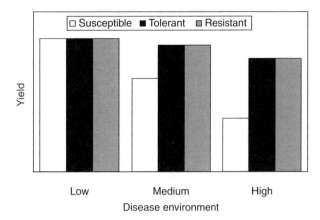

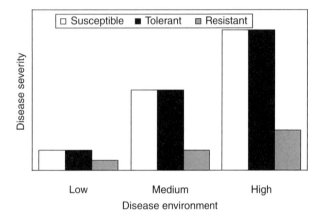

FIGURE 31.9 Susceptible, tolerant, and resistant cultivars yield similarly in environments with low disease pressure because disease severity is low. Under moderate or high disease environments, disease is equally severe on susceptible and tolerant cultivars but yield is affected less adversely on tolerant cultivars. Under moderate or high disease environments, yield of resistant and tolerant cultivars are similar but disease is less severe on resistant cultivars.

Tolerance frequently has a somewhat different meaning when used in seed catalogs and other literature produced by the commercial seed industry. Often, disease tolerance is used to identify cultivars that perform well in spite of adverse conditions, including severe disease pressure. When a cultivar yields well in spite of being infected as severely as a poor-yielding, susceptible cultivar, the cultivar may have true tolerance. More likely, the cultivar has a level of quantitative resistance that allows it to perform better than the susceptible cultivar under disease-conducive conditions because it is less severely infected than the susceptible cultivar (Figure 31.9). In this case, "tolerance" is due to partial resistance and the two terms are being used synonymously.

INFORMATION ABOUT RESISTANCE AND TOLERANCE IN SEED CATALOGS AND OTHER SOURCES

Definitions of resistance, susceptibility, and tolerance vary somewhat among seed catalogs from different companies and among other materials that report on disease reactions of cultivars. Similarly, methods of reporting the disease reactions of cultivars vary among different sources and even among different crops within the same company. Some people view this as confusing and problematic. However, it should be obvious that the types and degree of host resistance to diseases can be as varied as responses for all other host traits. It would be unreasonable to think that all resistant or tolerant reactions could be categorized easily by a single system. Hence, definitions and descriptions differ to some extent because all types of resistance are not the same. Nevertheless, a cultivar that is resistant or tolerant should sustain less damage than a susceptible cultivar grown under similar conditions.

The disease reaction of the cultivar being grown is one of the primary factors that determines whether or not disease will be severe and economically important. The more information that is known about the type and degree of resistance or tolerance in a cultivar, the easier it is to predict the performance of the cultivar under various conditions. Under similar conditions, a cultivar that is moderately resistant will be less severely infected than one that is moderately susceptible, and a tolerant cultivar will perform better than a susceptible cultivar. Knowing the reaction of the cultivar being grown allows resistance to be integrated more effectively with other disease control tactics.

REFERENCES

Adipala, E., P.E. Lipps and L.V. Madden. 1993. Reaction of maize cultivars from Uganda to *Exserohilum turcicum*. *Phytopathology* 83: 217–223.

Clifford, B.C. and R.B. Clothier, 1974. Physiologic specialization of *Puccinia hordei* on barley. *Trans. Br. Mycol. Soc.* 63: 421–430.

Daub, M.E. 1986. Tissue culture and the selection of resistance to pathogens. *Annu. Rev. Phytopathol.* 24: 159–186.

Jones, S.S., T.D. Murray and R.E. Allan. 1995. Use of alien genes for the development of disease resistance in wheat. *Annu. Rev. Phytopathol.* 33: 429–443.

Lenne, J.M. and D. Wood. 1991. Plant diseases and the use of wild germplasm. *Annu. Rev. Phytopathol.* 29: 35–63.

Pataky, J.K., L.J. du Toit and N.D. Freeman. 2000. Stewart's wilt reactions of an international collection of *Zea mays* germplasm inoculated with *Erwinia stewartii*. *Plant Dis.* 84: 901–906.

Shaner, G., E.L. Stromberg, G.H. Lacy, K. R. Barker and T. P. Pirone. 1992. Nomenclature and concepts of pathogenicity and virulence. *Annu. Rev. Phytopathol.* 30: 47–66.

Smith, D.R. 1977. Monitoring corn pathogens, pp. 106–121 in *Proceedings of the 32nd Annual Corn and Sorghum Research Conference,* H.D. Loden and D. Wilkinson (Eds.), December 6–8, 1977, Chicago.

Smith, D.R. 1984. Monitoring corn pathogens nationally, in *Proceedings of the 20th Annual Illinois Corn Breeders School,* March 6-8, 1984, Alexander, D.E. (Ed.), University of Illinois, Champaign.

Van der Plank, J.E. 1963. *Plant Diseases: Epidemics and Control.* Academic Press, New York, 349 pp.

Wynne, J.C., M.K. Beute and S.N. Nigam. 1991. Breeding for disease resistance in peanut (*Arachis hypogea*). *Annu. Rev. Phytopathol.* 29: 279–303.

SUGGESTED READINGS

Hooker, A.L. 1977. A plant pathologist's view of germplasm evaluation and utilization. *Crop Sci.* 17: 689–694.

Hulbert, S.H. 1997. Structure and evolution of the *rp1* complex conferring rust resistance in maize. *Annu. Rev. Phytopathol.* 35: 293–310.

Leonard, K.J., Y. Levy and D.R. Smith. 1989. Proposed nomenclature for pathogenic races of *Exserohilum turcicum* on corn. *Plant Dis.* 73: 776–777.

National Academy of Sciences. 1972. *Genetic Vulnerability of Major Crops.* NAS, Washington, D.C., 307 pp.

Parlevliet, J.E. and A. von Ommeren. 1985. Race-specific effects in major-gene and polygenic resistance of barley to barley leaf rust in the field and how to distinguish them. *Euphytica* 34: 689–695.

Schafer, J.F. 1971. Tolerance to plant disease. *Annu. Rev. Phytopathol.* 9: 235–252.

32 Cultural Control of Plant Diseases

Gary Moorman

CHAPTER 32 CONCEPTS

- The way plants are grown and maintained can influence whether disease develops and how severe it becomes. Where and when plants are grown, their proximity to other hosts of certain pathogens, and the removal of infected plants or plant parts have an effect on disease spread.

- Some things can be done to prevent pathogens from coming in contact with susceptible plants. Pathogens can be reduced or eliminated before or during crop production without the use of fungicides, bactericides, or nematicides. In other words, the way plants are grown can be manipulated to reduce or avoid some plant diseases or make it very likely that disease will develop and become severe.

Cultural practices for growing plants include soil preparation methods, propagation techniques, fertilization and irrigation regimes, where or how seeds or transplants are planted, and how plants are harvested. Each cultural practice may influence the amount of disease that occurs during the growing season. Sometimes, cultural practices can be chosen or modified in ways that reduce the amount of disease. To be effective, a cultural control practice must fit into the production system used by the farmer or the system must be modified to accommodate the practice. Because changing a cultural practice can be expensive or time-consuming, it must be effective enough in suppressing disease to justify its use. In other words, the cost of a cultural control practice has to be recovered through an increase in yield or a suppression of crop loss. In that sense, cultural control practices are no different than using chemicals or biological controls. An advantage of cultural controls is that they may reduce or eliminate the need for other types of control measures. The pros and cons of each cultural practice selected must be weighed, not only in terms of their expense and ease of implementation, but also in terms of how they will influence the most important pathogens that could affect the crop.

IRRIGATION (WATER MANAGEMENT)

Irrigation methods, amounts, and frequency can have profound effects on disease incidence. Most plant pathogenic fungi, almost all plant pathogenic bacteria, and foliar nematodes require a film or droplet of water on the plant surface in order to invade the plant. Spores landing on a dry leaf surface do not, in most cases, germinate or penetrate the leaf. Irrigation water applied directly to the soil via furrow flooding or drip and trickle irrigation, and kept off the plant surface, creates an environment unfavorable to many plant pathogens. Overhead irrigation applied through sprinklers, center pivot arrangements, or water cannons wet the foliage of plants and create a favorable environment for spore germination and infection. Many pathogens are splashed from leaf to leaf during overhead irrigation. In dry climates, overhead irrigation may provide moisture necessary for spore germination and infection not available to pathogens through rainfall. An advantage of drip irrigation is that water is placed near a plant's root system without wetting the foliage. This is useful for preventing foliar diseases of horticultural crops.

Some pathogens are inhibited by free water on foliage and these include the fungi that cause powdery mildew of roses and other plants. When the spores of powdery mildews land on a wet leaf, they do not germinate. Before fungicides were readily available, greenhouse rose growers routinely misted or syringed the leaves during the day in order to suppress spore germination. Thus, the cultural practice of misting helped control powdery mildew on roses. Eventually, fungicides effective for rose powdery mildew became widely available and misting was stopped. One adverse side effect of this change in cultural practice was an increase in mite problems because mites thrive under dry conditions and are suppressed under wet conditions. Also, some fungicides used for powdery mildew management may kill naturally occurring beneficial fungi that parasitize mites. So, all aspects of a cultural practice should be examined before implementation.

Irrigation methods that place water on the soil rather than on the plant tend to inhibit most foliar pathogens but can favor soilborne pathogens. For example **ebb and flow** or flood and drain irrigation systems in greenhouses and nurseries recycle unused water back to a reservoir. If pathogens such as *Pythium* or *Phytophthora* are already in the crop and enter the returning water or contaminate the reservoir from elsewhere, then recycling the water may

0-8493-1037-7/04/$0.00+$1.50

result in inoculation of a large portion of the crop with each watering.

Regardless of the method of irrigation employed, care must be taken in the amount and frequency of watering. Excessively wet soil is low in oxygen. Roots can be damaged directly by this. *Pythium* and *Phytophthora* are termed water molds because their activity is favored by excessive moisture and low oxygen content in soil. Most species of these pathogens produce a swimming spore stage (zoospore) when soil moisture is high. Zoospores are attracted to roots by chemicals, such as simple sugars, exuded from roots. Roots damaged by excessive soil moisture, excess fertilizer, and other factors tend to leak more such chemicals than do healthy roots.

POTTING MEDIA

Over the years, greenhouse operators and nurserymen have moved from using field soil as the growing medium for plants in pots and other containers to **soilless** components. Field soil may harbor plant pathogenic fungi, bacteria, and nematodes as well as weed seeds and therefore should be treated with heat or chemicals before use. This coupled with the facts that field soil is a nonrenewable resource and it makes containerized plants heavy to move within the operation as well as heavy to ship long distances has made the use of lighter, pathogen-free potting media attractive. Perlite, vermiculite, rockwool, and styrafoam are free of pathogens because of the way they are made. These materials can be used in combination or to amend peat moss or composts to produce a potting medium with the desired physical characteristics needed for healthy root development. Their major role, however, is to physically support the plant and they provide few or no nutrients. All nutrients must be applied in a solid or liquid form. It should be noted that although peat moss is usually pathogen-free, it sometimes harbors *Pythium* and another root rotting fungus, *Thielaviopsis*, and should be treated as if it were field soil. In general though, the use of soilless potting media has resulted in fewer crop losses in containerized plants than was the case when field soil was used extensively.

Composted tree bark and other composted organic materials, if prepared properly, will be free of pathogens and weed seeds and their microbial communities will greatly suppress root-infecting pathogens. Compost preparation must be monitored carefully because all parts of the pile being composted must reach the proper temperature and must be kept at the proper moisture level so that the organic matter is thoroughly colonized by the desired organisms. If parts of the pile are not composted properly or if the pile becomes too hot or too dry, the beneficial bacteria and fungi may not colonize the organic matter or may be killed.

MULCHING

Mulching can also be used to create an environment favorable to the plant and unfavorable to the pathogen. For example, plastic sheeting can be applied to the soil after initial cultivation and seedlings transplanted through it. Soil covered with black, but especially clear plastic, reaches a higher temperature during the day than bare soil. Verticillium wilt of eggplant is suppressed when plastic mulch is used. The fungus is not killed, however. By the end of the season, many plants exhibit symptoms and infected plants yield less than healthy plants. However, mulched plants yield more fruit and show symptoms less rapidly than infected plants grown on bare ground.

Organic mulches, such as straw, protect roots from drying and keep rain and overhead irrigation systems from splashing soil onto the plant. When soil that harbors a pathogen, such as *Phytophthora parasitica* that attacks tomatoes, is splashed onto leaves, stems, and fruits, rot can occur. Care must be taken in selecting the organic mulch material so that unwanted crop (oats, wheat, etc.) or weed seeds are not introduced into the growing area. Some growers use pine needles as mulch for that reason.

SOLARIZATION

When moderately moist soil (neither excessively wet or dry) is heated, disease-causing organisms are killed at various temperatures. For example, many water molds (e.g., species classified in the Oomycota — Chapter 20) are killed at 115°F (46°C), nematodes at 120°F (49°C), and many pathogenic fungi and bacteria at 140°F (60°C). At these temperatures, many beneficial organisms survive and provide competition to any pathogens that may arrive after treatment. For many decades, greenhouse operators have used steam, aerated steam (steam from which liquid water is removed), and electric systems to heat potting soil in order to eliminate pathogens. In some areas, steam and open flames have been used outdoors to treat soil. The heat of the sun, **solarization**, can also be used to treat soil. How well pathogens are reduced in the treated soil is determined by the actual temperature in the soil and the length of time the soil is held at a particular temperature. Some of the pathogen death is caused by the heat itself, some from chemicals released from decomposing organic matter, and some from the enhanced activity of beneficial organisms at elevated temperatures.

To solarize soil, clear plastic mulch is used because soil temperatures reach higher levels than with black plastic. Plant pathogens in dry soil are generally in a dormant, inactive state that is very resistant to heat, cold, or chemicals. Therefore, the cultivated soil should be moistened to a level that would support excellent seed germination. It must not be excessively wet or dry. The soil should then be covered with the clear plastic sheeting in the summer and kept covered for as long as possible. Several weeks of treatment are better than one week. After the solarization, the plastic can be left in place and slits cut into it for the placement of transplants. Or, the plastic can be removed. Pathogens will eventually return to the treated area through the movement of pathogen-containing soil on tools, shoes, or by wind and rain.

SANITATION

Sanitation practices are cultural methods that can play a role in suppressing disease outbreaks, especially initial disease development. For example, if potato tubers infected with *Phytophthora* are left in the field or are discarded into piles near the field, those tubers act as reservoirs for the initial inoculum of late blight. Initial inoculum may be reduced if the infected tubers are buried or destroyed.

Once disease has begun, removing infected plant parts or entire plants can reduce the amount of inoculum present and lessen the possibility that nearby plants will become infected. For example, removing dead flowers from ornamentals can help to reduce the amount of gray mold (*Botrytis*) inoculum (see Plate 7D following page 80). Similarly, removal of branches infected by fire blight (*Erwinia amylovora* bacteria) from crabapples, hawthorns, and other rosaceous plants helps to alleviate the disease on those plants. The removal from the greenhouse of entire plants (**roguing**) infected with *Impatiens necrotic spot virus* is a crucial step in managing that pathogen in potted crops. Roguing of weeds in the greenhouse is important because weeds can act as a reservoir for viruses and other pathogens and for the insects, particularly thrips and aphids known to move (vector) viruses from plant to plant.

DISINFECTION

Tools used to prune trees and shrubs may become contaminated with fungal spores or bacterial cells and spread the pathogen with each subsequent pruning cut. Thorough disinfection of tools between cuts or between trees being pruned can lessen this means of pathogen spread. Fire blight spreads in pear orchards on contaminated pruning tools from cankers on infected trees to healthy trees. Likewise, the canker stain fungus, *Ceratocystis*, on sycamore trees is readily spread from tree to tree on contaminated pruning tools and other equipment. *Tobacco mosaic virus* is spread very efficiently in sap from infected plants on almost anything that comes in contact with that plant. Another bacterial disease spread via contaminated tools is crown gall, caused by *Agrobacterium*. This bacterium survives in plant tissue and in soil. In crops where grafting is done, knives must be disinfected on a regular basis so that *Agrobacterium* is not introduced into the cuts made for grafting.

In addition to sanitation measures applied to plant material, pots, flats, and other containers, the greenhouse structure such as walkways and anything that may come in contact with the plants or places where plants may be grown should be disinfected in order to eliminate pathogens in the greenhouse and nursery. Floors and benches that are flooded during ebb and flow irrigation and the reservoirs where this water is stored between irrigations should be thoroughly dried, swept clean, and disinfected between crops. Such measures are particularly important in managing the soil-borne pathogens *Pythium*, *Phytophthora*, and *Thielaviopsis*. Outdoors, it may be important to power wash farm or forestry equipment and vehicles after working in an area known to harbor soilborne pathogens and before moving that equipment elsewhere. Therefore, sanitation of all tools and equipment coming in contact with infected plants or pathogen-harboring soil is an important cultural control technique used to reduce the spread of certain pathogens.

CROP ROTATION

Crop rotation is used to suppress or avoid certain diseases. Crop rotation is most effective for soilborne pathogens that reside in soil or plant residue left in and on the soil after harvest. One rotation strategy is to plant a crop that does not support the development of a pathogen that is a problem on the next crop to be grown in that particular field. For example, neither wheat nor corn is susceptible to *Sclerotium rolfsii*, the cause of southern blight of soybeans. Therefore, soybeans can be rotated with grain crops. Sometimes, a noncrop species is used between years of planting the crop. For example, tall fescue planted in rotation with tobacco can result in reduced losses of tobacco to blackshank, caused by *P. nicotianae*. However, the land used for such a rotation will yield little or no revenue during the fescue-growing year.

Plant parasitic nematodes obtain their nutrients from live plants. If the crop in the field is not a suitable host for the nematode, many nematodes die and their population level plummets. This is typically the case when lesion nematode (*Pratylenchus*) is present. The following season, the initial nematode population may be so low that nematodes do not adversely affect the new crop even if it is a suitable host. However, the nematode population may rebound on the suitable crop to a level that would make it unwise to plant that same crop again the following year. Crop rotation can allow you to continue to grow a crop off and on over a period of years despite the presence of serious pathogens at that particular site. Rotation is a very important strategy in managing diseases caused by the root-knot nematode (*Meloidogyne* species) on vegetables and the cyst nematode (*Heterodera* species) on soybeans and some other field crops (Chapter 8). Usually the farmer continues to grow the susceptible crop each year, but in a different field each year so that at least part of the farm is producing the main revenue-generating crop.

In some cases, it is best to continuously grow a particular crop because doing so tends to create an environment in the soil that inhibits certain pathogens. For example, the continuous culture of wheat may suppress take-all disease caused by *Gaeumannomyces graminis*. Although during the first few years of continuous wheat production the amount of take-all increases, disease incidence and severity decrease for the next several years. Apparently there is a build-up of beneficial organisms in the soil associated with wheat residue that suppress the pathogen.

REMOVAL OF NONCROP HOSTS AND ALTERNATE HOSTS

Another cultural control method is the removal of noncrop host plants of the pathogen. Black knot on plum and cherry trees, caused by the fungus *Apiosporina*, is often prevalent on wild plum and wild cherry trees along the edges of the forest and in hedgerows. By removing these trees, the amount of inoculum available to attack orchard trees is reduced. Similarly the removal of alternate hosts, required by the pathogen to complete its life cycle, is important such as in the case of cedar-apple rust on apple trees. The fungus, *Gymnosporanium juniperi-virginianae*, overwinters in galls on *Juniperus* species that are often found in abandoned pastures, in hedgerows and along roadsides. By removing junipers from the vicinity of the orchard, the number of spores that spread to infect apple fruit and leaves in the spring is reduced. An example of a forest tree disease that is managed through the removal of the alternate host is white pine blister rust, caused by *Cronartium ribicola*. By removing the alternate host, *Ribes* species, the life cycle of the fungus is disrupted and pines are less likely to become infected.

PLANTING DATE

How plants are established can greatly influence subsequent disease development. Tomatoes and many other high-value crops can be seeded directly into soil or can be started in a greenhouse and then transplanted into the field. Seeds and young seedlings are susceptible to damping-off caused by *Pythium* species, especially in cold, wet soils. To have tomatoes for sale early in the season, farmers may plant when soil temperatures favor development of **damping-off** due to *P. ultimum*. Because more mature plants tend to be much less susceptible to *Pythium* than young plants, using transplants avoids exposing seeds and

small seedlings to the fungus. Seeds can be planted into pathogen-free soilless media in containers in a greenhouse where soil temperature and moisture can be adjusted. More mature plants can be transplanted to the field after the field soil has reached temperature and moisture levels less conducive to damping-off.

INDEXING

Plants propagated vegetatively via cuttings, tubers, or rhizomes are generally less susceptible to disease than young seedlings. However, plant material propagated vegetatively is likely to be infected by pathogens from the mother (stock) plant. For example, bacterial blight of geraniums caused by *Xanthomonas* is carried in practically all cuttings taken from an infected mother plant. To avoid this problem, specialty propagators of geraniums thoroughly test each mother plant to ensure that it is free of blight bacteria. Most propagators that specialize in geranium production also test for viruses that affect geraniums. This process of testing for specific pathogens is called **indexing**. Culture indexing refers to testing for pathogens that can be grown in media (agar or broth), separate from the plant such as bacteria and most fungi. Indexed plants are not necessarily **disease-free** or **pathogen-free**. They may harbor pathogens other than the ones for which they are being tested. Plant indexing and propagating from indexed plants is an important cultural control method for strawberries, geraniums, carnations, chrysanthemums, and other high-value crops.

CROP PLACEMENT

Cucumber mosaic virus (CMV; Chapter 4), vectored by aphids, is an important pathogen of fresh market bell peppers in some regions. If infection occurs during flowering, severe fruit spotting develops. As few symptoms develop on fruit if infection occurs after flowering, strategies to delay infection can suppress disease losses. The initial onset of disease can be influenced by the crops placement in relation to where aphids can be expected to enter the field. When a very CMV-susceptible crop is planted near a hedgerow or other obstruction that can slow the wind, in the same manner as a snow fence, CMV-carrying aphids may settle out of the air and onto the crop at that location. Aphids are not strong fliers. The aphids may have overwintered on woody plants in the hedgerow, acquired CMV from nearby weeds in the spring, and then moved to the adjacent crop. Once in the field, the viruliferous aphids move from plant to plant, down the row. By planting the CMV-susceptible crop at a distance from where aphids first enter a field, the onset of CMV can be delayed and at least some crop losses avoided.

PLANT NUTRITION

Plant nutrition can have a dramatic impact on disease susceptibility and development. However, very few generalities can be stated concerning fertilizer type and rate in relation to subsequent disease development. Each crop and pathogen combination must be examined separately. High rates of nitrogen fertilization encourage succulent growth of pear trees that is very susceptible to fire blight. Similarly, high fertilization rates tend to favor Verticillium wilt development in trees and shrubs in the landscape. On the other hand, take-all disease severity increases in plants deficient in any of the major elements. Pythium root rot of the ornamental foliage plant, *Peperomia*, is most severe when fertilizer rates are low.

In some host and pathogen combinations, the form of the nutrient plays a role in disease susceptibility. For example, high nitrate fertilization favors Verticillium wilt in snapdragons and tomatoes, whereas high ammonium fertilization suppresses Verticillium wilt. In other diseases, nitrate may suppress disease.

Soil pH can be used to suppress disease but results may be erratic. Soil chemistry is very complex. Organic and inorganic compounds, living organisms, soil moisture and temperature, and many other factors all interact to influence soil chemistry. The result of altering the pH of soil with chemicals varies with each soil being amended. The suppression of clubroot of crucifers, caused by *Plasmodiophora brassicae*, through the addition of lime to raise the soil pH has been practiced for well over 200 years. In certain soils where the pH can be adjusted to 6.9 or higher, control can be very good. However, control may be poor at the same pH level in different soils. Many factors in the soil are influenced by pH adjustments and it is difficult to predict what will occur in a particular location without trying the technique.

From the examples provided in this chapter, it is clear that each pathogen/host combination needs to be looked at as a separate case when considering the use of cultural practices for disease control. A cultural practice that works with one host and pathogen may not work for a different host of that same pathogen. However, many cultural practices can be used to avoid or reduce disease development. The practice will only be used if it fits into the overall crop production system, is not too expensive to implement, and is effective.

EXERCISE

The effects that cultural practices have on disease development in a crop can be dramatic. For example, the concentration of fertilizer used on geraniums or poinsettias exposed to the root rotting fungus *Pythium* determines disease severity. Little disease develops at low fertilizer rates, yet good plant quality is obtained. A high fertilizer rate increases Pythium root rot losses. The presence of high amounts of nitrogen (N), phosphorus (P), or both N and P predisposes plants to attack by *Pythium*. The effect of fertilizer rate on disease severity is not merely due to damage from the high concentration of salts and associated high conductivity of the soil moisture.

EXPERIMENT 1. EFFECTS OF FERTILIZATION RATE ON PYTHIUM ROOT ROT OF GERANIUMS (*PELARGONIUM* × *HORTORUM*) OR POINSETTIAS (*EUPHORBIA PULCHERRIMA*).

This experiment will require 7 to 10 weeks to complete.

Materials

The entire class will require the following materials:

- Geranium seeds (or 120 poinsettia rooted cuttings)
- A culture of either *Pythium ultimum* or *P. aphanidermatum*. This can be obtained from the American Type Culture Collection. However, APHIS Permit No. 526 must be obtained in order for ATCC to ship the culture. Alternatively, a plant pathologist at a Land Grant university within your state may be able to supply a culture, in which case no permit is required.
- Plastic pots (4-inch diameter; 6-inch diameter pots for poinsettias)
- Commercial soilless potting mix (preferably one that does not contain composted components)
- Fertilizer to be applied as a liquid with each irrigation
- Conductivity meter for measuring soluble salts
- 250-ml beaker for collecting leachate
- Corn meal agar, V8 agar, or potato dextrose agar in 10-cm diameter petri dishes

Follow the protocols listed in Procedure 32.1 to complete the experiment.

Procedure 32.1 Effect of Fertilization Rate on Pythium Root Rot of Geranium (*Pelargonium* × *hortorum*)

1. Plant geranium seed in a flat or pot indoors or transplant poinsettia rooted cuttings in pots (6-inch diameter).
2. Transplant (approximately 10 to 14 days after planting) 120 geranium seedlings into pots (4-inch diameter) filled with a potting mix that does not contain

compost. (Many composts have *Trichoderma* in them as a natural part of the flora. *Trichoderma* can be a very efficient biocontrol of *Pythium*. Alternatively, the potting mix can be steamed or autoclaved to eliminate *Trichoderma*.)

3. To 35 plants, apply a liquid fertilizer (such as 15–15–15) at a rate of 150 ppm N. Apply the fertilizer to a second 35 plants at a rate of 300 ppm N and at 600 ppm N to the remaining 35 plants. Directions for mixing the fertilizer to achieve the desired concentrations are printed on the product container. Apply the fertilizer to the soil surface to thoroughly soak the potting mix so that about 25% of the moisture added leaches out the bottom of the pot. Repeat this each day for 4 or 5 days in order to establish three different fertilizer levels. After that, fertilizer-amended water should be used each time irrigation is necessary and enough should be applied each time so that about 10% of what is applied leaches through the potting mix. Such leaching helps to maintain the desired fertilizer level and lessens the build-up of salts to higher than desired levels. Measure the conductivity of each fertilizer that was prepared and use these readings as a guide to maintaining the three separate levels desired (See step 5).

4. Inoculate corn meal agar, V8 agar, or potato dextrose agar in standard 10-cm diameter petri dishes about a week before inoculating the plants (see Step 6 for timing). Inoculate a minimum of five dishes.

5. Four or five days after beginning the fertilization, take one pot from each fertilizer level and hold it over a beaker while pouring 100 to 200 ml of distilled water onto the soil surface. Measure the soluble salt level of the leachate by a conductivity meter to be certain that three different levels have been established. If the levels seem too low or too high as compared to those in Step 3, apply fertilizer twice a day for 2 days and repeat this step.

6. Seven to 10 days after transplanting and beginning the fertilizer treatment, inoculate 15 plants at each fertilizer level as follows. Scoop the entire contents of *Pythium*-inoculated, colonized petri dishes of agar into a blender and add 100 ml of tap water per dish. Very briefly homogenize this mixture in order to obtain a suspension of mycelium and agar. Apply 10 ml of suspension to the surface of each

of the 15 pots in each fertilizer level. Label these plants "inoculated." Label 15 additional plants "not inoculated." Use one of the remaining five plants per treatment to check the fertilizer level in each of the three treatments as noted in Step 5 each week.

7. Each week, note the number of days it has been since inoculation and record the number of dead plants in each inoculated and not-inoculated treatment. Discard the dead plants. Calculate the percentage of plants that have died.

8. Graph the total number of dead plants (*Y* axis) vs. days after inoculation (*X* axis) for each treatment.

Anticipated Results

Pythium root rot kills 600-ppm N-fertilized plants at a faster rate than at 300 ppm N. Usually, no plants are killed in the 100-ppm N treatment. Some noninoculated plants in both the 300- and 600-ppm N treatments will die as a result of the high salt levels. Although other types of plants can be used, note that begonias and impatiens generally do not react as dramatically as geraniums or poinsettias. Some plants, such as *Peperomia*, may have fewer losses due to root rot at high fertilizer levels than at low levels.

Questions

- What is the advantage of using lower fertilizer amounts in greenhouse plant production?
- How might *Pythium* get into a crop of geraniums in a greenhouse? From where would it have come?
- What can be done to prevent *Pythium* from getting into a crop in the greenhouse?
- What can be done to a greenhouse crop of potted plants once *Pythium* is found on the roots of some of the plants?

SUGGESTED READINGS

Engelhard, A.W. (Ed.) 1989. *Soilborne Plant Pathogens: Management of Diseases with Macro- and Microelements*. APS Press, St. Paul, MN, 217 pp.

Horst, R.K. 1983. *Compendium of Rose Diseases*. APS Press, St. Paul, MN, 50 pp.

33 Chemical Control of Plant Diseases

Alan S. Windham and Mark T. Windham

CHAPTER 33 CONCEPTS

- A fungicide is a chemical compound that kills or inhibits the growth of fungi.

- A bactericide is a chemical compound that kills or inhibits the growth of bacteria.

- A nematicide is a chemical compound that kills or disrupts the feeding or reproductive behavior of nematodes.

- A disinfectant is a chemical agent that is used to eliminate plant pathogens from the surface of plants, seed, or inanimate objects such as greenhouse benches, tools, pots, and flats.

Agricultural chemicals have been used to control diseases for centuries, even before microorganisms were associated with various plant maladies. Most pesticides targeted at plant diseases are used to minimize losses and are applied prior to infection. Although fungicides can be very efficacious against fungal pathogens, bacterial plant pathogens are much more difficult to control. Few fungicides have the capacity to halt the infection process once the pathogen is firmly established inside a plant.

Many farmers and gardeners equate disease control with the use of pesticides, such as **fungicides**, **bactericides**, **nematicides**, or **disinfectants**. Chemical control is one of several strategies used in plant disease management. The use of pesticides for disease control should be a component of an integrated system (Chapter 35) that uses cultural (Chapter 32) and biological controls (Chapter 34). In many cases, chemical control should be the strategy reserved for last when all other methods have been used.

PESTICIDE NAMES

Fungicides, as all pesticides, have the following three names: a chemical name of the active ingredient, a common name, and a trade name. While the chemical and common names remain constant, the fungicide may be marketed under several trade names depending on the target crop or manufacturer. For instance, a fungicide may be marketed under one trade name to specialty crops such as turf and ornamentals and under a different trade name to fruit, vegetable, and field crop markets.

An example:

Chemical name: Ethyl hydrogen phosphonate
Common name: Fosetyl-al
Trade name: Aliette

PESTICIDE LABELS

Pesticide labels list valuable use and safety information for the person mixing and applying the pesticide. At the top of each label, the trade, common, and chemical names are listed with the percentage of the active ingredient. **Signal words** such as Caution, Warning, or Danger appear in bold print, depending on the toxicity of the product (Table 33.1). An EPA registration number, which ensures that the product has met all regulatory standards, should be prominently displayed on the label. The precautionary statement lists hazards to humans and wildlife; worker protection standards include necessary personal protection equipment and reentry intervals for workers that may come into contact with fungicide residue. The label will also include directions for use, instructions for mixing and application, and potential environmental hazards. Pesticide labels generally contain information on pesticide storage and disposal.

MATERIAL SAFETY DATA SHEET (MSDS)

The **material safety data sheet** (MSDS) contains important information about pesticides. These sheets should be available for inspection by employees mixing, loading, applying, and working in areas where pesticides have been used. At the top of each MSDS, the pesticide trade or product name, chemical name, and common name are listed. Also listed on these sheets is information about adverse effects of overexposure, first-aid measures, firefighting measures, how to handle a spill or leak, proper storage conditions, and information about worker protection standards including required personal protection equipment. MSDS also list emergency numbers to call in the event of a spill or accidental poisoning.

0-8493-1037-7/04/$0.00+$1.50
© 2004 by CRC Press LLC

TABLE 33.1
Toxicity Classes for Pesticides

Hazard Indicators	Categories			
	I	**II**	**III**	**IV**
Signal word	Danger/poison	Warning	Caution	Caution
Oral LD_{50}	0–50 mg/kg	50–500 mg/kg	500–5,000 mg/kg	+5,000 mg/kg
Inhalation LC_{50}	0–0.2 mg/l	0.2–2 mg/l	2–20 mg/l	+20 mg/l
Dermal LD_{50}	0–200 mg/kg	200– 2000 mg/kg	2000–20,000 mg/kg	+20,000 mg/kg
Skin effects	Corrosive	Severe irritation at 72 h	Moderate irritation at 72 h	Mild or slight irritation at 72 h
Lethal dose to human adults (oral)	Few drops to 5 ml (1 teaspoon)	5–30 ml (1 teaspoon to 2 tablespoons)	30–486 ml (1 ounce to 1 pint)	486 ml (>1 pint)

FUNGICIDES

HISTORY OF FUNGICIDES

In 1802, lime sulfur was first used to control mildew on fruit trees. A few years later in 1807, Prevost used copper sulfate as a seed treatment to prevent bunt of wheat. In the latter part of the century, **Millardet**, a professor of botany at Bordeaux University, observed that grapes treated with a mixture of copper sulfate and lime to make them unappetizing to travelers were free of downy mildew (Chapter 20). In 1885, he demonstrated the effectiveness of this "Bordeaux mixture" in controlling downy mildew of grape.

Organomercurial fungicides were used as seed treatments to control bunt of wheat in 1913. The use of mercury and other heavy metals is now restricted by legislation due to animal toxicity. In the 1930s, the organic fungicides called dithiocarbamates were developed. Dithiocarbamates such as maneb, ziram, and thiram were used to prevent foliar diseases of plants. Although some of the original fungicides of this group are no longer available, dithiocarbamates are currently some of the most widely used fungicides around the world. By 1950, antibiotics such as streptomycin sulfate and cyclohexamide were introduced to control bacterial or fungal diseases of plants. From 1960 to 1970, widely used protectant fungicides such as captan and chlorothalonil were first used for foliar diseases of plants.

One of the first systemic fungicide groups, the benzimidazoles, was introduced to combat a wide variety of diseases such as powdery mildew, stem rots, and leaf spot diseases around 1970. Other systemic fungicides were introduced in the following decade. The SBIs (sterol biosynthesis inhibitors) were true systemic fungicides that controlled diseases such as powdery mildew and leaf spots. The acylalanines were the first systemic fungicides used to combat root rot diseases caused by species of *Pythium* and *Phytophthora*. By the end of the 20th century, a new group of fungicides, the strobilurins, were introduced to prevent a long list of diseases such as mildews, leaf spots, and stem and root rots. The strobilurins are similar to antifungal compounds produced in nature by wood decay fungi. They are widely used in row and specialty crops.

FUNGICIDE FORMULATIONS

Fungicide **formulations** consist of an active ingredient (a.i.) and inert ingredients, and may contain wetting agents, emulsifiers, or stickers. Wetting agents and stickers help to distribute the product over the crop and slow the weathering process, respectively. Prepackaged mixtures of two a.i.s may be used to broaden the spectrum of fungicidal activity or to prevent or slow the development of resistance. For example, mancozeb is often mixed with systemic fungicides at risk to resistance. Different formulations of a fungicide may have different concentrations of the a.i. depending on its use and target organisms.

Wettable Powders (WP)

A wettable powder is a finely ground fungicide powder that does not dissolve when added to water, but remains suspended in the spray tank. Agitation of the spray solution is normally needed to prevent settling. A disadvantage of a wettable powder is dust during measuring and weighing. Many of the older fungicides are formulated as a wettable powder.

Water-Dispersable Granules (WDG)

Dispersable granules have larger particles than a wettable powder. These particles break up rapidly when added to water and go into suspension. An advantage of a dispersible granule is less dust during measuring and weighing.

Water-Soluble Pouch (WSP)

Water-soluble pouches make mixing pesticides easier because a premeasured amount of pesticide is sealed in a bag that dissolves on contact with water. These pouches are usually designed to be added to 50 to 100 gal of water. An advantage of this formulation is that it minimizes exposure of the person mixing the pesticide. A disadvantage is that the pouches are preweighed for large volumes of spray and are not convenient for small quantities.

Flowable (F)

Flowable formulations have concentrated amounts of fungicide particles suspended in a liquid form. They are convenient because they are measured by volume rather than by weighing as in the dry formulations. A disadvantage of flowable formulations is that settling may occur if they are stored for long periods.

Emulsifiable Concentrate (EC)

An emulsifiable concentrate, like a flowable fungicide, is a liquid formulation. The active ingredient of this formulation is dissolved in an organic solvent. Emulsifiable concentrates are measured by volume.

Granule (G)

A granular formulation is made up of dry pellets or granules of a fungicide, often in low concentrations. These granules may break down or release the fungicide after coming into contact with water after application by a broadcast or drop spreader. Granular formulations are generally used for soilborne pathogens.

Dust (D)

Older fungicides such as sulfur are sometimes applied to foliage of plants as a fine dust. Dust formulations are often shaken or blown onto foliage and redistributed by dew or rainfall.

FUNGICIDE APPLICATION

Fungicides may be applied in many different ways. The application technique depends on the target pest, the crop, and available equipment. Most pesticides are mixed with water and applied by hand-held sprayers for small areas or by hydraulic sprayers or mist blowers for larger areas. Hydraulic sprayers are used where larger volumes of water are used to apply pesticides to the canopy of crops. Hydraulic sprayers may use as much as 100 to 200 gal of water per acre to apply pesticides to the foliage of crops. Air blast or mist blowers use much less water to apply the same amount of pesticide per acre. An air blast sprayer in a nursery or orchard may use as little as 15 to 20 gal of water per acre to apply pesticides to the canopy of crops.

Fungicides are often applied as seed treatments or in-furrow applications to prevent seed rots and damping-off.

Vegetable and field crop seed are often treated with fungicides to prevent seed rots caused by fungi such as species of *Pythium* and *Fusarium* species or damping-off caused by *Rhizoctonia solani*. A brightly colored dye is added to the treatment to indicate that the seed has been treated with a fungicide and should not be used for food or animal feed.

Drench applications of pesticides are used in specialty crops for certain soilborne pathogens. Fungicides labeled for stem or root rot diseases are applied in large volumes of water to containers or soil beds, often through a proportioner connected to the hose or irrigation system. Drench applications are intended to saturate the media or soil and may take the place of an irrigation cycle.

Broadcast application of fungicide granules are used when drench applications are impractical. Granules may be applied over the top of containers, propagation beds, or landscape beds with a rotary spreader. Drop spreaders are often used for precise application of granules to turfgrass. Irrigation or rainfall is needed after granules are applied to release the active ingredient.

Fungicide granules are sometimes incorporated into bulk soilless media prior to planting. Granules are thoroughly mixed into the media along with other amendments such as fertilizer or lime. Fungicides incorporated into media generally combat damping-off and stem and root rot diseases.

TYPES OF FUNGICIDES

Fungicides are often grouped as contact or eradicants. **Contact fungicides** are most often applied as foliar sprays to protect above-ground plant parts from infection; they may also be used as seed treatments. Contact fungicides must be applied uniformly over the leaf surface. As these fungicides are on the outer plant surface, they are subject to weathering and photodegradation. Contact fungicides do not protect new growth, so they must be applied frequently. **Eradicant fungicides** may be applied as seed treatments or on growing plants. They are usually systemic and move upward though the plant in the xylem tissue. Eradicant fungicides, like contact fungicides, work best if they are in place prior to infection; however, they will halt fungal growth if applied shortly after infection.

Systemic fungicides usually have one or more of the following characteristics: the ability to enter through roots or leaves, water solubility to enhance movement in the vascular system, and stability within the plant. Systemic fungicides may move varying distances in plants after being applied. An advantage of systemic fungicides is that they provide control away from the site of application. They are inside the plant and are not affected by weathering. Locally systemic (mesostemic) fungicides are translaminar and will protect the undersides of leaves after being applied to the upper leaf surface. Locally systemic

TABLE 33.2
Chemical Classes of Fungicides

Fungicide Class	Common Name
Inorganic fungicides	
Sulfur	Sulfur, lime sulfur
Copper	Copper sulfate, copper pentahydrate, Bordeaux mixture
Dithiocarbamate	Mancozeb
Aromatic compounds	Pentachloronitrobenzene, chlorothalonil
Oxathiins	Carboxin, oxycarboxin
Benzimidazole	Benomyl, thiophanate methyl
Dicarboximide	Iprodione, vinclozolin
Acylalanine	Metalaxyl, mefenoxam
Organophosphate	Fosetyl-al
Sterol biosynthesis inhibitor	Fenarimol, myclobutanil, propiconazole, triadimefon, triflumizole
Strobilurin	Azoxystrobin, kresoxim-methyl, trifloxystrobin
SAR stimulant	Acibenzolar-S-methyl

fungicides may provide protection for short distances on the leaf surface by vapor action. Most truly systemic fungicides move upward in the apoplast (xylem) to protect new plant growth from infection. Movement in the apoplast is usually passive as the fungicide moves upward in the transpiration stream. Arborists take advantage of this flow when they inject fungicides into the trunk of an elm to protect the tree from Dutch elm disease. An example of a fungicide that moves in the apoplast is azoxystrobin, a strobilurin fungicide. Few systemic fungicides move downward in the symplast (phloem) after they are applied to foliage. Systemic fungicides that move in the symplast are transported in phloem sieve tubes to the root system along with carbohydrates manufactured during photosynthesis. One of the few fungicides documented to move in the symplast is fosetyl-al.

FUNGICIDE RESISTANCE

If a fungicide is used continuously, there is always the possibility that certain individuals of a fungal population may become less sensitive to the fungicide. This decrease in sensitivity or resistance of a fungal population may be the result of a genetic mutation either present or induced in the population and the subsequent selection and multiplication of resistant individuals. Environmental conditions, disease pressure, and fungicide application frequency affect resistance development.

Fungicides that attack specific sites in the fungal cell may become vulnerable if the fungus becomes less sensitive to the fungicide with one mutation. Fungi that produce a large number of spores are more likely to become resistant to fungicides. The best chance for resistance development occurs when thousands or millions of fungal spores representing numerous individuals are exposed to one fungicide continuously. A single gene mutation in a

few spores may lead to a lack of control if the individual spores are also highly pathogenic. Fungi that cause diseases of floral or vegetable crops in greenhouses such as powdery (Chapter 14) and downy mildews (Chapter 20) and botrytis blight have been reported to acquire resistance to some fungicides.

Cross resistance may occur when a fungus becomes resistant to a particular fungicide active ingredient. For instance, *Botrytis* (gray mold) isolates that are insensitive to iprodione are often tolerant of the fungicide vinclozolin, which is in the same chemical class and targets the same site. Also, powdery mildews that are resistant to benomyl are often resistant to thiophanate methyl, both benzimidazole fungicides. If a resistant strain of a fungus is present, it is often appropriate to choose a fungicide of a different chemical class (Table 33.2).

The development of fungicide resistance can be slowed by the following strategies:

- Using labeled rates for pesticide applications. The use of less than labeled rates may accelerate the development of resistance.
- Alternate or mix a fungicide of a different chemical class to slow resistance. Use fungicides vulnerable to resistance sparingly.
- Use fungicides as a part of an integrated disease management program that includes biological and cultural controls and host resistance (Chapter 31) when available.

NONTARGET EFFECTS OF FUNGICIDES

The use or repeated use of certain fungicides may have unexpected results. The application of some fungicides may inhibit the growth of some fungi while having no effect on others. Fungicides applied to soil may reduce the growth of soil fungi such as *Trichoderma* species, which are important competitors of nutrients and potential parasites of plant pathogens. The result is the increased incidence of diseases caused by plant pathogens that are tolerant of the fungicide. Some SBI fungicides are closely related to plant growth regulators. Use at higher than labeled rates or at shortened intervals may lead to shortened internodes and stunting of turfgrass; use of high rates of SBI fungicides as seed treatments may reduce the growth of small grains such as barley or wheat. Broad-spectrum fungicides that inhibit the growth of saprophytic fungi in turf areas may lead to increased thatch layers.

BACTERICIDES

Bactericides are used to protect plants from bacterial plant pathogens. Antibiotics such as streptomycin sulfate and inorganic metal compounds such as basic copper sulfate and copper hydroxide are used to combat bacterial diseases. Streptomycin is produced by the actinomycete *Streptomyces griseus*. It is most commonly used to prevent fire blight of pome fruit trees and some ornamental hosts during bloom and is often ineffective during the shoot blight phase of fire blight. Copper compounds are used as protectant treatments for foliar bacterial diseases. The copper ion is toxic to bacterial cells. Fixed copper compounds, such as copper hydroxide, are relatively insoluble in water, but release enough copper ions to inhibit the growth of bacteria. Overuse of streptomycin or copper compounds may lead to resistance.

DISINFECTANTS

Disinfectants are used to eliminate bacteria, fungi, and algae from the surfaces of greenhouse benches, pots, flats, tools, and equipment. Common disinfectants include phenolic compounds, alcohol, sodium hypochlorite, quarternary ammonium compounds, and hydrogen peroxide. Disinfectants have little residual activity and dissipate shortly after use, especially if they are exposed to organic matter. Items to be disinfected should be relatively free of soil or organic matter.

CHEMICAL-INDUCED SYSTEMIC ACQUIRED RESISTANCE (SAR)

Plants, unlike animals, do not have an immune system to deter disease. Plants do have defense mechanisms that may be triggered by infection by plant pathogens. Pathogens that cause localized lesions on leaves may activate a systemic defense reaction that protects the entire plant from infection by other plant pathogens. This systemic acquired resistance (SAR) (Chapter 28) may also be acti-

vated by chemicals such as acibenzolar-*S*-methyl (ASM). Chemicals that induce SAR are generally not toxic to plant pathogens, but work by stimulating the defense system of the plant. As ASM is transported systemically, it is distributed uniformly throughout the plant.

Chemicals that induce SAR have been used successfully to protect plants from diseases such as blue mold (tobacco), downy mildew (lettuce), powdery mildew (wheat), and bacterial speck (tomato). Activity varies depending on disease and plant species. In general, chemicals that induce SAR should be applied several times at weekly intervals prior to possible infection. Traditional fungicides provide short-term protection whereas the chemical that induces SAR provides long-term control. Chemically induced SAR may be used to reduce the frequency and quantity of fungicides applied to crops. Because these chemicals do not act directly on plant pathogens, they are at lower risk for inducing pathogen resistance.

Chemical control of plant diseases is an important management strategy that works well when integrated into an overall program that includes host resistance and biological and cultural controls. The use of chemicals to control plant diseases generally works well if they are applied according to label directions as preventative treatments. This does not mean that fungicides should be applied indiscriminately. Consider the cropping history of the field, landscape bed, or turf area. Have certain diseases been observed causing significant damage each year for several years? Have you mapped areas of a field or noted certain golf greens that are disease prone? Can you change the environment in a greenhouse or increase air circulation on a golf green? What cultural practices or inputs could you modify to decrease the incidence or severity of disease? Many factors can affect the use of chemicals in the management of plant diseases, including societal and environmental concerns.

Fungicides are valuable tools in crop protection, but should only be used to complement other disease control tactics. Fungicides should not be used to cure diseased plants, but to keep healthy plants healthy or to minimize the damage of plant diseases. They are not a panacea or a "silver bullet"; however, fungicides are useful tools for plant protection.

EXERCISES

EXPERIMENT 1. READING AND COMPREHENDING A PESTICIDE LABEL AND MATERIAL SAFETY DATA SHEET

Pesticide labels contain essential information about the safe use of pesticides. Additional facts about mixing, application, and storage may be found on the label. Information on the label should be reviewed before purchasing, mixing, applying, storing, or disposing of the pesticide. Pesticide applicators or farm workers that are exposed to pesticides should be familiar with information on how to handle an accidental poisoning or spill prior to the incident. The goal of this exercise is to familiarize students with the safety and use information found on a pesticide label and material safety data sheet (MSDS).

Materials

- Provide each student with a sample pesticide label from a product guide or agrichemical company website
- MSDS sheets may be found in the back of product guides or agrichemical company website

Follow the instructions in Procedure 33.1 to complete the study.

Procedure 33.1 Reading Pesticide Labels and MSDS Sheets

1. Find the following information on a sample pesticide label:
   Trade name
   Manufacturer
   Common name
   Formulation
   Percent active ingredient
   EPA Registration number
   Signal word
   Precautionary statements: (Are there any hazards to humans or wildlife?)
   Does the pesticide list Worker Protection Standards (WPS)?
   What is the reentry interval (REI)?
   What Personal Protective Equipment (PPE) is required?
   List any environmental hazards
   List any restrictions
   How is the pesticide to be applied
   List emergency numbers (Chemtrec, etc.)
2. Find the following information on the MSDS sheet:
   Pesticide trade name
   Common name
   Toxicological information
     Acute effects of exposure
     Chronic effects of exposure
     Carcinogen (yes or no)
     Teratogencity (birth defects)
     Reproductive effects
     Neurotoxicity
     Mutagenicity (genetic effects)

Accidental spill or leak information
Ecological information
Handling and storage requirements

Questions

- Who would you call for questions about pesticide use or safety?
- What would you do if a pesticide spill occurred at your business?
- How should pesticides be stored?
- What does it mean if a pesticide is a carcinogen? Mutagenic?

EXERCISE 2. EFFECT OF NOZZLE SIZE ON WATER OUTPUT OF A SMALL HAND SPRAYER

Sprayer calibration is essential for the safe, economical, and effective application of pesticides. Hand sprayers are used to spray small areas of lawns, shrubs, or small trees and may be used to apply fungicides, insecticides, or herbicides. The volume of water applied to a given area is often determined by the pesticide and target pest. For example, many turf herbicides are applied in 1 gal of water/1000 sq ft, whereas most turf fungicides are applied in 3 gal of water/1000 sq ft because good coverage is necessary to protect the foliage.

The amount of water output of most agricultural sprayers over a given area is determined by travel time (in the case of hand sprayers, walking speed), pressure, and nozzle type and size. With most hand sprayers, output is controlled by nozzle size because pressure is constant if the sprayer is continually pumped. Constant pressure is easier to maintain with backpack sprayers that can be pumped during application. Calibrate the sprayer in a parking lot or other area where the spray pattern of the nozzle on the ground is easily observed.

Materials

Each student or team of students will require the following items:

- Carboy of water or easily accessible source of water
- One to three gallon compressed air hand sprayer or a Solo backpack sprayer
- Flat fan nozzles (Delevan D1, D3, D5 or Spraying Systems TK1, TK3, TK5)
- 1-l graduated cylinder

Follow the protocols listed in Procedure 33.2 to complete the exercise.

Procedure 33.2 The Effect of Nozzle Size on Output of a Small Hand Sprayer

1. Measure a test area of 500 sq ft (e.g., 20 ft × 25ft) on a parking lot (where it is easy to see the coverage and spray pattern) or a grassy area.
2. Fill sprayer with a measured volume of water or to a marked level on the sprayer.
3. Uniformly spray the test area. Make sure that the surface is evenly wet.
4. Release compressed air and determine the amount of water used by measuring the water remaining in the sprayer or the amount of water needed to raise the water level to the initial level.
5. Calculate the application rate of the sprayer.

   Method 2 — 128th of an Acre Calibration Method:

6. Measure a test area of 340 sq ft (c.a. 18.5 x 18. 5ft).
7. Repeat Step 2 to Step 4.
8. Another method to determine water output is to record the amount of time that it takes to spray the test plot. Spray the test plot three times and average the spray times.
9. Fill sprayer with water, pressurize, and catch the output in the 1-l graduated cylinder for the average time required to spray the test plot. The output in fl oz is equal to the number of gallons the sprayer would apply to an acre.

Anticipated Results

Flat fan nozzles should give uniform coverage. Water output should have increased as nozzles with larger orifice sizes were tested. If adequate coverage is not provided at a normal walking speed, then nozzle size should be increased or walking speed decreased.

Questions

- How would you adjust water output of a compressed air hand sprayer? A tractor-mounted hydraulic sprayer?
- What types of nozzles are used to apply pesticides?

SUGGESTED READINGS

Agrios, G.N. 1997. *Plant Pathology*, 4th ed. Academic Press, New York, 635 pp.

Bohmont, B.L. 1990. *The Standard Pesticide User's Guide.* Prentice Hall, Englewood Cliffs, NJ.

Hassall, K.A. 1990. *The Biochemistry and Uses of Pesticides.* VCH Publishers, New York.

Hopkins, W.L. 1996. *Global Fungicide Directory.* Ag Chem Information Services, Indianapolis.

Jeffers, S.N., R.W. Miller and C.C. Powell, Jr. 2001. Fungicides for ornamental crops in the nursery, pp. 409–416 in *Diseases of Woody Ornamentals and Trees in Nurseries*, Jones, R.K. and D.M. Benson (Eds.), APS Press, St. Paul, MN.

Köller, W. 1999. Chemical approaches to managing plant pathogens, in *Handbook of Pest Management*, J.R. Ruberson (Ed.), Marcel Dekker, New York.

Matthews, G.A. 2000. *Pesticide Application Methods*, 3rd ed. Blackwell Science, London.

Marsh. R.W. 1977. *Systemic Fungicides.* Longman Press, New York.

Parry, D.W. 1990. *Plant Pathology in Agriculture.* Cambridge University Press, Cambridge.

Simone, G.W. 2001. Bactericides and disinfectants, pp. 417–422 in *Diseases of Woody Ornamentals and Trees in Nurseries*, Jones, R.K. and D.M. Benson (Eds.), APS Press, St. Paul, MN.

Staub, T., W. Kunz and M. Oostendorp. 2001. Chemical activators of disease resistance in crop protection, in *Encyclopedia of Agrichemicals* John Wiley & Sons, New York.

Waller, J.M., J.M. Lenne and S.J. Waller. 2002. *Plant Pathologist Pocketbook.* CABI Publishing, New York.

34 Biological Control of Plant Pathogens

Bonnie H. Ownley and Mark T. Windham

CHAPTER 34 CONCEPTS

- Plant diseases can be controlled by living microorganisms that are antagonistic to plant pathogens.

- The phenomenon of suppressive soils involves a build-up of microorganisms that are antagonistic to disease-causing plant pathogens.

- Mechanisms of biological control of plant pathogens include antibiosis, parasitism, competition, induced systemic resistance, cross protection, and hypovirulence.

- Biological control of plant diseases often involves multiple mechanisms.

Biological control is the use of natural or modified organisms, genes, or gene products, to reduce the effects of undesirable organisms such as plant pathogens and to favor desirable organisms such as crops (Research Briefings, 1987). This definition is broad and includes genetic modification (genetic resistance) of the host plant. However, the main focus of this chapter will be on natural and modified organisms as biological control agents of plant pathogens. Biocontrol agents are known also as **antagonists** and **antagonism** is the generalized mechanism that they use to reduce the survival or disease-causing activities of plant pathogens. Antagonism is actively expressed opposition and includes antibiosis, competition and parasitism. Biological control of plant diseases with antagonists is accomplished by destroying existing pathogen inoculum, excluding the pathogen from the host plant, or suppressing or displacing the pathogen after infection has occurred (Cook and Baker, 1983).

DISEASE SUPPRESSIVE SOILS

Research on the biological control of plant pathogens followed the discovery of **disease suppressive soils**. These are soils in which either pathogens cannot establish, they establish but fail to produce disease, or they establish and cause disease at first, but disease becomes less important with continued culture of the crop (Baker and Cook, 1974). Disease suppression can be characterized as general or specific. General suppression results from the activity of the total microbial biomass in soil and is not transferable between soils; specific suppression results from the activity of individual or select groups of microorganisms

and is transferable (Weller et al., 2002). One of the best-documented examples of disease suppressive soils is the take-all decline phenomenon. Take-all root disease, caused by the soilborne fungus *Gaeumannomyces graminis* var. *tritici*, is a root and crown (basal stem) rot disease of wheat (*Triticum aestivum*) and barley (*Hordeum vulgare*) that occurs worldwide in temperate regions. In some soils, take-all decline occurs naturally following 5 to 7 years of growing continuous wheat with severe take-all disease and poor yields. Then the disease becomes less severe and yields recover. Increases in specific populations of microorganisms have been associated with take-all decline. In the Pacific Northwest of the U.S., an increase in populations of antibiotic-producing pseudomonad bacteria has been correlated with take-all decline. In Australia, increases in populations of the fungus *Trichoderma* are thought to be responsible for take-all suppressive soils.

WHY IS BIOLOGICAL CONTROL POPULAR?

Scientific interest in biological control of plant pathogens has been spurred by growing public concerns over the potentially harmful effects that some chemical pesticides pose to human health and the environment. There is also a need to control various diseases for which there are currently no controls or only partial control because there is little or no genetic resistance in the host, crop rotation is impractical or not economically feasible, or reliable, economical chemical controls are not available. For example, no practical or economical chemical control for crown gall disease was replaced when biological control with *Agrobacterium radiobacter* K84 was developed (Cook,

1993). In addition, registered biological controls are generally labeled with shorter reentry times and preharvest intervals than conventional chemical pesticides. This gives growers greater flexibility in balancing their operational and pest management procedures (McSpadden Gardener and Fravel, 2002).

PROBLEMS WITH BIOLOGICAL CONTROL

Although an array of microorganisms has been shown to protect crop plants from disease under experimental conditions, commercial development of many antagonists has been hampered due to inconsistent performance between field locations and seasons. Variation in performance of biological control agents has been attributed to many factors. These include compatibility of the host plant and the biocontrol agent due to host plant genotype, agricultural practices, mutation of the biocontrol organism resulting in a loss of effectiveness, resistance of the pathogen to biocontrol mechanisms, vulnerability of the biocontrol agent to defense mechanisms of the pathogen, and effects of the environment on survival and effectiveness of the biocontrol agent.

Antagonists are living organisms and whether applied directly to the host plant, to field soil, or to growth medium in a greenhouse, the antagonists will be active only if their growth and reproduction are favored by the environment. A change in environmental conditions during the growing season may have profound effects on the ability of a biocontrol agent to control a plant pathogen, whereas the same environmental changes may have little effect on the ability of a chemical pesticide to control the pathogen. Preparation and storage of antagonist inocula and application of the antagonist often have exacting requirements. Inocula of biocontrol agents usually cannot be stored at extreme temperatures that might be satisfactory for storing a wettable powder fungicide. In addition, shelf life of antagonist inocula is not as long as that of a chemical pesticide; therefore, growers cannot stockpile large quantities of antagonist inocula for later use.

MECHANISMS OF ANTAGONISM

Antagonists used for biological control of plant pathogens include bacteria, fungi, nematodes, protozoa, and viruses. With some exceptions, antagonists often are not pathogen specific; instead their effect on plant pathogens is coincidental (Cook and Baker, 1983). For example, an antagonist that aggressively colonizes roots and inhibits a wide range of microorganisms may protect roots against pathogens but the effect against any particular pathogen is probably coincidental. However, there are also antagonists that have a true parasitic relationship with their microbial host.

Antagonists interfere with plant pathogens through antibiosis, competition, and parasitism. These mechanisms are not mutually exclusive. An antagonist may use multiple mechanisms to detrimentally affect a plant pathogen or may use one mechanism against one type of pathogen and a different mechanism against another. For example, control of *Botrytis* on grapes (*Vitus*) with the fungal antagonist *Trichoderma* involves competition for nutrients and parasitism of sclerotia. Both mechanisms contribute to the suppression of the pathogen's capability to cause and perpetuate disease (Dubos, 1987).

ANTIBIOSIS

Antibiosis is the inhibition or destruction of one organism by a metabolite produced by another organism. Antagonists may produce powerful growth inhibitory compounds that are effective against a wide array of microorganisms. Such compounds are referred to as broad-spectrum **antibiotics**. On the other hand, some metabolites, such as **bacteriocins**, are effective only against a specific group of microorganisms. The bacterial antagonist *A. radiobacter* K84 produces agrocin 84, a bacteriocin that is effective only against bacteria that are closely related to *A. radiobacter*, such as the crown gall pathogen *A. tumefaciens*. Antagonists that produce antibiotics have a competitive advantage in occupying a particular niche and securing substrates as food sources because their antibiotics suppress the growth or germination of other microorganisms.

Antibiosis can be an effective mechanism to protect germinating seeds. For example, the bacterial antagonist *Pseudomonas fluorescens* Q2–87 can protect wheat roots against the take-all pathogen, *G. graminis* var. *tritici*, when coated onto seed. As the seeds germinate, the bacteria multiply in the rhizosphere, using exudates from the roots as a food source. The **rhizosphere** is the thin layer of soil that adheres to the root after loose soil has been removed by shaking and is directly influenced by substances that are exuded by the root into soil solution. The antibiotic 2,4-diacetylphloroglucinol, which is produced by *P. fluorescens* Q2–87, is effective against the take-all pathogen in minute quantities and has been recovered from the wheat rhizosphere (Bonsall et al., 1997). However, the effectiveness of antibiotics in soil can be variable as they can become bound to charged clay particles, degraded by microbial activity, or leached away from the rhizosphere by water.

Antibiotics

Many antibiotics manufactured by antagonists are not produced to target a specific pathogen and a specific antagonist may not produce the same antibiotics under different environmental conditions. Some antagonists produce several bioactive compounds that are effective against different plant pathogens. The bacterial antagonist *P. fluorescens* Pf-5 produces multiple antibiotics, including pyoluteorin, pyrrolnitrin, and 2,4-diacetylphloroglucinol. Pyoluteorin inhibits *Pythium ultimum,* a common cause of seedling disease in cotton (*Gossypium hirsutum*); however, it has little effect on other cotton seedling pathogens, such as *Rhizoctonia solani, Thielaviopsis basicola*, and *Verticillium dahliae* (Howell and Stipanovic, 1980). Pyrrolnitrin inhibits *R. solani, T. basicola*, and *V. dahliae,* but is not active against *P. ultimum* (Howell and Stipanovic, 1979). Examples of bacterial biological control agents that produce antibiotics include *Bacillus, Pseudomonas*, and *Streptomyces.* Fungal antagonists that produce antibiotics include *Gliocladium* and *Trichoderma.*

Volatile Compounds and Enzymes

Several volatile substances have a role in biocontrol of plant pathogens. These include ammonia (produced by the bacterial antagonist *Enterobacter cloacae* against the plant pathogens *P. ultimum, R. solani*, and *V. dahliae*), alkyl pyrones (produced by *T. harzianum* against *R. solani*), and hydrogen cyanide (produced by *P. fluorescens* against *T. basicola*, which causes black root rot). Although the biocontrol mechanism of parasitism involves many enzymes, there are enzymes that are involved in antibiosis only. For example, the fungal biocontrol agent *Talaromyces flavus* Tf1 is effective against Verticillium wilt of eggplant (*Solanum melongena*). *Talaromyces flavus* produces the enzyme glucose oxidase; hydrogen peroxide is a product of glucose oxidase activity and kills microsclerotia of *Verticillium* in soil (Fravel, 1988). The enzyme alone does not kill microsclerotia.

COMPETITION

Competition is the result of two or more organisms trying to utilize the same food (carbon and nitrogen) or mineral source, or occupy the same niche or infection site. The successful competitor excludes the others due to a faster growth or reproductive rate or is more efficient in obtaining nutrients from food sources. *Pseudomonas fluorescens* produces a **siderophore**, pseudobactin, which deprives pathogens such as *Fusarium oxysporum* of iron. Siderophores are extracellular, low-molecular-weight compounds of microbial origin with a very high affinity (attraction) for ferric iron. Chlamydospores of *F. oxysporum* require an exogenous source of iron to germinate. Although *F. oxysporum* also produces siderophores, the siderophores of *P. fluorescens* are more efficient in binding iron. If *P. fluorescens* is active in soil, chlamydospores of *F. oxysporum* remain dormant and cannot germinate due to low iron conditions.

Biological control of annosus root rot of conifers is another example of using competition to control a plant disease. Annosus root rot is caused by the fungus *Heterobasidion annosum*, which can survive for many years in stumps and logs, and causes extensive damage in managed forests or plantations of pure stands. To control annosus root rot, freshly cut stumps are inoculated with the fungal antagonist *Phlebia gigantea*. The mycelium of *P. gigantea* physically prevents *H. annosum* from colonizing stumps that it would use as a food base for attacking young pine (*Pinus*) trees (Cook and Baker, 1983).

Cross protection is a form of competition in which an avirulent or weakly virulent strain of a pathogen is used to protect against infection from a more virulent strain of the same or closely related pathogen. This term originated in virology to describe cases where infection of a cell by one virus lessened the likelihood that a second virus would damage the cell (Chapter 4). There is no host response in cross protection (see induced systemic resistance) and it is not transmissible (see hypovirulence). Inoculation with a small satellite-like self-replicating RNA molecule, known as CARNA 5, can be used to reduce virulence of *Cucumber mosaic virus* (CMV) in some vegetable crops. However, it is important to remember that weakly virulent strains can become more virulent or there may be unexpected synergistic effects that enhance disease. For example, CARNA 5 reduces disease caused by CMV in squash (*Cucurbita pepo*) and sweet corn (*Zea mays*), but enhances disease severity caused by CMV in tomato (*Lycopersicon esculentum*) (Cook and Baker, 1983).

PARASITISM

Parasitism is the feeding of one organism on another organism. As a mechanism of biocontrol, parasitism can be successfully used to reduce inoculum of sclerotia-forming fungi or prevent root rots, but may be less effective in protecting germinating seeds because establishing a parasitic relationship between the antagonist and pathogen may take more time than the time needed for the seed to become infected by the pathogen. Fungi that parasitize other fungi are termed **mycoparasites**. Parasitism by the fungal antagonist *Trichoderma* often begins with detection of the fungal host (plant pathogen) from a distance. The hyphae of *Trichoderma* grow toward a chemical stimulus given by the pathogen. This is followed by recognition, which is physical or chemical in nature, and attachment of *Trichoderma* hyphae to the host fungus that coils around the pathogen hyphae (Figure 34.1). *Trichoderma* produces lytic enzymes that degrade fungal cell walls. In some cases, cell-wall-degrading enzymes and antibiotics act synergistically in the biocontrol process (Chet et al., 1998). Examples of mycoparasites include *T. hamatum, T. harzianum, T. koningii, T. virens, T. viride, Pythium nunn*, and *P. oligandrum.*

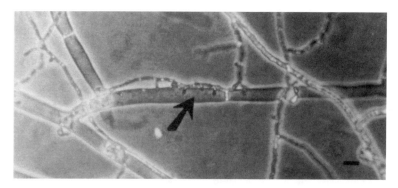

FIGURE 34.1 The smaller hypha of *Trichoderma virens* has penetrated the surface of the *Rhizoctonia solani* hypha at the arrow. Black line on lower right = 10 µm. (Photomicrograph reprinted from Howell, C.R. 1982. *Phytopathology,* 72: 496–498. With permission.)

OTHER MECHANISMS OF BIOLOGICAL CONTROL

HYPOVIRULENCE

Biological control by **hypovirulence** occurs when a hypovirulent (weakly virulent) strain of a fungal pathogen fuses (**anastomosis**) with a virulent strain of the pathogen and transmits the hypovirulent condition to the virulent strain. Anastomosis is fusion of touching hyphae and represents vegetative compatibility. Transmissible hypovirulence results from the infection of the pathogen with one or more dsRNA of viral origin. The classic example of hypovirulence is biocontrol of chestnut blight caused by the fungal pathogen *Cryphonectria parasitica* with hypovirulent strains of the fungus. In Europe, it was noted that cankers on chestnut (*Castanea*) trees were healing or were only superficial. Atypical strains of *C. parasitica* were isolated from the healing cankers. They had reduced pigmentation and sporulation. These atypical strains were also less virulent than strains from cankers that did not heal. In addition, when hyphae from a virulent strain were allowed to fuse with hyphae of a strain from a healing canker, the virulent strain became hypovirulent. Essentially, the hypovirulent strain converted the virulent strain to hypovirulent by the transfer of dsRNA via hyphal anastomosis (Heiniger and Rigling, 1994).

INDUCED SYSTEMIC RESISTANCE

Colonization of plants with nonpathogenic **plant-growth-promoting rhizobacteria** (PGPR) can cause **induced systemic resistance** (ISR) in the host plant (Chapter 28). ISR is a plant-mediated mechanism of biocontrol in which the biocontrol agent and the pathogen do not come in contact with one another. In ISR, host plant defenses are stimulated and plants are protected systemically. The level of

host response is modulated by jasmonic acid and ethylene (van Loon et al., 1998). Because many PGPR also produce antibiotics, iron-chelating siderophores, or lytic enzymes, their ability to suppress disease may involve more than one mechanism. PGPR may mediate biological disease control or plant growth promotion or both. Several strains or combinations of strains are available as commercial products for crop protection (McSpadden Gardener and Fravel, 2002).

INCREASED GROWTH RESPONSE

Microorganisms used in biological control, including PGPR and certain fungal biocontrol agents, can be associated with enhanced plant growth. In some cases, increased growth of the host plant is due to a reduction of viable inoculum of undetected pathogens, such as root-infecting *Pythium* species, which cause only slight reductions in vigor or yield. In other cases, enhanced plant growth, particularly in the absence of pathogens, may be due to plant-growth-promoting compounds of microbial origin (Chet, 1993).

DELIVERY, APPLICATION, AND FORMULATION OF BIOCONTROL PRODUCTS FOR PLANT DISEASES

There are various delivery and application methods for applying biocontrol products, including foliar sprays, soil or soilless mix treatments, seed treatments, root dips, and postharvest applications to fruit as a drench, drip, or spray. Formulations of commercially available products include dusts, dry and wettable powders, and dry and water-dispersible granules and liquids (Desai et al., 2002). Foliar sprays are most often used to protect above-ground plant parts from a foliar pathogen. Incorporating inocula of an antagonist into soil is a common approach to biological control of soilborne pathogens. Usually the antagonist is added in a dormant condition, often with a food base so that it can become established quickly in the soil. Commercially available soilless mixes have been marketed in the horticultural industry, that contain one or more biological control agents established in the mixes. The mixes are suppressive to pathogen populations under a prescribed set of environmental conditions. Adding organic amendments such as ground shrimp, crab shells, or bean leaf powder to greenhouse mixes or field soil has been used also to stimulate the growth of antagonistic organisms at the expense of plant pathogens.

One of the more successful methods of introducing biocontrol agents into an agricultural system involves treating seeds with antagonists. This approach delivers antagonists of soilborne pathogens as close to the target as possible. Seed treatments place the antagonist in the infection court (the seed coat surface) at planting, before the seed can be attacked by the pathogen. Formulated products can be applied directly at the time of planting as powders or liquids without stickers. Alternatively, antagonists can be precoated onto seed as dry powders or liquid-based formulations. Additives are used to prolong the survival of biocontrol agents on seeds. Once the seeds dry, they can be stored with the antagonists until planting. In some biocontrol products, bacterial antagonists and chemical fungicides are applied in combination to seeds (Warrior et al., 2002).

COMMERCIALLY AVAILABLE PRODUCTS

In the past ten years, there has been a significant increase in the number of commercially available biological control products for plant disease management. In 1993, there were only seven products registered with the U.S. Environmental Protection Agency (Cook, 1993); currently there are almost 30 registered products available in the U.S. (McSpadden Gardener and Fravel, 2002). On the worldwide market, there are approximately 50 biological control products for plant diseases (Desai et al., 2002). Increased understanding of the complex interactions that occur within the microbial community of agricultural soils and improved formulations of biocontrol agents to ensure their survival between production and use are needed for future development of more efficacious, cost-effective products.

CASE STUDIES

The following examples of biological control agents illustrate the diversity and complexity of the interactions between biocontrol agents and their target plant pathogens (Table 34.1 to Table 34.3).

TABLE 34.1
Case Study — *Agrobacterium radiobacter* strain K1026

Biocontrol agent — *Agrobacterium radiobacter* K1026, a Gram-negative bacterium.
- *Target pathogen* — *A. tumefaciens*, a Gram-negative soilborne bacterium.
- *Disease* — Crown gall of fruits, nuts, and ornamental nursery stock.
- *Mechanisms of biocontrol* — *A. radiobacter* produces the bacteriocin agrocin 84, which is effective against *A. tumefaciens*. It also competes with the crown gall pathogen for nutrients and attachment sites on plant roots.
- *Application and formulation* — Cells of *A. radiobacter* produced in culture medium are suspended in nonchlorinated water and applied to seeds, seedlings, cuttings, roots, stems, and as a soil drench (McSpadden Gardener and Fravel, 2002).

TABLE 34.2
Case Study —— *Coniothyrium minitans* CON/M/91–08

Biocontrol agent — *Coniothyrium minitans*, a Deuteromycete (an imperfect fungus).
- *Target pathogens* — *Sclerotinia sclerotiorum* and *S. minor*, soilborne ascomycete fungi.
- *Diseases* — Sclerotinia diseases of many crops including white mold of snap beans, pink mold of celery, stem rot of sunflowers and soybeans, lettuce drop, white mold of canola, and Sclerotinia blight of peanuts.
- *Mechanism of biocontrol* — *C. minitans* is a parasite of *Sclerotinia*. It attacks and parasitizes the **sclerotia**, or survival stage of the pathogen. This leads to a reduction in the amount of *Sclerotinia* inoculum available to cause disease in the following crop.
- *Application and formulation* — *C. minitans* is formulated as a water dispersible granule and sprayed onto agricultural soils before planting or after harvest (McSpadden Gardener and Fravel, 2002).

TABLE 34.3
Case Study — Trichoderma harzianum KRL-AG2 (T-22)

Biocontrol agent — *Trichoderma harzianum* KRL-AG2 (T-22), Deuteromycete (an imperfect fungus).

- *Target pathogens* — Several soilborne fungal pathogens, including *Pythium* species, *Rhizoctonia solani*, *Fusarium* species, *Sclerotinia* species, and *Thielaviopsis basicola*.
- *Diseases* — Damping-off and root rot diseases of many different plants, including ornamentals and food crops.
- *Mechanisms of biocontrol* — *T. harzianum* is a parasite of other fungi and can rapidly colonize plant roots, thereby out-competing pathogens for nutrients and space. *Trichoderma harzianum* also promotes plant growth in the absence of pathogens.
- *Application and formulation* — *T. harzianum* is formulated as a granule that is incorporated into soil or potting mix prior to planting (McSpadden Gardener and Fravel, 2002).

EXERCISES

The laboratory exercises are designed to demonstrate the biocontrol mechanisms of antibiosis and parasitism, and the suppressive effect of biological control agents on disease caused by plant pathogens.

GENERAL CONSIDERATIONS

These experiments are designed for teams of two or more students, depending on space and resources. The bacteria and fungi used in these exercises can be purchased from the American Type Culture Collection (ATCC; http://www.atcc.org/). An approved USDA application and permit are required to purchase or transport *Rhizoctonia solani* (http://www.aphis.usda.gov/forms/). Permits are not required for nonpathogens, such as *Pseudomonas fluorescens* and *Trichoderma virens*. It is very important to maintain aseptic conditions when conducting Experiment 1 and Experiment 2 and preparing cultures of micro-organisms.

EXPERIMENT 1. SUPPRESSION OF MYCELIAL GROWTH OF *RHIZOCTONIA SOLANI* BY ANTIBIOTIC-PRODUCING *PSEUDOMONAS FLUORESCENS* PF-5

Pseudomonas fluorescens Pf-5 was originally isolated from the surface of a cotton root. It produces several antibiotics, including pyrrolnitrin, pyoluteorin, and 2,4-diacetylphloroglucinol. It also produces the toxin hydrogen cyanide and two siderophores, pyochelin and pyoverdine, which are involved in competition for iron in low-iron environments. *Pseudomonas fluorescens* Pf-5 is an antagonist of the soilborne fungal pathogen *R. solani*, which causes seedling diseases of cotton and many other plants. Antibiosis plays a major role in the biocontrol of *R. solani* on cotton seedlings with *P. fluorescens* Pf-5.

Materials

Each team of students will require the following items:

- Microbiological media: Using dehydrated media, follow the manufacturer's instructions for preparing Difco tryptic soy agar (TSA) and Difco nutrient broth (NB). Prepare dilute potato dextrose agar (dilute PDA) with 4.4 g Difco potato dextrose broth and 20 g granulated agar. Follow manufacturer's instructions for sterilization of media.
  - Four TSA 10-cm deiameter dishes petri
  - Sterile noninoculated tube with 5 ml of NB
- 5-day-old cultures of *R. solani*; one culture on dilute PDA per team. Several isolates of *R. solani* from cotton are available from the ATCC. The ATCC numbers of these isolates are 14011, 18184, 28268, 38922, 60734, 90869, and MYA-986.
- 24-hour-old cultures of *P. fluorescens* Pf-5; one tube with a 5-ml culture in nutrient broth per team. This bacterium is available from the ATCC (ATCC number BAA-477).
- Transfer needle with stiff bent (90° angle) wire to cut 5-mm^2 mycelial plugs of *R. solani*
- Bunsen burner or alcohol lamp and 95% ethanol for sterilizing the transfer needle
- Sterile, disposable 1-ml pipettes and bulb; two pipettes per team
- Metric ruler
- Sharpie® marker
- Four 2-cm × 10-cm sections Parafilm®

Follow the protocols listed in Procedure 34.1 to complete Experiment 1.

Procedure 34.1 Suppression of Growth of *R. solani* by Antibiotic-Producing *P. fluorescens*

1. Label the bottom of the TSA petri dishes with your name, date, and "antibiosis assay."
2. Maintain aseptic techniques throughout this experiment.

3. Cut four 5-mm² plugs of mycelium from the actively growing edge of the *R. solani* culture on dilute PDA with a flame-sterilized and cooled transfer needle.

4. Remove lid from TSA dish and use the transfer needle to place one mycelial plug of *R. solani* in the center of the dish. Repeat for the three remaining TSA dishes. Incubate *R. solani* at room temperature (25°C to 28°C) for 24 h.

5. On the bottom side of the four TSA dishes, mark two spots about 10 mm from the edge of the dish, on opposite sides of the *R. solani* plug (Figure 34.2). In two of the TSA dishes, use the sterile 1.0-ml pipette to place a 0.1-ml drop of *P. fluorescens* broth culture on the agar over each marked spot. In the two remaining TSA dishes, use another sterile pipette to place a 0.1-ml drop of noninoculated sterile nutrient broth over the marked spots. The latter dishes will serve as controls.

6. Allow the liquid drop of bacteria or broth to dry on the TSA before moving the dishes.

7. Seal the top and bottom edges of each TSA dish together with Parafilm® to prevent moisture loss and maintain sterility.

8. Incubate the "antibiosis assay" dishes at 25°C to 28°C for 4 to 5 days. Observe daily and do not allow the edge of the fungal colony to reach the edge of the petri dish in the control dishes before making measurements (Step 9).

9. Measure and record the width of the inhibition zone between the leading edge of the *R. solani* culture and the *P. fluorescens* colony or the sterile broth droplet.

Anticipated Results

A clear measurable zone of inhibition should be evident between the bacterial colony and the leading edge of the *R. solani* mycelium (Figure 34.2). Antibiotics such as pyrrolnitrin produced by *P. fluorescens* Pf-5 diffuse into the TSA medium and inhibit mycelial growth of *R. solani*. No zone of inhibition should be observed between the drop of nutrient broth and the mycelium of *R. solani*.

Question

- What environmental factors may affect suppression of a soilborne pathogen, such as *R. solani*, by an antibiotic-producing antagonist?

EXPERIMENT 2. PARASITISM OF *RHIZOCTONIA SOLANI* BY THE FUNGAL ANTAGONIST *TRICHODERMA VIRENS*

Applications of the fungal antagonist *Trichoderma virens* (formerly named *Gliocladium virens* GL-21) can protect plants against damping-off and root rot pathogens, such as *R. solani*. *Trichoderma virens* is the active ingredient of the commercially available biological control product, SoilGard™. To control *R. solani*, SoilGard, a granular formulation of *T. virens*, is incorporated into soil or soilless potting mix prior to seeding. The biocontrol mechanisms involved in the interaction between *T. virens* and *R. solani* include parasitism (through production of enzymes and toxins), antibiosis (production of the broad-spectrum antibiotic gliotoxin), and competition for space and nutrients. Recently, seed treatment of cotton with another isolate of *T. virens* was reported to induce a plant host resistance response against *R. solani* (Howell et al., 2000).

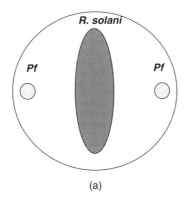
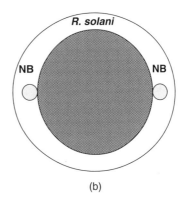

(a)   (b)

FIGURE 34.2 Schematic of antibiosis assay. (A) Inhibition of *Rhizoctonia solani* by *Pseudomonas fluorescens* Pf-5 (*Pf*) on tryptic soy agar. (B) No inhibition of mycelial growth of *R. solani* by droplet of sterile nutrient broth (NB).

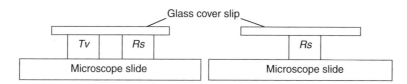

FIGURE 34.3 Schematic of parasitism assay. (left) Paired mycelial plugs of *Trichoderma virens* (*Tv*) and *Rhizoctonia solani* (*Rs*). (right) Control slide with one plug of *Rs*. Both slides are placed inside separate petri dishes containing dilute PDA.

Materials

Each team of students will require the following items:

- Microbiological media: Dehydrated Difco PDA; prepare dilute PDA as described for Experiment 1. Four dilute PDA petri dishes.
- 2-day-old cultures of *T. virens* (*Tv*) and *R. solani* (*Rs*), one culture of each organism on dilute PDA. Isolates of *R. solani* are available from the ATCC (see previously); *T. virens* GL-21 can be cultured on dilute PDA from granules of the commercial product SoilGard.
- Transfer needle with stiff bent (90° angle) wire to cut 5-mm² mycelial plugs of *R. solani* and *T. virens*
- Bunsen burner or alcohol lamp and 95% ethanol for sterilizing the transfer needle
- Eight glass microscope slides (25 mm × 75 mm)
- Four glass microscope cover slips (24 mm × 50 mm)
- Sharpie® marker
- Metric ruler
- Four 2-cm × 10-cm sections of Parafilm®
- Compound microscope with 10× ocular lens and 10× and 40× objective lenses

Follow the protocols listed in Procedure 34.2 to complete Experiment 2.

Procedure 34.2 Parasitism of *Rhizoctonia solani* by the Fungal Antagonist *Trichoderma virens*

1. Label the bottom of dilute PDA dishes with your name, date, and "parasitism assay."
2. Maintain aseptic techniques throughout this experiment.
3. Remove lid from dilute PDA dish and place a flame-sterilized, cooled microscope slide on the agar surface in the center of the dish. Repeat for remaining three dishes of dilute PDA.
4. Cut four 10-mm² plugs of mycelium from the actively growing edge of the *R. solani* culture with a flame-sterilized, cooled transfer needle.
5. Remove the lid from two of the dilute PDA dishes and use a transfer needle to place one mycelial plug of *R. solani* on the microscope slide, approximately 20 mm from one edge of the slide (Figure 34.3). For the two remaining dilute PDA dishes, place a plug of *R. solani* in the center of the microscope slide.
6. Flame-sterilize and cool the transfer needle; cut two 10-mm² plugs of mycelium from the actively growing edge of the *T. virens* culture.
7. Remove lids from the two dilute PDA dishes that have a plug of *R. solani* 20 mm from one edge of the microscope slide. Use the transfer needle to place a mycelial plug of *T. virens* on the microscope slide, 20 mm from the opposite edge of the slide from the *R. solani* plug. Place a flame-sterilized, cooled glass cover slip over the pair of plugs of fungal mycelium in both dishes.
8. In the two remaining dilute PDA dishes that have only one plug of *R. solani*, place a flame-sterilized, cooled glass cover slip over the plug.
9. Seal the top and bottom edges of each dilute PDA dish together with Parafilm® to prevent moisture loss and maintain sterility.
10. Incubate the "parasitism assay" dishes at 25°C to 28°C for 3 to 5 days.
11. Observe assay dishes daily to determine when hyphae of *T. virens* and *R. solani* have contacted one another. About 48 hours after contact, remove the glass cover slip from each "parasitism assay" and place it fungal side down on a few drops of immersion oil onto a clean microscope slide.
12. Observe slides with light microscopy at 100× and 400× magnification. Make observations on the appearance and size of the fungal hyphae on slides.

Anticipated Results

Hyphae of *R. solani* are significantly larger than those of *T. virens* and typically form right angles at the branches. Smaller hyphae of *T. virens* penetrate the surface and parasitize hyphae of *R. solani* (Figure 34.1). The smaller hyphae of *T. virens* are not observed in the control slides.

Question

- Which enzyme would be important for penetration and parasitism of *R. solani* hyphae by *T. virens*?

EXPERIMENT 3. BIOCONTROL OF DAMPING-OFF OF COTTON SEEDLINGS CAUSED BY *RHIZOCTONIA SOLANI* WITH *BACILLUS SUBTILIS* GB03

Bacillus subtilis GB03 is a PGPR and is the active ingredient in the commercially available biological control product Kodiak® that is used to control seedling diseases caused by soilborne pathogens such as *R. solani* in cotton, legumes, wheat and barley. The biocontrol activity of *B. subtilis* involves induced systemic resistance (ISR) in the host plant, competition for space and nutrients, and production of antibiotics. In the absence of pathogens, *B. subtilis* GB03 promotes plant growth.

Materials

Each team of students requires the following items:

- Cornmeal sand inoculum of *R. solani* prepared by instructor. Isolates of *R. solani* are available from the ATCC (see Experiment 1). Prepare cultures of *R. solani* on Difco dilute PDA (see Experiment 1) and incubate for 5 days at 25°C to 28°C. Prepare cornmeal sand medium using 97% quartz sand, 3% cornmeal, and 15% deionized water by weight. Place mixture into 500-ml Erlenmeyer flasks and cover opening with nonabsorbent cotton plug and aluminum foil. Autoclave the slurry for 90 min on two consecutive days, then inoculate each flask with twenty 10-mm² mycelial plugs from the edge of the *R. solani* cultures growing on dilute PDA. Incubate the flasks at 25°C to 28°C for 14 days. Shake flasks periodically to redistribute the mycelia of *R. solani* in the cornmeal sand medium.
- Twenty cotton seeds treated with *B. subtilis* (Kodiak® concentrate) per team. Instructor should treat the seeds as per manufacturer's instructions.

- Forty untreated cotton seeds
- Three stakes for flats
- Three plastic greenhouse flats
- Potting soil to fill three greenhouse flats
- Sharpie® marker

Follow the protocols listed in Procedure 34.3 to complete Experiment 3.

Procedure 34.3 Biocontrol of Damping-Off of Cotton Seedlings Caused by *Rhizoctonia solani* with *Bacillus subtilis*

1. Fill one plastic greenhouse flat with moist potting soil and label with your name, date, "biocontrol assay," and "HC." This flat serves as the untreated **healthy control** and does not receive either *R. solani* or seed treated with Kodiak®. Plant 20 untreated cotton seeds in four rows of five seeds in the "HC" flat.

2. Fill two greenhouse flats with moist potting soil. Remove the soil and weigh. Place the potting soil in a plastic bag and thoroughly mix with the cornmeal sand inoculum of *R. solani* at the rate of 2% cornmeal sand inoculum by weight. For example, if the potting soil weighs 1000 g, then add 20 g of *R. solani* inoculum.

3. Refill one greenhouse flat with half of the infested potting soil. Label the flat with your name, date, "biocontrol assay," and "IC." This flat serves as the untreated **infested control**. Plant 20 untreated cotton seeds in four rows of five seeds in the "IC" flat.

4. Fill the third greenhouse flat with the remaining infested potting soil. Label the flat with your name, date, "biocontrol assay," and "T." This serves as the **biocontrol treatment** with treated seeds planted into infested potting soil. Plant 20 Kodiak®-treated cotton seeds in four rows of five seeds in the "T" flat.

5. Place flats on a greenhouse bench and water as needed for 14 days.

6. At 5, 10, and 14 days after planting, observe the flats and record the number of living seedlings emerged.

7. Calculate the percentage of living seedlings emerged from each flat at each date by the equation:

$$(\text{Number of surviving seedlings}/20) \times 100$$

8. Make a line graph with three lines, each line representing the data from one flat. Plot "Days after planting" (*X*-axis) versus "Percentage of surviving

seedlings" (*Y*-axis) at 0, 5, 10, and 14 days after planting.

Anticipated Results

The untreated healthy control (HC flat) is expected to have nearly 100% seedling survival. The untreated infested control (IC flat) is expected to have the lowest percentage seedling survival, typically about 50% or less. The percentage of diseased seedlings will vary with environmental conditions in the greenhouse. The percentage of surviving seedlings from seed treated with *B. subtilis* GB03 (T flat) is expected to be significantly greater than the IC flat.

Question

- Additional information could have been gained from including more controls in this experiment. Suggest two additional controls that would help a grower make a decision on whether or not to use this product.

REFERENCES

Baker, K.F. and R.J. Cook. 1974. *Biological Control of Plant Pathogens.* W.H. Freeman, San Francisco, 433 pp.

Bonsall, R.F., D.M. Weller and L.S. Thomashow. 1997. Quantification of 2,4-diacetylphloroglucinol produced by fluorescent *Pseudomonas* spp. *in vitro* and in the rhizosphere of wheat. *Appl. Environ. Microbiol.* 63: 951–955.

Chet, I. (Ed.). 1993. *Biotechnology in Plant Disease Control.* Wiley-Liss, New York, 373 pp.

Chet, I., N. Benhamou and S. Haran. 1998. Mycoparasitism and lytic enzymes, pp. 153–172 in *Trichoderma and Gliocladium, Vol. 2: Enzymes, Biological Control and Commercial Applications*, Harman, G.E. and C.P. Kubicek (Eds.), Taylor & Francis, London.

Cook, R.J. 1993. Making greater use of introduced microorganisms for biological control of plant pathogens. *Annu. Rev. Phytopathol.* 31: 53–80.

Cook, R.J. and K.F. Baker. 1983. *The Nature and Practice of Biological Control of Plant Pathogens.* American Phytopathological Society, St. Paul, MN, 539 pp.

Desai, S., M.S. Reddy and J.W. Kloepper. 2002. Comprehensive testing of biocontrol agents, in *Biological Control of Crop Diseases,* Gnanamanickam, S.S. (Ed.), Marcel-Dekker, New York, pp. 387–420.

Dubos, B. 1987. Fungal antagonism in aerial agrobiocenoses, pp. 107–135 in *Innovative Approaches to Plant Disease Control,* Chet, I. (Ed.), Wiley-Liss, New York.

Fravel, D.R. 1988. Role of antibiosis in the biocontrol of plant diseases. *Annu. Rev. Phytopathol.* 26: 75–91.

Heiniger, U. and D. Rigling. 1994. Biological control of chestnut blight in Europe. *Annu. Rev. Phytopathol.* 32: 581–599.

Howell, C.R. 1982. Effect of *Gliocladium virens* on *Pythium ultimum, Rhizoctonia solani,* and damping-off of cotton seedlings. *Phytopathology* 72: 496–498.

Howell, C.R., L.E. Hanson, R.D. Stipanovic and L.S. Puckhaber. 2000. Induction of terpenoid synthesis in cotton roots and control of *Rhizoctonia solani* by seed treatment with *Trichoderma virens. Phytopathology* 90: 248–252.

Howell, C.R. and R.D. Stipanovic. 1979. Control of *Rhizoctonia solani* on cotton seedlings with *Pseudomonas fluorescens* and with an antibiotic produced by the bacterium. *Phytopathology* 69: 480–482.

Howell, C.R. and R.D. Stipanovic. 1980. Suppression of *Pythium ultimum*-induced damping-off of cotton seedlings by *Pseudomonas fluorescens* Pf-5 and its antibiotic, pyoluteorin. *Phytopathology* 70: 712–715.

McSpadden Gardener, B.B. and D.R. Fravel. 2002. Biological control of plant pathogens: Research, commercialization, and application in the USA. Online. *Plant Health Progress* doi:10.1094/PHP-2002–0510–01-RV.

Research Briefings. 1987. *Report of the Research Briefing Panel on Biological Control in Managed Ecosystems.* National Academy Press, Washington, D.C.

van Loon, L.C., P.A.H.M. Bakker and C.M.J. Pieterse. 1998. Systemic resistance induced by rhizosphere bacteria. *Annu. Rev. Phytopathol.* 36: 453–483.

Warrior, P., K. Konduru and P. Vasudevan. 2002. Formulation of biological control agents for pest and disease management, pp. 421–441 in *Biological Control of Crop Diseases,* Gnanamanickam, S.S. (Ed.), Marcel-Dekker, New York.

Weller, D.M., J.M. Raijmakers, B.B. McSpadden Gardener and L.S. Thomashow. 2002. Microbial populations responsible for specific soil suppressiveness to plant pathogens. *Annu. Rev. Phytopathol.* 40: 309–348.

35 Integrated Pest Management

Clayton A. Hollier

CHAPTER 35 CONCEPTS

- IPM has been defined as "A sustainable approach to managing pests by combining biological, cultural, physical and chemical tools in a way that minimizes economic, health and environmental risks" (Anonymous, 1994).

- There are many definitions of IPM. Most make the following assumptions: (1) presence of a pest does not necessarily indicate a problem; (2) a cropping system is part of an ecosystem; and (3) all pest management options should be considered before choosing a pest management strategy (Dufor and Bachmann, 2001).

- The first step in IPM implementation is to identify the pest.

- Analysis of information collected in the scouting and identification process is essential to determine pest management needs.

- A successful IPM program is one that has been planned and is never considered a "quick fix."

- Biocontrol tactics fit very well in the management of ecologically sensitive systems.

Integrated pest management (IPM) is not a new concept. In reality, the ancient people used IPM as a means to grow crops for food and fiber. Although their cropping methods were determined by trial and error, they realized that crops should be cultivated in certain ways to protect against pests. Some of their techniques seem foreign to us today, such as the offering of sacrifices to the "rust gods" in order to have grains free of diseases, or more understandable in a modern context, such as planting particular crops in certain soil types at a certain time of year.

DEFINITION AND IMPORTANCE

IPM has been defined as "A sustainable approach to managing pests by combining biological, cultural, physical and chemical tools in a way that minimizes economic, health and environmental risks" (Anonymous, 1994). Although modern concepts and practices of IPM began in the 1950s because of the pest management needs in apple production, it expanded into cotton production in the 1960s and today plays a role in all major crop production systems in the U.S. (McCarty et al., 1994). In this chapter, **pest** will be used to describe an insect pest or **vector** and plant **pathogens**.

Pest management represents a significant portion of a grower's crop production costs. Even with current IPM strategies, there are still 10% to 30% crop losses (Jacobson, 2001). In the mid 1940s, pesticides were seen as tools needed for all existing and future pest problems. Dependence on pesticides was so heavy that many alternative nonchemical strategies were abandoned (McCarty et al., 1994). Even today, pesticides play an important role in managing pests; however, consumer concerns, environmental regulation, pest resistance to pesticides, and development costs of products have reduced the availability and use of new pesticides. Thus, producers and the society in general are seeking new and adequate pest management strategies to protect production and the environment (Jacobson, 2001).

COMPONENTS OF AN IPM PROGRAM

There are many definitions of IPM. Most make the following assumptions: (1) presence of a pest does not necessarily indicate a problem; (2) a cropping system is part of an ecosystem; and (3) all pest management options should be considered before choosing a pest management strategy (Dufor and Bachmann, 2001). Many strategies can be used as part of IPM programs.

SCOUTING

Scouting is the key to successful IPM programs. It involves a comprehensive, systematic check of fields at regular intervals to gather information on the crop's progress and pest level. Plant pathologists use scouting as an essential part of the IPM process for managing plant diseases. Scouts that monitor crops for disease look for the presence of pathogen signs (spore masses, bacterial oozing) or symptoms (wilting, leaf spots, damping off, etc). Various patterns are used to scout fields so that a representative sample of the crop is examined. Scouts often walk fields in an S-, V-, W-, X- or Z-pattern and look for pests or crop damage or take samples for laboratory assays. Just as pattern and techniques vary depending on the pest, scouting methods vary by crop. The more frequently a crop is monitored, the more information is collected for IPM decision-making. The frequency of scouting depends on the crop, the pest, and the environmental influences that exist during the growing cycle.

PEST IDENTIFICATION

The strategies for IPM implementation vary in complexity. The first step in IPM implementation is to identify the pest. Most land-grant universities have disease and insect clinics that specialize in plant pest identification (Chapter 36). Many of these clinics use digital imaging to speed pest identification. Rather than shipping a plant or insect specimen to the lab, county agents or other practitioners post digital images of the pest or plant damage on the clinic web page for specialists to view, diagnose, and make management recommendations. Digital diagnostics allows farmers to react faster to pest outbreaks as the pests are often identified the same day as the images are posted on the website. Once the pest is identified, useful information can be gathered by studying the life cycle of the insect or pathogen. Pest identification is essential before selecting an appropriate pest management tactic. Misidentification can lead to the use of inappropriate strategies and further lead to a loss of time and money.

PEST SITUATION ASSESSMENT

The **damage threshold** (DT) is the amount of crop damage that is greater than the cost of management measures. Because growers want to protect their crop before a population of pest reaches DT, IPM programs use the concept of an **action threshold** (AT). These two levels are related but the AT is the point at which management tactics should be applied to prevent pest population increases and injurious levels to the crop (Zadoks and Schein, 1979). The value of the crop determines the AT.

Analysis of information collected in the scouting and identification process is essential to determine pest management needs. Has the AT been reached? It must be kept in mind that DTs or ATs have not been determined for every crop and those that have are generally developed for average conditions. In times of stress, such as drought, thresholds may have to be altered (McMullen, 2001).

IMPLEMENTATION AND EVALUATION

Once appropriate management strategies have been determined, they should be implemented in a timely manner. Late implementation usually will result in ineffective management and a loss of time and money. To determine success, the IPM strategy is chosen by comparing the pest activity before and after implementation of the IPM program. Review the following: (1) the pest identification procedures to determine the accuracy of the identification, (2) the field sampling procedures to eliminate bias and ensure appropriate representation, (3) the criteria for choosing the IPM appropriate strategy, and (4) the changes that should be made to the system to make it better.

DEVELOPING AN IPM PROGRAM

A successful IPM program is one that has been planned and is never considered a "quick fix." Quick fixes and IPM are not compatible strategies. In developing an IPM program, several technical and informational needs must be met (McMullen, 2001). Before looking at these classes, consider whether an IPM program tailored to the problem already exists. To determine this, check with state or county extension agents for the existence of the program, and if so, obtain the latest information for a particular crop/pest complex. If an IPM program does exist for a particular pest, formulation of a program includes the DTs for the pest and if the DTs change with time or crop development stage and how this influences management practices.

DEVELOPMENT STEPS

All steps are important in the process of developing an IPM program; however, there are issues from the beginning that need to be agreed on to make the program run as efficiently as possible. Define who is responsible for the monitoring process, what is its purpose, what are the costs involved, what expertise is necessary, and whether there are any potential conflicts of interest. Deciding who will scout the field(s) is extremely important. Monitoring is important to determine what pests are present; what their numbers and stage of development are; host development; what damage is being done, including incidence and severity, sampling methodology, soil type and moisture, dew periods, nutrient status, temperature, and relative humidity. Field books or software packages for the computer are places to store such information along with historic but retrievable information, including field maps and history of pests, soil types, soil profiles, where problems have occurred, previous crops, and cropping systems.

Accurate pest identification is necessary, but determining who is available to help identify the pests being monitored varies from state to state. University extension/research faculty and consultants traditionally have been employed to identify pests (Chapter 36). Accuracy in the identification will influence the success of the IPM program being developed. Inaccuracy can contribute to financial and ecological disasters.

Other considerations such as variety selection, crop rotation, plant populations, cultivation/tillage practices, planting/harvesting dates, and fertility levels influence the success of implementation of an IPM program. For example, use of **resistant** or **tolerant** varieties (Chapter 31) is the most economical and most widely used method to manage pests. Unfortunately, in some cases there is little or no resistance, but where available, resistance or tolerance cultivars should be employed. An example of success in this area is the use of resistant varieties to reduce the negative impact on yield and quality of wheat due to leaf rust development (Chapter 18).

Cultural practices (Chapter 32) such as crop rotation and cultivation/tillage have been widely used historically to reduce inoculum levels of pathogens. For example, crop rotation has been extensively used to reduce disease inoculum by growing nonhosts. Subsequent hosts then benefit from the short-term reduction in pest populations, thus reducing their overall impacts on the yield. An example of this practice is the reduction of *Sclerotinia* species (white mold) by rotation to nonsusceptible host. Additionally, cultivation has been used to decrease weed populations and their competition to crops. The weeds that are alternative hosts for pathogens are destroyed and thus reduce pathogen populations. With the advent of herbicides and their widespread use and the increase of conservation practices, cultivation has played a diminished role in crop production. However, as an IPM practice, it is still viable tactic and can reduce pathogen populations.

Temporal and **spatial** dimensions of crop management also can influence pathogen levels. Variation of planting or harvesting dates can reduce or eliminate certain pests as a factor in causing damage or reducing yield. These shifts in dates should be more conducive to the host development and less so for the pest. The following are some examples where these factors have reduced the inoculum load: early planting of corn to reduce late-season southern rust development, or late planting of ornamental bedding plants allowing for warming of the soil and reducing several cool-season soilborne plant pathogens.

Appropriate plant spacing allows for adequate growth of the host but also allows for air circulation or ventilation-reducing moisture on leaves of the canopy. This reduces a microenvironment that is conducive to disease development. Conversely, narrower row spacing or dense canopies favor development of plant diseases that thrive in high relative humidity and long dew periods.

Healthy seeds or propagating material are the backbone of successfully growing any type of plant. If seeds or propagating material are damaged or infected, the germination rate will be reduced, increasing the chances of replanting. Cost of production will increase, reducing the profit margin for the grower. An example of successful propagation is the planting of disease-free seed or using seed treated with fungicides that will protect seed before and after germination from seedling blight and other pathogens.

The use of biocontrol as a tool in IPM has increased during the last decade as the understanding increases in this vitally important area (Chapter 34). Introducing mycoparasites and antagonists as enemies of pathogens has reduced the inoculum level in several cases. Biocontrol reduces the use of the traditional pesticide, thus gaining support of many environmentally concerned groups. Ecologically, biocontrol fits well in sensitive systems because nontarget organisms, such as beneficial insects, are not impacted. Many within these groups feel that IPM should not include the use of pesticides to aid in the pest management process. In reality, the management of pests often does include the judicious use of pesticides in conjunction with the previously outlined strategies. Pesticides are a tool in a complete IPM system (Chapter 33).

AN EXAMPLE OF A PEST MANAGEMENT SYSTEM: WHEAT DISEASE MANAGEMENT IN THE SOUTHEASTERN U.S.

Disease management is a key component of high-yielding wheat. Some diseases, such as take-all disease, must be managed proactively and cannot be affected once they are established. Other diseases, such as foliar diseases caused by fungi, can often be managed by the timely application of fungicides. Generally, wheat producers place too much emphasis on disease management by foliar fungicides only. Most diseases are best managed through the use of multiple tactics, both proactive (crop rotation, delayed planting, resistant varieties, proper fertility, and seed treatment fungicides) and reactive (application of foliar fungicides).

Scouting for Diseases

Scouting for diseases is important for two reasons. First, yearly scouting helps to build an on-farm database that can be used to select appropriate disease management tactics for future crops. Second, scouting helps to determine if and when to apply fungicides. Once fields have been properly scouted, these data are helpful to determine disease management options. For help with this, contact local extension agricultural agents or crop consultants for the latest recommendations. An appropriate course of action can begin only when up-to-date, accurate information is obtained.

How Preplant Decisions Affect Diseases

Wheat producers have a significant portion of their total disease management program in place once the seed is in the ground. By that time, decisions have been made about crop and cultivar selection, method of tillage/seedbed preparation, variety selection, seed quality, seed treatment, planting date, seeding method and rate, and fall fertility. Individually and collectively, these decisions can play an important role in influencing which diseases develop, their severity, and their effect on crop yield. Because preplant and planting decisions are so important in the management of wheat diseases, understanding how they affect disease is necessary.

Variety Selection

Decisions relating to variety selection are perhaps the most important decisions that can be made in managing diseases. Every commercially available wheat variety has a unique range of reactions to diseases common in the region. Which and how many varieties are planted determine the potential for certain diseases. Failure to consider the ramifications of variety selection in managing diseases is a costly mistake. Selecting two or three varieties with the greatest amount of available resistance to the diseases most common on a farm or in the community is important. To do this, some idea about the disease history on the farm is necessary. If that information is not readily available, extension agricultural agents or crop consultants will be of help. This information will not be as good as actual data from the grower's own farm, but it is far better than basing decisions on no information. It is important to plant more than one variety for this key reason — it is common for a single disease to severely damage a single variety. When multiple varieties are planted, the risks of more significant losses are reduced. Planting more than one variety, especially when different maturities are represented, can help with the logistics of harvesting and soybean planting, if this crop is used in rotation.

Crop Rotation

Crop rotation helps to manage wheat pathogens that survive between wheat crops in wheat residue. When a crop other than wheat is grown in a field, levels of pathogens specific to wheat decline. This occurs simultaneously as the residue of previous wheat crops deteriorates. Lower levels of pathogens can translate into less disease pressure the next time wheat is produced. Crop rotation is helpful in the management of hidden diseases, such as Pythium root rot, and destructive diseases, such as take-all. In fact, rotating fields out of wheat is the only practical means of controlling take-all. Rotation also can reduce infections by the fungal species *Stagonospora* and *Septoria*. However, pathogen spores introduced from neighboring fields can negate the beneficial effects of rotation on these diseases. In some areas of the soft red winter wheat-growing region, wheat is planted following corn. Corn is generally a good nonhost crop to grow in nonwheat years. There has been some evidence that planting conventionally tilled wheat before corn increases the chances for a Fusarium head blight problem in wheat because the head blight fungi also attacks corn, causing stalk and ear rots. Observations have shown, however, that planting wheat behind corn, even in a no-till environment, does not increase the amount of head blight in wheat to any great extent. Thus, head blight inoculum is produced and blows around readily and, as long as conditions are favorable for head blight development, the disease will be a problem regardless of the rotation scheme.

Tillage

Tilling wheat stubble hastens the breakdown of residue that harbors certain disease organisms. This can help reduce levels of take-all and foliar diseases, such as Septoria leaf blotch and tan spot. "Help" is the operative word here because it is unlikely that tillage will be of much good in the absence of other management methods. For fields in a wheat/double-crop soybean/corn rotation, tillage prior to planting corn should cause a significant decline in surviving wheat stubble. The year between wheat crops in this rotation also helps, except where high levels of the take-all fungus exist. In those cases, two or more years between wheat crops may be required to reduce inoculum.

Planting no-till wheat following corn is not a problem except that this practice can slightly increase the risk for Fusarium head blight to occur in borderline situations. However, no data suggest that planting wheat directly into corn residue will be the difference between a serious head blight problem or an insignificant one.

Seed Quality, Seed Fungicides, Seeding Rate, and Planting Method

All these factors can influence stand establishment and seedling development. To achieve the highest possible yields, sufficient stands are necessary. There must be excellent seed germination followed by emergence and development of seedlings to ensure the desired stands. The use of high-quality (certified) seed treated with a broad-spectrum fungicide and good planting techniques foster good stand establishment. Excess stands, however, are undesirable because they encourage foliar and head diseases by reducing air circulation and light penetration into the canopy later in the season. Lodging of plants because of weakened stems is also a problem caused by excessively dense stands.

PLANTING DATE

The trend in recent years has been to plant wheat earlier and earlier each year. Early planted wheat, defined as wheat planted before the Hessian fly-free planting date, is at greater risk of damage caused by *Barley yellow dwarf virus* (BYDV), take-all disease, and Hessian fly than is later-planted wheat. If logistic considerations cause a grower to plant some wheat acres before the fly-free date for that area, those acres should have been well rotated and planted to a variety that can tolerate some BYDV. Planting all wheat acreage before the fly-free date is extremely risky and is not recommended.

NITROGEN FERTILITY

Too much nitrogen in the fall can encourage excessive fall growth that can increase problems with BYDV and most foliar diseases caused by fungi. Increased problems with BYDV have to do with an extended period of activity by aphids that transmit the virus when stands are dense in the fall. The same situation encourages infection and over-wintering of pathogens causing foliar diseases, such as leaf rust, powdery mildew, and leaf blotch complex. Excessive spring nitrogen results in lush stands that promote disease in a manner similar to that associated with excessive seeding rates. Lodging may increase too, resulting in higher moisture within the canopy of fallen plants and creating favorable conditions for fungal diseases.

FUNGICIDE SEED TREATMENTS

Obtaining and keeping a good stand of wheat is a key component of high yields. One management strategy many wheat producers use to attain excellent stands is to treat seed with a fungicide. Wheat seed treatment fungicides accomplish the following:

- Encourage good stand establishment — Wheat is planted at a time of the year that can be hostile to germinating and emerging seedlings. Excessively cool, wet, or dry soils, seed planted too deep or too shallow, and no-till plantings where seed-to-soil contact is poor slow the germination process and predispose developing seedlings to infection by seed- and soilborne fungi. Species of *Pythium* are probably the main pathogens in excessively wet and warm soils. The other conditions mentioned favor infection of seedlings by species of the fungi such as *Fusarium, Rhizoctonia, Septoria,* and *Stagonospora,* among others. Infection can result in fewer emerged seedlings and reduced vigor of the seedlings that do emerge. Most producers who plant wheat according to recommended guidelines have little difficulty achieving dense,

vigorous stands of wheat seedlings. The fact that a good percentage of the wheat seed planted in the region is treated with a fungicide probably has something to do with this situation. Routine use of high-quality, high-germ seed is probably a key contributing factor. In fact, historically, most small-plot seed treatment research done has shown that treating high-quality seed with a fungicide only rarely results in stand or yield increases. Nonetheless, it is still advisable to treat seed with a good general-use fungicide to protect seedlings from adverse soil conditions if they develop after planting. In this regard, seed treatment fungicides should be seen as a form of low-cost crop insurance. Environmentally, seed treatment fungicides are desirable because of their low toxicity, low use rates, rapid breakdown, and target application strategy.

- Enhance germination of marginal-quality seed lots — If seed quality is marginal because of fungi such as *Fusarium,* seed treatment can be used to bring seed germination/survival up to acceptable levels. In many cases, seed testing laboratories can provide tests that indicate whether fungicides will enhance germination of seed. Poor response of low-to-moderate germination seed lots to fungicides is indicative of a high percentage of dead seed, mechanical damage, or some factor apart from disease. Seed lots of low germination are not likely to be helped by any seed treatment fungicide and should not be used where high yield is a primary goal.

- Control loose smut — Loose smut control is probably the main reason seed treatment fungicides are so widely accepted and used in the southeastern U.S. Before seed certification, loose smut was a serious problem for both seed and grain producers. The use of seed treatment fungicides, such as carboxin, allowed seed lots to be "cleaned up" by eliminating loose smut from infected seed. Carboxin is still considered by many to be the standard for loose smut control. It is inexpensive and highly effective in most situations. However, sporadic, reduced activity of carboxin caused by poor application procedures or soil conditions that caused the active ingredient to be washed off the seed supported the development of a new generation of seed treatment materials that are highly effective against loose smut. All of these new-generation fungicides are extremely active at low use rates. In fact, because of the excessively low use rates of these products, application can be done only by seed conditioners with the experience and equipment to do the job properly. This eliminates the option of on-farm seed treatment by producers and increases the cost in many cases.

- Control foliar diseases — The new generation of systemic, sterol-inhibiting seed treatment fungicides (Chapter 33) can provide fall management of several fungal diseases, including Septoria and Stagonospora leaf blotches (leaf blotch complex), leaf rust and powdery mildew. Occasionally, management of these diseases extends into early spring as a result of reduced inoculum levels of the causal fungi in fields planted with treated seed. In some cases, this activity can be quite substantial, as is the case with triadimenol and powdery mildew. Management of powdery mildew through head emergence the following spring is not uncommon. Nonetheless, seed treatment fungicides should not be considered as a total replacement for spring-applied foliar fungicides because no seed treatment provides season-long control of foliar diseases.

In general, sterol-inhibiting fungicides are not highly effective against many common soilborne pathogens because of their specific mode of action. For this reason, most are marketed as a mixture with either thiram or captan, both of which provide at least moderate activity against a wide range of soil-borne fungi. Newer formulations may include metalaxyl to control soilborne *Pythium* species.

When contemplating the use of wheat seed treatment fungicides, consider the following factors. Costs of materials and disease control vary widely, so it is critical to assess cost–benefit ratios of the various fungicides. The main consideration is to determine why to use a seed treatment fungicide. Specifically, what diseases are to be managed and what is to be accomplished by seed treatment use? Fungicide labels should be read and followed completely. These labels refer to the specific diseases the products manage. Once these determinations have been made, selection of the most appropriate material (best disease control for the money) can be adopted. For example, triadimenol seed treatment is relatively expensive. If powdery mildew is an extensive problem on a farm, triadimenol may negate the need for an early spring foliar application of a fungicide on a mildew-susceptible variety. Taken in this context, the economics of triadimenol seed treatment become more favorable. If loose smut or general soil-borne pathogens are the main concern, and the risk of powdery mildew is minimal, triadimenol is not the most economical choice because less expensive materials are as good as triadimenol at managing these diseases.

Although fewer options are available to producers regarding on farm treatment of seed, some hopper-box treatments are still available. If a hopper-box treatment is to be attempted, it is important to note that complete coverage of all seeds is essential. Poor coverage equals poor results and, perhaps, a waste of money. Even distribution can be accomplished on farm only with considerable effort and planning. Having seed treated by a professional eliminates potential problems of poor fungicide distribution and uneven rates on seeds.

FOLIAR FUNGICIDES FOR WHEAT

Deciding whether to apply foliar fungicides to wheat is one of the most difficult decisions a producer or crop consultant has to make because of the many variables that influence the need for and effectiveness of foliar fungicides. First, fungicides must be applied in the early stages of a disease epidemic to be very effective (Chapter 30). Applying fungicides too far in advance of an epidemic or waiting too long to apply results in poor disease management and little or no economic benefit. Similarly, there is no economic gain from using foliar fungicides if yield-reducing levels of disease fail to develop or if crop yield potential is too low to cover costs. Finally, foliar fungicides manage only certain foliar and head diseases caused by fungi. They do not manage diseases caused by bacteria, viruses, or nematodes, and they have no effect on some fungal diseases, such as loose smut and take-all disease. Take the following steps when making decisions to use foliar fungicides.

Step 1. Commit to Scouting Fields

When considering the use of foliar fungicides, certain questions must be answered. Is there a commitment to scouting fields to determine the need to apply a fungicide? (If there is uncertainty about how to answer this question, refer to Step 3 through Step 6 for specific field scouting requirements.) If there is no commitment to field scouting (whether it is done by the grower or by a consultant), there is a question as to the decision to even consider fungicide use. Ultimately, a grower will need to decide how important wheat is to the total farming operation. If wheat is important to the profitability of the farm, it is advisable to make both time and monetary commitments to produce the best crop possible. If wheat is of only secondary importance relative to other farm operations, perhaps management of diseases by resistance and cultural practices should be considered.

Step 2. Determine the Number of Potential Fungicide Applications

Once the commitment has been made to scout wheat fields, the next significant determination is how many fungicide applications are you willing to make? Nearly every producer says only one and few indicate two. The answer to this question is important because it determines the approach to fungicide use. If a grower is going to make only one fungicide application, timing of the application is crucial. Research and experience show that a single application made during heading performs at least as well as and usually better than a single application made at flag leaf emergence in most situations.

The problem with single applications at flag leaf emergence (regardless of the fungicide used) is that they frequently allow late-season disease pressure to build to excessive levels. As a result, the crop is damaged even though early diseases may have been kept in check. Heading applications, on the other hand, usually limit disease build-up on the flag leaf (F), the second leaf down (F-1) and the head, although disease is allowed to develop unchecked early in the season. Protecting the F and F-1 leaves and the head is much more important to yield and grain quality than is protecting lower leaves. The risk in making a single application is waiting too late to apply the heading treatment. Fungicides are of little or no value once the flag leaf and head are severely diseased. The best way to limit this risk is to start scouting operations during flag leaf extension. It is unlikely that significant disease will have developed on the F and F-1 leaves by this time.

Crops receiving two fungicide applications (an early application followed by a late application) often yield significantly more than crops getting even the best single application. The question, however, is whether the economic benefit that results from the additional treatment is greater than its cost. As a general rule, the extra treatment at least pays for itself if early disease pressure is moderate to heavy and crop prices are good. If early disease pressure is minimal or crop prices are low, it probably would not be an economically justified treatment.

Step 3. Know the Disease Reaction of the Wheat Variety Planted

Typically foliar fungicides are not necessary on wheat varieties rated as resistant or moderately resistant to a particular fungal disease. Careful scouting and observation are the keys. Leaf rust and powdery mildew can adapt to and attack a formerly resistant variety. This can happen in a single season, so growers and consultants need to be vigilant and still scout those crops.

Step 4. Estimate Crop Yield Potential

Does the field have sufficient yield potential to justify a foliar fungicide application? Spraying with fungicides protects only yield already built into a crop; fungicides do not increase yield. Although various techniques can be used to estimate yield potential, most producers can look at a crop after green-up and know intuitively if the crop is worth protecting. In most cases, there will be a need to harvest an additional three to eight bushels per acre (depending on grain price and chemical cost) to offset the cost of a fungicide application. The higher the yield potential of a crop the more likely the economic benefit is to be realized from applying a foliar fungicide if disease becomes a problem.

Step 5. Know the Disease(s)

As indicated earlier, fungicides manage a relatively small number of fungal diseases. Fortunately, the diseases controlled are those that commonly reduce yields of soft red winter wheat crop in the southern U.S., such as leaf rust, powdery mildew, leaf blotch complex, and glume blotch. Other diseases, except tan spot, which is rarely a problem, are not managed with foliar fungicides. Thus, proper identification of the disease is critical in developing a control strategy.

Step 6. Scout Fields

Scouting wheat fields to determine crop growth stage and current disease situation is critical to making good fungicide-use decisions. When scouting fields, observe the entire field. Decisions should not be based on what is found along the edges or what is seen from the seat of a moving vehicle. The key is to make a decision based on the average disease situation in a field. This requires assessing disease levels in eight to ten randomly selected sites within the field.

Once in a field, it is important to determine the growth stage of the crop for the following two reasons: (1) All fungicides must be applied within specific growth-stage restrictions. Tilt, for example, cannot legally be applied once the flag leaves in a crop are fully expanded (Feekes 9 or decimal scale 39). Some states have received exemptions that allow for heading application of Tilt™ fungicide. Other fungicides, such as Dithane M-45™, have well-defined days-to-harvest restrictions. (2) Fungicides provide the greatest benefit when plants are protected from disease between flag leaf emergence and soft dough stage. In much of the southeastern U.S., the most critical stage is typically from mid-head emergence through flowering. This is the period in which fungicide applications are often most beneficial.

Step 7. Determine Disease Levels

To be effective, fungicides must be applied early in an epidemic. Too often, fungicides are applied too early, before any disease is visible. This approach results in no economic benefit if disease pressure remains low. Waiting too late to apply fungicides, although common, is equally ineffective. For leaf blotch complex including *S. nodorum*, if F-2 and lower leaves have symptoms and rain has been recent, there is a good chance the flag leaf and head are already infected (with first symptoms being 7 to 12 days away). Therefore, examination of symptoms on lower leaves can help determine when to apply a fungicide. Herein lays the greatest obstacle to effective, economical use of foliar fungicides — how much disease is enough disease to justify a foliar application of a fungicide? There are no absolutes, but many states have developed various threshold guidelines to help producers make informed fungicide-use decisions. Thresholds must be used along with some common sense. For example, if a specific threshold is reached for powdery mildew, application of a fungicide would not be recommended if an extended period of hot, dry weather is predicted. The threshold indicates that yield loss caused by one or more of the above diseases is likely; however, they do not mean losses will definitely occur. Weather can always intervene and impact the development and progression of a disease epidemic. There is no way to develop disease thresholds that are appropriate for all situations.

Step 8. Select a Fungicide

Product labels provide detailed use instructions and product limitations. Apply all pesticides according to label specifications.

Step 9. Understand the Risks

One problem is the inability to determine whether disease-favorable conditions will persist after a fungicide is applied. Fungicides are valuable only if yields and test weights are threatened by disease. Similarly, fungicides are of limited value if other diseases develop that are not managed by those chemical treatments. Examples of organisms not responding to fungicides are all virus and bacterial diseases and other fungal diseases including take-all, loose smut, and head scab. Lastly, fungicides may be of limited value if yields and test weights are reduced by nondisease factors such as a spring freeze, lodging, delayed harvest, or poor grain fill period. Unfortunately, these situations are always a risk to the fungicide user. Monitoring crop development throughout the season can reduce some risk by ascertaining yield-limiting factors that indicate fungicides would not be warranted. Of

course, this is a moot point once a fungicide is applied. In all instances where fungicides are used, check the response of the crop to the treatment by leaving a non-treated strip in the field for comparison (Hollier et al., 2001).

This example takes into consideration the discussions from the beginning sections of the chapter. It is important to review every aspect of how a crop is grown and look at a broad picture of the influences on the crop.

Integrated pest management is undergoing a tremendous transition as needs arise and the expectation of society demands change in the way food supply is grown, handled, and stored. IPM offers a flexible tool for satisfying those demands. Conceptually, IPM is valuable, but in practice that value increases. IPM takes a closer look at the way in which agricultural products are grown and offers the producer the underlying reasons of why and how.

REFERENCES

Anonymous. 1994. Integrated pest management practices on fruit and nuts. *RTD Updates: Pest Management*, USDA-ERS, 8 pp.

Dufor, R. and J. Bachmann. 2001. *Integrated Pest Management*. Appropriate Technology Transfer for Rural Areas (ATTRA), Fayetteville, AR, 41 pp.

Hollier, C.A., D.E. Hershman, C. Overstreet and B.M. Cunfer. 2001. *Management of Wheat Diseases in the Southeastern United States: An Integrated Pest Management Approach*. Louisiana State University AgCenter, Baton Rouge, 44 pp.

Jacobson, B. 2001. USDA integrated pest management initiative. In: Radcliffe, E.B. and W. D. Hutchinson (Eds.), *Radcliffe's IPM World Textbook*. URL: http://ipmworld.umn.edu, University of Minnesota, St. Paul.

McCarty, L.B., M.L. Elliott, D.E. Short, R.A. Dunn, G.W. Simone and T.E. Freeman. 1994. Integrated pest management strategies for golf courses, pp. 93–102 in *Best Management Practices for Florida Golf Courses*, McCarty, L.B. and M. L. Elliott (Eds.), University of Florida, Gainesville.

McMullen, M. 2001. Integrated pest management (IPM). URL: http://www.ag.ndsu.nodak.edu, North Dakota State University, Fargo.

Zadoks, J.C. and R.D. Schein. 1979. *Epidemiology and Plant Disease Management*. Oxford University Press, Oxford, 427 pp.

36 Plant Disease Diagnosis

Jackie M. Mullen

CHAPTER 36 CONCEPTS

- Disease diagnosis is the process of disease identification. Plants should be initially examined in the field, landscape or garden setting, and the site, plant and problem history should be determined and recorded.

- If the plant disease or other problem cannot be diagnosed in the field setting, a plant sample should be collected, packaged, and mailed so that the sample arrives at a diagnostic laboratory or clinic in a fresh condition with an adequate quantity of symptomatic tissue for examination and testing. The sample must be representative of the problem.

- Plant disease clinics or laboratories may be private, state-supported, or land-grant university supported.

- Clinics use a variety of techniques to diagnose a plant problem. Many clinics charge for services, which consist of the diagnosis and disease control recommendations.

- Initial diagnostic techniques may involve visual study, use of references, soil pH analysis, total soluble salt analysis, and microscopy. Additional specialized procedures used (including culture work, serology and molecular testing) depend on the disease suspected, the value of the crop, and the client. A diagnosis is typically based on more than one procedure.

Disease **diagnosis** is the act or process of biotic or abiotic disease identification. Disease may be biotic or abiotic. Some plant disease agents cause visible symptoms distinct enough to allow for disease identification to be made relatively quickly on the basis of a visual study of only plant appearance. For example, crown gall caused by the bacterium *Agrobacterium tumefaciens,* induces irregular galls to form on the lower portions of stems and upper sections of roots of infected plants. These galls are often distinctive and allow immediate diagnosis. However, in many situations, visual appearance of the diseased plant is not plain enough to allow for an exact diagnosis of the problem. In these cases, diagnosis depends on one or more assays or tests in addition to a visual inspection of the plant. These added studies may involve soil pH testing, soil analysis (total soluble salt and fertility and nematode assays), light and electron microscopic study of the damaged plant tissues, cultural and physiological studies of the isolated pathogen, molecular studies of the pathogen [**serology, gel electrophoresis, gas chromotography,** PCR (polymerase chain reaction), and DNA probe identification], plant tissue analysis, and soil or pathogenicity studies. The exact procedures used to diagnose a plant disease depend on the suspected disease and the plant or crop situation. The crop situation or the crop owner often dictates the level of specificity needed in the diagnosis. For a backyard garden, the identity of a corn (*Zea mays*)

leaf spot as a *Cercospora* species leaf spot may be sufficient, but to a plant breeder concerned with the leaf spot damage in a field situation, the identity of the leaf spot needs to be more specific, and the leaf spot would be identified by genus and species, such as a *Cercospora zeae-maydis* leaf spot.

The process of disease diagnosis has existed, in some form, since diseases were first recognized. The Romans were aware of rust diseases of small grains as early as 310 BC. The actual cause of rust diseases and other plant diseases was not known or objectively studied until after the development of the compound microscope in 1675. Pasteur and other early scientists in the mid 1800s dispelled the earlier belief of spontaneous generation and proved that microorganisms were present in our environment and were responsible for many diseases of plants and animals. DeBary in 1861 was the first to scientifically prove that the fungus *Phytophthora infestans* caused the disease of Irish potato called late blight, which had caused a widespread famine in Ireland in 1845 to 1846 (Agrios, 1997 — Chapter 1 and Chapter 30). Since that time many biotic plant diseases have been identified and most are caused by fungi (various chapters), bacteria (Chapter 6), viruses (Chapter 4), and nematodes (Chapter 8). The abiotic diseases (Chapter 23) are caused by such factors as temperature extremes, moisture extremes, low or high soil

pH, fertilizer excesses or deficiencies, pesticide damage, pollution effects, or weather/soil problems.

The hundred years following DeBary's first plant disease identification were very active in the area of new disease diagnosis. The procedure used to confirm the existence of a new plant disease was developed by Robert Koch in 1876 and is known as Koch's postulates or proof of pathogenicity. This procedure is still used today for the diagnosis and identification of a new (previously unreported) plant disease. The procedure involves the following four steps:

1. The pathogen must found with all symptomatic plants.
2. The pathogen must be isolated and grown in pure culture and its characteristics described. If the pathogen is a biotroph, it should be grown on another host plant and have the symptoms and signs described.
3. The pathogen from pure culture or from the test plant must be inoculated on the same species or variety as it was originally described on and it must produce the same symptoms that were seen on the originally diseased plants.
4. The pathogen must be isolated in pure culture again and its characteristics described exactly like those observed in Step 2.

This procedure was amended by Erwin F. Smith to include reisolation of the pathogen from the inoculated, symptomatic plant. If both cultures contain the same microorganism, then this organism is considered to be the pathogen or cause of the disease. Today, Koch's postulates or proof of pathogenicity procedures include the last step added by E. Smith (Agrios, 1997).

In diagnostic labs or clinics today, Koch's postulates are performed infrequently, except when the disease agent is suspected to be new and previously unreported. Most of the plant disease diagnoses done today involve identification of plant diseases that have been previously described and named. Several techniques may be performed to determine the identity or causal agent (pathogen) of a disease. These diagnostic procedures are completed by or under the direction of plant pathologist diagnosticians. These procedures or analyses may include many of the ones mentioned earlier in this chapter. PCR techniques are not routinely used in clinics, but recent advances in molecular methods, such as kits for PCR DNA hybridization, may be available for clinical use in pathogen detection and identification in the near future (Schaad et al., 2001). The PCR methodology will allow for pathogen identification from very small samples. Visual studies of symptoms and signs, microscopy, culture media studies, and serology techniques are the most frequently used techniques in diagnostic clinics. Disease diagnosis may involve one or more procedures, depending on the disease, the client, and the planting situation.

This chapter describes the basic steps involved in plant disease diagnosis from the time the abnormality is noticed in the field or landscape until the diagnosis is accomplished. The chapter sections are as follows: (1) field diagnosis and observations, (2) procedures for sample collection, packaging, and mailing, (3) the overall operation of a diagnostic clinic, and (4) diagnostic methods used in clinics — a somewhat chronological listing of methods with a brief description and discussion of each procedure. The laboratory exercises in Chapter 37 emphasize practical experience with disease diagnosis.

FIELD DIAGNOSIS

Diagnosis begins in the field or backyard. In many cases, growers or homeowners are able to diagnose their own problem by a field inspection and knowledge of the history of the field and plant problem. Knowledge of recent extreme weather conditions allow the grower to identify wilted plants as abiotic (drought conditions) or biotic diseases that are likely to develop under these conditions. Some diseases may be diagnosed by the presence of visible symptoms or signs and a previous experience may allow the grower/homeowner to recognize the malady by a field inspection only. For example, corn smut (Chapter 18), caused by *Ustilago maydis*, is a disease that can be easily recognized and diagnosed in the field by its large fleshy galls on the ears and stalks and by the black, powdery, sooty masses of spores contained inside the galls, which will crack open when mature.

Field diagnosis requires that the grower/homeowner be a good observer. Knowledge of the normal appearance of the plant and requirements for growth is paramount in the diagnosis process. Foliage may be inspected and found to have an abnormal appearance, but the inspection should proceed to include more than the foliage. Stems, fruit, and roots should be observed if possible. In many cases, a dieback of twigs and branches may relate to a stem, trunk, or root problem. Soil also should be considered as a factor that may contribute to a dieback or other foliage problem. Many plant problems involve more than one causal agent, so it is important to consider the condition of the whole plant and soil when trying to make a diagnosis. Also, consideration should be given to the site history, fertilization, lime and pesticide applications, and recent weather conditions. Not all plant problems are attributable to micropathogens (fungi, bacteria, and viruses) — consider the possibility of an insect or nematode infestation. Also, pay attention to the distribution of the disease. Biotic diseases often exhibit a scattered occurrence or incidence, whereas abiotic problems may occur uniformly in an area or in patterns of rows or edges of a field, for example.

COLLECTING, PACKAGING, AND MAILING PLANT SAMPLES

If diagnosis is not possible by inspection in the field, then it will be necessary to seek assistance. County extension agents are a good source for such assistance. They may be able to visit the planting or crop area and help arrive at a field diagnosis. If it is not possible for them to visit the site, it will be necessary to collect a plant sample. A few considerations should be given to collecting samples. First, the sample collected should be representative of the problem. If a leaf spot and fruit rot of apple (*Malus sylvestris*) are observed as the only type of abnormality present, then leaves and fruit should be collected showing early, middle, and late stages of the damage. An accurate diagnosis depends on a sample that is representative of the situation. Second, the sample should be fresh. Ideally, the specimens collected should be recently infected. Samples should be collected for diagnosis when disease development is first noticed and during initial stages of disease spread. These types of recently infected plant tissues are most easily diagnosed because the pathogen is actively developing. For example, a disease that develops in rainy weather may be dependent on those wet conditions for disease activity and spread. Collecting the sample during a later dry period may result in a sample where disease is no longer active. The symptoms may be dramatically altered, signs may not be present, and secondary decay bacteria and fungi may develop and prevent an accurate diagnosis. Collection of fresh, recently infected tissues is usually required for an accurate diagnosis. Third, the sample should be adequate in size. How large is an adequate sample? This answer varies with the sample and problem. For a leaf spot problem, usually about 20 leaves providing about 20 leaf spots are adequate for the diagnosis procedures, which may involve microscopy, culture work, and possibly other testing and study by more than one specialist. Of course, the exact size of an adequate sample will vary with the plant and disease agent. Large leaves with many spots will require fewer leaves for diagnosis than will small leaves with fewer spots. For galls and cankers, usually four or five specimens are sufficient to complete the diagnostic process. Whole plants are generally needed for all problems that are not leaf spots, galls, or cankers. In general, turfgrass samples should be 7 to 8 square inches and 3 inches deep and taken from the edge of the damaged area. About half of the sample should contain damaged plants; the other half should contain healthy plants. The pathogen is usually located at the margin of the damage area and diagnosis is often not possible if all the turfgrass in the sample is dead. For seedlings, generally 25 plants or more are needed. With larger plants, fewer plants are needed in the sample; for example, one mature tomato plant is sufficient to diagnosis most problems. Determination of tree problems not involving leaf spots, cankers, or galls can be difficult as collecting the whole plant is obviously not possible. In these cases, sections of the plant must be sampled.

Never collect dead plant samples. When plant tissues die, plant pathogens may become overrun with saprophytic microorganisms that make diagnosis impossible. Collect soil samples around damaged and healthy plants as soil pH or nutrients may play a key role in plant problems. When it is not possible to collect a whole plant, it is sometimes not possible to give an exact diagnosis. It is not unusual for samples from tree limbs showing leaf scorch to be diagnosed as "suspect the problem/disease involves the roots or lower trunk."

Once the sample is collected, it must be packaged to keep it as fresh as possible until a specialist examines it. Most samples should be placed into a plastic bag to prevent desiccation. Damp paper towels are not needed to preserve a specimen; however, a dry paper towel can be placed in the bag to absorb excessive moisture and prevent rot in transit to the laboratory. If moisture is added, additional decay and increased populations of bacteria and fungi could make diagnosis difficult. Fruits and vegetables should be packaged in several layers of newspaper instead of plastic. The newspaper allows the vegetables and fruit to have some air exchange and at the same time, samples will not dry out. Soil should be packaged separately in a plastic bag. Samples should be refrigerated as soon as possible after collections are made to help prevent secondary decay bacteria and fungi from developing on the sample.

Samples to be mailed should be placed into a padded envelope or box with ample packing material (styrofoam chips or other such material) to prevent damage during transit. Packages should be addressed clearly. Incorrectly addressed packages that stay in transit for long periods of time are usually not fresh enough for accurate diagnosis. It is always best to mail a sample during the first part of the week, rather than late in the week when there is a risk that the sample will stay in transit for the weekend. If you have any question about mailing procedures for your sample, call the clinic to ensure that the sample arrives promptly and is processed in a timely manner.

Almost every state has a plant disease clinic associated with the land-grant university college of agriculture. Usually, these clinics are associated with the plant pathology department or another unit of the agriculture college. Also, there are private labs in some states.

OVERVIEW OF THE OPERATION OF A DIAGNOSTIC CLINIC

ORGANIZATION/SUPPORTING AGENCY

Diagnostic clinics may be privately owned and entirely supported by their fee structure or they may be supported or partially supported by the state land-grant university system. Also, some diagnostic laboratories are associated with the department of agriculture in some states.

Private diagnostic clinics usually receive samples from homeowners and commercial operations. They are typically well equipped for a variety of testing services, including soil minerals and pH analysis, disease diagnosis, and nematode analysis. Fees for services vary, but generally charges are higher than at a land-grant university diagnostic laboratory.

Most state land-grant university systems support one or more plant diagnostic laboratories/clinics and are usually listed on the web page of the university. Every state has a land-grant university that offers research, teaching, and extension activities for agriculture and human sciences. Depending on the specific state, the diagnostic clinics may be listed under the extension component of the college or school of agriculture or under a specific department such as plant pathology, botany, or possibly entomology. Ruhl et al. (2001) have recently prepared a table listing all university-related diagnostic clinics.

State department of agriculture diagnostic laboratories usually accept samples only from state inspectors who examine commercial samples to certify plants free of disease (or free of specific diseases) before they are shipped out of the state. Not all states have a department of agriculture diagnostic lab. In some states, agriculture inspectors cooperate with university clinics when laboratory diagnostic procedures are needed.

SUBMITTING AND RECEIPT OF PLANT SAMPLES

Land-grant university diagnostic clinics usually receive plant samples from county extension agents, homeowners, commercial growers, consultants, golf course superintendents, business operations, and researchers. When a plant sample arrives at a diagnostic clinic, it is recorded in a logbook or in a computer database and assigned a tracking number. Many clinics use their databases for rapid information retrieval on samples. Also, annual reports are compiled with disease listings for each crop received.

PLANT DISEASE DIAGNOSES

Plant diagnostic clinics provide plant disease diagnosis of both biotic and abiotic diseases. Problems diagnosed include fungal, bacterial, viral, and nematode diseases as well as damage caused by nutritional imbalances, abnor-

mal soil pH and total soluble salts, and pesticide, environmental, mechanical and air pollution damage. Clinics may also provide soil nematode analysis, but in some states, nematode analysis is handled separately and may be administered by another department. Integrated clinics provide plant disease diagnosis, insect damage identification, and sometimes plant/weed identification and insect identification, and possibly soil nematode analysis. Integrated clinics usually are staffed by or cooperate with pathologists, entomologists, weed scientists, and possibly nematologists. Botanists, agronomists, and horticulturists may also be consulted on some samples.

In many diagnostic clinics, about 50% of the disease samples are diagnosed with biotic diseases and the remaining samples are found to have problems caused by abiotic factors. These percentages vary from clinic to clinic and from year to year. In 2000, when a severe drought occurred during the summer in most of the Southeast, about 65% of the samples submitted to clinics in the region exhibited damage caused by drought or drought-related factors. In more typical years, about 80% of the biotic diseases seen in these clinics are caused by fungal plant pathogens. Bacteria and viruses together comprise about 15% or more. Plant parasitic nematode problems account for less than 5% of the plant problems submitted for diagnosis. Plants with insect damage are also submitted to many integrated clinics. Insect damage is considered to be an injury rather than a biotic or abiotic disease problem. Some clinics have an entomologist on their staff or refer their samples to the entomologist(s) at a different location(s).

RESPONDING TO THE CLIENT

The response from the clinic to the client consists of the diagnosis and the control recommendation. The name of the disease and the name of the pathogen are provided. The name of the disease may or may not include the genus name of the pathogen. Because the disease name is usually specific for a particular pathogen, the scientific name of the pathogen is supplied to unequivocally identify the cause of the disease and avoid ambiguity. As an example, consider the disease name of "common leaf spot" of strawberry. In the U.S., most people familiar with strawberry diseases will know that "common leaf spot" is caused by the fungus *Mycosphaerella fragariae*. However, the disease name "common leaf spot" may not provide adequate information to one who is unfamiliar with strawberry diseases. The scientific name of the pathogen always identifies the disease and pathogen.

Control recommendations vary with the disease, the disease severity, and the cropping situation. Control recommendations given to a homeowner may not be practical or economical for a commercial grower. For example, removal of the infected plants and plant foliage in the fall may be the most practical and economical recommendation for control of common leaf spot of strawberry in a small garden where the leaf spotting is severe on only a few plants. For a commercial grower with a moderate amount of leaf spot throughout the planting, a regular fungicide spray program in the early spring of the following year may be the most practical and economical control recommendation.

SERVICE CHARGES

Clinics were typically begun as free services and were supported entirely by extension or university funding, or both. This situation has changed in many states during the past 15 to 20 years. Today many of the university-associated diagnostic clinics charge for their diagnostic services and fees vary from state to state. Plant disease diagnostic services may vary from $5 to $25, but some highly technical and expensive analyses, such as DNA identification, may require additional charges. Soil nematode analysis fees also vary depending on the specific state. Charges usually range between $5 and $15. When out-of-state samples are sent to clinics, higher charges are usually imposed.

DISTANCE DIAGNOSIS BY DIGITAL IMAGES

With the development of e-mail and capabilities to send images electronically, many clinics have adopted a program for receiving digital images of plant disease situations and samples from county extension agents. These programs have been designed so that transmitted images of plant damage showing the field situation, close-up views of the plant damage, and microscopic views of damaged tissues. Additionally, a digital diagnostic submission form may be transmitted from cooperating county offices or other university locations to the diagnostic clinic or pathology extension specialists. All responses from the clinic are via e-mail, shortening the diagnostic response time by 2 to 3 days. When the disease is one that may spread rapidly, a quick diagnosis and control recommendation reply will facilitate a timely implementation of control practices, which may make the difference between a large or small amount of crop lost to disease. The digital image diagnosis program allows for the images, submission form, and the diagnosis to be kept in an electronic file for future educational or reference uses.

Not all samples may be diagnosed by images. Although fungal diseases that produce visible symptoms and macroscopic or microscopic signs are ideal for this system of sending digital images for diagnosis, this method is often not suitable for bacterial and viral diseases. When microscopic structures of fungal pathogens are not present, a plant sample usually must be sent to the clinic for culture work or other types of analyses. Digital

diagnosis is also dependent on the photographic skills of the cooperator sending the images and collection of vital information about the problem.

CLINIC DIAGNOSTIC PROCEDURES: CHRONOLOGY OF INITIAL TESTS AND SPECIALIZED ANALYSES, AND DISCUSSION OF INDIVIDUAL DIAGNOSTIC PROCEDURES

Initial diagnostic procedures followed in most laboratories or clinics are fairly standard and these procedures usually follow in roughly the same order as listed:

1. Visual examination for symptoms and signs
2. Review of the information sent with the sample
3. Consultation of plant disease compendia or host indices, or both
4. Microscopic examination with stereo-dissecting microscope
5. Microscopic examination with compound microscopes
6. Possible further consultation with reference books
7. Soil pH determination

If visible symptoms and signs are distinct enough to be diagnostic of a particular disease, then the diagnosis may be completed after Procedure 1 to Procedure 3. If an experienced diagnostician is making the diagnosis and the disease is common, then consultation with compendia or host indices may not be necessary. If visible symptoms and signs are not diagnostic, then microscopy (Procedure 4 to Procedure 6) will usually follow. Usually both the stereo-dissecting and compound microscopes are used and consultation with references will usually follow unless the microscopic structures are well known to the diagnostic worker. The evidence may be sufficient for diagnosis of the disease if microscopic fungal spores are observed or bacterial cells are observed oozing out of damaged tissues.

If soil is sent with the sample, the soil pH is usually measured. Some symptoms of damage, such as scorch, dieback, or poor growth, may directly relate to pH levels that are too alkaline or too acidic. An extremely acidic soil pH may cause some minor element toxicities. The pH of soil is measured as well — inappropriate pH may predispose a plant to certain diseases.

If visible symptoms are not diagnostic, microscopic structures are not observed, and pH determination does not account for the damage, then other diagnostic testing procedures are used. The exact procedures used and the order followed in using each diagnostic procedure depends on symptoms and the possible agents that could cause those symptoms. For example, if microscopy does not unequivocally confirm a disease and if the suspected pathogen is fungal or bacterial, then culture isolations usually follow. If an isolated plant pathogen that has been associated with the symptoms is observed on the specimen, the diagnosis may be considered complete. If culture results are not conclusive, then other analyses may follow, depending on the suspected pathogen.

Diagnosis of a disease is usually based on results of more than one type of study or analysis. Many diagnoses are based on visual symptoms and possibly visual signs, accompanying information, microscopic evidence of signs, and literature descriptions of diseases. When trying to arrive at a disease diagnosis, review the following information: notes on symptoms and signs, the problem description and site/plant history, information from the host index, descriptions of possible diseases, microscopic evidence of pathogen structures, and results of other tests. The following specialized analyses are used to identify specific pathogens:

1. Measurement of electrical conductivity as a criterion of total soluble salts in a soil sample. This is very useful in determining whether excess fertilizer caused the damage.
2. Moist-chamber incubations of plant materials to help determine the presence of a fungal pathogen.
3. Culture isolation techniques aid in the identification of a fungal or bacterial pathogen.
4. ELISA or other serology methods. These techniques detect fungal, bacterial, and viral pathogens.
5. Specific physiological tests depending on bacteria suspected of causing the disease.
6. Gas chromatography of bacterial fatty acids.
7. Gel electrophoresis for identification of pathogen species or strain by specific enzyme or protein identification.
8. DNA identification by various molecular methods to identify specific pathogens.
9. Proof of pathogenicity or Koch's postulates protocol (Chapter 2) when fungal or bacterial agents are suspected to be the pathogen.

The diagnostic techniques used will vary according to the particular disease(s) suspected. However, an overall pattern and chronology is often followed with the initial six studies listed next. The results of the initial studies will help determine which of several subsequent specialized analyses may be performed. Other factors that determine the diagnostic tests conducted are the value of the crop and the grower. For a home gardener, it may not be necessary to identify the exact species of the fungus (e.g., *Alternaria*) that is causing a leaf spot on a shade tree. However, nursery growers or plant breeders may request an identification of the species of *Alternaria* on their containerized shade trees.

Visual Symptoms and Signs

Initially, the plant sample is examined for visual symptoms and signs of disease. For example, black spot (*Diplocarpon rosae*) on rose is often diagnosed on the basis of visible symptoms, as the leaf spots are distinctively irregular with a feathery margin. Cedar-apple rust causes distinctive galls to form on juniper and spore-producing structures to form on apple leaves. The presence of the symptoms and signs of this rust disease are sufficient for disease diagnosis.

Review of the Information Sent with the Sample

Every diagnostic clinic/lab has a diagnostic information sheet or questionnaire that should be completed and should accompany the plant sample. If the client does not have access to these questionnaires, then a letter should accompany the sample to describe the damage, the distribution of the damage, the history of the site, pesticides and fertilizers applied, and recent weather events. A description of the damage as it appeared when the sample was collected can be very helpful, especially if additional decay occurs in transit to the clinic. A description of the development of the problems also is helpful to the diagnostician. Site history is important in diagnosis, especially if a herbicide carryover may be causing the damage. Pesticide or fertilizer applications may be directly or indirectly involved in problem development. Excess lime, fertilizer, or pesticides may cause damage or affect disease development.

Refer to Disease Host Indices and Crop Compendia

At this point, references are often used to develop a list of possible problems or diseases, based on visual symptoms and signs with supporting information. Disease compendia and disease host indices (Farr et al., 1989) are especially helpful. As indicated previously, some diseases produce visual symptoms and signs that are diagnostic. However, most disease diagnosis requires further study before diagnosis can be completed. References provide clues of diseases or pathogens that may be involved. Additional assays or tests confirm the absence or presence of plant pathogens.

Study with a Stereomicroscope

When diagnostic structures are not visible to the naked eye, damaged plant tissue is examined with a stereomicroscope (sometimes called a dissecting microscope) at a magnification of 0.5× to 60× (Figure 36.1). Low magnification is useful for viewing fungal fruiting bodies and large fungal spores. Stereo-dissecting microscopes are also useful for finding fruiting bodies or spore masses to

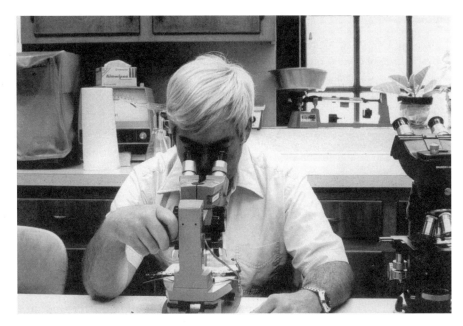

FIGURE 36.1 Examination of damaged plant tissue by a stereomicroscope.

transfer to a glass slide for examination at a higher magnification with a compound microscope.

If a fungal infection is suspected, but fungal fruiting bodies are not found, it is possible that the fungus has not yet produced any fruiting bodies or does not produce fruiting bodies at all, but rather produces spores on individual conidiophores on the plant tissue surface (Chapter 16). This tissue should be examined for the presence of spores by using a compound microscope.

Study with a Compound Microscope

A compound microscope is useful for examining fungal fruiting bodies and spores at 45× to 1000×. Many fungal diseases are identified based on spore morphology. This microscope is also used to view masses of bacteria associated with plant diseases as they ooze out of infected tissue. Occasionally, viruses may be identified with a compound microscope. Tissue may be stained with a protein stain or a nucleic acid stain, highlighting the protein or nucleic acid component of virus bodies. Virus inclusion bodies consist of aggregations of virus particles or virus products. The bodies are visible at 450× and are specific for individual virus groups, but not individual viruses (Christie and Edwardson, 1986).

Microscopic examination of plant parts and pathogen structures is an integral part of the diagnostic process at plant disease clinics. For example, Septoria leaf spot, a common disease on tomato, is usually identified by such visual symptoms as small (approximately 2.6–mm diameter) gray, circular spots with dark brown or black borders; the presence of fruiting bodies called pycnidia, which are observed in the stereomicroscope as small (100- to 200-

µm diameter), usually black, flask-shaped structures with a small hole or ostiole at the top or apex of the body and the presence of very long (about 67 by 3.2 µm), thin, filiform spores (Jones et al., 1991). The diagnosis is based on three pieces of information, two of which involve microscopic study. Sometimes microscopic study does not reveal any diagnostic fungal or bacterial structures. In these situations, additional studies/procedures are usually performed.

Consultation with Reference Books

Even if observed structures are familiar to the diagnostician, references may be checked to confirm that structures viewed are exactly the same size, shape, or arrangement and are diagnostic for a particular fungal disease. Small variations in spore morphology may indicate that a different or new fungal species is present and possibly the cause of the problem. Disease compendia are excellent references for identifications as they contain descriptions of visible symptoms and signs, microscopic structure descriptions and comments on other tests that may be necessary for diagnosis of particular common pathogens of a specific crop or crop group. However, not all crops and landscape plants are covered by the compendia series. Many ornamentals are not included and other references must be consulted along with mycology texts and fungal genus keys and monographs. Disease host indices are often very helpful in disease diagnosis. They include listings of all disease agents reported on a particular plant. Some disease and pathogen descriptions are only available in manuscripts in research journals where the disease was first described.

Soil pH Determination

If soil is sent with the plant sample, it is always a good idea to check the soil pH (level of acidity or alkalinity of the soil). If the soil pH is inappropriate for a particular plant, stress could predispose the plant to a variety of weak disease agents. Also, the stressed plant might become more susceptible to some moderately pathogenic disease agents. Leaf spot diseases may be more severe on plants weakened by soil pH that is too acidic for normal plant growth. Some root diseases may be more prevalent and damaging when plants are weakened by a number of environmental stresses, including soil that is too acidic or too alkaline. If the environmental stress problem is not corrected, the plant will continue to be highly susceptible to certain diseases and as soon as disease control treatments conclude the plant may very quickly become reinfected and develop disease symptoms again. To determine soil pH, a small amount of soil is mixed with an equal volume of filtered water, allowed to stand for about 30 min and the solution pH measured.

Total Soluble Salt Measurements Based on Soil Electrical Conductivity

An electrical conductivity meter determines total soluble salts level in a sample of soil or soilless media. Measurements are given in units of electrical conductivity, mmhos/cm. The usual procedure is to mix 40 ml of soil with 80 ml of filtered water. After mixing is complete, the soil is allowed to settle and the water solution is poured into a graduated cylinder. The electrical probe is submersed into the solution and the total soluble salt reading is recorded. Charts are available for the specific ratio of soil to water (in most cases, 1:2) and for the type of soil (sandy soil mix, heavy soil mix, and soilless mix) to give an indication of whether the levels of salt are low, moderate, acceptable, or too high for most plants.

Low readings may indicate nutrient deficiencies. Very high readings indicate that roots may have been injured by high concentrations of salts. Root injury from high salts often cause new foliage to show severe levels of leaf scorch. Sometimes new growth is entirely brown and dead as a result of root injury from salts. High salt levels kill root hairs and effectively mimic drought stress.

Moist-Chamber Incubations

Tissue incubation in a moist environment may stimulate fungi to produce spores within developing fruiting bodies or on specialized hyphae. Once spores and fruiting bodies are produced, diagnosis is possible.

Moist chambers should be set up only after visual symptoms and signs are documented because these characteristics are usually destroyed in moist chambers. Sometimes moist chambers are not helpful in the diagnostic process as secondary fungi completely overgrow the tissues, making it impossible to detect the pathogen spores and structures (Waller et al., 1998; Shurtleff and Averre, 1997). A moist chamber consists of a plastic bag or other closed container (not completely sealed from air exchange) that contains plant tissue and a moist paper towel. The goal is to provide an environment with high relative humidity without exposing the specimen to free water. High humidity favors the development of fungal and bacterial pathogens and saprophytes present on decaying tissues. Tissues in a moist chamber should be examined daily for 1 to 5 days for the development of fungal structures. Microscopic examination of tissues after incubation may reveal several types of spores of which one may be the pathogen. Therefore, it is important to be familiar with common genera of plant pathogenic and saprophytic fungi.

Isolation of the Plant Pathogen on Culture Media

When fungal disease is suspected and microscopic evidence is not present, moist chamber incubations are often used first. Culture work is used if the moist chamber technique is not successful for fungal identification or if bacterial pathogen identification is desired. Tissue used in isolations in culture media must be fresh, consisting of recently infected tissue areas bordering on healthy tissue. Tissue to be cultured is usually surface sterilized with a 10% (v/v) solution of household bleach. The duration of surface **sterilization** ranges from a few seconds to 2 min. Immediately after soaking the tissue pieces in the bleach solution, the tissue should be rinsed in sterile distilled or filtered water. Tissue should then be blotted dry (clean paper towels usually work fine for blotting). Tissue sections should be aseptically cut into small pieces (2- to 3-mm diameter) and placed into sterile culture medium in sterile petri dishes (Figure 36.2 and Figure 36.3). Dishes are usually maintained at room temperature for 3 to 7 days and examined daily for fungal growth. Potato dextrose agar acidified with lactic acid to retard bacterial growth is often used as a general-purpose medium for the culture of fungal foliage pathogens. When fresh, recently infected tissues are cultured from marginal areas of infection, cultures should produce a consistent type of fungal growth. Fungal isolates should be transferred to a sterile petri dish with sterile medium to produce a single fungal isolate in pure culture. The growth of certain fungal pathogens is very distinctive in culture, aiding in identification of the pathogen. Conclusive evidence for fungal identification is the development of distinctive fungal spores or fruiting bodies. If tissue decay is advanced, several fungi may grow out in culture; each isolate should be examined to determine which ones are plant pathogens. In some cases, secondary pathogens or decay fungi may grow faster than the pathogen and the medium in the dish is overgrown by

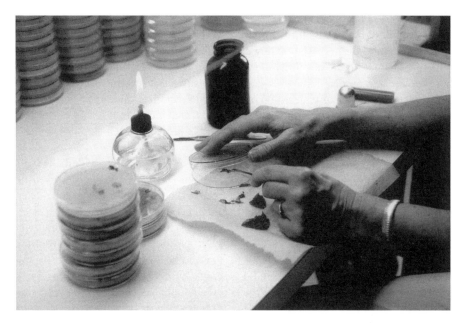

FIGURE 36.2 Surface-sterilized tissue pieces cut from the edge of symptomatic tissues are placed into a sterile culture dish containing medium growth using aseptic technique.

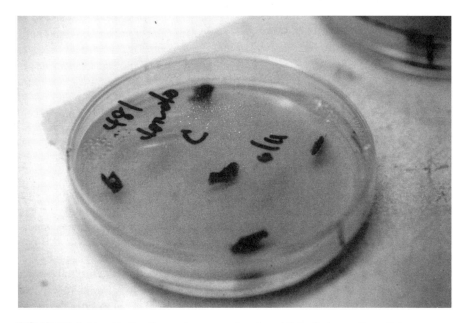

FIGURE 36.3 Usually five tissue pieces are placed equidistantly in a culture dish.

these undesirable fungi, preventing identification of the primary pathogen (Waller et al., 1998; Shurtleff and Averre, 1997; Fox, 1993).

If fungus fails to grow from the cultured plant tissues, surface sterilization may have been too severe. It is often a good idea to prepare two to four culture dishes by using a range of surface sterilization times. The optimum sterilization time varies with the pathogen, the plant tissue type, and the secondary and decay organisms present.

If a bacterial disease is suspected, a different method is used for isolation in culture. After surface sterilization

of the tissue, a piece of the damaged tissue is placed into a sterile plate and cut up or crushed in a drop of sterile filtered water. The macerated tissue remains in the water for 3 to 5 min, and then a small amount of water is streaked across a culture dish by a bacteriological transfer loop. A general-purpose medium for bacterial growth is tryptic soy agar. Most diagnostic labs will streak the bacteria in four quadrants in a prescribed manner so that the bacterial concentration is diluted by the time the last quadrant streak is streaked. The objective of the streak technique is to dilute the bacteria enough so that single cells will be separated in the last quadrant. The resulting colonies from single cells may be easily transferred to another culture dish, which is then a pure culture derived from a single bacterial cell (see Figure 7.1). Bacteria are not easily identified as they do not produce microscopic structures that are diagnostic. Instead, bacteria are typically identified by specific physiological reactions or by specific molecular characteristics. Therefore, the isolation of bacteria in pure cultures is just the first step toward other procedures to identify the bacteria by genus and species (Waller et al., 1998; Shurtleff and Averre, 1997; Fox, 1993).

Baiting

Baiting to isolate a pathogen typically involves placing a piece of infected or damaged tissue into a healthy plant part where the specific pathogen will be stimulated to grow. Usually baits involve selective stimulation of growth for certain fungal pathogens. Carrot roots are baits for the fungal pathogen *Thielaviopsis basicola,* which causes black root rot of some plants. Green apples and green pears are used as baits for growth of the fungus *Phytophthora*. If culture methods are not available, successful baiting is an alternative strategy for isolating and identifying pathogens (Fox, 1993; Shurtleff and Averre, 1997; Waller et al., 1998).

Use of Serology Techniques (Ouchterlony and ELISA)

Serology, the study of immunological reactions, derives its name from serum, which is blood fluid after all the blood cells are removed. There are several types of tests that involve immune reactions, that is, several tests that involve the reaction of **antigens** (substances foreign to the body, usually proteins) and **antibodies** (specific molecules produced by mammals in response to the presence of the foreign protein or substance). One of the first serology tests to be conducted was known as the Ouchterlony test, after the scientist who designed the procedure. In this test, the antigen–antibody reaction takes place in water agar within a petri dish. The antigen or foreign protein is placed into a well in the center of the plate. This antigen is tested against several known antibodies, which are individually placed into wells around the edge of the agar dish. The antigen and antibodies diffuse out of the wells and move through the agar medium. As these substances come in contact with each other, about midway from the center and edge of the plate, an arc of white precipitate is formed between the antigen and its antibody. There is no precipitate formed between the antigen and the other nonreacting antibodies (Chapter4; Figure 5.5).

Today the most common serology test used by most diagnostic clinics/labs is known by the acronym ELISA, enzyme linked immunosorbent assay (Chapter 4; Figure

4.2). The immune-reacting molecules (antigen and antibody) are adsorbed onto the wells of a plastic multiwell microtiter dish. One of the reacting antibodies is linked to an enzyme that allows for a color reaction to indicate a positive test reaction. The type of ELISA may be direct or indirect. The direct test is often called a double-antibody sandwich and it is the shorter procedure of the two. The indirect ELISA is a longer procedure, but the longer protocol allows for a more reactive antibody–enzyme component to combine with the antigen. The indirect test is often considered to be more sensitive than the direct method. For the purposes of this general plant pathology text, the shorter, simpler, direct ELISA will be described (Figure 36.4). The wells of a multiwell plate are coated with a specific antibody (AB). After this AB is allowed to dry and be adsorbed onto the wells, the unknown antigen (usually the sap expressed from the infected tissues) is added to the test wells. At the same time control wells are prepared by adding sap expressed from a healthy plant of the same age, variety, and location as the infected test plant (a negative, healthy plant control); sap expressed from a plant known to contain the specific antigen or pathogen that is suspected to be present in the unknown test plant (a positive control); and buffer or water only (a negative control). The antibody (AB)–antigen (AG) reaction mixture is allowed to incubate for a period of time that varies depending on the specific antigen test. After the incubation time is complete, the wells (test and controls) are washed several times (usually six times) and then allowed to drain. If the AG reacted with the AB, the bound mixture of AB–AG remains attached to the test and positive control wells. If the test plant sap did not contain the specific antigen reactive with the added AB, then no binding took place and only the AB remains attached to the wells. The next addition to the wells is the AB preparation attached to an enzyme, which is usually peroxidase or alkaline phosphatase. The AB-enzyme (AB-E) will attach to those wells that contain the AG bound to the originally added AB. This three-component mixture of AB–AG–AB-E is allowed to incubate for the recommended period of time. After the incubation time is completed, the wells are again washed several times (usually six times) to remove any unbound AB-E from wells. Next, the substrate (S) for the enzyme (hydrogen peroxide or *p*-nitrophenylphosphate, respectively, for the aforementioned enzymes) is added and allowed to react with the mixture. A color change indicates that the enzyme has bound to the substrate and is a positive test for the presence of the AB in question (Figure 36.4). If positive control wells and negative control wells react or not as they should, then the completed ELISA provides very valuable and specific information as to the identity of the pathogen causing the plant damage (Fox, 1993; Matthews, 1993; Schaad et al., 2001). ELISA results usually are ready in a matter of hours or after an overnight reaction, so the

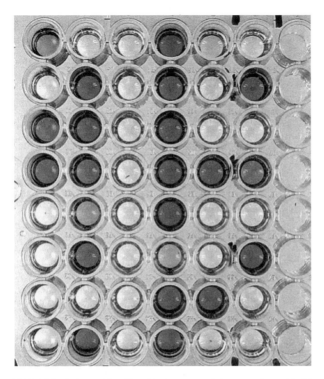

FIGURE 36.4 Multiwell strip shows positive reactions (dark) for samples tested for *Phytophthora* by the ELISA technique.

clients may obtain an indication of the cause of their problem in less than 24 h. This test procedure is very valuable for commercial growers where a rapid implementation of control measures could save the crop from widespread damage and economic loss.

Specific Physiological Testing to Determine the Identity of Bacterial Pathogens

Bacteria (Chapter 6) do not produce many characteristic microscopic structures that can be used as a tool for identification. Most plant pathogenic bacteria are short rods. Bacteria do differ in their capability for specific physiological characteristics. An initial analysis is the determination of cell wall structure by the Gram stain or by bacterial reaction with 3% potassium hydroxide. Bacteria are generally divided into two groups on the basis of their cell wall structure. Either one of these two tests will identify the bacteria as having a tight cell wall structure (Gram positive or KOH viscosity negative) or having a loose cell wall structure (Gram negative or KOH viscosity positive; Chapter 6). Many of the physiological characteristics involve the ability of bacteria to utilize specific carbohydrates or other carbon or protein molecules. Some tests determine the presence or absence of a particular enzyme capable of bringing about a particular chemical reaction. The identity of a bacterial genus may depend on the presence (or absence) of growth and the appearance of the growth on a specific agar or broth medium. Some bacterial

growth media are referred to as differential media because the bacteria may be differentiated from each other on the basis of the type of growth on the media. Yeast Dextrose Carbonate agar (YDC) and Pseudomonas Agar F (PAF) are common bacterial differential media. Xanthomonads typically grow as yellow, mucoid colonies on YDC and can be distinguished from many other types of bacteria growth on the basis of appearance. However, not all yellow colonies on YDC are xanthomonads. Pseudomonas agar F is a differential medium for the group of pseudomonad bacteria that have the ability to produce a fluorescent pigment. Fluorescence of bacterial colonies under ultraviolet light on this medium is characteristic of a group of bacteria that cause plant diseases. Many different types of differential media help categorize and identify bacterial genus and species. Prior to the development of more rapid and accurate molecular testing, bacterial identification was based on numerical taxonomy; that is, the genus groupings and species were identified on the basis of results of multiple tests. To identify a particular bacterial species, 10 to 20 different physiological tests might be necessary. Selective media are available as aids for identification of some bacteria. Depending on on the degree of selectivity of the media, bacterial species may be identified with fewer tests (Schaad et al., 2001).

BIOLOG

BIOLOG is a combination of "numerical taxonomy," molecular analysis, and the use of a computer software library program. The analysis system consists of a multiwell dish where each well is coated with a different carbon or food source. A metabolically active bacterial preparation is added to each well. Positive and negative controls must be included with each test. After 1 to 2 days incubation, the mixture in the wells changes color if a degradation (oxidation) of the substrate took place. The optical densities of the wells are measured by a spectrophotometer, and the readings are compared via a computer program where the pattern of positive reactions in the multiwell plate is compared to the known pattern for plant pathogenic bacteria that have been recorded in the software library program. Results are given as a percentage of similarity to the known bacterium most similar to the test bacterium. This analysis system is considered to be fairly accurate for genus identification and sometimes accurate for species (Fox, 1993; Schaad et al., 2001).

Gas Chromatography of Bacterial Fatty Acids to Determine the Identity of Bacterial Pathogens

A gas chromatographic system has been designed to separate and quantify the fatty acids present in a bacterial preparation (Schaad et al., 2001) and results in a fatty acid profile. Studies of the plant pathogenic bacteria and other bacteria have shown that fatty acid profiles may be characteristic and specific for individual bacterial genera, species, and pathotypes. The results of the gas chromotography are transmitted to a specially designed computer program where the fatty acid profile of the test bacterium is compared to a library of fatty acid profiles of other bacteria. Results are given as a similarity coefficient that expresses the degree or percentage of agreement between the test organism and the most similar bacterium in the library system. A very high coefficient, such as 0.925, indicates very good agreement and the probability that the identification is accurate to the genus, species, and pathotype taxa. A lower coefficient, such as 0.456, usually indicates the identification is accurate to the genus and species level. Not many diagnostic clinics perform bacterial fatty acid gas chromatography due to the expense of the equipment — some research labs will cooperate with clinics and perform identifications for diagnostic purposes. Before the analysis, the bacterial preparation in broth must be treated to disrupt or break the cell walls and release the fatty acids, which are then methylated so that they become volatile and move easily in the gas medium. The fatty acids are separated on size and identified by their rate of movement as compared to the movement rates for known fatty acids (Schaad et al., 2001).

Gel Electrophoresis Used to Identify Specific Enzymes or Proteins of Specific Pathogens

Gel electrophoresis is a method of identifying proteins or enzymes by the rate at which they move through an electric field. The rate of their movement through the gel depends on their size and their electric charge. The rate of movement of a particular protein is compared to the rate of movement for known proteins. If production of a unique protein is specific for a specific pathogen or pathogen subspecies, the electrophoresis method can be used as a diagnostic method. This method was used to identify an enzyme specific to A2 mating types of metalaxyl-resistant *Phytophthora infestans* in the early 1990s. Gel electrophoresis assisted pathologists in advising growers whether they should use metalaxyl or one of the newer, more expensive fungicides for late blight control.

The analysis involved lysing the fungal cells, extracting the proteins by centrifuging, and organic phase separations. Identification of the protein bands in the gel involved a specific staining technique. The known proteins (positive controls) and negative controls must always be included with the test protein sample (Goodwin et al., 1995).

DNA Probe Molecular Methods to Determine the Identity of Pathogens

The most accurate method of organism identification is through DNA analysis. Identification is always done by determining compatibility (or hybridization) of the test DNA single strand to the known DNA single strand. The "probe" refers to the marker attached to the known DNA single strand. The presence of the marker allows for visual identification of the double-stranded hybridized DNA. The attached probe might be radioactive P-32, biotin, a serology component of an ELISA reaction, or other markers. Basically the test involves extraction of the DNA, heating the DNA to cause separation of the double strands into single strands, addition of the known single-stranded DNA with attached probe, and cooling to facilitate reassembly (annealing) of the single-strands of DNA into double strands where the probe containing DNA is hybridized to the unknown DNA if the two DNA strands are compatible. If the unknown DNA is of the same identity as the known probe-connected DNA, then the probe will allow for detection of the double-stranded result. If the probe DNA is not compatible (does not hybridize) with the unknown DNA, then double strands will not form with the probe. The single-stranded probe DNA will be removed from the test solution. P-32 probes are detected by x-ray film and biotin is detected by a visible color reaction as is the case for an ELISA probe system (Duncan and Torrance, 1992; Matthews, 1993; Fox, 1993; Schaad et al., 2001).

PCR is not being used now in clinics, but recent advances indicate that methodology will soon be available to allow a rapid and easy procedure for polymerizing very small amounts of a DNA sample so that small plant samples may be used for the DNA probe detection method (Schaad et al., 2001).

Proof of Pathogenicity

If the organism suspected to be the pathogen has not been reported to cause the symptoms and signs observed, then it is necessary to isolate the suspected pathogen in culture and inoculate a healthy plant of the same cultivar and age and nutritional status. If the original symptoms develop and the original organism can be isolated in culture again, then pathogenicity is considered to be proven (Baudoin, 1988).

Plant disease diagnosis is the process or procedure of determining the cause of a plant disease. Initial steps in the diagnostic procedure for most diseases are very similar. Initially, the plant is examined visually for symptoms and signs, plant sample information is examined, and reference materials are consulted. Next, microscopy often follows. Moist chambers and cultures may later follow, depending on the suspected problem cause(s). When abiotic problems are suspected, some of the initial procedures given may be followed so as to help confirm the abiotic cause through the negative evidence obtained from microscopy and culture work. Once the diagnostician completes the initial diagnostic procedures, a list of possible pathogens can be formulated and then set out to prove or disprove the involvement of these disease agents. Methods used may depend on the cropping situation and client interest. An exact species identification of a pathogen may not be necessary in many situations.

REFERENCES

Agrios, G. N. 1997. *Plant Pathology.* Academic Press, New York, 635 pp.

Baudoin, A.B.A.M. 1988. Diagnosis of disease and proof of pathogenicity (Koch's postulates) in *Laboratory Exercises in Plant Pathology: An Instructional Kit,* Baudoin, A.B.A.M. (Ed.), APS Press, St. Paul, MN, 213 pp.

Christie, R.G. and J.R. Edwardson. 1986. Light microscopy techniques for detection of plant virus inclusion. *Plant Dis.,* 70: 273–279.

Duncan, J.M. and L. Torrance. 1992. *Techniques for the Rapid Detection of Plant Pathogens.* Blackwell Scientific, Cambridge, MA, 235 pp.

Farr, D.F., G.F. Bills, G.P. Chamuris and A.Y. Rossman. 1989. *Fungi on Plants and Plant Products in the United States.* APS Press, St. Paul, MN, 1252 pp.

Fox, R.T.V. 1993. *Principles of Diagnostic Techniques in Plant Pathology.* CAB International, Wallingford, U.K., 213 pp.

Goodwin, S.B., R.E. Schneider and W.E. Fry. 1995. Cellulose-acetate electrophoresis provides rapid identification of allozyme genotypes of *Phytophthora infestans. Plant Dis.* 79: 1181–1185.

Jones, J.B., J.P. Jones, R.E. Stall and T.A. Zitter (Eds). 1991. *Compendium of Tomato Diseases.* APS Press, St. Paul, 73 pp.

Matthews, R.E.F. (Ed.). 1993. *Diagnosis of Plant Virus Diseases.* CRC Press, Boca Raton, FL, 374 pp.

Ruhl, G., J. Mullen and J. Williams-Woodward. 2001. Plant problem diagnosis and plant diagnostic clinics, pp. 442–450 in *Diseases of Woody Ornamentals in Nurseries,* Jones, R. K. and D.M. Benson (Eds.), APS Press, St. Paul.

Schaad, N.W., J.B. Jones and W. Chun. (Eds.). 2001. *Laboratory Guide for Identification of Plant Pathogenic Bacteria,* 3rd ed. APS Press, St. Paul, MN, 373 pp.

Shurtleff, M.C. and C.W. Averre, III. 1997. *The Plant Disease Clinic and Field Diagnosis of Abiotic Diseases.* APS Press, St. Paul, MN, 245 pp.

Waller, J.M., B.J. Ritchie and M. Holderness. 1998. *Plant Clinic Handbook,* IMI Technical Handbooks No. 3. CAB International, Wallingford, U.K., 94 pp.

SUGGESTED READINGS

McCain, J.W. 1988. Use and care of the light microscope, in *Laboratory Exercises in Plant Pathology: An Instructional Kit,* Baudoin, A.B.A.M. (Ed.), APS Press, St. Paul, MN, 213 pp.

Nelson, B.D. 1988. Plant disease diagnosis in practice, in *Laboratory Exercises in Plant Pathology: An Instructional Kit,* Baudoin, A. B. A. M. (Ed.), APS Press, St. Paul MN, 213 pp.

37 Diagnostic Techniques and Media Preparation

Jackie M. Mullen

Plant disease diagnosis is a process that begins when a plant problem is first noticed at a field, garden, nursery, or greenhouse. A home gardener, farmer, or county agent observes the damaged plant(s) and the entire field and notes the obvious visible symptoms and signs. The distribution of the damage should be documented. Scattered or clumped distribution of the plant problem is often characteristic of a biotic disease whereas a row or obvious spray pattern is often indicative of pesticide injury or a cultural problem (Chapter 32). When entire fields or sections of fields are damaged with symptoms of uniform age, the problem is often due to an abiotic factor related to weather such as freeze, frost, or drought injury (Chapter 23). Information should be gathered on the history of the problem and site or planting area. All recent pesticide or fertilizer application dates and rates of application should be noted. Insect injury should be considered.

If the problem cannot be diagnosed at the site, then a sample must be collected, packaged, and sent for examination by a diagnostic specialist. The sample should consist of whole plants if possible, as foliage damage may relate to root and/or soil problems. Care should be taken to select plants showing a variety of damage stages or development and should be representative of the damage seen in the field. If the damage consists of leaf spots and cankers, then the sample collected should include leaves and stems with each symptom. A diagnosis is based on the quality of the sample submitted to the clinic. If the sample is not representative of the problem, then the diagnosis will be inaccurate and the control recommendations of little value. If plants are small, then many plants should be collected. For example, when bedding plant plugs or seedlings are sent for diagnosis, a flat of 1-inch tall seedlings is usually sufficient for diagnosis. If 5-inch tomato transplants are sent for diagnosis, usually five to ten plants are collected. When small shrubs are damaged, one or two whole plants are usually sufficient. Trees are not usually collected as whole plant specimens, but occasionally clinics do receive whole plants. In general, submit the portion of the tree exhibiting symptoms: leaves with leaf spots, anthracnose, rust, or mildew; branch segments with cankers or galls; etc. Roots are usually not requested unless the tree is small. Root disease is not easily diagnosed on large trees and root samples are often not requested until all other problems are eliminated. Also, send 1 to 2 pints of soil for analysis. Sending a soil sample from areas with healthy plants is very helpful for comparison.

Plant disease diagnosis may involve several procedures, depending on symptoms noted by the diagnostician. Initially, all samples are examined for symptoms and signs that usually give an indication of the causal agent. Depending on whether the suspect pathogen is fungal, bacterial, or viral, the diagnosis will proceed to the next level or block of diagnostic tests. The following three exercises are described in this chapter: diagnosis of a fungal leaf spot, preparation of culture media, and diagnosis of a root rot disease by isolating the pathogen in culture and the complete process of disease diagnosis including field observations, collection and packaging of a sample, and the performance of diagnostic techniques for diagnosis of the disease. The last exercise will require students to locate a plant problem in their home landscape. The plant and the area should be studied. The plant diagnostic questionnaire should be filled out, noting as much information as possible on the problem, its development, treatments made, recent weather, etc.

A sample representative of the problem must be collected and packaged appropriately so as to keep the sample as fresh as possible. The sample will be diagnosed with the help of the instructor. At the end of the laboratory session, students should present the disease they collected and justify their diagnosis. Diagnostic reports should be turned in to the instructor.

EXERCISES

EXPERIMENT 1. DIAGNOSIS OF FUNGAL LEAF SPOTS

Fungal leaf spot identification is accomplished by associating a fungus with the symptoms on the foliage. Fungi are usually identified by spore morphology and spore arrangement on conidiophores or by spore morphology and fruiting bodies observed in the diseased plant tissue. If spores are present on the symptomatic tissues, then diagnosis may be completed with light microscopy. If spores are not present, then a moist chamber may be prepared to stimulate spore development and maturation. After 1 to 7 days, tissues are examined for signs of sporulation. If spores are not produced in a moist chamber, then diagnosis proceeds to fungal isolation in culture. After approximately 3 to 7 days, cultures are examined microscopically to observe fungal spore development. (If possible, culture dishes should be examined daily during the 3- to 7-day period.) In addition to or instead of cultures, ELISA methods may be used to detect the presence of the pathogen protein. If the previous methods are not successful, other methods such as gel electrophoresis or DNA probe analysis can be used.

Materials

Each student or team of students will require the following materials:

- Plant samples showing leaf spots. Some common leaf spot diseases that might be used for this lab exercise are early blight on tomato, Septoria leaf spot on tomato, black spot or Cercospora leaf spot on rose, anthracnose on cucumber or other plants, Alternaria or Entomosporium leaf spot on Photinia.
- Dissecting and compound microscopes
- Microscope slides and coverslips
- Dropper bottle with water and cotton blue stain (0.5%)
- Plastic bag and paper towel

Follow the protocols listed in Procedure 37.1 to complete this experiment.

Procedure 37.1 Diagnosis of a Fungal Leaf Spot Disease by Visual Symptoms, Microscopic Signs, and a Moist-Chamber Incubation

1. Select a sample (provided by instructor) for diagnosis. Leaf spots of varying size and color should be included in the sample.
2. Study and record observations of the symptoms present. Describe the leaf spot color, shape, size, texture, etc.
3. Select some leaf spots of varying appearance and observe with the aid of a dissecting microscope. Look for evidence of small black or other colored bodies scattered over the leaf spots. Record observations.
4. If fruiting bodies are observed, cut out tissue sections with bodies and place one or two sections on a microscope slide, add a drop of water and cover

slip. Observe mount (wet mount) for presence of fruiting bodies and spores.
5. Make a second wet mount after attempting to cut through some of the fruiting bodies. Again, examine slide wet mounts for presence of spores in fruiting bodies.
6. If fruiting bodies are not observed with the dissecting microscope, make a slide mount of two sections of leaf spots. Also prepare two tape mounts of the leaf spot tissue by applying the sticky side of cellophane tape to the leaf spot. Remove the tape and place it onto a microscope slide containing a drop of water. (The tape is used like a cover slip.) Observe the tape mount at 100× and 450× for the presence of spores.
7. Take a few leaves with leaf spots and place them onto a plastic bag containing a damp paper towel. Incubate for 7 days at room temperature. Check the leaves for the presence of additional fruiting bodies and spores.
8. Refer to the *Illustrated Genera of Imperfect Fungi*, 4th ed. (Barnett and Hunter, 1986) to identify the fungus. Consult host disease indices and compendia of crop diseases to determine whether the fungus has been associated with a previously reported leaf spot disease.
9. Make the diagnosis.

Anticipated Results

Fungal fruiting bodies with spores or spores on conidiophores not within a fruiting body should be observed at 100× and 450× with the compound microscope. The fungus should be identified by spore morphology, fruiting bodies, or arrangement of spores on conidiophores. The moist chamber should stimulate sporulation of the causal fungus. After referring to a disease host index and a compendium or similar reference, determine whether the symptoms and signs observed have been previously described on the host selected. Comments should be made on the diagnosis of the problem.

Questions/Activities

- Distinguish between recently developed leaf spots and old leaf spots.
- How many types of spores did you see when examining the tissues microscopically before moist chamber incubation and after incubation? Why this difference? Draw the spores you observed.
- When you see more than one spore type, how can you determine which one is a plant pathogen?

- When would a tape mount be more effective than a wet mount?
- Why should the whole specimen not be placed in a moist chamber?

EXPERIMENT 2. CULTURE MEDIA PREPARATION AND ITS USE IN DISEASE DIAGNOSIS

Isolation of fungi and bacteria in culture is an important component of diagnosis of many diseases. Diagnostic clinics often use several culture media depending on the pathogen to be isolated. Lists of culture media for isolation of specific pathogens or pathogen groups are available in several references. A general-purpose medium is used for isolations when the exact fungal or bacterial pathogen is unknown. Potato dextrose agar and nutrient agar or tryptic soy agar are most often used as general-purpose media to isolate fungal and bacterial pathogens, respectively. These media may be prepared from individual ingredients as they were initially formulated or they may be purchased pre-mixed in a dry form. A commercially available medium is measured according to directions on the container and mixed with filtered water. The medium is then sterilized by heat at 121°C under pressure for 15 to 20 minutes so it does not boil. Usually the medium is sterilized in an **autoclave**, but sterilization may be performed in a pressure cooker. The high temperature will kill all bacteria and fungi and will inactivate viruses. When the medium cools to approximately 50°C, it can be poured into sterile petri dishes. **Aseptic** technique should be used when pouring media into petri dishes to prevent contamination from airborne bacteria or fungi. Once the poured dishes are cool and the medium has solidified, the culture dishes may be used for isolation of pathogens from plant tissues.

Plant tissue for fungal isolations should be cut from the edge or margin of the damaged area. Four to five pieces of tissue should be surface sterilized, rinsed with sterile water, blotted with clean paper towel, and placed equidistantly from each other on the medium in the dish. Again, aseptic technique must be used, and care taken to sterilize the scalpel or other instruments used to move the small tissue pieces into the petri dish. A 10% solution of household bleach is often used for surface sterilization. The time allotted depends on the tissue. Thin, fragile tissue is surfaced sterilized for 0.5 to 1.0 min, whereas thicker tissue is sterilized for 1 to 2 min. In many situations, several sterilization times are used to see which time produces the best results.

When preparing bacterial isolations, whole leaf spots or tissue cut from the edge of the damaged areas may be used for culture purposes. Surface sterilization of this tissue is similar to the technique used with fungal diseases. After surface sterilization and rinsing, the leaf spots or other damaged tissues with some marginal area are cut up or crushed by a sterile technique in a sterile water droplet and allowed to stand for 5 min (or more) to allow bacteria to ooze out into the water. A sterile bacterial loop is used to lift up a small quantity of water containing bacteria and the suspension streaked over the surface of a general-purpose bacterial medium such as tryptic soy agar. Most clinics streak bacteria in a four-quadrant dish to dilute the bacteria concentration so that single cells may be isolated in the last quadrant streak. Colonies from single cells may be transferred to sterile culture dishes for future studies.

In this experiment, students will prepare potato dextrose agar, pour culture dishes with the acidified agar, and prepare tissue isolations of infected roots using the prepared PDAa (acidified potato dextrose agar) medium in order to diagnose the cause of a root rot disease of garden bean. With many fungal root rot diseases, symptoms are not distinct enough to allow for a symptom-based diagnosis. In some situations, the causal agent may be observed microscopically with a compound microscope. When diagnostic fungal structures are not present, root segments must be placed into a culture medium to allow for growth and identification of the fungal root pathogen. Some fungal root pathogens are isolated with specialized culture media that contain antibiotics. Diseased root tissue may be placed on three to four types of media, each specific for different fungal pathogens.

Materials

Each pair of students will require the following materials:

- Potato dextrose agar, commercially prepared and dehydrated
- 250-ml Erlenmeyer flask
- Balance with weighing paper or cups
- Magnetic stir plate
- Autoclave or pressure cooker
- 10-cm diameter petri dishes
- 25% lactic acid in a dropper bottle
- Alcohol burner
- Plants with root rot disease
- Dissecting and compound microscope

Follow the protocols listed in Procedure 37.2 to complete this experiment.

Procedure 37.2 Preparation of Acidified Potato Dextrose Agar and its Use in Root Disease Diagnosis

1. Prepare potato dextrose agar, following directions on the container. Weigh out enough powder to prepare 150 ml of medium. Place powder into a 250-ml Erlenmeyer flask and add 150 ml of distilled water.

2. Add a magnetic stir bar to the flask of medium and place on a stirrer unit for 3 min to disperse media clumps and dissolve powder in the water.

3. Place nonabsorbent cotton as a plug at the top of the flask and cover with aluminum foil and autoclave (or pressure cook) medium at 121°C for 20 min.

4. Allow the unit to cool while slowly reducing the pressure. When pressure is reduced to normal and the internal temperature of the unit is below 100°C, remove the flask and allow to cool to about 50°C. Add 5 to 7 drops of 25% lactic acid and swirl the contents of the flask to mix.

5. Pour the medium into ten sterile 10-cm diameter petri dishes by sterile technique in a laminar flow hood. If a hood is not available, select a clean table surface away from traffic flow and air currents. Wipe the table with 70% ethanol or other disinfectant and allow to dry. Begin by tilting the erlenmeyer flask at an angle and removing the aluminum foil and cotton plug. Flame the top of the flask opening and allow it to cool. While keeping the flask tilted, pour 15 ml of medium (about half the depth of the petri dish bottom) into each of five dishes. When pouring medium, hold the dish lid at an angle so that there is very little opportunity for microorganisms in the air to fall into the dish. Reflame the mouth of the flask and to allow to cool. Pour the remainder of the medium into the other five petri dishes. Allow the medium in dishes to cool to room temperature.

6. Observe plants with root decay. Record symptoms and damage.

7. With the dissecting microscope, observe the rotted roots for mycelia and/or spore masses.

8. Cut roots into small segments (1 cm) and place on a microscope slide. Prepare a wet mount. Observe tissues at 100× and 450× magnification. Make notes of your observations.

9. Select some small root pieces that show slight discoloration and some that show a darker discoloration.

10. Surface sterilize these small roots with a 10% solution of household bleach for 1 min. Rinse with sterile water and blot dry with a clean paper towel.

11. Cut the surface sterilized roots showing slightly discolored (SD) and moderately discolored (MD) regions into smaller lengths using sterile technique. Place three pieces of small roots in a clump onto one location on the PDAa dish by an aseptic technique. When opening the petri dish, crack the cover of the dish so as to just allow space for placement of the root pieces. Do not take the lid completely off the bottom of the petri dish. Repeat this procedure placing three or four more root clumps onto the agar surface, spacing tissues equidistant from each other. Prepare two PDAa isolation dishes and label each with your name and date. Label one dish "SD" and the other "MD."

12. Incubate the tissue isolation dishes for 7 days at room temperature and observe fungal growth. Record observations.

13. Prepare wet mounts of the fungal growth and observe at 100× and 450×. Record observations.

14. Consult with your instructor if you need assistance identifying fungal isolate(s).

15. Refer to disease host indexes and compendia of crop diseases to determine whether the fungi isolated have been associated with a previously reported root rot disease.

16. Report your conclusions.

Anticipated Results

After 3 days, fungal growth should be visible in culture dishes. After two or three more days, fungal growth should be examined using dissecting and compound microscopes. Identify the fungus isolated by using references provided (host indexes, crop disease compendia, etc.) Fungal pathogens should be consistently isolated from root segments.

Questions

- What would you conclude if fungal growth originated from all over the dish rather than from tissue placed into the dish?
- What would you conclude if no fungal or bacterial growth occurred in culture isolation plates?
- What is your conclusion if several fungi grew out of isolations in your culture plates?
- Most fungal root pathogens are identified by the characteristic spores they produce. What is one fungal root pathogen that is identified by its characteristic hyphae? Does identification of this fungus in culture require more or less time than the other fungal root pathogens that are identified by spore characteristics? Explain your answer.

EXPERIMENT 3. PLANT DISEASE DIAGNOSIS: THE COMPLETE PROCESS OF FIELD OBSERVATIONS, SAMPLE COLLECTION, SAMPLE PACKAGING, AND LABORATORY DIAGNOSIS

Successful diagnosis of plant problems often depends on observation of symptoms, history of the problem, and the site and knowledge of recent fertilizer, pesticide, or other chemical applications. The grower should record all this information (see the diagnostic questionnaire). Collect and package samples so as to provide representative, fresh samples of adequate quantity. If possible, collect whole plants with disease symptoms in early, middle, and late stages of development.

Materials

Each student will require the following materials:

- Plants with damage. Herbaceous plants such as bedding plants or vegetables are recommended as samples, because whole plants can usually be collected. Also, plants showing leaf spot, blight, or crown or root lesions are suggested. Some common diseases that might be used for this exercise are anthracnose of pansies; *Sclerotium rolfsii* — crown rot of pepper or other plants; *Fusarium* or *Rhizoctonia* — lower stem rot of bean; Alternaria leaf spot of zinnia, marigold, cabbage, or tomato; and Botrytis blight of rose or poinsettia.
- Diagnostic questionnaire
- Plastic bags and twist ties
- Stereomicroscope and compound microscopes
- Microscope slides and coverslips
- Plastic bags and paper towels
- Two 10-cm diameter petri dishes of acidified potato dextrose agar

Follow the protocols listed in Procedure 37.3 to complete this experiment.

Procedure 37.3 Plant Disease Diagnosis: The Complete Process of Plant and Site Observations, Information Gathering, Sample Collection, Sample Packaging, and Laboratory Diagnosis

1. Inspect a garden, field, or landscape and locate damaged or diseased plants. For purposes of this exercise, fungal leaf spot diseases are preferred, but some lower stem (crown) or root rot diseases may also be successfully incorporated into the experiment.
2. Thoroughly inspect the plant(s) for symptoms and signs. Make notes about the distribution of the disease — isolated, scattered, or the presence of a pattern.
3. Obtain information on the site and problem history. Determine dates and rates of fertilizers, lime, and pesticide applications and note recent weather conditions in the area. Complete a plant diagnostic questionnaire.
4. Collect a sample of the problem. Collect whole plants if possible. If plants are small, collect several and if plants are 2-ft tall or more, one or two plants are sufficient. If it is not possible to collect a whole plant, collect several pieces of the plant showing varying stages of plant damage and symptom development. Do not collect dead plants and ideally damage should be recent. The sample should be representative of the problem, and the sample size must be large enough to subdivide for two to four different analyses and examinations.
5. Package the sample in plastic or several layers of newspaper as soon as possible after collection and place into a refrigerator or cool ice chest. Most samples may be placed into a plastic bag. Fruits and vegetables keep best if wrapped in newspaper.
6. In the laboratory, make notes of the appearance and disease symptoms and signs of the sample.
7. Observe the damaged tissues for pathogen signs with dissecting and compound microscopes. If diagnostic structures are not seen with microscopy, prepare a moist chamber and culture isolations (see Procedure 37.2).
8. After incubating the specimen in moist chambers and culture dishes for 7 days, examine the plant material/cultures for mycelia, spores, and fruiting bodies. Consult references to identify fungal structures and fungi and to determine whether the disease can be confirmed by symptoms and microscopic structures.
9. Complete a diagnostic report form and prepare a report including all notes on plant problem, appearance in field, pertinent site/plant information, symptoms and signs, and microscopic fungal structures observed in moist chambers or cultures. Conclude the report with a disease identification citing key diagnostic criteria.

Anticipated Results

Most common plant diseases caused by fungi are identified by symptoms and signs observed with dissecting and compound microscopes. Many leaf spot diseases, powdery mildews, downy mildews, and rusts can be confirmed via light microscopy. Acidified PDA is very useful for isolating fungal pathogens from root and leaf tissue. If a fungal pathogen is the causal agent of a leaf spot or blight, it is usually consistently isolated from diseased tissue. Obligate parasites such as powdery and downy mildews will not grow on culture media used to isolate most fungal pathogens. Moist chambers are helpful for stimulating sporulation of most fungal foliage pathogens. However, saprophytic fungi, which often grow rapidly and sporulate profusely in moist chambers, may make identifying the causal agent difficult. Host indexes and disease compendia are very helpful for information on potential pathogens, symptomatology, and conditions that favor disease development.

Questions

- How would you prove or disprove a diagnosis of pesticide spray damage?
- How could recent weather have affected the disease and diagnosis?
- What would you conclude if you consistently isolated a *Colletotrichum* species from a leaf spot on cucumber, but the leaf spot symptoms did not resemble symptoms described for this disease?

REFERENCE

Barnett, H.L. and B.B. Hunter. 1986. *Illustrated Genera of Imperfect Fungi,* 4th ed. Burgess, Minneapolis, MN, 218 pp.

SUGGESTED READING

Baudoin, A. B. A. M. (Ed.). 1988. *Laboratory Exercises in Plant Pathology: An Instructional Kit.* APS Press, St. Paul, NM.

Part VI

Special Topic

38 In Vitro Plant Pathology

Subramanian Jayasankar and Dennis J. Gray

CHAPTER 38 CONCEPTS

- Plants can be regenerated from single cells. Thus, a large number of single cells with potential to become plants can be grown in a container and subjected to intense screening with appropriate selection agents. Surviving cells are regenerated to produce resistant phenotypes. This process is called in vitro selection.

- To determine whether resistance is actually induced immediately after selection process, a number of small tests that use the surviving cell or callus culture and the selection agent (or the organism that produces the selection agent) can be performed under laboratory conditions. These bioassays are very effective especially when working with plants with long regeneration cycles.

- When plant cells are subjected to intensive in vitro selection, a number of defense genes are induced; however, only those cells whose defense mechanism is activated quickly and in a sustained manner remain viable and survive recurrent in vitro selection. Plants regenerated from such cells exhibit certain native defense genes in a constitutive manner and thus they are more resistant than their parental material.

- This approach of activating innate resistance in the plant by in vitro selection constitutes a viable, noncontroversial biotechnological approach of generating resistant plants.

In vitro culture and selection of plant cells and tissues has been used effectively as a tool for developing novel, disease-resistant genotypes. The first demonstration of this technique was to produce wildfire-resistant tobacco plants through in vitro selection by using methionine sulfoximine, a structural analog of wildfire toxin. Since then, plant cells have been successfully selected against an array of pathogenic microorganisms and regenerated into plants with enhanced disease resistance. In addition, plant cell and tissue culture has become an important tool in the study of plant–pathogen interactions at the cellular and molecular levels. Plant cells react to certain biotic and abiotic stresses in a manner similar to that of an intact plant. This makes such cell cultures ideal candidates to understand the resistance responses and changes occurring at the cellular and subcellular levels, when infected with pathogenic organisms (Jayasankar, 2000). Some of the common responses that are well documented include changes in permeability of plasma membrane and triggering the synthesis of new biochemical compounds, especially defense-related enzymes (such as the pathogenesis-related proteins; see Chapter 28). Plant cell cultures provide an ideal population of homogeneous genetic material. A single flask of embryogenic cell suspension culture theoretically represents millions of plants that can be effectively screened. For instance, a suspension culture of grape proembryogenic masses (Figure 38.1) contains enough totipotent cells to regenerate plants for hundreds of acres of vineyard. Furthermore, the culture is very useful to perform a number of genetic tests such as **bioassays** and can also be useful to culture biotrophic pathogens. This chapter will address how in vitro culture can be best used to study and understand plant pathogen interactions.

DEVELOPING DISEASE-RESISTANT GENOTYPES OF CROP PLANTS

Although cell cultures of most species are homogeneous for a specific trait, subtle differences at the subcellular level do exist among individuals in the population. To isolate such cells with slight genetic differences out of a huge population, we have to devise a method to select or screen the population with a selection agent. Sometimes such selection agents can also cause minor but specific mutations, thereby altering their genetic makeup. Plants regenerated from such mutant cells exhibit altered phenotypes. Among the most common selection agents that are used to generate disease-resistant phenotypes are metabolites such as **phytotoxin** produced by the pathogen, crude culture filtrate, and, rarely, the pathogen itself (Daub, 1984, 1986). Table 38.1 summarizes the important work done in this field.

FIGURE 38.1 Suspension culture of grapevine (*Vitis vinifera*). These suspension cultures contain highly embryogenic cell aggregates called proembryogenic masses, which are ideal for in vitro selection against plant pathogens.

The use of phytotoxins as selection agents has received the most attention mainly because of the ease in exposing a population of cells to controlled, sublethal doses of the isolated phytotoxin. In addition, toxin selection is easier to handle than pathogen selection, because it is a purified chemical compound that can be incorporated into the culture medium at measurable doses. In several instances, cells that were selected against the toxin retained significant levels of resistance to the toxin as well the pathogen when regenerated into plants. In a few studies, such resistance following selection against the toxin was transmitted to the progeny as well. However, this approach has a number of pitfalls as well. First, toxin selection has worked well only against host-specific toxins such as the HMT-toxin produced by *Cochliobolus heterostrophus* race T (syn. *Helminthosporium maydis*). Second, there are only a handful of characterized phytotoxins, whereas the number of pathogenic microbes is vast. Because selection is targeted against a particular pathogen, the regenerated plants often exhibit resistance only to the particular pathogen in question. When these plants are planted in the field, they are susceptible to other pathogens, thus making evaluation very difficult.

These pitfalls perhaps can be overcome by the use of crude culture filtrates, instead of purified phytotoxin,

obtained by growing the pathogen in nutrient broth solutions. Culture filtrates of both pathogenic bacteria and fungi have been successfully used to establish resistant cultures and regenerate resistant plants (Table 38.1). The use of culture filtrates, though not considered the best approach by many (Daub, 1986), has its own advantages. The culture filtrate approach more closely approximates aspects of the host–pathogen interaction than toxin selection, because culture filtrates have the entire spectrum of compounds (including the toxin) produced by the pathogen. In addition, recent studies have shown that some compounds (e.g., proteins) produced by pathogens in culture that are not a critical factor in the disease process are often implicated in evoking a general disease-resistance response in the host plant (Strobel et al., 1996). Similarly, culture filtrate selection has also evoked broad-spectrum resistance in grapevine (Jayasankar et al., 2000).

Whether it is purified toxin or crude culture filtrate, the mode of selection is very important. In some studies, solid medium (medium that was gelled with agar or similar compounds) was used for effecting such selections, incorporating measured doses of toxin or culture filtrate in the medium. Plants regenerated from such selection schemes quite often showed epigenetic responses that faded away with time. The probable reason is that in a

TABLE 38.1
In Vitro Selection for Disease Resistance

Crop	Pathogen	Selection Agent	Results
Alfalfa (*Medicago sativa*)	*Colletotrichum gloeosporioides*	Culture filtrate	Enhanced disease resistance
Asparagus (*Asparagus offcinalis*)	*Fusarium oxysporium* f. species *asparagi*	Pathogen inoculation	Increased disease resistance
Barley (*Avena sativa*)	*Helminthosporium sativum*	Partially purified culture filtrate	Plants with increased disease resistance and transmitted to progeny
Chinese cabbage (*Brassica campestris* species *pekinensis*)	*Erwinia carotovora*	Culture filtrate combined with UV irradiation	Plants with increased disease resistance and transmitted to progeny
Coffee (*Coffea arabica*)	*Colletotrichum kahawae*	Partially purified culture filtrate	Plants with increased resistance
Eggplant (*Solanum melangena*)	*Verticillium dahliae*	Culture filtrate	Increased disease resistance
Grapevine (*Vitis vinifera*)	*Elsinoe ampelina*	Culture filtrate	Plants with increased resistance
Grapevine (*Vitis vinifera*)	*Colletotrichum gloeosporioides*	Culture filtrate	Plants with increased resistance
Maize (*Zea mays*)	*Helminthosporium maydis*	T-Toxin	Phytotoxin-resistant cells and plants
Mango (*Mangifera indica*)	*Colletotrichum gloeosporioides*	Colletotrichin and culture filtrate	Resistant embryogenic cultures
Peach (*Prunus persica*)	*Xanthomonas campestris* pv. *pruni*	Culture filtrate	Resistant clones to the disease
Potato (*Solanum tuberosum*)	*Phytophthora infestans*	Culture filtrate	Phytotoxin-resistant plants
Rice (*Oryza sativa*)	*Xanthomonas oryzae*	Culture filtrate	Filtrate-resistant plants
Strawberry (*Fragaria* species)	*Fusarium oxysporum* f. species *fragariae*	Fusaric acid	Resistant shoots
Sugarcane (*Saccharum officinarum*)	*Helminthosporium sacchari*	Culture filtrate	Disease-resistant clones
Tobacco (*Nicotiana tabaccum*)	*Pseudomonas tabaci*	Methionine sulfoximine	Phytotoxin-resistant plants
Tomato (*Lycopersicon esculentum*)	*Fusarium oxysporum*	Fusaric acid	Plants with elevated resistance

solid medium, the cultured explants are not in full contact with the medium and often a gradient is established from the top to bottom of the explant (Litz and Gray, 1992). This results in **acclimatization** of those cells at the top of the gradient to the selection agent and the plants regenerated from such cells are generally false positives, exhibiting **epigenetic** resistance. It is possible to circumvent this gradient factor by plating and growing the cells as a thin layer, but several plant species require a mass of small cells rather than single cells for optimum regeneration. In a suspension culture system of selection, cell masses are completely and rapidly immersed in toxin-containing medium. This not only helps to avoid the gradient factor, but the cell masses do not have sufficient time to elicit an epigenetic response. Hence, suspension cultures are better for such selections.

PROBLEMS ENCOUNTERED IN SELECTION

Selection against toxin or culture filtrate is not possible under certain circumstances and against certain diseases. To give a few examples, it is very difficult to select against fungal pathogens that cause powdery mildew (Chapter 14), because none of these fungi grow in culture. Some bacterial pathogens (Chapter 6), such as the xylem-limited bacteria (e.g., *Xylella fastidiosa* of grapevine that causes the Pierce's disease), kill the plant by physical means rather than chemical action. These bacteria are so restricted in their growth that isolation and culture of these bacteria and addressing resistance through selection is often futile. To date, successful selection has not ever been carried out against viral diseases, because the only way possible to select against virus is to grow the pathogen in its hosts and look for resistance. To impart resistance under such conditions, one has to use gene or genome transfer techniques.

BIOASSAYS

In vitro bioassays are very useful tools to determine the level of resistance in a breeding program or for screening a population for sensitivity to a pathogen or pathogen-derived metabolites. Assays that use intact plants (seedlings) or plant parts (leaves or shoots) against pathogens or metabolites are being used routinely for screening purposes. More recently, **callus** or cell cultures are also used extensively for such screening. Barring a very few exceptions, in vitro response exhibits a direct correlation to *in planta* response. Hence, these assays, when combined with a selection program (as described previously in this chapter) are very effective, especially in perennial crops where the regeneration cycles are long. Another added advantage of using cell or callus cultures for screening against pathogen/pathogen-derived metabolites is that they facilitate studying the underlying subcellular mechanisms in those interactions (detailed in a later section of this chapter).

Germination of seeds in a toxin solution is a very easy and common method of testing phytotoxicity. Both host-specific and nonspecific toxins can be assayed in this manner. Typically, a wide range of concentration is used to determine the LD 50 levels and sublethal doses of toxin, based on germination inhibition. Additional parameters such as inhibition of root growth (especially for soilborne pathogen), malformation of cotyledons, and chlorosis in the emerging leaves of intact plants can also be used to determine toxicity. The most common way to test a live pathogen is to spray a measured quantity of spore or bacterial suspension on clean leaves and let it incubate under ideal conditions for 24 to 72 h. Alternatively, a drop of spore or bacterial suspension can be placed over the leaf lamina, which, after sufficient soaking, is incubated and observed for disease development. In some cases, vacuum-infiltrate or a prick with a needle may be necessary to facilitate disease development. It is very important to include appropriate controls in these studies. The easiest method is by soaking the same leaf or on a similar leaf from the same plant that is used for pathogen infection in water or buffer in which the spores were suspended. These tests are also equally effective when a pathogen-derived metabolite is tested (see Jayasankar et al., 1999 for details). For assaying "wilt-toxins" (toxins involved in wilt diseases) or pathogens that cause wilt diseases, the best method is to place a live and clean cutting of the plant into a solution containing the metabolite or the pathogen for a predetermined time and subsequently incubating the cuttings in sterile water for disease symptom development. Parameters such as inhibition of shoot growth, chlorosis, or necrosis of leaves and time required for the plant organ to wilt as compared to the appropriate untreated control will serve to determine the resistance/susceptibility of the plant in question.

Callus cultures are routinely used for in vitro screening of multiple genotypes to assess their sensitivity to a pathogen. For instance, Nyange et al. (1995) screened nine genotypes of coffee against *Colletotrichium kahawae*, the fungal pathogen causing coffee berry disease. Early studies (before plant tissue culture was very common) involving plant tissue cultures aimed only at assessing the effect of phytotoxin. However, our current knowledge of plant tissue culture and plant–pathogen interactions at the molecular level has provided a powerful tool in understanding the disease process. A list of crop plants where tissue culture-based assays were used to study plant–pathogen interactions is furnished in Table 38.2. An effective variant among the bioassays based on tissue culture is the dual culture assay. Other screening procedures are based on the growth of pathogen, whereas dual culture assays are based on inhibition of the pathogen by plant cell culture (Figure 38.2). This assay is an effective tool where the objective is to induce disease resistance using

TABLE 38.2
In Vitro Culture Based Bioassays Involving Plant–Pathogen Interaction

Plant Species	Pathogen	Type of Bioassay	Purpose
Alfalfa (*Medicago sativa*)	*Verticillium albo-atrum*	Viability staining of cell cultures	Evaluation of phytotoxicity of culture filtrate
Alfalfa (*Medicago sativa*)	*Fusarium* species	Intact seedlings in hydroponic culture containing culture filtrate	Evaluation of phytotoxicity of culture filtrate
Coffee (*Coffea arabica*)	*Colletotrichum kahawae*	Growth of fungus on callus cultures	Screening of genotypes for resistance
Dogwood (*Cornus florida*)	*Discula destructiva*	Growth of callus on medium containing culture filtrate	Screening of genotypes for resistance
Grapevine (*Vitis vinifera*)	*Elsinoe ampelina*	Dual culture using proembryogenic mass	Assessing resistance after in vitro selection
Lemon (*Citrus limon*)	*Phoma tracheiphila*	Dual culture using callus	Assessing resistance after in vitro selection
Mango (*Mangifera indica*)	*Colletotrichum gloeosporioides*	Dual culture using proembryogenic mass	Assessing resistance after in vitro selection
Pine (*Pinus taeda*)	*Cronartium quercuum* f. species *fusiforme*	Inoculation of embryos with live fungus	Assessing in vitro resistance of embryos
Soybean (*Glycine max*)	*Fusarium solani* f. species *glycines*	Viability staining of cell cultures; stem cutting assay	Evaluation of phytotoxicity of culture filtrate
Tomato (*Lycopersicon esculentum*)	*Alternaria alternata* f. species *lycopersici*	Detached leaf bioassay	Evaluation of phytotoxicity of toxin
Wheat (*Triticum aestivum*)	*Microdochium nivale*	Detached leaf and young seedlings	Screening of cultivars for resistance

in vitro selection, as shown in some perennial crops such as lemon, grapevine, and mango (Jayasankar et al., 2000).

MOLECULAR STUDIES

In vitro plant–pathogen interaction studies provide an ideally controlled environment in which to study events occurring at the molecular level. Synthesis and accumulation of defense-related proteins have been observed following a pathogen infection, and in several species genes encoding these proteins have been identified, cloned, and characterized. Such studies lead to the identification of an important group of proteins, termed **pathogenesis-related proteins (PR proteins)** (Linthorst, 1991). Examples of PR proteins include chitinases, glucanases, osmotin, and thaumatin-like proteins. To date, several PR-proteins have

been identified and grouped into 11 "families" based on their serological affinity.

One of the earliest responses that plant cells exhibit following pathogen attack is the rapid increase in reactive oxygen species (ROS), which is also referred to as the oxidative burst (Chapter 28). This initial oxidative burst can be detected within a few minutes after infection occurs regardless of the host's resistance or susceptibility. In a resistance reaction, a second oxidative burst develops a few hours later and is sustained for longer periods. Such elevated levels of ROS can directly or indirectly inhibit the invading pathogen and also serve as intermediates in the activation of other defense responses (Baker and Orlandi, 1995).

Several days after these initial responses, other defense mechanisms are activated. Among these are the **hypersensitive responses (HR)** that result in the rapid

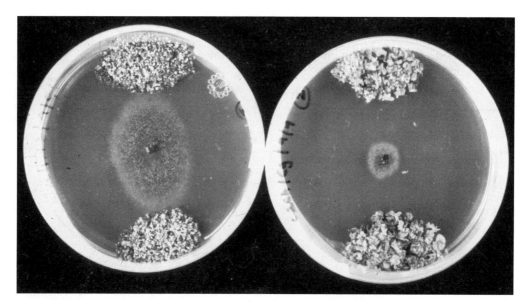

FIGURE 38.2 Dual culture assay. Embryogenic cultures of grapevine that were selected *in vitro* with culture filtrate of *Colletotrichum gloeosporioides* and tested for resistance against the fungus. Fungal mycelia grew uninhibitedly in the dishes containing nonselected control (left), whereas their growth was severely inhibited by the *in vitro* selected cultures (right).

and localized death of a small group of plant cells surrounding the site of pathogen infection in a programmed manner. As this programmed cell death proceeds around the infection site, the invading pathogen is killed, thereby preventing any further spread of necrosis (Hammond-Kosack and Jones, 1996). As these cells prepare to die, they also set forth a series of other defense responses, which include the stimulation of genes encoding PR proteins. These responses culminating in the expression of PR-proteins are also seen in other noninfected tissues of the plant and are termed **systemic acquired resistance (SAR)** (Ward et al., 1991).

In the past ten years, SAR has become one of the most widely researched areas in plant–pathogen interactions. These studies lead to the finding of several interesting secondary signals such as ethylene, jasmonates, and salicylic acid (SA) that have crucial roles in mediating the induction of plant defenses. Although it has been over 20 years since White (1979) showed that application of SA or (acetyl) salicylic acid (aspirin) to tobacco leaves increased its resistance to TMV infection, molecular studies elucidating the role of SA in plant defense have been done only recently. SA has been shown to be an endogenous plant signal that has a central role in plant defense against a variety of pathogens including viruses (Dempsey et al., 1999). SA treatment also triggers the same set of PR-proteins that are expressed as a result of SAR. Plants treated with SA and those exhibiting SAR have heightened resistance to virus as well, whereas none of the PR pro-

teins themselves have been shown to exhibit antiviral activity.

CONCLUSION

It is clear that in vitro plant pathology has come a long way over the past 20 years or so. What was once designed to study the plant–pathogen interaction in a controlled environment has slowly evolved into a decisive tool in generating disease-resistant crop plants and, in turn, greatly enhanced our understanding of these interactions at the molecular level. It is crucial that we understand the mechanisms by which plants defend against pathogen attack so that it will be possible for us in future to activate their own defenses, instead of using transgenes to confer resistance. Induced resistance conferred by "native genes" will be more stable and will also help us to generate "environmentally friendly" genotypes of crop plants.

EXERCISES

The following laboratory experiment will provide students with some experience in performing experiments related to in vitro plant pathology. The experimental subjects provided in these exercises are primarily based on ease of availability. However, these approaches are very broad based and can be adapted or modified to any other similar situations. As with any biological experiment, details often differ with plant and fungal species. Although the examples provided in these experiments are based on plant–fungal interactions, they can be easily adapted to address other plant–microbe interactions also.

GENERAL CONSIDERATIONS

Preparation of Media

The agar media for growing fungal pathogens are quite simple and are usually available as ready-to-use powders from many biochemical suppliers. Resuspend the required amount of powdered medium in distilled water in 500-ml Pyrex bottles and autoclave for 20 min at 121°C. After cooling to about 65°C (can be held in hand without too much discomfort), dispense the medium in petri dishes in a laminar flow hood. Avoid condensation in petri dishes by either stacking the dishes into columns of ten or so immediately after pouring or by keeping the lids of the dishes ajar inside the laminar flow hood until the medium solidified and cooled to room temperature. Medium in petri dishes can be sealed in polythene sleeves (reuse the sleeves in which the petri dishes are supplied) and kept in the refrigerator for 3 to 4 months.

Handling of the Pathogen and Phytotoxins

Most of the plant pathogenic fungi are harmless to humans and generally do not require extra precaution while handling. In most instances, the same laminar flow hood may have to be used for both plant and pathogenic cultures. If the same laminar flow hood is used for both plant and fungal/bacterial cultures, at least 1 h should be allowed between working with fungal and plant cultures. If possible, the pathogen cultures should be done at the end of the day, so that the laminar flow hood can be cleansed thoroughly and allowed to run overnight.

Some phytotoxins may be harmful to human beings even at very low concentrations. If commercially available phytotoxins are used for experiments, read the instructions very carefully before using them. Follow appropriate precautions such as wearing gloves and mask and using fume hood as directed.

Materials that are used in Experiment 3 (SDS-PAGE) contain chemicals, such as acrylamide, mercaptoethanol, and TEMED, which have been listed as carcinogenic chemicals. These chemicals should be handled with extra caution as indicated.

EXPERIMENT 1. EFFECT OF PHYTOTOXIN ON TOBACCO SEED GERMINATION

Generally, phytotoxins will have some adverse effect on seed germination. This is more conspicuous when non-specific phytotoxins are used. Tobacco seeds are convenient models to assay the efficacy of phytotoxins, because they are small, easy to handle, and germinate rather

quickly. Although this experiment uses a commercially available phytotoxin as a model, any fungal (or bacterial) product, such as concentrated culture filtrate or partially purified extracts, can also be tested effectively. This experiment will require approximately 2 weeks.

Materials

Each team of three students will require the following materials:

- Twenty disposable 10-cm diameter petri dishes
- Twenty sterile, Whatman No. 3 filter papers
- Five 100-ml beakers
- Ten 1-ml serological pipettes

The whole class will require the following materials:

- Approximately 500 mg of tobacco seeds. Although cultivar is not a constraint, it will be good to use different cultivars with different teams, so that a comparison of the response can be made between the cultivars. These seeds are available with most seed vendors.
- *Alternaria* toxin stock solutions. Several toxins from *Alternaria* are commercially available from chemical companies. Prepare stock solutions altertoxin I of at least five strengths (e.g., 0, 5, 10, 50, and 100 µM). Each team will need 20 ml of each stock; prepare stocks accordingly.

Follow the protocols in Procedure 38.1 to complete the experiment.

Procedure 38.1 Inhibition of Tobacco Seed Germination by *Alternaria* Toxin

1. Preparation of tobacco seeds for germination assay — Wash the seeds with 10% commercial bleach containing few drops of Tween 20 by vortexing in a microcentrifuge tube. Centrifuge briefly and remove the bleach solution by a pipette. Rinse the seeds three times with sterile distilled water. Transfer the seeds to a sterile filter paper and allow it to dry in the laminar flow hood. The seeds are now ready for assay.

2. Germination — Place a sterile filter paper in each petri dish and label four dishes for each treatment. Wet the filter paper with 5 ml of appropriate toxin solution. Place 100 tobacco seeds in each petri dish. Close and carefully (gently) seal the petri dishes with parafilm. Incubate at 28°C ± 2°C in a growth room with a 16-h photoperiod.

3. Observations — After 1 week, count the number of seeds that have germinated in each dish. Repeat the count after 2 weeks. During the second week, count any abnormalities such as nonemergence of cotyledonary leaves and malformation of germinated seedlings in each plate separately.

Anticipated Results

There will be a decline in germination percentage as the concentration of toxin increases. At higher concentrations, for example, 50 μM or more, there may be noticeable abnormalities among the germinated seedlings. Some abnormalities that can be readily identified include albino types and variegated cotyledons or young leaves.

Question

- Why does the germination capacity of tobacco seeds decline when germinated in toxin solution?

EXPERIMENT 2. EVALUATION OF ANTHRACNOSE RESISTANCE IN COMMON BEANS BY A LEAF DISC BIOASSAY

Anthracnose is a very common disease of several crop plants, usually caused by the fungus *Colletotrichum*. In common beans (*Phaseolus vulgaris*), anthracnose is caused by *C. lindemuthianum*. The disease is easily identified by spots, mostly on the underside of the leaf. Red-brown spots and streaks also develop on stems, petioles, and leaves. A characteristic symptom of the disease is veins that turn brick-red to purple and eventually black on the underside of the leaf. In resistant varieties, small specks of infection can be noticed and usually the disease spread is arrested. This experiment will provide the students with an opportunity to evaluate the leaf disc bioassay, which is the most common assay when several cultivars have to be evaluated for disease resistance. Usually, the results can be seen within a week.

Materials

Each team of students will require the following materials:

- Four 50-ml beakers
- Six- to eight-week-old bean plants, five of each variety
- Cork borer of 1-cm diameter
- 20 sterile disposable 10-cm diameter petri dishes
- 20 sterilized Whatman No. 3 filter papers
- Two pairs of forceps
- 500-ml sterile, distilled water

The whole class will require the following materials:

- Culture of *C. lindemuthanium* (several accessions are available from ATCC, e.g., accession numbers 16341 or 16342) on lima bean agar medium (Difco 0024) for 1 week or until orange to pinkish clusters of spores can be seen. Culture two dishes per team. Collect the spores in distilled water and adjust the spore density to 100,000 spores/ml. Each team needs 20 ml of spore suspension. Obtain APHIS permit to transport.

Follow the protocols outlined in Procedure 38.2 to complete the experiment.

Procedure 38.2 Evaluation of Anthracnose Resistance in Common Beans with a Leaf Disc Bioassay

1. Preparation of leaf discs for bioassay — Wash young, fully expanded bean leaves thoroughly in tap water and rinse well with deionized water. Gently blot the leaves dry with paper towels. Cut 30 1-cm diameter leaf discs by a clean cork borer. Keep the discs on a moist paper towel.

2. Inoculation with fungal spores — Soak 20 leaf discs in fungal spore suspension for at least 5 min in a 5-ml beaker. Make sure that all the leaf discs are completely immersed in spore suspension by gently swirling the beaker. After 5 min, gently decant the spore suspension into another beaker. As a control, soak ten leaf discs in sterile deionized water.

3. Incubation and observations — Place wet, sterile filter paper in a standard petri dish. Carefully place the leaf discs on the wet filter paper, five per petri dish, with adequate spacing. Gently wrap the petri dishes with parafilm and incubate in a growth room at 28°C ± 2°C in a growth room with a 16-h photoperiod. Observe the leaf discs for any symptom development after 3 days. After 1 week, count the number of discs showing symptoms in each variety. Also try to grade the intensity of the symptom on a suitable scale (e.g., 1 to 5, from 1 representing no symptom to 5 representing very pronounced and widespread symptoms).

Anticipated Results

Most of the leaf discs will start developing lesions and in some cases will turn necrotic within 1 week. There will also be differences in sensitivity between the resistant and susceptible varieties. Resistant varieties may develop lesions much later than the susceptible varieties do. Hypersensitive cell death around the sites of fungal infection may be present. Other saprophytic fungi and mold may grow on the leaf discs.

Questions

- Define resistance and susceptible based on your test results.
- Describe hypersensitive response.

EXPERIMENT 3. INDUCTION OF PATHOGENESIS-RELATED (PR) PROTEINS IN TOBACCO AND *ARABIDOPSIS*

As discussed in Chapter 26, plants respond to pathogen attack in a number of ways. Researchers have also discovered that certain chemicals such as salicylic acid can also evoke similar responses. Induction of PR proteins is one such common response. This experiment will enable the students to see whether such responses can be induced in tobacco or *Arabidopsis* by polyacrylamide electrophoresis (Laemmli, 1970). Nucleic acid and proteins are negatively charged and hence when subjected to an electric field will migrate toward the cathode. All electrophoresis techniques are based on this principle. This experiment will suit graduate students studying plant–pathogen interactions and will take approximately 10 days to complete.

Materials

Plants for this experiment should be germinated at least 6 weeks before the planned dates of experiment. If there are several groups, they can be split to test tobacco and *Arabidopsis*. Two members are suggested for each team and both should be encouraged to collect and analyze their own samples. The students will require the following materials:

- Six sets of axenically-grown plants, consisting of three to four uniform plants grown in tall jars
- Sterile pestle and mortar
- 30 sterile, 1.5-ml microcentrifuge tubes

The whole class will require the following materials:

- Micropipettes and sterile pipette tips
- Liquid nitrogen
- Spore suspension — Collect spores from a fresh plate of *C. destructivum* (or *A. alternata*) culture and suspend in 5 ml distilled water. After determining the density with a hemocytometer, adjust the spore suspension to 10,000 spores/ml and make 200 ml (or the desired quantity) of suspension.

- 200 mM salicylic acid solution. Dissolve 13.81 g of salicylic acid in 500 ml distilled water.
- Protein extraction buffer — Several buffers are used successfully to extract total proteins from leaves. A common buffer to extract proteins from leaves is 100 mM Tris (pH 6.8), 100 mM sodium phosphate, 2.5% sodium dodecyl sulfate (SDS), 5% glycerol, and 10 mM dithiotreitol (DTT). It is safe to add protease inhibitors such as phenyl-methyl sulfonic acid (10 mM) or leupeptin (10 mM) to the buffer.
- Polyacrylamide gel electrophoresis units and required power supplies — Mini-gel units sold by several commercial vendors such as Bio-Rad or Hoefer Scientific can be used. Prepare 10% or 12% SDS-PAGE gels. These gels can be prepared in the lab if gel casters are available or ready gels can be purchased from the makers. Refer to Sambrook et al. (1989) for casting gels in the lab.
- Protein molecular weight standard — Available from several companies such as Bio-Rad and Sigma. Reconstitute as per manufacturer's directions.
- Protein sample-loading buffer — Dissolve 2.4 ml of 0.5 M Tris (pH 6.8), 2.0 ml glycerol, 4.0 ml of 10% (w/v) SDS, and 1.0 ml of 0.1% (w/v) bromophenol blue in 9.6 ml of sterile distilled water. This can be kept at room temperature for 3 months. Just before use add 25 µl ß-mercaptoethanol to 475 µl of the buffer. Use this working buffer within 48 h.

Follow the protocols described in Procedure 38.3 to complete the experiment.

Procedure 38.3 Induction and Analysis of PR Proteins by SDS-PAGE

1. Induction — Spray 5 ml of spore suspension or salicylic acid solution by an atomizer to two sets of plants each and label them. As mock inoculation, spray sterile distilled water on one of the remaining two sets of plants. The sixth jar of plants serves as "absolute control." Seal the jars with parafilm and incubate in a 16-h photoperiod.
2. Protein extraction — One week after spraying, collect young leaves from different sets of plants. Weigh 500 mg of leaves and wrap in aluminum foil. Immediately, plunge in liquid nitrogen and grind the frozen leaves to a fine powder in a mortar and pestle. Add 1 ml of protein extraction buffer to the fine powder and homogenize further. Transfer the resulting slurry to a 1.5-ml microcentrifuge

tube. Keep this on ice until all four samples are extracted. Centrifuge the slurry at 12,500 rpm for 10 min at 4°C and carefully pipette 750 µl of the supernatant into a fresh tube. This constitutes a crude protein extract. Add ammonium sulfate to this supernatant to 80% saturation and centrifuge again at 12500 rpm for 10 min. The proteins will precipitate as a small white pellet. Resuspend this pellet in 100 µl of sterile water. This is the protein sample to be analyzed.

3. Protein estimation — Determine the protein concentration in the samples by a protein estimation kit (commercially available from several vendors). Readjust the protein sample volume to a final concentration of 1 or 0.5 µg/µl, depending on the reading. Keep this sample at –20°C, if it is not used immediately.

4. Loading and running the gel — After setting up the gel apparatus and the gel in the assembly (follow manufacturer's directions), load the protein samples carefully in appropriate wells using a gel loading pipette tip or Hamilton syringe. Take care to avoid cross-contamination of wells. Load a standard marker to one of the wells to serve as a convenient reference. Connect the electrodes appropriately and run the gel at the recommended voltage.

5. Staining — Many commercial products based on the principles of Coomassie staining of proteins and peptides are available and are easy to use. Any of the stains (e.g., GelCode blue from Pierce Endogen) may be used to stain and visualize the protein bands. These stains are fast, very sensitive, and only a small volume is required.

Anticipated Results

Proteins and peptides can be seen as blue bands of varying intensities like a ladder in each lane. There should be additional bands or higher expression (more intensely stained) of certain bands in the lanes containing samples that were sprayed with spores or salicylic acid in comparison with untreated control and mock-inoculated control. It is likely that these differences will be seen in the 20- to 45-kDa region, which can be referenced by using the marker standard.

Questions

- What will happen if the protein samples are allowed to remain at room temperature (not kept on ice, until denaturing) for extended period?
- What are pathogenesis-related (PR) proteins? Give examples of PR proteins and describe their role in plants.

ACKNOWLEDGMENTS

We sincerely thank Dr. Richard Litz, Tropical Research and Education Center (TREC), University of Florida, Homestead, for his encouragement and critical suggestions in the preparation of this chapter. This contibution is Florida Agricultural Experiment Station Journal series number R-08485.

REFERENCES

Baker, C.J. and E.W. Orlandi. 1995. Active oxygen in plant pathogenesis. *Annu. Rev. Phytopathol.* 33: 299–321.

Daub, M.E. 1984. A cell culture approach for the development of disease resistance: Studies on the phytotoxin cercosporin. *HortScience* 19: 382–387.

Daub, M.E. 1986. Tissue culture and the selection of resistance to pathogens. *Annu. Rev. Phytopathol.* 24: 159–186.

Dempsey, D.A., J. Shah and D.F. Klessig. 1999. Salicylic acid and disease resistance in plants. *Crit. Rev. Plant Sci.* 18: 547–575.

Hammond-Kosack, K.E. and J.D.G. Jones. 1996. Resistance gene-dependent plant defense responses. *Plant Cell* 8: 1773–1791.

Jayasankar, S. 2000. Variation in tissue culture, pp 386–395 in *Plant Tissue Culture Concepts and Laboratory Exercises*, 2nd ed., Trigiano R.N. and D.J. Gray, (Eds.). CRC Press, Boca Raton, FL.

Jayasankar, S., Z. Li and D.J. Gray. 2000. In vitro selection of *Vitis vinifera* "Chardonnay" with *Elsinoe ampelina* culture filtrate is accompanied by fungal resistance and enhanced secretion of chitinase. *Planta* 211: 200–208.

Jayasankar, S., R.E. Litz, D.J. Gray and P.A. Moon. 1999. Responses of embryogenic mango cultures and seedling bioassays to a partially purified phytotoxin produced by a mango leaf isolate of *Colletotrichum gloeosporioides* Penz. In Vitro *Cell. Dev. Biol.-Plant.* 35: 475–479.

Laemmli, U.K. 1970. Cleavage of structural proteins during the assembly of the head of bacteriophage T4. *Nature* 227: 680–685

Linthorst, H.J.M. 1991. Pathogenesis-related proteins of plants. *Crit. Rev. Plant Sci.* 10: 123–150.

Litz, R.E. and D.J. Gray. 1992. Organogenesis and somatic embryogenesis, pp. 3–34 in *Biotechnology of Perennial Fruit Crops,* Hammerschalg, F.A. and R.E. Litz (Eds.). CAB International, Wallingford, U.K.

Nyange, N.E., B.Williamson, R.J. McNicol and C.A. Hackett. 1995. In vitro screening of coffee genotypes for resistance to coffee berry disease (*Colletotrichum kahawae*). *Ann. Appl. Biol.* 27: 251–261.

Sambrook, J., E.F. Fritsch and T. Maniatis. 1989. *Molecular Cloning: A Laboratory Manual.* Cold Spring Harbor Laboratory Press, New York.

Strobel, N.E., C. Ji, S. Gopalan, J.A. Kuc and S.Y. He. 1996. Induction of acquired systemic resistance in cucumber by *Pseudomonas syringae* pv *syringae* 61HpZpss protein. *Plant J.* 9: 431–439.

Ward, E.R., S.J. Uknes, S.C. Williams, S.S. Dincher, D.L. Wiederhold, D. Alexander, P. Al-Goy, J.P. Metraux and J.A. Ryals. 1991. Coordinate gene activity in response to agents that induce systemic acquired resistance. *Plant Cell* 3: 1085–1094.

White, R.F. 1979. Acetylsalicylic acid (aspirin) induces resistance to tobacco mosaic virus in tobacco. *Virology* 99: 410–412.

Glossary

Abiotic disease. Stress or injury to plants caused by cultural practices or environmental factors.

Abrasive. Any of a group of chemicals or compounds used to cause wounding of plant cells.

Abscission. Loss of leaves, fruit, or other plant parts from the plant due to a breakdown of the specialized cells at the base of the structure; referred to as the abscission layer or zone.

Absorption. The movement of water and solutes from the exterior environment into the various locations of the plant, including into cells.

Acclimatization. The process of becoming accustomed to a specific condition (e.g., the environment or the presence of a phytotoxin).

Acervulus (pl. acervuli). A mass of conidiophores and conidia that are initially covered by either the host cuticle or epidermis.

Acrylamide (polyacrylamide). A gel-like medium used to electrophoretically separate different weights of DNA fragments (amplicons) or proteins.

Active defense. Defense mechanisms that are produced by the plant after pathogen recognition.

Advisory. A report giving information, especially a warning; with regard to plant disease, an advisory is used to alert for the need for a control measure such as a fungicide application.

Aeciospore. Vegetative, dikaryotic (*N* + *N*), nonrepeating spore of rusts formed following dikaryotization of receptive hyphae by pycniospores (spermatia).

Aerobic. Requires oxygen for respiration.

Agarose. A highly purified fraction of agar.

Aggressiveness. Amount of disease caused by an isolate of a pathogen.

Agonomycetes. See **Mycelia Sterilia**.

Alkaloid. A nitrogenous organic compound with alkaline properties, usually bitter in taste, poisonous to animals, and produced by certain plants.

Allele. One of a pair of genes at a single locus.

Alloinfection. Pathogen infection of a host plant by inoculum from another individual plant.

Aminopyrrolizidine. An alkaloid produced by some plants that is toxic to animals including humans.

Amplicon. A portion of DNA amplified or copied during the polymerase chain reaction.

Amylase (amylolytic). Enzymes that degrade starch to maltose and D-glucose.

Anaerobic. Does not require oxygen for respiration (fermentive).

Analysis. The resolution of problems by representing component processes through mathematical equations.

Anamorph. Conceptually, the asexual aspect of a fungus; contrast with **holomorph** and **teleomorph**.

Anastomosis. Fusion of cell walls between two hyphae.

Anastomosis groups (AG). Refers to *Rhizoctonia* isolates having three or more nuclei per hyphal cell (multinucleate).

Angiosperms. Flowering plants.

Annealing temperature. The temperature at which rejoining of single strands of DNA into the double helix occurs or temperature at which primers pair with DNA.

Annulus. The ring around the stipe that is a remnant of the inner veil in the developing mushroom that once enveloped only the gills or pores.

Antagonism. An association between organisms in which one or more of the participants is harmed or has its activities limited by the other through antibiosis, competition, parasitism, or predation.

Antagonist. An organism that harms another organism through its activities.

Antheridium (pl. antheridia). Specialized cell in which nuclei are formed that function as male gametes, and is, therefore, a male gametangium; contrast with gynecium.

Antibiosis. The inhibition or destruction of one organism by a metabolite produced by another organism.

Antibiotic. Chemical produced by one organism that is harmful to another organism.

Antibody. A new or modified protein produced by an animal's lymphatic system in response to a foreign substance (called an antigen). The antibody binds with the antigen, and this binding causes the antigen to become inactive.

Antigen. Usually a foreign protein (occasionally a lipid, polysaccharide, or nucleic acid) that is introduced into animal tissue that reacts by producing antibodies that specifically bind and inactivate the protein.

Antisense RNA. An RNA strand that is complementary or antisense to a positive sense RNA strand.

Antiserum (pl. antisera). The serum fraction of the blood from an animal that has been immunized with an antigenic substance; the serum contains a number of different immunoglobin types and reactivities.

Apothecium (pl. apothecia). An ascomycete fruiting structure that is cup-shaped or based on a cup shape in which exposed asci form; the structure develops as a result of nuclear exchange, typically involving an ascogonium and antherdium.

Appressorium (pl. appressoria). Terminal or intercalary cell of a germ tube of a fungus used for attachment to the host.

Arbitrarily amplified DNA (AAD). Collective term that describes those **nucleic acid** amplification techniques that use arbitrary primers. AAD includes several well-established techniques that provide genetic information from a plurality of sites (loci) in a genome or nucleic acid molecule, such as **randomly amplified polymorphic DNA (RAPD)**, arbitrarily primed PCR (AP-PCR), **DNA amplification fingerprinting (DAF)**, and AFLP.

Arbitrary primers. Generally, short (7 to 12 base pairs) segments of DNA that do not correspond to known genomic sequences and thus may pair with (anneal to) many sites (arbitrary). They are used in PCR methods such as RAPD and DAF.

Arbuscule. Special, dichotomously branched haustorium that develops intracellularly and serves as the site of nutrient and carbon exchange in glomalean endomycorrhizae.

Area under curve. The region of a graph bounded by axes and a line, where the line depicts disease levels in time.

Area under the disease progress curve (AUDPC). See **area under curve.**

Arthropods. Any member of the phylum Arthropoda; arthropods have distinct body segments, exoskeleton, jointed appendages, and bilateral symmetry.

Arthrospore. An asexual spore formed when hyphal fragments break into unicellular sections.

Ascogenous hyphae. Dikaryotic hyphae that will give rise to asci.

Ascogonium (pl. ascogonia). Specialized cell that functions to accept nuclei from an antheridium and is therefore a female gametangium.

Ascoma (pl. ascomata). Fungal tissue in or on which asci are formed; a fruiting body, syn. **ascocarp**.

Ascomycota (Ascomycetes). A diverse group (phylum) of fungi that are characterized by production of asci usually containing eight ascospores.

Ascospores. Fungal spores produced in a sac-like structure called an **ascus**, most often following sexual recombination.

Ascus (pl. asci). Specialized cell in which nuclei fuse, undergo meiosis, and are subsequently incorporated into ascospores.

Aseptic. Free of all living organisms.

Asexual reproduction. Vegetative reproduction of an organism accompanied by mitotic divisions of nuclei in which the chromosome number is not reduced by meiosis.

Autoclave. A chamber that sterilizes materials by using pressurized steam; a temperature of 121°C for 15 min is required to kill most microbes.

Autoinfection. Pathogen infection of a host plant by inoculum arising from that same individual plant.

Auxins. Plant growth regulators (hormones) affecting shoot elongation and other functions.

Avirulence. Inability of a pathogen to cause a compatible (susceptible) reaction on a host cultivar with genetic resistance.

Axenic. Free from all microbial contamination (describing a pure culture).

Bactericide. A chemical compound that kills or inhibits the growth of bacteria.

Bacteriocin. Antibiotics that inhibit specific groups of microorganism that are closely related to the microorganism that produced the chemical.

Baiting. The use of a living organism, instead of artificial culture media, to promote the growth of a particular microorganism.

Bark splitting. Localized death of the cambium layer of tree trunk, which leads to bark pulling away from the sapwood.

Basidiocarp. Fruiting structure in the Basidiomycota.

Basidiomycota (Basidiomycetes). A diverse group (phylum) of fungi that is characterized by the production of basidia and basidiospores.

Basidiospore. Ephemeral, thin-walled, haploid (N) spore of basidiomycetes following nuclear fusion and meiosis in basidia. In rusts and smuts formed on germination of teliospores.

Basidium (pl. basidia). Specialized cell in the hymenium where basidiospores are formed following karyogamy and meiosis.

Bioassay. A technique that uses a plant cell, tissue, or organ as a source to test its response against another organism or chemicals under controlled conditions. May also be used to detect chemical agents.

Biofilms. Bacterial cells embedded in extracellular polysaccharides. Zoogloea are examples of biofilms occurring among plant cells.

Biological control. The use of natural or modified organisms, genes, or gene products to reduce the effects of undesirable organisms such as plant pathogens and to favor desirable organisms such as crop plants.

Biotic disease. A disease caused by living pathogens including viruses.

Biotroph(ic). An organism that can live and multiply only on or within another living organism (i.e., obligate parasite).

Bitunicate ascus. Having a double wall.

Blastospores. Spores produced by budding.

Bordeaux mixture. Copper sulfate and lime; one of the first fungicides.

Callus (pl. calli). Tissue mass (usually not uniform) arising from disorganized proliferation of cells and tissues.

Canker. Localized infection of the bark on the stems or trunk.

Capsid. The protein that encapsulates or encloses a viral genome.

Cellulase (cellulolytic). A collection of hydrolytic enzymes capable of degrading native and modified celluloses to D-glucose.

Cellulose. ß-(1–4) linkages of D-glucose; major component of the cell wall of the Oomycota and plants.

Central vacuole. The large membrane-bound fluid-filled body found in plant cells. It functions in storage of chemicals, enzymes, and maintenance of osmotic potential.

Ceratobasidium anastomosis group (CAG). *Rhizoctonia* isolates having two nuclei per hyphal cell (binucleate).

Character. An observable feature of an organism that can distinguish it from another.

Chelating agent. A chemical with the ability to bind divalent ions such as calcium or magnesium.

Chemotaxis. Swimming orientation of a zoospore in response to an external chemical stimulus.

Chitin. ß-(1–4) linkages of *n*-acetylglucosamine; chief structural component of the cell wall of higher fungi (kingdom Fungi).

Chlamydospore. A thick wall resting spore characteristic of some fungi.

Chloroplast. Specialized membrane bound cellular organelle containing the chlorophyll pigments that are involved in the photosynthetic processes of the plant.

Chlorosis. Abnormal yellowing or fading of the normal green color of leaves due to loss of chlorophyll production or increase in chlorophyll degradation.

Chytridiomycota. Fungal phylum where a true mycelium is lacking, uniflagellate zoospores are present, and asexual reproduction occurs when the mature vegetative body transforms into thick-walled resting spores; in the kingdom Fungi.

Cibarial pump. The muscularized chamber in Homopterans that serves to intake fluid during feeding.

Cirrus (pl. cirri). A mass of spores, usually in a gelatinous matrix, that forms a droplet or ribbon as it is forced from the fungal fruiting body.

Clade. A monophyletic group (single lineage or origin inferred through cladistic analysis) that are each other's closest relatives.

Clamp connection. A structure connecting two hyphal cells and involved in continuation of the dikaryotic condition for some basidiomycetous fungi.

Clavate. Club-shaped.

Cleistothecium (pl. cleistothecia). Sexual fruiting body of an ascomycete that lacks a natural opening for spore discharge. Spores are usually dispersed following breakage by freezing, thawing, or wear and tear.

Coelomycetes. An informal class of mitosporic (Deuteromycota) fungi where spores are produced in conidiomata.

Coenocytic. Hyphae that lack, or mostly lack, cross walls; aseptate.

Colony (fungal). Discrete or diffuse, circular collections of hyphae or spores, or both, that arise from one cell or one grouping of cells.

Columella. Swollen tip of a conidiophore that bears asexual sporangiospores (or mitospores); in the Zygomycota.

Competition. A process by which two or more organisms try to utilize the same food or mineral source or occupy the same niche or infection site.

Competitors. Chemical compounds that are used to slow a reaction by competing with the substrate of interest for enzymatic binding sites or for reactive chemicals.

Conidiomata. Asexual fruiting structure (e.g., acervulus, pycnidium, or synnema).

Conidiophore. Simple or branched hyphae where mitosis occurs and conidia are produced.

Conidium (pl. conidia). Asexually produced spore borne on a conidiophore.

Contact fungicides. Chemicals applied as foliar sprays to protect above-ground plant parts from infection; they may also be used as seed treatments.

Crop rotation. Growing various plants that differ in disease susceptibility in a given area.

Cross protection. A form of competition in which an avirulent or weakly virulent strain of a pathogen (e.g., virus) is used to protect against infection from a more virulent strain of the same or closely related pathogen (e.g., virus).

Cultural practices. Methods used in planting, maintaining crops, and ultimately suppressing plant pathogens.

Culture. To grow a microorganism or other cells or tissue on an artificially prepared nutritional medium.

Culture media. An artificially prepared solid or liquid media on which microorganisms or other cells are grown.

Cytokinins. Plant growth regulators (hormones) affecting cell division and other functions.

Damage threshold. The amount of crop damage that is greater than the cost of management measures.

Damping-off. Rotting of seeds in the soil or the death of seedlings soon after germination.

Denaturing temperature. Temperature at which there is loss of three-dimensional structure of proteins and DNA; in PCR reactions, effectively strand separation of DNA.

Diagnosis. The process or procedure for determining the cause of a plant disease or other problem; the process of disease identification.

Dicots. Plants having two cotyledons (as beans).

Dikaryotic. Having two haploid nuclei per cell ($N + N$: a condition found in most basidiomycete cells and the ascogenous hyphae of ascomycetes).

Diploid. Having a single nucleus containing 2 sets of chromosomes ($2N$), one from each parent.

Disease complexes. Diseases caused by a combination of pathogens such as fungi and nematodes.

Disease cycle. The series of events, processes, and structures involving both host and pathogen in development of a disease, repeated occurrence of the disease in a population, and injurious effects of the disease on the host.

Disease diagram. A pictorial representation of a disease level on a leaf or plant.

Disease-free (pathogen-free). A loosely used phrase to denote that specific plant pathogens are not present on or in selected plants. Other pathogens may or may not be present.

Disease progress. The development of disease, usually in time.

Disease suppressive soils. Soils in which pathogens cannot establish, they establish but fail to produce disease, or they establish and cause disease at first, but disease becomes less important with continued culture of the crop.

Disinfectant. A chemical agent that is used to eliminate plant pathogens from the surface of plants, seed, or inanimate objects such as greenhouse benches, tools, pots and flats.

Dispersal. To separate or move in different directions.

DNA amplification fingerprinting (DAF). An arbitrarily amplified DNA technique that uses one or more synthetic oligonucleotide primers typically 5 and 15 nucleotides in length and high primer-to-template ratios. Amplification products are generally visualized by **polyacrylamide gel electrophoresis** and silver staining and fingerprints are highly complex; syn. **DNA profile**.

DNase. Enzyme that degrades or destroys DNA.

DNA polymerase. Enzyme that makes the DNA polymer.

DNA polymorphism. A variation in a DNA sequence detected through DNA sequencing or DNA analysis (fingerprinting, profiling, etc.).

Dolipore septum. A complex septum found in hyphae of basidiomycetes that prevents movement of organelles between cells.

Downy mildews. Diseases caused by obligate fungal-like parasites classified in the phylum Oomycota and the Peronsporaceae.

Durable resistance. Resistance that remains effective during a prolonged and widespread use in an environment favorable for disease.

Ebb and flow irrigation (flood and drain). Process by which potted plants are placed in a tray or nonporous floor and water is used to fill the tray or floor area. After the potting medium reaches the desired moisture level, the excess water is drained away. Often, the excess water is returned to a reservoir and used again at the next watering.

Ectomycorrhizae. Symbiotic, thought to be mutualistic, association between a plant root and fungus. The fungus forms a layer or mantle around the root and grows inward between the outer cortical root cells. The complex of hyphae around and in the root is referred to as a Hartig net. Fungus receives carbohydrates from the plant and the plant gains greater access to water and mineral nutrients and some protection from root pathogens.

Ectoparasite. In nematology, parasitic nematodes that feed on roots from outside.

Egestion. The expulsion of food materials through the aphid stylet; aphids alternate between ingestion (the intake of food materials) and egestion during feeding. Egestion may serve to ensure that the stylet is not blocked during feeding.

Electrophoresis (gel). Movement of charged molecules, such as DNA and proteins, through a supporting medium (a gel) that is usually composed of starch, agarose, or acrylamide. Molecules are usually separated by weight and/or change.

Endemic. Prevalent in or peculiar to a specific locality or region.

Endobiotic. Living entirely in the interior of a host; in the Plasmodiophoromycota.

Endoparasites. In nematology, nematodes that penetrate the root and feed on tissue inside the root.

Endophyte. Plant or fungus developing and living inside another plant.

Endophytic (prokaryotes). Endophytes live within plants; some are pathogens and others apparently are nonpathogenic inhabitants. The degree of endophytic intimacy varies. Some endophytes live within cells (liberobacteria, phytoplasmas, and spiroplasmas), and others colonize the surface of living cells within the plant (pseudomonads and xanthomonads).

Environment. The conditions, influences, or forces that influence living forms.

Enzyme. A protein that is a catalyst for chemical reactions without being altered or degraded during the process; in fungi and bacteria, exoenzymes.

Enzyme linked immunosorbant assay (ELISA). A serological technique that employs binding antibodies or antigens to a polystyrene microtiter plate and uses antibodies linked to an enzyme by gluteraldehyde for detection.

Epidemic. Any increase in the occurrence of disease in a population.

Epidemiology. The study of factors that lead to a change in disease.

Epidermis. The outer layer of cells on a plant.

Epigenetic. Environmentally induced changes in appearance or response of an organism that are not permanent.

Epinasty. Downward bending of the leaf from its normal position.

Epiphytic. Growing on a plant without harming it.

Eradicant fungicides. Chemical compounds that halt fungal growth if applied shortly after infection.

Eradicate. To completely eliminate a pathogen from a plant or field.

Ergot. A sclerotium produced by *Claviceps purpurea* and related species in the floret of infected grains.

Etiolation. Excessive elongation of internodes in plants usually in response to low light or disease.

Eukaroytic. Organisms that have a nucleus as well as other double-membrane organelles.

Exponential model. A mathematical equation used to represent disease development in time; this model allows infinite increase in disease.

Extension temperature. Temperature in PCR cycle that is optimum for the activity of DNA polymerase (cf. 72°C).

Extracellular enzyme. A protein (enzyme) produced with the cytoplasm of fungi and bacteria that is transported across plasmalemma to act on a substrate in the environment, for example, **cellulases**. May be involved in pathogenesis or nutrition of the organism or both.

Facultative fermentive. An organism that can derive energy using oxygen (**aerobic**) or fermentive (**anaerobic**) processes.

Facultative parasite. An organism that normally lives as a saprophyte but under certain conditions can become a parasite.

Facultative saprophyte. An organism that normally lives as a parasite but under certain conditions can become a saprophtye.

Flagellum (pl. flagella). A motile appendage projecting from a zoospore. Some flagella have hairs (e.g., straminipilous flagellum), whereas others are hairless.

Fluorometer. A single wavelength spectrophotometer used to measure DNA concentration.

Forecast. To estimate in advance.

Forma(ae) specialis(es). Special forms of a pathogenic fungal species, separable based on morphology or host. For bacteria, pathovan or subspecies is equivalent.

Formulation. Pesticides consist of an active ingredient and inert ingredients in a dry or liquid form such as granules or emulsifiable concentrates.

Fungicide. A chemical compound that kills fungi.

Fungus (pl. fungi). Eukaryotic, heterotrophic, absorptive organism that develops a microscopic, diffuse, branched, tubular thread called a hypha.

Gall. An abnormal, somewhat spherical growth produced by the plant in response to some insects, nematodes, fungi, or bacteria.

Gametangial contact. Sexual reproduction in the oomycota involving the contact of **gametangia** and resulting in the formation of oospore.

Gametangial copulation. Sexual reproduction in the Zygomycota involving the fusion of **gametangia** and resulting in the formation of a zygospore.

Gametangium (pl. gametangia). Sex organ that contains gametes.

Gametes. Sexual cells, for example, sperm and eggs, or their functional equivalents.

Gas chromatography. The movement of molecules through a gas medium so that molecules of different molecular weight are separated; used to identify bacteria by profiles of bacterial fatty acids.

General resistance. Resistance effective against all biotypes of a pathogen.

Genetic engineering. The addition of genetic material to the genome of the organism.

Genomic DNA. Total DNA of the organism. May include DNA from plasmids, chloroplasts, and mitochondria in addition to chromosomal DNA.

Genus. Taxonomic classification below that of family that includes related species.

Gill. Plate on the undersurface of the pileus that supports the production of basidia and basidiospores.

Glucans. Long chains of glucosyl residues; with chitin, a structural component of the cell wall of higher fungi (kingdom Fungi).

Glycocalyx. Cell wall; composed chiefly of polysaccharide that is slimy as in slime molds or firm as in most other fungi.

Glycogen. A storage polysaccharide used as a major carbon reserve compound in fungal cells.

Glycolysis. Enzymatic breakdown of glucose and other carbohydrates to form lactic acid or pyruvic acid and the production of energy in the form of adenosine triphosphate (ATP).

Gradient. The rate of increase or decrease of a variable.

Hartig net. Intercellular network of fungal hyphae in the root cortex that does not penetrate cortical cells and that serves as an exchange site for nutrients in certain groups of mycorrhizae; connected to the mantle.

Haustorium (pl. haustoria). Specialized branch that extends from an intercellular hypha into a living host cell and serves as a nutrient absorptive organ for the fungus; absorptive organ of parasitic plants.

Hemibiotroph(ic). A plant pathogenic fungus that maintains its host cells alive for a few hours or days and then kills them for nutrition.

Hemiendophytic. Living partly within plant tissue.

Hemolymph. The circulatory fluid produced by arthropods.

Herbicide injury. Plant injury caused by the misapplication, drift, or residue of a herbicide on nontarget plants.

Heterokaryon. Hyphal cells that have two or more genetically different nuclei.

Heterothallic. Self-sterile; a complementary mating type is needed for sexual reproduction.

Heterotroph(ic). An organism living only on organic food substances as primary sources of energy.

Holocarpic. Entire thallus matures to form thick-walled resting spores (Chytridiomycota).

Holomorph. Collective concept that refers to both the sexual and asexual aspects of a fungus; contrast with **anamorph** and **teleomorph**.

Homothallic. Self-fertile; only one mating type required for sexual reproduction.

Hormone. Organic compound produced by organisms at minute levels and is generally transported to other parts of the organism to control various growth and physiological processes of the organism; in plants also referred to as plant growth regulator.

Host growth stage. Particular physical or physiological point in plant host development.

Host range. The variety of plant genera and species that a pathogen can infect.

Hymenium (pl. hymenia). Layer of hyphae where sexual reproductive cells, such as asci and basidia, and other sterile cells occur.

Hyperauxiny. Increased levels of indole acetic acid in host tissues that results in tumorous tissue development and hypocotyl elongation.

Hyperparasitism. Parasitism of parasites by other organisms.

Hyperplasia. Increased or abnormal frequency in division of cells.

Hypersensitive response or reaction (HR). Active defense mechanism that limits the progression of the infection by the rapid death of a few host cells. The HR is a manifestation of host recognition of the pathogen avirulence gene product. Typically, an HR includes signal transduction, programmed cell death, increased activation of defense-related genes, and a distant induction of general defense mechanisms that serve to protect the plant.

Hypertrophy. Abnormal enlargement of plant cells.

Hypha (pl. hyphae). Thread-like filaments that form the mycelium or vegetative body of a fungus that may or may not have crosswalls.

Hyphomycetes. An informal class of mitosporic fungi (Deuteromycota) where spores are produced on separate conidiophores.

Hypovirulence. Reduced level of virulence in a strain of pathogen resulting from genetic changes in the pathogen or from the effects of an infectious agent on the pathogen.

Immunobinding. The technique of binding proteins to membrane sheets and using antibodies for their detection.

Immunoglobulins. Blood serum proteins of vertebrates formed in response to antigenic regions or epitopes of foreign proteins or other substances. See also **Antibodies**; **ELISA**.

Incidence. The proportion of a population affected by a disease.

Indexing. The examination and testing of plants in order to detect the presence of certain specific pathogens (see **disease-free**).

Indolediterpenes. Mycotoxins of a diverse range of chemical structures that share a common biosynthetic origin.

Induced systemic resistance (ISR). An active defense mechanism in which a root-colonizing bacterium causes distant induction of general defense mechanisms that serve to protect the plant. In ISR, jasmonic acid and ethylene signal distal portions of the plant and antimicrobial proteins are not produced.

Infectivity. The ability of a pathogen (virus) to establish infection.

Infectivity curve. The comparison of diluted virus inoculum and the number of infections caused by the dilution.

Inflorescences. Groupings or arrangements of flowers on a common axis.

Inhibitors. Compounds with the ability to slow or stop a specified reaction.

Initial inoculum. Parts or forms of a pathogen that are able to infect after survival through time periods when a host is absent.

Inoculation. Placement of the propagule of a pathogen on or near a host cell or at a site where it can it can infect the host.

Inoculum. Sexual or asexual spores, vegetative mycelium, or other infectious bodies that serve to initiate infection in a host organism (being restricted to only specific spore stages in rusts and smuts).

Integrated pest management (IPM). A system for controlling disease, mite, and insect problems by multiple strategies (cultural, biological, and chemical).

Intensity. The level or degree of occurrence.

Intercalary. A structure formed as an interruption of a continuous strand, as with certain fungal spores (chlamydospores).

In vitro selection. A process by which cells or tissues are selected by screening with an appropriate selection agent in a container or under laboratory conditions.

Isolates. Any propagated culture of a virus (or other organism) with a unique origin or history.

Karyogamy. Fusion of two haploid nuclei (N) to form a diploid nucleus (2N); final step in fertilization.

Larva (pl. larvae). Juvenile organism that undergoes several growth stages prior to reaching adult development.

Lesion (local). Abnormal appearance in a localized area due to a wound or disease.

Life cycle. All the successive stages, morphological and cytological, in an organism that occur between the first appearance of a stage and the next appearance of that same stage, and including all the intervening stages.

Lignin. Complex organic material derived from phenylpropane that imparts rigidity and strength to woody tissues.

Lipase (lipolytic). Enzyme that hydrolyzes ester bonds between fatty acids and glycerol (or sorbitol in Tween) backbone found in fats.

Locule. A small cavity.

Lodge. A falling over of a plant generally at the soil line.

Logistic model. A mathematical equation used to represent disease development in time; this model limits increase in disease.

Loline. Fungal secondary metabolite present in grasses symbiotic with endophytes in the genera Epichlöe and Neotyphodium.

Lolitrems. **Indolediterpenes** that are potent neurotoxins; cause tremors in mammals and deter insect feeding.

Maceration. The blending, grinding, or pulverizing of tissue until a smooth homogenous mixture results; may also refer to soft rots caused by enzymatic degradation of host cell walls.

Mantle. Fungal sheath enveloping the outside of infected feeder roots; in ectomycorrhizae and some ericaceous mycorrhizae.

Marginal. Along the edge (e.g., leaf).

Material safety data sheet (MSDS). Contains important information about pesticides, such as adverse effects of overexposure, first-aid measures, fire-fighting measures, how to handle a spill or leak, proper storage conditions, and information about worker protection standards, including required personal protection equipment.

Mating types. "Sexes"; complementary mating types are needed for sexual reproduction in **heterothallic** fungi.

Mechanical inoculation. The ability to inoculate a virus into a host without the natural vector. For example, abrasives cause wounding that enable the mechanical inoculation of sap extracts.

Mechanical transmission. The transmission of a virus into a host without a vector.

Meiosis. A sequence of cellular events in sexual cell division leading to the division of the nucleus into four nuclei that are genetically dissimilar due to chromosome recombination.

Melanin. Dark brown or black pigment found in fungal cell walls.

Mesophile(ic). An organism that grows well between 10°C and 40°C.

Migratory. Moving from one host to another or an individual nematode that feeds from many sites along a root.

Millardet. Professor of botany at Bordeaux University, who studied and promoted the use of a mixture of copper sulfate and lime (Bordeaux mixture) to manage downy mildew of grape and other diseases.

Mitochondrion (pl. mitochondria). Cytoplasmic organelle that generates energy in the form of ATP.

Mitosis. A sequence of cellular events in asexual cell division leading to the division of the nucleus into two genetically similar nuclei.

Model. Any physical, mathematical, or theoretical representation of some process.

Molecular marker. A distinguishing molecular feature that can be used to identify a segment or region of a genome, chromosome, or genetic linkage group.

Mollicutes. Prokaryotes lacking cell wall thriving in high osmoticum habitats protecting the integrity of their plasma membranes from lysis. Mollicutes include the polymorphic phytoplasmas and helical spiroplasmas. These organisms are nutritionally fastidious, growing only in phloem sap in their plant and hemolymph in their insect vectors.

Monocots. Plants having a single cotyledon (as grasses).

Monocycle. Relative to plant disease, a process of pathogen infection, growth, production of symptoms, and reproduction without repetition.

Monogenic resistance. Resistance conditioned by a single gene.

Monokaryotic. A cell having genetically identical haploid nuclei.

Monophyletic. Descendants from a single common ancestor.

Multigene resistance. Resistance encoded by several to many genes; may be similar to horizontal resistance.

Mutualism. A form of symbiosis in which two or more organisms of different species live together for the benefit of both or all.

Mycelia Sterilia. A group of fungi in the Deuteromycota that is characterized by not producing spores (see **Agonomycetes**).

Mycelial fan. Fan-shaped aggregation of hyphae found under the bark of trees that invades and kills the cambial tissues of the host; common in *Armillaria* species.

Mycelium (pl. mycelia). Mass of highly branched thread-like filaments (hyphae) that constitutes the vegetative body of a fungus.

Mycoparasitism. One fungus being parasitic on another fungus.

Mycorrhizae. Fungus-root; a mutualistic relationship between certain fungi and the roots of plants; also **mycorrhizal fungi**.

Mycotoxin. Toxin produced by a fungus in infected grain or legumes that is poisonous to humans or animals when consumed.

Naked ascus. Ascus not produced in an ascocarp.

Necrosis. Death of plant tissue in a localized area resulting in brown or black lesion.

Necrotroph(ic). A plant pathogenic fungus that kills its host cells for nutrition.

Nematicides. Chemicals used to kill or interrupt the life cycle of plant-parasitic nematodes.

Nematology. The study of nematodes.

Nucleic acid. A single- or double-stranded polynucleotide containing either deoxyribonucleotides (e.g., DNA) or ribonucleotides (e.g., RNA) linked by $3'$-$5'$ phosphodiester bonds.

Obligate aerobes. Require oxygen for respiration and metabolism.

Obligate anaerobes. Are poisoned by oxygen; they use compounds other than oxygen as electron acceptors for respiration and metabolism.

Odontostyle. A stylet with or without flanges and found in dagger and needle nematodes.

Oidium (pl. oidia). Asexual spore that splits off in succession from the tip of a short branch or oidiophore; for example, powdery mildews.

Oligogenic resistance. Resistance conditioned by a few genes.

Onchiostyle. A tooth-like stylet found in stubby-root nematodes.

Onset. The beginning or early stages.

Oogonium. Female gametangium of the fungi-like **Oomycota** species.

Oomycota. Water molds; a fungus-like phylum where sexual reproduction results in production of oospores; in the kingdom Stramenopila.

Oospore. A thick-walled resting spore that develops from the fertilization of an **oogonium(ia)** by one or more **antheridium(a)**.

Ostiole. An opening or pore in a **pycnidium** or **perithecium**.

Oxidative phosphorylation. Metabolic processes where energy in the form of adenosine triphosphate (ATP) is stored during the transfer of electrons through the electron transport pathway.

Papilla (pl. papillae). Deposit of callose and other material on the inside of a plant cell wall that are produced in response to pathogen ingress.

Parasite(ic or ism). An organism, virus, or viroid living with, in, or on another living organism and obtaining food from it.

Partial resistance. Resistant reactions in which some symptoms develop or pathogen reproduction occurs but the amount is less than on a fully susceptible reaction. Partial resistance implies nothing about inheritance, gene expression, or race specificity of the resistance.

Passive defense. Defense mechanisms that are present before pathogen recognition.

Pathogen. An organism or agent capable of causing a disease on a host plant.

Pathogenesis-related proteins (PR proteins). Synthesis and accumulation of defense-related proteins following pathogen infection of plants.

Pathogen fitness. The ability of the pathogen to survive or compete in the environment where the host is grown.

Pathogenicity. The ability of a pathogen to cause disease on a specific host species.

Pectinase (pectinolytic). A complex battery of enzymes capable of degrading pectin to a number of products including glacturonic acid.

Peloton. An intra- or intercellular hyphal coil in the roots of some endomycorrhizae.

Penetration peg. Specialized, slender fungal hypha that penetrates the cell wall of a host plant through a combination of enzymatic and mechanical force; usually produced from the undersurface of an appressorium.

Pericarp. Outermost wall of a ripened ovary or fruit; the seed coat, that is, the outer layer of the karyopsis in cereals and other grasses.

Perithecium (pl. perithecia). A flask-shaped structure with a hole (**ostiole**) at the top of the flask in which asci form in a single layer (hymenium); the structure develops as a result of nuclear exchange, typically involving an ascogonium and antheridium.

Permanent wilt. Wilt due to dysfunction of vascular tissues.

Permeability. The ability of a membrane to control the passage of materials through it.

Pest. An organism that injures or has a harmful effect on a plant.

pH. The measure of acidity or alkalinity within a range of 0 to 14, pH 7 being neutral. The pH scale is logarithmic, so a change of one unit is equal to a tenfold change of hydrogen ion concentration.

Phenol(ic). Compound having one or more hydroxyl groups substituted in single or multiple benzene rings

Photosynthesis. Process conducted by plants that transfers light energy with carbon dioxide and water into a chemical form of energy, carbohydrates, releasing oxygen in the process.

Phyllosphere. The aerial surfaces of plants (of leaves, stems, fruit, and flowers) that serve as habitats for phytopathogenic prokaryotes during the resident phase or pathogenesis.

Phylogenetic analysis. The use of hypothesis of **character** transformation to group taxa hierarchically into nested sets and then interpreting these relationships as a **phylogenetic tree.** Phylogenetic relationships are reconstructed by distance, parsimony, and maximum likelihood methods.

Phylogenetic tree. A hypothesis of genealogical or evolutionary relationship among a group of taxa.

Phylum (pl. phyla). A taxonomic rank between the kingdom and class (formerly referred to as division).

Physiology. Study of basic activities conducted within the cells and tissues of a living organism as they are related to the chemical and physical processes of the organism.

Phytoalexins. Active defense antimicrobial low-molecular-weight compounds that are produced after infection or elicitation by abiotic agents.

Phytoanticipins. Passive defense antimicrobial low-molecular-weight compounds that are stored in the plant cell or are released from a glucoside.

Phytotoxin. A chemical produced by a plant pathogenic microbe in the process of disease development.

Pileus. The cap of a mushroom basidiocarp.

Plasmalemma. Cell membrane that bounds the protoplast; found inside the cell wall in pathogens other than viruses and plants.

Plasmid. A self-replicating piece of circular DNA that is present in some bacteria and fungi and is not a part of the chromosomal DNA. Its replication is not associated with cell division.

Plasmodiophoromycota. A phylum within the kingdom Protists characterized by organisms that have cruciform nuclear division and produce plasmodia.

Plasmodium (pl. plasmodia). A cell without a wall (naked protoplast), usually consisting of a mass of protoplasm with many nuclei.

Plasmogamy. Fusion of protoplasts to form a dikaryotic cell; first step in fertilization.

Pollution. Contamination of air, water, or soil by particulates or chemical substances.

Polycyclic. Relative to plant disease, the repeating process of pathogen infection, growth, production of symptoms, and reproduction.

Polyetic. Taking place over more than one year.

Polygenic resistance. Resistance conditioned by several genes.

Polymerase chain reaction (PCR). An in vitro nucleic acid amplification technique capable of increasing the mass (number) of a specific DNA region (fragment). The method uses two synthetic oligonucleotide primers (10 to 30 nucleotides long) that specifically hybridize to opposite DNA strands flanking the region to be amplified. A series of temperature cycles involving DNA denaturation, primer annealing, and the extension of annealed primers by the activity of a thermostable DNA polymerase amplify the target region many millionfold.

Polymorphism. More than one form. In DNA fingerprinting, the presence or absence of a specific band or PCR product at a specific locus.

Polyphenol oxidase. An enzyme that oxidizes adjacent hydroxyl groups on phenolic compounds.

Powdery mildew. A plant disease caused by a fungus in the order Erysiphales, generally causing white, powdery patches on the surface of leaves, stems, or flower parts.

Predispose. To make more susceptible (to disease).

Programmed cell death (PCD). An active defense mechanism commonly associated with reproductive and xylem tissue development. In PCD, the attacked cell and several plant cells around it die in response to chemical signals. The sacrifice of these cells isolates the pathogen.

Prokaryotes. Bacteria and mollicutes that differ from eukaryotes (higher plants, fungi, and animals) in that they lack organelles (membrane-bound nuclei, mitochondria, and chloroplasts) and their proteins are translated from mRNA exclusively on 70S ribosomes.

Proprioceptor. Sensory receptor located deep in tissues such as skeletal muscles and tendons.

Protein subunits. The individual proteins composing the viral **capsid**.

Pseudothecium (pl. pseudothecia). A flask-shaped structure with a hole (**ostiole**) at the top of the flask in which **asci** form in a single layer (hymenium); the structure develops before any nuclear exchange involving an ascogonium and **antheridium**.

Psychrophile(ic). An organism that grows well at colder temperatures (<10°C).

Puccinia path. The path, some 600 to 900 miles (1000 to 1300 km) wide, extending from northern Mexico and southern Texas to the Prairie Provinces of Canada, over which urediniospores of stem and leaf rust fungi annually blow northward and successively reproduce, extending those diseases from Mexico and Texas to Canada.

Pycnidium (pl. pycnidia). Conidiomata that are flaskshaped and contain exposed conidia and conidiophores.

Pycniospore. Thin-walled, monokaryotic (N) gamete produced in a pycnium (spermatium), the nuclei of which pair, but not fuse, with compatible nuclei in receptive, flexuous hyphae to establish the dikaryotic (N + N) condition of the rust fungus life cycle.

Qualitative resistance. Resistance that can be placed into distinct categories that form the basis of Mendelian ratios.

Quantitative resistance. Resistant host reactions that are not differentiated into distinct classes because there is continuous variation from resistant to susceptible phenotypes.

Quarantine. The exclusion of pathogens that are not established in an area; quarantine is a form of legal control for plant diseases.

Race. Population of pathogen isolates that have the same virulence.

Randomly amplified polymorphic DNA (RAPD). Amplified DNA polymorphism uncovered by the use of arbitrarily amplified DNA.

Rating scale. A progressive classification, as of size, amount, importance, or disease grades.

Reaction mixture (slang: cocktail). The components of a PCR reaction; usually includes buffer, primer, nucleotides, template DNA, magnesium ion, DNA polymerase, and water.

Reducing agent. A chemical that donates electrons to reduce another chemical reaction.

Reducing sugar (compound). Any sugar (or sugar-acid) that has an anomeric carbon; one in which the hydroxyl group around a carbonyl carbon can exist in different isomeric forms (e.g., D-glucose).

Regurgitant. The fluid mixture produced by beetles during feeding; it contains high levels of ribonucleases, deoxyribonucleases, and proteases.

Repeating spore. A spore stage that gives rise to the same type of mycelium and subsequent type of spore production as that from which it developed.

Resident phase. Epiphytic colonization of plants by phytopathogenic prokaryotes without apparent disease production. Among phytopathogenic prokaryotes lacking resistant spores, resident phases are important for survival of the pathogen until conditions favor pathogenesis.

Resistance. Ability of the host to reduce the growth, reproduction, or disease-producing activities of the pathogen; in pathogens, lack of sensitivity to a pesticide, acquired through a genetic change.

Resistant. A condition where plants are not hosts for a specific pathogen and the pathogen cannot cause a disease on that plant; often used to describe varieties of a plant species on which the pathogen cannot cause disease.

Respiration. Metabolic processes by which carbohydrates and lipids are converted into energy with the uptake of oxygen from the environment.

Resting spore (or cyst). A spore that functions as a survival structure under conditions of extreme temperature and moisture.

Restriction fragment length polymorphisms (RFLP). The variation in the length of a DNA fragment produced by a specific restriction endonuclease and generally detected by Southern hybridization with a nucleic acid probe.

Rhizoid. A branched fungal hypha that resembles a root and penetrates a substrate; in the Zygomycota.

Rhizomorph. Root-like structure composed of thick strands of somatic hyphae; facilitates the dispersal of some fungi to new substrates.

Rhizoplane. The plant surface/soil interface on roots.

Rhizosphere. The volume of soil that is physically, chemically, or biologically affected by the presence of living roots.

Ribozyme. Self-cleaving RNA strand that can specifically bind and cleave other RNA strands.

RNase. Enzyme that destroys or degrades RNA.

Rogue (rouging). Removal and destruction of undesired plants.

Saline. Containing salt or salts.

Salivary glands. The organs in aphids that produce the salivary excretions.

Sanitation. Control methods used to prevent the introduction or spread of plant pathogens.

Saprophyte(ic). An organism that obtains its food from a nonliving (dead) host.

Sclerotium (pl. sclerotia). Spherical resting structure 1 mm to 1 cm in diameter with a thick-walled rind and a central core of thin-walled cells with abundant lipid and glycogen reserves; individual hyphae have lost identity.

Scouting. Comprehensive, systematic check of fields at regular intervals to gather information on the crop's progress and pest level.

Secondary metabolite. A compound not directly associated with the processes that support growth but that presumably plays some other role (e.g., defense compounds).

Seed certification. The practice of providing disease-free seed by verifying that source plants used in seed production or the seed themselves are free of infection or contaminating pathogens.

Selection. The process of isolating and preserving individuals or characters from a group of individuals or characters.

Sensillum (pl. sensillia). An epithelial sense organ composed of one or a few cells with a nerve connection and taking the form of a spine, plate, rod, cone, or peg.

Serology. The study of antigen and antibody reactions; a method used to identify antigens and the organisms that produce them.

Seta (pl. setae). A sterile, bristle-like hair.

Severity. The proportion of an individual (leaf, plant, single field) affected by disease.

Sexual dimorphism. Exhibition of distinct morphological forms by the male and female of a nematode species.

Sexual reproduction. Reproduction of an organism requiring the union of compatible nuclei, forming the diploid chromosomal number, followed by a meiotic division that reduces the chromosomal number to the haploid state, followed by the subsequent formation of gametic cells that unite, reestablishing the diploid condition.

Siderophore. Compounds released by an antagonist that inhibit the growth of other organism by sequestering iron.

Sign. The presence of the pathogen or its parts on a host plant.

Signal recognition/transduction. Active defense mechanism by which the plant recognizes the pathogen and begins the production of induced defenses. Ion fluxes, oxidative bursts, protein phosphorylation, and signal molecules are involved in signal recognition and transduction.

Signal words. Words such as Caution, Warning, or Danger that appear on pesticide labels depending on the toxicity of the product.

Simple sequence repeats (SSRs). One of many DNA sequences dispersed throughout fungal, plant, and animal genomes, composed of short (2 to 10 base pairs) sequences that are repeated in tandem and are usually highly variable. Also know as microsatellites.

Single gene resistance. See **monogenic resistance**.

Soilless media. Material used in pots, flats, or other containers in which plants are grown. Components of soilless media may include peat moss, sand, vermiculite, perlite, rockwool, styrafoam, compost, bark, or other organic materials. Soil test laboratories usually consider a potting medium soilless if it is composed of less than 25% field soil.

Solarization. Use of sunlight to heat soil to temperatures lethal to plant pathogens and other biological contaminants.

Sorus (pl. sori). A group of fruiting bodies that contains resting spores or sporangia, or both.

Spatial. Of, or pertaining to, existing in space.

Species. A group of individuals genetically or morphologically distinct from other organisms; specific epithet.

Specific resistance. Resistance effective against certain biotypes or races of a pathogen but ineffective against other biotypes or races; syn. **race-specific resistance**.

Spermatia. Spores that function as male gametes in the Ascomycota and Basidiomycota.

Sporangiophore. The hyphal stalk that bears a sporangium.

Sporangiospore. Motile or nonmotile asexual spore produced in a sporangium.

Sporangium (pl. sporangia). A sac-like structure that is internally converted into spores or a single spore (downy mildews).

Spore. The cellular reproductive unit of many organisms (similar in function to seeds of plants).

Sporocarp. Fruiting structure that bears spores.

Sporodochium (pl. sporodochia). Conidiomata with short conidiophores, which are clustered into a rosette and produced on a superficial mycelial layer.

Sporophore. Stalk that supports a spore (see **conidiophore** and **sporangiophore**).

Sterilization. The process for killing all forms of life.

Stipe. The stalk of a mushroom **basidiocarp** that supports the **pileus**.

Stolon. Branch of hyphae that skips over the substrate — a runner; in the Zygomycota.

Strain. A virus isolate that differs from the type isolate of the species in a definable character, but does not differ enough to be a new species.

Stylet. A needle-like structure found in the anterior end of nematodes and used to penetrate cell walls for feeding and movement inside roots. Also, a type of piercing-sucking mouthpart found in Homoptera.

Subcuticular. Beneath the waxy cuticle.

Sun scald. Injury that occurs to fruit or leaves when exposed to a sudden increase in light intensity.

Susceptibility (susceptible). Inability of the host to reduce the growth, reproduction, or disease-producing activities of the pathogen when environmental conditions are favorable (also susceptible).

Symptom. The response of the plant to the presence of a pathogen or adverse conditions.

Syncytium. A specialized, multinucleate feeding cell induced by cyst nematodes inside plant roots.

Synnema (pl. synnemata). Conidiomata with compact or aggregated clusters of erect conidiophores with conidia formed at or near the apex.

Systemic acquired resistance (SAR). An active defense mechanism in which a necrotizing pathogen causes distant induction of general defense mechanisms that serve to protect the plant. Salicylic acid or methyl salicylate is produced as a primary and secondary response, and antimicrobial proteins are produced.

Systemic fungicides. Fungicides that have one or more of the following characteristics: the ability to enter a plant through roots or leaves, water solubility to enhance movement in the plant's vascular system, and stability within the plant. Systemic fungicides are often more vulnerable to the development of resistance in the target fungus population.

Tannins. Polyphenolic compounds occurring in plants that have the ability to bind and precipitate proteins.

Teleomorph. Conceptually, the sexual aspect of a fungus; contrast with **holomorph** and **anamorph**.

Teliospore. Thick-walled, overwintering or resting spore of rusts and smuts, in which karyogamy (nuclear fusion) occurs and from which the basidium arises.

Temporal. Of or limited by time.

Thallus. The vegetative (nonreproductive) body of a fungus.

Thermocycler. The machine that is programmed for automatic temperature changes (cycles) required for the PCR reaction. Typically has set points for denaturing, annealing, and extension temperatures.

Thermophile(ic). An organism that grows well at warmer temperatures (>40°C).

Thigmotropism. Contact response to a solid or rigid surface that results in orientation of an organism or one of its parts.

Tolerance. Ability of a cultivar to perform well under adverse conditions. "True tolerance" occurs if one cultivar sustains less damage than another cultivar when the amount of infection is the same for both cultivars. Tolerance also is used synonymously with partial or general resistance in which case the improved performance under adverse conditions is the due to a lesser amount of disease on the "tolerant" cultivar (i.e., partial resistance).

Transcription. The transfer of genetic information associated with deoxyribonucleic acid (DNA) to messenger ribonucleic acid (mRNA).

Translation. The transfer of genetic information associated with the messenger ribonucleic acid (mRNA) into a sequence of amino acids forming a polypeptide chain ultimately forming a protein.

Translocation. Movement of materials, including water, minerals, and organic materials through the vascular system of the plant.

Transpiration. Loss of water in the vapor phase from leaves and other above-ground plant parts.

Trehalose. Disaccharide of glucose used as a carbon reserve compound in fungal cells.

Trichomes. Plant hairs.

Tylosis (pl. tyloses). Occurs in xylem tissue; outgrowth of a parenchyma cell that partially or completely blocks the lumen of the vessel; often associated with wilts.

Unitunicate ascus. An ascus having a single wall.

Urediniospore. Dikaryotic ($N + N$), repeating, summer spore stage of rust fungi.

Vector. A nematode, insect, or other organism that can transmit a virus or other agent into or onto a plant.

Vegetative compatibility. Vegetative hyphae that are able to fuse and maintain the heterokaryotic genetic state.

Veraison. Period of grape ripening during which berry growth slows down and coloration develops.

Vesicle. Intra- or intercellular, ovate to spherical structure that contains storage lipids and may also serve as a propagule in some endomycorrhizae. Also, the thin walled, balloon-like structure in which zoospores of *Pythium* species (Oomycota) differentiate.

Viroid. Small particle (250 to 400 nucleotides) of circular, single-stranded RNA whose genome is too small to code for proteins and thus has no protein coat.

Virulence. Ability of a pathogen to cause a compatible (susceptible) reaction on a host cultivar with genetic resistance.

Virulence formula. Identification of isolates of pathogens based on effective and ineffective host genes.

Viruliferous. A vector that has acquired virus and is capable of transmitting it.

Virus. A pathogen that is composed of either RNA or DNA and is enclosed in a protective protein coat.

Viscin. A sticky mucilaginous pulp that coats the seed of dwarf mistletoes.

Volva. At the base of the stipe, a remnant of the universal veil that once enveloped the entire developing mushroom.

White rusts. Diseases caused by *Albugo* species (Oomycota, Albuginaceae).

Witches' brooms. Distortions of normal shoot growth in which a loss of apical dominance creates a bunched growth of shoots.

Wood decay. Enzymatic and chemical process by which microoganisms, such as basidiomycetes, degrade plant cell walls to obtain nutrients for growth and reproduction. Generally, white and brown rots are recognized.

Zoospore. A spore produced during reproduction capable of moving in water (i.e., swimming spore).

Zygomycota. Fungal phylum where sexual reproduction results in the production of zygospores; includes common bread mold and decay fungi; in the Kingdom Fungi.

Zygospore. A diploid, thick-walled spore that results from sexual reproduction; in the Zygomycota.

Appendix I

Careers in Plant Pathology

Alan S. Windham and Mark T. Windham

Plant pathologists are the ultimate plant doctors. Just as your family doctor and medical researcher diagnose and study human diseases, respectively, and a veterinarian diagnoses and studies animal diseases, a plant pathologist studies diseases of plants. Just as there are specialists in animal and human medicine, there are specialists in plant pathology who study the causes of plant disease, research management strategies, diagnose plant problems, breed and select new cultivars of plants, and work in many related fields.

New and emerging diseases that threaten food, fiber, and ornamental crops are a constant danger to our agricultural economy. Professionals with training in plant pathology are needed to regulate the movement of these pathogens, educate the general public and agricultural professionals about the threat of these diseases, and conduct research on the nature and management of these diseases.

If you have decided on a career in plant pathology, it is important to choose college courses that will increase your knowledge and be valuable in your career path. As an undergraduate, you may wish to take a diverse number of courses that cover disciplines such as genetics, mycology, botany, entomology, microbiology, virology, chemistry, botany, biochemistry, agronomy, soils, and horticulture. If you decide to specialize in an advanced degree, you can tailor your curriculum to meet your specific goals. For example, if you are interested in nematology, you may wish to fashion an academic program that explores invertebrate biology, soils, ecology, genetics, and molecular biology. For research degrees such as M.S. and Ph.D., you will design an original research project on some aspect of nematology.

Professionals with undergraduate training in plant pathology may be employed as growers or pest management specialists at greenhouses or nurseries, as county extension agents with the Cooperative Extension Service, research associates or technicians with land grant universities, and in agrichemical companies or businesses specializing in biotechnology. Graduates with a master's degree in plant pathology may work as a sales or technical representative for seed or agrichemical companies. There are also opportunities to work as research technicians with land grant universities, federal agencies such as the United States Department of Agriculture – Agricultural Research Service, agrichemical companies, or businesses specializing in biotechnology. There are also survey and regulatory positions available with USDA APHIS (Animal and Plant Health Inspection Service) and PPQ (Plant Protection and Quarantine). Each land grant university has a plant disease clinic that often hires graduates in plant pathology to diagnose plant problems. Diagnosticians serve as a valuable link to the general public. There are also teaching positions available at community and four-year colleges in the biological sciences.

Graduates with a doctoral degree in plant pathology may work as extension specialists for the Cooperative Extension Service, join the faculty of colleges and land grant universities, conduct research for the USDA Agricultural Research Service, the US Forest Service, seed companies, agrichemical companies, or companies specializing in biotechnology. There are also private practitioners that specialize in the diagnosis of plant diseases. These diagnosticians frequently work with farmers and growers of specialty crops such as greenhouse and nursery crops and turfgrass.

If you are interested in a career in plant pathology, contact a plant pathology department at a land grant university for educational and career opportunities. You may also contact the American Phytopathological Society. Find out more about plant pathology at www.apsnet.org.

Appendix II

Fantastic Plant Pathology Websites

Appendix II lists some excellent websites that provide supplemental materials to the book chapters. The URLs have been checked as of February 1, 2003 and all are functional. Websites in Appendix II will be updated periodically and may be accessed at http://dogwood.ag.utk.edu/ and by clicking on "Fantastic Plant Pathology Websites."

Chapter	Website	Comments
1 and 2	http://apsnet.org/	Home for American Phytopathological Society — website has feature articles on plant pathological topics, modules on plant diseases, and other information of interest to people concerned about plant diseases
	http://www.bspp.org.uk/	The British Society of Plant Pathology — excellent web site and links to subjects on plant pathology
	http://www.vms.utexas.edu/~jdana/history/famine.html	Late blight of potato historical site
3	http://forestpathology.org/index.html	Forest and Shade Tree Pathology (all types of pathogens)
	http://helios.bto.ed.ac.uk/bto/microbes.htm#top	The Microbial World
4 and 5	http://tulane.edu/~dmsander/garryfavweb.html	All about virology on the World Wide Web
	http://image.fs.uidaho.edu/vide/	Plant Viruses Online
	http://micro.msb.le.ac.uk/335/viorids/html	Viroid website
	http://www.bidogi.e.,uni-hamburg.de/b-online/e35/35.htm	Plant viruses and viroids
6 and 7	http://www.hcs.ohio-state.edu/hcs300/bact.htm	Bacteria web page
	http://wheat.pw.usda.gov/~lazo/docs/xmal	*Xanthomonas* web page
8 and 9	http://nematode.unl.edu	University of Nebraska nematology website
	www.wcrl.ars.usda.gov/cec/teaching/nema.htm	USDA listing of nematology sites on WWW
10–21	http://botit.botany.wisc.edu/toms_fungi/	Tom Volk's Fungi — excellent, general web page about fungi
	http://www.masfungi.org/	The Mycological Society of America
	http://www.britmycolosoc.org.uk/	The British Mycological Society
	http://mycology.cornell.edu/	The WWW Virtual Library: Mycology
	http://mycolog.com/fifthtoc.html	The Fifth Kingdom
	http://www.suddenoakdeath.org/	Web site on the destructive disease of sudden oak death
	http://helios.bto.ed.ac.uk/bto/microbes/biotroph.htm	Biotrophic plant pathogens
	http://zoosporic-fungi.dmc.maine.edu	Excellent site for fungi and fungal like organisms that have zoospores
22	http://www.science.siu.edu/parasitic-plants/index.html	University of Southern Illinois at Carbondale; parasitic plant site
	http://www.rms.nau.edu/mistletoe	The mistletoe center — site of the USFS Rocky Mountain Research Station

Chapter	Website	Comments
24–25	http://cropsci.uiuc.edu/faculty/gca/kaffe/index.htm.	Excellent website for comparing different molecular techniques
26–31	http://www.scri.sari.ac.uk/MBN/Erwinia/chartxt.htm	Plant resistance mechanisms site for *Erwinia*
32–35	http://www.cnr.umn.edu/FR/extension/foresthealth/foreststandshadetreehealthpage.htm	Forest and Shade Tree Health (University of Minnesota)
	http://www.cas.psu.edu/docs/CASDEPT/PLANT/ext/bact_dis.html	Penn State Fact Sheets
	http://www.ca.uky.edu/agcollege/plantpathology/PPAExtesnion/Ppalinks.htm	University of Kentucky Fact Sheets
	http://ohioline.osu.edu	Ohio State Fact Sheets
	http://www.aces.edu/dept/plantdiagnosticlab	Auburn University Plant Diagnostic Lab
36–37	http://www.sciencemedcentral.com/bpo/general/home.htm	Biological Procedures Online
	http://www.agdia.com	Agdia's Diagnostic Site
	http://www.caf.wvu.edu/Kearneysville/wvufarm6.html	University of West Virginia Diagnostic keys to major tree fruit diseases in the mid Atlantic State

Index

A

Abiotic diseases, 201
Abiotic stresses, 6
Abscisic acid, in plant diseases, 273
Acervulium, 135
Acervulus, 11, 80
Acetobacter, 41
Acidovorax, 41
Actinomyces, nocardioform, 43
Action threshold, 338
Active defense strategies of plants, 261
Active mechanisms to present pathogen penetration, 262–263
Active penetration, 50
Acylalanines, 320
Adapted cultivars, 303
Agaricales, 166
    habitat of, 167
    wood rots caused by, 170
Agarose plates, preparation of, 37
Aggressiveness, 301
Aglycone, 50
Agonomycetes, 87
Agrobacterium, 41, 315
    cytokinin production by, 245
    experiment to determine effect on normal plant growth, 277–278
    indole-3-acetic acid production by, 245
    radiobacter
        K1026, 331
        K84, 327
    rhizogenes, 48
    tumefaciens, 8, 45, 48, 244
        production of plant growth hormones by, 272
    tumorigenic, type IV secretions of proteins from, 45
Air pollution
    as an abiotic stress, 6
    injury due to, 205–206
Albuginaceae, 174
Albugo, 174
    candida, 180, 272
        observation of morning glory leaf tissue infected with, 190–191
    ipomoeae-panduranae, 181
Alder
    Brennaria pathogenicity of, 43
    Phyllactinia pathogenicity of, 118
Alfalfa
    crown wart of, 93
    rust, 158
    stem and bulb nematodes and, 66
    wilt caused by *Clavibacter michiganense,* 45, 48
Algae, 5, 8, 15, 199
Alkaline phosphatase, use of in ELISA, 28
Alkyl pyrones, use of for biological control, 329
Alloinfection, 286
Almond
    deformation of fruit by *Taphrina deformans,* 114–115

scorch of caused by *Xylella fastidiosa,* 42
Alternaria
    alternata, 273
    solani, 135
        use of TOM-CAST for control of, 291
Alternate host elimination, for control of rusts, 160
Alyssum, as host for *Plasmodiophora brassicae,* 94
AM-toxin, 273
Amanita vaginata, 168
American Phytopathological Society, 6
Aminopyrrolizidines, 239
Ammonia, use of for biological control, 329
Amoracia, as host for *Plasmodiophora brassicae,* 94
Ampelomyces quisqualis, use for biocontrol of powdery mildew, 125
Amplicons, 218
Amplification
    exercise for, 221–224
    of internal transcribed spacers of rRNA, exercise for, 229–231
Amplification fingerprinting of DNA, 211
    data collection and analysis exercise, 227–229
    electrophoresis and, exercise for, 224–227
    exercise for, 220–221
Amplification-based techniques for generating molecular markers, 211
Anamorph, 81
Anastomosis, 77
    groups, 170–171
Angiosperms, *Taphrina* parasites of, 112
Anguina tritici, 66
Angular leafspot, 49
Annosum root rot, 169
Antagonism, 239
    mechanisms of, 328–330
Antagonists, 328
Antheridia, 14
Anthracnose diseases
    effects of on photosynthetic competence, 270
    experimental evaluation of resistance in beans by leaf disc
        bioassay, 374–375
Antibiosis, 328
Antibiotics
    production of, 235–236
    use of as antagonists for biological control, 328–329
Antibodies, 28
Antimicrobial compounds, 263
Antimicrobial proteins, 264
Antisera, 28
Aphelenchoides, 66
Aphids, as vectors of viruses, 25–26
Aphyllophorales, 166
Apothecia, 86
Apothecia, 13, 130
Apple
    alternaria leaf blotch of, 273
    discolorations caused by *Gluconobacter,* 42
    disease of caused by *Erwinia amylovora,* 45
    fire blight caused by *Erwinia amylovora,* 45

post harvest rot caused by *Mucor piriformis*, 86
    scab, 81, 132, 239–240, 283
Appresoria, 239
Apricots, *Taphrina* diseases of, 112
Arabidopsis
    experimental induction of pathogenesis-related proteins in, 375–376
    thaliana, 94
Arbitrarily amplified DNA, 211
Arceuthobium, 193
    as a limiting factor in crop production, 5–6
Area under disease progress curve, 290
Armillaria, 166
    mellea, 82, 86, 168
    tabescens, 168
Arthrobacter, 43
Artificially induced variation, 303–304
Ascocarps, 86
Ascomata, 130
Ascomycetes
    description of, 127
    nomenclature for, 128
    reproduction of, 128
    taxonomy of, 130
Ascomycetous yeasts, 111
Ascomycota, 12, 86
Ascospores, 12–13, 81
Ascostroma, 86
Ascus, 13
Asparagus, autoecious rusts of, 158
Aspergillus, 134–135
    flavus, 133
Assays
    bio, 27–28
    dot immunobinding, 39
    ELISA, 28, 39
    infectivity, 27
    pathogenicity experiment, 55–56
Assembly, viral replication by, 22
Auriscalpium, 167
Autoecious rusts, 158
Autoinfection, 286
Auxin, 272
Avirulent pathogens, 7, 209, 300
Azaleas
    leaf gall, 283
    parasitic algae infection of, 199

B

Bacilliform viruses, 24
Bacillus, 43, 46
Bacillus, as an antibiotic used for biological control, 329
Backcross methods, of resistance incorporation, 304
Bacteria, 5, 8, 14–15. *See also* prokaryotes
    case study for wilt, 265
    experiment for physiological testing of, 57–59
    experiment isolate from plant tissue, 54–55
    experiment to determine physical defenses of potato against soft
        rotting, 267–268
    experiment to observe in plant tissue, 53–54
    identification of plant pathogenic, 44
    phloem-limited, 44
    physiological testing to determine identity of, 355
Bactericides, 323
Badiospores, 13

Baiting, 354
Bakanae, elongation of internodes due to, 272
Balansia, 87
Ball moss, 199
Banana
    Brennaria pathogenicity of, 43
    Granville wilt of caused by *Ralstonia solanacearum*, 42, 45, 48
Bark splitting, 202
Barley
    smut diseases of, 151–153
    stripe mosaic virus, transmission of by pollen, 27
    take-all root disease of, 327
    yellow dwarf virus, 341
        persistent circulative nonpropagative transmission of, 26
Basidiomycota, 86–87, 151
    morphology of, 165–166
    non-fleshy, 170–171
    taxonomy of, 166–167
    wood rots caused by, 169–170
Basidiomycota, 13
Basidiophora, 178
Basidiospores, 81
Basidium, 13
Bean
    autoecious rusts of, 158
    experiment on root rot caused by *Pythium* species, 183–184
    haloblight of caused by *Pseudomonas syringae*, 48
    leaves, colonization of by ice-nucleation active *Pseudomonas*
        syringae, 45
    pod mottle virus, 24
        experiment for mechanical transmission of, 35
        Ochterlony double diffusion experiment for detection of, 37–39
    rust, experiment, 161–162
Beet
    curly top virus, transmission of by leafhoppers, 26
    necrotic yellow vein virus, transmission of by *Polymyxa*, 27
Beetles, as virus vectors, 26
Begomovirus, transmission of by whiteflies, 26
Belonolaimus, 65
Benzimadiazoles, 320
Betulaceae, 167–169
Bioassays
    experiment to evaluate anthracnose resistance using leaf disc
        procedure, 374–375
    identification of viruses by, 27–28
    in vitro, 370
    to determine levels of pesticides, 205
    use of plant cell cultures in, 367
Biocontrol products, delivery, application, and formulation of, 330–331
Biofilms, 49
BIOLOG, 355
Biological control, 327–331
Biotic diseases, 201
Biotrophs, 8, 77
Biotypes, 306
Bipolaris, 135
Bird's nest fungi, 165
Bizarre tulips, 22
Black knot, 316
Blackrot, caused by *Xanthomonas campestris*, 48
Blights
    chestnut, 5, 76, 281–282
    definition of, 9
    due to abiotic stresses, 6
    early, forecasting, 291
    fire, 45

late, 4, 76, 281
northern corn leaf, 296
Pythium, 174
Sclerotinia, 285
southern corn, 5, 133, 273, 282
tomato, 135
BLITECAST, 291
Blueberry, galls on caused by *Nocardia vaccinii,* 43, 48
Blumeria graminis, 122–123
Blumeriella jaapii, 270
Bordeaux mixture, 84, 178–179, 320
Botryosphaeria, 201
Botrytis, 315
chemical attack by, 244
cinerea, 82, 120
Bradyrhizobium, 41–42
binding of by soybean, 46
japonicum, rhizobitoxin produced by, 47
nitrogen-fixing nodules on legume roots caused by, 48
Branch death, due to ball moss, 199
Brassica, as host for *Plasmodiophora brassicae,* 94
Breeder's stock, 303
Bremia, 174, 178
Bremiella, 178
Brennaria, 43
Broad-spectrum antibiotics, 328
Broccoli, *Plasmodiophora brassicae* as a pathogen of, 93–94
Broomrapes, 198
Brown rots, 170
Brussell sprouts, *Plasmodiophora brassicae* as a pathogen of, 93–94
Bud and leaf nematodes, 66
Buffalo nut, 193
Bulk method, of developing cultivars, 305
Bunt
common, 153
dwarf, 153
Karnal, 151–153
Burenia, 111–112
Burkholderia, 42
cycstic fibrosis caused by *cepacia* species of, 42
Burl, definition of, 9
Bushy stunt, 24
Butt rot, definition of, 9
Bymovirus, transmission of by fungi, 27

C

Cabbage
blackrot of caused by *Xanthomonas campestris,* 48
Plasmodiophora brassicae as a pathogen of, 93–94
Xanthomonas pathogenicity of, 42
Callus cultures, use for in vitro screening to assess pathogenic sensitivity, 370
Calvin cycle, 270
Camelina, as host for *Plasmodiophora brassicae,* 94
Candida albicans, 134
Candidiosis, 134
Canker
definition of, 9
fungi causing, 202
stain fungus, 315
Canola, *Plasmodiophora brassicae* as a pathogen of, 93–94
Capsella, as host for *Plasmodiophora brassicae,* 94
Capsids, 22
Captan, 320

Carbon fixation, 270
Carmovirus, transmission of by thrips, 26
Carrots
galls of caused by *Rhizobacter daucus,* 42
leaf gall of wild, 112
soft rot of caused by *Claustridium,* 43
Cassava, empty root of, 199
Cauliflower
mosaic virus, semipersistent transmission of, 25
Plasmodiophora brassicae as a pathogen of, 93–94
Cedar-apple rust, 158, 316
Cell permeability, alternation of by plant pathogens, 273
Cell walls, 45
experiment for determining structure of, 56–57
Cellular association, 22
Cellulases, 175
bacterial, plant pathogenesis by, 46
Cellulolytic activity, quantitative measurement of by experiment, 251–259
Cellulose, decomposition of dead organic matter by metabolism of, 237
Central America, early plant diseases in, 4
Cephaleuros, 199
Ceratobasidiales, 166
Ceratobasidium, 165, 171
Ceratocystis, 315
Cereals
powdery mildew diseases of, 122
roots
diseases of caused by *Polymxya graminis,* 92
Herbaspirillum rubrisubalbicans pathogenicity of, 42
smut diseases of, 151–153
Ceylon, coffee rust in, 4
Chemical attack of pathogenic organisms, 244–246
Chemical control, of bacteria, 51
Chemical-induced systemic acquired resistance, 323
Chenopodium quinoa, use of as an indicator host, 27
Cherries
black knot, 316
leaf spot, 270
Taphrina diseases of, 112
Cherry, rasp leaf virus, 65
Chestnut blight, 5, 281–282
Chinese cabbage, *Plasmodiophora brassicae* as a pathogen of, 93–94
Chitinases, 51
Chlamydospores, 12, 80
Chlorophyll
experiment to determine effects of plant pathogen on, 274–276
viral symptoms based on changes in, 23–24
Chlorophyllum, 170
Chlorosis
as a symptom of flagellate protozoa infection, 199
caused by *Heterodera glycines,* 63
definition of, 9
due to abiotic stresses, 6
due to drought stress, 202
due to flooding, 202
due to nutrient deficiencies, 203
of local lesions, 23
Chlorthalonil, 320
Chromista, 84. *See also* Stramenopila
Chytridiomycota, 85, 92
Chytrids, 92–93
experiment of plant infection by gall-inducing, 102–106
Cirrhus, 11

Citrus
    disease of caused by *Sprioplasma citri*, 44
    greening caused by *Liberobacter asiacticum*, 49
    infections of by *Eremothecium coryli*, 111
    leaf blight, caused by *Xylella fastidiosa*, 42
    stubborn disease caused by *Spiroplasma citri*, 48
    tristeza virus, cross protection as a means to control, 29
Clam connections, 165
Clavibacter, 14, 43
    michiganense
        alfalfa wilt caused by, 45
        tomato canker caused by, 48
Claviceps purpurea, 4, 81, 86, 131. *See also* ergotism
Claviciptaceae, 87
Cleistothecia, 86
Cleistothecia, 13, 82, 120
Clones, 212
Closteroviruses, transmission of by whiteflies, 26
Clostridium, 43, 46
Cochliobolus
    carbonum, 273
    heterostropus, 5, 273 (*See also* southern corn blight)
Coconut palm, heartrot of, 198–199
Coelomycetes, 87
    conidiomata, 137–138
Coffee
    infection of by *Eremothecium*, 111
    phloem necrosis of, 199
Coffee rust, 4–5, 76, 158
Cole crops, *Xanthomonas* pathogenicity of, 42
Collards, *Plasmodiophora brassicae* as a pathogen of, 93–94
Colletotrichum, 131
    acervulus of, 12
    coccodes, 291–292
    lagenarium, 137
    lindemuthianum, 78, 239
Common bunt, 153
Comovirus, 19
    transmission of by beetles, 26
Competition, 329
Complete resistance, 299
Condia, 11
Conidiophores, 11, 80, 117, 134–135
Conifers, rusts of, 158
Coniothyrium minitans CON/M/91-08, 331
Conks, 13–14
Conopholis americana, 193
Contact fungicides, 321
Contagium fluidium vivium, 23
Control measures, 288–290
Coprinus, 170
Corky root, of lettuce caused by *Rhizomonas suberifaciens*, 42
Corn, 133
    brown spot of, 93
    northern leaf blight, 296
    pathogenicity of mushroom-forming genera on, 166
    smut diseases of, 151–155
        diagnosing, 346
    southern blight, 133, 273, 282
        in hybrid seeds, 5
    Stewart's wilt caused by *Pantoea stewartii*, 43
    stubby-root nematodes and, 65–66
    stunt caused by *Sprioplasma kunkelii*, 44
Cornus. See also flowering dogwood
    florida, 5
    nuttallii, 5

Corynebacterium, 14
Cotton
    antibiotics used for biological control of pathogens attacking, 329
    Fusarium wilt of, 136
    infections of by *Eremothecium coryli*, 111
    lance nematodes and, 66
    ring nematodes and, 66
    root-knot nematodes of, 61
Cowpea mosaic virus, 19
Cricket-bat willow, *Brennaria* pathogenicity of, 43
Criconemella, 66
Criniviruses, transmission of by whiteflies, 26
Cronartium ribicola, 283
Crop compendia, 350
Crop failure, 4
Crop losses, 290
    resulting from nematodes, 62
Crop management, 339
Crop placement, 316
Crop rotation, 315–316
    as disease control measure, 289, 340
Cross protection, 329
Cross-pollination, effect on disease development, 284
Crown gall, 6, 8, 272, 315
    caused by *Agrobacterium tumefaciens*, 48
Crown rot, 84
Crown rust fungus, 308
Crucifers
    Albugo candida as a pathogen of, 180
    blackrot of caused by *Xanthomonas campestris*, 48
    clubroot of, 84, 92
    experiment on susceptibility of to *Plasmodiophora brassicae*, 99–102
Cryphonectria parasitica, 5. *See also* chestnut blight
Cryptonectria parasitica, 283
Cucumber
    angular leafspot of, 49
    mosaic virus, 271, 283, 316
    powdery mildew of, 125
Cucurbita pepo, 271
Cucurbits
    powdery mildew disease of, 124–125
    yellow vine caused by *Serratia marcesens*, 43
Cultivar mixtures, 308
Cultivated germplasm, 303
Cultural control
    of prokaryotes, 51
    of rusts, 160
    of viruses, 30
Culture indexing, 316
Culture media preparation, experiment for disease diagnosis, 361–362
Curtobacterium, 43
Cuscuta, 193–195
Cutinases, bacterial, plant pathogenesis by, 47
Cyst nematodes, 63
Cytokinins, 272
    production of, 245
Cytoplasm, 46

D

Damage threshold, 338
Damping off, 84, 175, 316
    experiment on biocontrol of, 335–336
Date palms

rot of terminal buds caused by *Paenibacillus,* 43
smut diseases of, 151–153
Deformations, caused by viruses, 24
Deuteromycota, 87
    conidiomata, 135–137
    experimental identification of, 141–143
Deutromycota, description of, 133–135
Diagnosis of disease, 348–357
Dieback, definition of, 9
Dill, galls caused by *Protomyces,* 112
Discomycetes, 131
Discula
    destructiva, 5, 212 (*See also* dogwood anthracnose)
    umbrinella, 212
Disease
    aggressiveness, 301
    analysis or models for epidemics, 287–290
    assessment exercise, 292–293
    complexes, associated with lesion nematodes, 64–65
    complications, 287
    control, 50–51
    cultural control of, 313
    cycle, 8
    diagnosis, 348–357
        experiment for culture media preparation, 361–362
        experiment on process of, 362–364
        process of, 345
    diagrams, 286
    forecasting and management, 291–292
    gradient, 287
    host indices, 350
    incidence of, 286
    onset of, 285
    pressure, 296
    progress exercise, 293
    rating scales, 286
    resistance, 295–300
        breeding for, 303–306
    severity, 286
    suppressive soils, 327
    tolerance, 309–310
Disease-resistant genotypes, in vitro development of, 367–370
Diseases, chemical control of, 319–323
Disinfectants, 323
Disinfection, 315
Dissemination, 8
Distortion, as a symptom of viral infection, 24
Dithiocarbamates, 320
Ditylenchus dipsaci, 66
DNA, 209
    amplification, exercise for, 221–224
    amplification fingerprinting, 211–212, 217
        data collection and analysis exercise, 227–229
        exercise for, 220–221
    arbitrarily amplified, 211
    comparison of profiles, 217
    probe molecular methods for identifying pathogens, 356
    randomly amplified polymorphic, 211
    selective amplification, 217
DNases, 219
Dodder, 193
    as a virus vector, 27
    control of, 195
    disease cycle, 194
Dogwood anthracnose, 5
Dolichodorus heterocephalus, 66

Dolipore septum, 165
Dot immunobinding assays, 39
Downy mildews, 8, 84, 174, 178–180
    hyperauxiny in diseases caused by, 245
    of grapes, 76
    pathogens causing, 10
Drip irrigation, 313
    as disease control measure, 289
Drought, 8
    infection of canker causing fungi due to stress from, 201
    stress, 202
Durable resistance, 299
Dust formulations of fungicides, 321
Dutch elm disease, 82, 137
    decrease of water flow in the xylem due to, 271
Dwarf bunt, 153
Dwarf mistletoe, 8, 193, 195–196
    as a limiting factor in crop production, 5–6
Dwarfing, as a symptom of viral infection, 24

E

Early harvest, as disease control measure, 289
Ebb and flow irrigation, 313
Ectendomycorrhizal fungi, 89
Ectomycorrhizal fungi, 88, 167–169
Ectoparasites, 61
Edema, 202
Eelgrass
    observations of wasting disease, 106–110
    wasting of caused by *Labyrinthula zosterae,* 92, 96–97
Egestion, 25
Electrophoresis, 219
    and staining of DAF products, exercise for, 224–227
ELISA, 28, 39
    use for disease diagnosis, 354–355
Elm
    Dutch elm disease, 82, 137
        decrease in water flow through the xylem due to, 271
    scorch of caused by *Xylella fastidiosa,* 42
    wetwood caused by *Clostridium,* 43
Embden-Meyerhof pathway, 271
Empty root, due to flagellate protozoa, 199
Emulsifiable concentrates, 321
Enations, as a symptom of viral infection, 24
Endoglucanases, 46, 51
Endomycorrhizal fungi, 88–89
Endoparasites, 61
Endophytes, 77
    symbiosis with plants and microorganisms, 239
Endophytic fungi, 87
Endophytic plant pathogens, 48
    nutritional classes of, 48–49
Endothia parasitic. See Cryphonectria
Enterobacter, 43
    cloacae, as a bacterial antagonist, 329
Entomosporium, conidia of, 12
Environmental stresses, 6
Enzyme activities
    experiment to develop standard curves and measurements of, 254–259
    qualitative determination of by experiment, 247–250
Enzyme-linked immunosorbant assay. *See* ELISA
Enzymes
    bacterial, plant pathogenesis by, 46

use of for biological control, 329
Epichloë, 87, 239
Epicoccum, sporodochium of, 13
Epidemic, components of, 281
Epidemics
  disease analysis or models for, 287–290
  effects of environment on, 284
  effects of host population on, 284–285
  effects of pathogens on, 283–284
Epigenetic resistance, 370
Epinasty, 272
Epiphytes, 48
Eradicant fungicides, 321
Eradication, as a method to control viruses, 30
Eremothecium coryli, 111
Ergot alkaloids, 239
Ergotism, 4, 81
Ericaceous endomycorrhizae, 88–89
Eriophyid mites, transmission of viruses by, 27
Erwinia, 43
  amylovora, 283
    disease in apple caused by, 45
    extracellular polysaccharide production by, 245–246
    fire blight caused by, 45
  chemical attack by, 244
  herbicola, production of galls due to plant growth hormones produced by, 272
  soft rot pathogenesis by, 46
Erysimum, as host for *Plasmodiophora brassicae*, 94
Erysiphaceae, 117
Erysiphe
  cleistothecia of, 13
  necator, 120
Erysiphe pulchra, 125–126
Ethylene
  as a gaseous plant growth hormone, 272
  plant injury due to, 205–206
Etiolation, caused by fungi, 78
Exobasidium, 170
  vaccinii, 283
Exopolysaccharides, 271
Experiments
  Ouchterlony double diffusion technique, 37–39
  transmission of *tobacco mosaic virus*, 33–37
Exponential model of disease analysis, 287–289
Extracellular enzymes, use of by pathogens to attack plants, 244–245
Extracellular hydrolytic enzymes, 175
Extracellular polysaccharides, 45
  chemical attacks on plants by, 245–246

F

Facultative anaerobic gram-negative rods, 43
Facultative fermentives, 81
Facultative parasites, 8
  plant attacking strategies of, 238–239
  soil-inhabiting, 174
Facultative saprophytes, 8, 77, 239–241
Fagaceae, 167–169
Fairy rings, 170
  DNA amplification fingerprinting to study, 212
Ferns, *Taphrina* parasites of, 112
Fertilization, 313
Fescues, endophytic fungi associated with, 87
Field beans, *Xanthomonas* pathogenicity of, 42

Field diagnosis of disease, 346
Fire blight, 315
  caused by *Erwinia amylovora*, 45
  chemical control of, 51
  effect of nitrogen fertilization on, 317
Fission yeasts, 111
Flagella, 45
Flagellate protozoa, 15
Flagging, definition of, 9
Flax, autoecious rusts of, 158
Fleshy basidiomycetous fungi, 165–171. *See also* mushrooms
Flexuous rod viruses, 24
Flood and drain irrigation, 313
Flooding, 8, 202
Floral infection, 153
Flowable fungicide formulations, 321
Flowering dogwood, 5
  powdery mildew of, 125–126
Foliar disease, control of by fungicide seed treatment, 342
Foliar fungicidal sprays, 320, 330–331
  for rust, 160
Foolish seedling disease, elongation of internodes due to, 272
Forage grasses, smut diseases of, 151–153
Freeze injury, 201–202
Fruit
  abscission, caused by fungi, 78
  anthracnose, use of TOM-CAST to control, 291
  premature ripening of caused by ethylene, 273
  rot, 111
  tree pruning as disease control measure, 289
Fumonisins, 134
Fungal alkaloids, 239
Fungal growth and protein isolation, experiment for, 251–254
Fungal leaf spots, experiment for diagnosis of, 359–361
Fungi, 5, 8, 11–14
  as virus vectors, 27
  Ascomycota, 86
  Basidiomycota, 86–87, 165–171
  characteristics of, 76–77, 83
  Chytridiomycota, 85
  classification of, 82–87
  Deuteromycota, 87
  disease symptoms caused by, 78
  environment and, 81
  experiment to determine effects of light and temperature on, 143–144
  experiment to identify variability of growth on host materials and in culture, 144–145
  experimental effect of volatile candy flavorings on the growth of plant pathogenic, 264–267
  genome sequencing of, 210
  heterothallic, 81
  homothallic, 80
  imperfect, description of, 133–135
  infection of plants by, 77
  mitosporic, 87
  mycoparasites, 329
  reproduction of, 78–81
  stem rust, 239
  study of, 75
  survival and dispersal of, 81–82
  types of pathogens, 77–78
  zoosporic, 91
  Zygomycota, 85–86
Fungi imperfecti. *See* Deuteromycota
Fungicides
  application of, 321

formulations of, 320
history of, 320
nontarget effects of, 322–323
resistance to, 322
seed treatments, 341–342
selection pressure and, 307
use of to inhibit sterol biosynthesis, 289
Furovirus, transmission of by fungi, 27
Furrow flooding, 313
Fusarium oxysporum, 136, 238–239
    decrease of water flow in the xylem due to, 271

G

Gaeumannomyces, 201
    graminis, 327
Gall
    as a symptom of viral infection, 24
    azalea leaf, 283
    caused by *Agrobacterium,* 41
    caused by *Nocardia vaccinii,* 43
    caused by plant-parasitic nematodes, 62
    caused by *Protomyces,* 112
    crown, caused by *Agrobacterium tumefaciens,* 48
    definition of, 9
    development of due to uncontrolled cell division, 272
    of carrots caused by *Rhizobacter daucus,* 42
    Synchytrium macrosporum caused by, 94–95
Gametes, production of zygotes from, 12
Ganoderma applanatum, 237
Gas chromotography, for bacterial pathogen identification, 355–356
Gastromycetes, 166
Gel electrophoresis, identification of enzymes, proteins or
             pathogens by, 356
Geminiviruses, 244
Gene activation, as an active mechanism for plant defense, 262
Gene pyramiding, 308
Gene transfer techniques, 370
General resistance, 299
General virulence, 301
Genetic diversity, 306
Genetic engineering, 211–212
    control of viruses with, 29
Genetic resistance, to rusts, 160
Genetic uniformity, influence of disease development due to, 284
Gentamycin, use of to control fire blight, 51
Geranium, fasciation of caused by *Rhodococcus fascians,* 43
Germination enhancement, fungicide seed treatments for, 341
Germplasm, cultivated, 303
Giant ragweed, stem gall on, 112
Gibberella
    ascospores of, 13
    fujikuroi, elongation of internodes due to, 272
Gladiolus, *Burkholderia gladioli* pathogenicity of, 42
Globodera, 63
Glomalean endomycorrhizae, 88
Glomerella, 131
Gluconobacter, 42
Glycocalyx, 76
Glycosis, 271
Glycosylation, 50
Grafting, as a virus vector, 27
Grain sorghum, smut diseases of, 151–153
Grains, 76
Gram-negative bacteria

aerobic/microaerophilic rods and cocci, 41–42
    facultative anaerobic rods, 43
Gram-positive bacteria
    endospore-forming motile rods, 43
    fungus-like, 43
    irregular, nonsporing rods, 43
    nocardioform actinomyces, 43
Granular fungicide formulations, 321
Granville wilt, 45
    cellulases and, 46
Grapes
    bacterial necrosis and canker of caused by *Xylophilus ampelinus,* 42
    black rol, 132
    downy mildew of, 84, 178
    powdery mildew of, 120–122
Grapevine fan leaf virus, 65
Graphium
    synnema of, 13
    ulmi, 137
Grasses
    forage, smut diseases of, 151–153
    lesion nematodes and, 63
    pathogenicity of mushroom-forming genera on, 166
Gray mold, 315
Growth abnormalities, symptoms caused by, 24
Gymnosporagnium juniperi-virginianae, 283, 316

H

Hail injury, 203
Halos, 262
Harpin, role of in pathogenesis, 51
Haustoria, 77, 117, 239
Hazelnuts, infections of by *Eremothecium coryli,* 111
HC-toxin, 273
Heartrot, due to flagellate protozoa, 199
Hemibiotrophs, 78, 237
Hemicellulases, 175
Hemicelluloses, secretion of into the root cap, 49
Hemiendophytic mycelium, formed by powdery mildews, 118
Hemileia vastatrix, 4. *See also* coffee rust
Herbaceous rusts, 158
Herbaspirillum, 42
Herbicide injury, 204
Hessian fly, 341
Heterobasidium annosum, 168–169
Heterodera, 63
Heteroecious cereal rusts, 155–157
Heterothallic fungi, 81
Heterotrophic plant pathogens, 236
Hibiscus, infection of by *Eremothecium,* 111
Holly, American, bacterial blight of caused by *Arthrobacter iicis,* 43
Hollyhock rust, 158
Holomorph, 81
Holy Fire. *See* ergotism
Homothallic fungi, 80
Hoplolaimus, 66
Hops, rust, 158
Horizontal resistance, 299
Hormones, chemical attacks on plants by, 245
Horseradish peroxidase, use of in ELISA, 28
Host plant resistance
    categorization of, 296
    control of viral diseases by, 29
Host range experiments, plant species used in, 36

Host resistance, 50
Host tissue, replacement fo caused by fungi, 78
Human disease
    caused by *Burkholderia cepacia*, 42
    caused by *Enterobacter cloacae*, 43
    caused by *Klebsiella pneumoniae*, 43
    caused by *Serratia marcesens*, 43
Hybrid seed corn, southern corn blight in, 5
Hybridization-based techniques for generating molecular markers,
        210–211
Hydrogen cyanide, use of for biological control, 329
Hymenium, 166
Hymenomycetes, 166
Hyperauxiny, 245
Hyperplasia, 24
    caused by fungi, 78
Hypersensitive reaction, 51, 264, 299–300, 371. *See also* specific
        induced resistance
    experiment for, 59–60
    resistance to rusts, 158–160
Hypertrophy, 24
    caused by fungi, 78
Hyphae, 11, 76
Hyphomycetes, 87
Hypovirulence, 330

I

Iberis, as host for *Plasmodiophora brassicae*, 94
Icosahedral viruses, 24
Ilarvirus, transmission of by thrips, 26
Immunity, 299
Impatiens necrotic spot virus, 315
Impatient necrotic spot virus, 6
Imperfect fungi. *See also* deutromycota
    description of, 133–135
In vitro plant pathology
    development of disease-resistant genotypes with, 367–370
    molecular studies, 371–372
Inbred line cultivars, 305
Increased growth response, 330
Indexing, 316
Indian pipe, 193
Indicator hosts, 27
Indole-3-acetic acid production, 245
Indolediterpenes, 239
Induced nonspecific resistance systems, 50–51
Induced systemic resistance, 51, 264, 330
Infection, 8
    passive, 263
Infectivity assays, 27
Infectivity curves, calculation of, 35
Initial inoculum, 283–284
Inoculation, 8. *See also* mechanical inoculation
    antagonist, 328
Insecticides, selection pressure and, 307
Integrated pest management
    definition of, 337
    implementation and development of a program for, 338–339
    scouting as a component of, 337–338
Internal transcribed spacers of rRNA, exercise for amplification of,
        229–231
Internodes, shortened, due to drought stress, 202
Intumescence, 202
Invasion, 8

Iowa Stiff Stalk synthetic, 305
Iris, *Burkholderia gladioli* pathogenicity of, 42
Irish potato famine, 4
Iron deficiency, 203
Irrigation regimes, 313–314
Isolates, of viruses, 19

J

Jasmonic acid, induced systemic resistance signaled by, 51
Jelly fungi, 165

K

Kale, *Plasmodiophora brassicae* as a pathogen of, 93–94
Karnal bunt, 151–153
Karyogamy, 80
Klebsiella, 43
Koch's postulates, 9–10, 346
Kohlrabi, *Plasmodiophora brassicae* as a pathogen of, 94

L

Labyrinthula zosterae, 91, 92, 96–97
    life history, 97
    observations of eelgrass wasting disease, 106–110
Labyrinthulids, 92
Lalaria, 112
Lancaster Surecrop, 305
Lance nematodes, 66
Late blight of potato, 4, 281
    pathogens causing, 10
Leaf
    abscission, 272
        caused by fungi, 78
        due to ball moss, 199
    distortion, 203
    scorch, 49
        due to drought stress, 202
    spots
        caused by *Pseudomonas*, 42
        caused by *Xanthomonas campestris*, 283
        definition of, 9
        due to abiotic stresses, 6
        experiment to diagnose, 359–361
        Septoria, 291
Leafhoppers, as virus vectors, 26
Leafy mistletoe, 193, 196–198
Legumes, lesion nematodes and, 63
Lepidium, as host for *Plasmodiophora brassicae*, 94
Lepiota, 170
Lesion
    caused by plant-parasitic nematodes, 62
    definition of, 9
Lesion nematodes, 63–65
Lettuce
    corky root of caused by *Rhizomonas suberifaciens*, 42
    necrotic yellows virus, persistent circulative propagative
        transmission of, 25
Liberobacter, 44
Lichenized ascomycetes, 127
Light, abiotic diseases due to changes in, 201
Lightning injury, 203
Lignin

decomposition of dead organic matter by metabolism of, 237
  induced, 262
Line patterns, as a symptom of viral infection, 24
Lipopolysaccharides, 46
Lobularia, as host for *Plasmodiophora brassicae,* 94
Local lesions
    as symptoms of a virus, 23
    creation of in experimental conditions, 35
Loculoascomycetes, 132
Logistic model of disease analysis, 288–290
Lolines, 239
Lolitrems, 239
Longidorus, transmission of viruses by, 27
Loose smut, 153
    control of by fungicide seed treatments, 341
Love vine, 193. *See also* dodder
Low-temperature injury, 201–202
LSD, 4, 87
Lunaria, as host for *Plasmodiophora brassicae,* 94
Luteovirus, transmission of by aphids, 26

M

Macrophomina phaseolina, 285
Magnaporthe poae, 81
Maize
    dwarf mosaic virus, 23, 27
        transmission of by aphids, 30
    foliar diseases of, 236
    leaf spot disease in, 273
    Stewart's wilt of caused by *Pantoea stewartii,* 45, 48
    stripe virus, transmission of by leafhoppers, 26
Maneb, 320
Marasmius, 166, 168, 170
    oreades, 212
Material Safety Data Sheets, 319
    reading and comprehending, 323–324
Matthiola, as host for *Plasmodiophora brassicae,* 94
Mechanical inoculation, 22–23, 33
Meloidogyne, 62–63
    incognita, 136
Membrane-bound pigments, 46
Meristem infection, 153–155
Mesocriconema, 66
    disruption of absorption and translocation of water and nutrients
        due to, 271
Migratory ectoparasites, transmission of viruses by, 27
Mildew
    caused by fungi, 78
    powdery, 117–126
Mineral deficiency and toxicity, 203–204
Mistletoe
    dwarf, 8, 193
        as a limiting factor in crop production, 5–6
    leafy, 193
Mites
    as virus vectors, 27
    wheat curl, 30
Moist-chamber incubations, for disease diagnosis, 352
Moisture, effect on disease development of, 284
Molecular hybridization, 217
Molecular markers, 210–211
    in plant pathology, 211–214
Mollicutes, 5, 8, 15, 44
Monilinia, 131

Monitor plots, 307
Monitoring, 307–308
Monoclonal antibodies, 28
Monocyclic diseases, 8, 283–284
Monogenic resistance, 297, 307
Monotropa
    hypopiths, 193
    uniflora, 193
Morning glories
    Albugo ipomoeae-panduranae as a pathogen of, 181
    observation of tissue infected with *Albugo candida,* 190–191
Morphology, of viruses, 28
Mosaic
    as a symptom of viral infection, 23
    definition of, 9
Mosaic viruses
    barley stripe, transmission of by pollen, 27
    cauliflower, 25
    cowpea, 19
    cucumber, 271, 283, 316
    maize dwarf, 23, 27
        vector control of, 30
    pea enation, 24
    peach rosette, 65
    rose, 24
    tobacco, 6, 315
        experiment for transmission of, 33–37
        transmission of in laboratory studies, 23–24
        transmission of in seed coats, 27
    wheat streak, 23–24
        transmission of by mites, 27
Mosses, 15, 199
Motile spores, 11. *See also* zoospores
Mottles, as a symptom of viral infection, 23
Mucor, 86
Mulching, 314
Multigenic resistance, 285
    as disease control measure, 289
Multiline cultivars, 308
Mummy, definition of, 9
Mushrooms, 13–14, 86
    morphology of, 165–166
Mustard
    infections of by *Eremothecium sinecaudii,* 111
    Plasmodiophora brassicae as a pathogen of, 94
Mutualism, 239
Mutualists, 77, 87–89
Mycelia, 11
Mycelia Sterilia, 87, 138
    experimental identification of using hyphal characteristics, 145–146
Mycelium, 76
    production of zygotes from, 12–13
Mycetozoans, 91–92. *See also* slime molds
Mycology Society of America, 6
Mycoparasites, 329
Mycorrhizae, 167–169
Mycorrhizal fungi, 77, 87, 88–89
Mycotoxins, 75, 133

N

Nasturtium, as host for *Plasmodiophora brassicae,* 94
Necrosis
    caused by fungi, 78
    definition of, 9

due to abiotic stresses, 6
due to nutrient deficiencies, 203
of local lesions, 23
Necrotrophs, 8, 77
Nectarines, deformation of fruit by *Taphrina deformans,* 114–115
Nectria coccinea, perithecia of, 14
Nematodes, 5, 8, 15
    active mobility and energy dependency of, 242–243
    as virus vectors, 27
    bud and leaf, 66
    control of, 67
    cyst, 63, 315
    dagger, 65
    experiment to isolate plant-parasitic and free-living, 71–73
    lance, 66
    lesion, 63–65, 315
    physiology of plant-parasitic, 61
    reniform, 61, 66
    ring, 66, 271
    root-knot, 62–63, 136, 283, 315
        experiment on the effects of, 69–71
    seed gall, 66
    stem and bulb, 66
    sting, 65
    stubby-root, 65–66
Nematology Society, 6
Neotyphodium, 239
Neotyphodium, 87
Nepovirus, transmission of by nematodes, 27
Nepoviruses, 65
Nicotiana
    DNA sequences of geminiviruses in nuclear genome of, 244
    use of as an indicator host, 27
Nitrogen fertility, 341
Nitrogen fertilization, 317
Nitrogen-fixing nodules, caused by tumor-inducing bacteria species, 48
Nocardia, 43
    vaccini, galls on blueberries caused by, 48
Nocardioform actinomyces, 43
Noninfectious diseases, 201
Nonpersistent transmission of viruses, 25
Northern corn leaf blight, 296
Nucleic acids, 209
    viral identification with, 28–29
Nursery crops, stem and bulb nematodes and, 66
Nutrient deficiencies, 8
    as an abiotic stress, 6
Nutrients, disruption of absorption and translocation of, 271–272

O

Oak
    Brennaria pathogenicity of, 43
    scorch of caused by *Xylella fastidiosa,* 42
    sudden death, 84
Oats
    Acidovorax avenae pathogenicity of, 41
    blue dwarf virus, transmission of by leafhoppers, 30
    crown rust, 157–158
    smut diseases of, 151–153
Obligate endoparasites, 92
Obligate parasites, 22, 77, 174, 239–242
Oidium, 119, 283
Oil palm, sudden wilt of, 198
    gibberellin synthesis and, 272

Oleander, knot, caused by *Pseudomnas syringae,* 48
Oligogenic resistance, 297
Olive, knot
    caused by *Pseudomnas syringae,* 48
    gall formation due to auxin caused by pathogen, 272
Olpidium, 93
    brassicae, 85
    transmission of viruses by, 27
Onions
    bud and leaf nematodes and, 66
    Burkholderia cepacia pathogenicity of, 42
    smut diseases of, 151–153
    stem and bulb nematodes and, 66
Oogonia, 14
Oomycota, 8, 14, 76, 84
    description of, 173–174
Oospores, 81
Ophiostoma
    novo-ulmi, decrease of water flow in the xylem due to, 271
    ulmi, 82
Opines, 272
Orchidaceous endomycorrhizae, 88
Orchids, *Acidovorax avenae* pathogenicity of, 41
Organic materials, composted, 314
Organomercurial fungicides, 320
Ornamentals
    bud and leaf nematodes and, 66
    lesion nematodes and, 63
    powdery mildew disease of, 124
    sting nematodes and, 65
Orobanche
    minor, 198
    ramosa, 198
Ostioles, 11
Ouchterlony double diffusion technique
    experiment for detecting and identifying viruses by, 37–39
    use in disease diagnosis, 354
Overhead irrigation, 313
Oxidative phosphorylation, 271
Oxytetracycline, use of to control fire blight, 51
Ozone, plant injury due to, 205–206

P

Paecilomyces, 202
Paenibacillus, 43, 46
Pantoea, 43
    agglomerans, 46
        endophytic residence phase of, 48
    stewartii, 291
        Stewart's wilt of maize caused by, 45, 48
Papaya, brown discoloration of caused by *Enterobacter,* 43
Parasites
    definition of, 7
    facultative, 8
    obligate, 22
Parasitic seed plants, 5, 8, 15
Parasitic slime molds, 14. *See also* Plasmodiophoromycota
Parasitism, 329
Paratrichodorus, 65
Partial resistance, 300
Passive defense strategies of plants, 261
Passive mechanisms to prevent pathogen penetration, 262
Passive penetration, 50
Pathogen fitness, 7

Pathogen variation, 300–303
Pathogenesis-related proteins, 371
Pathogenicity, 300
    definition of, 7
    experimental assays, 55–56
    proof of, 356–357
Pathogens. *See also* specific pathogens
    archiascomycete, 111–116
    citrus-greening, 44
    Claviceps purpurea, 4
    Cochliobolus heterostropus, 5
    Cryphonectria parasitica, 5
    definition of, 7
    Discula destructiva, 5
    ecology of plant, 48–49
    enzyme-producing, 49
    hemiascomycete, 111
    Hemileia vastatrix, 4
    hypersensitive test for, 59–60
    in epidemics, 283–284
    Phytophthora infestans, 4
    races of, 283
    soft-rotting, 42
Pea
    early browning virus, 66
    enation mosaic virus, 24
    experiment on root rot caused by *Pythium* species, 183–184
    Rhodococcus fascians pathogenicity of, 48
    sweet, fasciation of caused by *Rhodococcus fascians,* 43
Peach
    deformation of fruit by *Taphrina deformans,* 114–115
    Monilinia fruticola disease of, 131
    post harvest rot caused by *Mucor piriformis,* 86
    ring nematodes and, 66
    rosette mosaic virus, 65
Peanut
    Granville wilt of caused by *Ralstonia solanacearum,* 42, 45, 48
    ring nematodes and, 66
    Sclerotinia blight of, 285
Pear
    discolorations caused by *Gluconobacter,* 42
    fire blight of caused by *Erwinia amylovora,* 43, 45
    post harvest rot caused by *Mucor piriformis,* 86
Pectic enzymes, bacterial, plant pathogenesis by, 46
Pectinase, 175
Pectins, secretion of into the root cap, 49
Pectobacteria, caused by *Erwinia amylovora,* 43
Pectobacterium, 43
    carotovora, 46
        endophytic residence phase of, 48
    soft rot pathogenesis by, 46
Pedigree method, of developing cultivars, 305
Pedigree methods, of resistance incorporation, 304
Penetration, 8, 262–263
Penicillium, 135
    digitatum, 272
Peperomia, effect of fertilization rates on, 317
Pepper, ringspot virus, 66
Peptidoglycans, 45
Peramine, 239
Perithecia, 86
Perithecia, 13, 82, 130
Perlite, 314
Peronospora, 174, 178
    tabacina, experimental coculture with tobacco callus, 187–190
Peronospora tabacina, 180

Peronosporaceae, 174, 178–181
Peronosporales, 174
Persistent circulative nonpropagative transmission of viruses, 25–26
Persistent circulative propagative transmission of viruses, 25
Pest identification, 338
Pesticide
    injury, 204
    labels, 319
        reading and comprehending, 323–324
        Material Safety Data Sheets, 319
    names, 319
Phaeolus schweinitzii, 170
Phaseolotoxin, 47
Phenolic fungitoxins, 239
Phloem necrosis, due to flagellate protozoa, 199
Phloem-limited bacteria, 44
Phoma, pycnidium of, 12
Phophodiester bonds, 209
Phoradendron, 193
Photosynthesis, disruption of, 269–271
Phyllactinia, 118
Phyllosphere, 49
Phylogenetic tree, 212
Physoderma maydis, 93
Phytoalexins, 263–264
Phytoanticipins, 263
Phytophthora, 14, 84, 174
    cinnamomi, 84
    experimental isolation of from plant tissues and soil, 184–186
    experimental production of sporangio and oospores by, 186–187
    in flood conditions, 202
    infestans, 4, 76, 84, 176–178, 239–242 (*See also* late blight of potato)
    parasitica, 314
    ramorum, 84
    sojae, 283
Phytophthra, disruption of absorption and translocation of water and nutrients due to, 271–272
Phytoplasma, 44
Phytoplasmas, 48
Phytotoxic rhizobitoxin, production of by *Bradyrhizobium,* 42
Phytotoxin, experiment to determine effect on tobacco seed germination, 373–374
Pigments
    membrane-bound, 46
    viral symptoms based on changes in, 23–24
Pili, 45
Pinaceae, 167–169
Pine needles, as mulch, 314
Pine-sap, 193
Pineapple
    discolorations caused by *Gluconobacter,* 42
    pink disease of caused by *Acetobacter diazotrophicus,* 41
Plant defense strategies, 261–264
Plant diseases. *See also* specific diseases
    causes of, 5–6
    coffee rust, 4–5
    cycle of, 8–9
    definition of, 7
    effects of, 4
    ergotism, 4
    Irish potato famine, 4
    monocyclic, 8
    polycyclic, 9
Plant growth hormones, 272–273
Plant nutrition, 317
Plant pathogens

biological control of, 327–331
chemical attack of, 244–246
ecology of, 48–49
experiment to determine effect on chlorophyll of, 274–276
heterotrophic, 236
isolation of on culture media for disease diagnosis, 352–354
Plant samples, collecting, packaging, and mailing, 347
Plant stresses, 6, 8
Plant surfaces, penetration of, 50
Plant virology, 19
historical principles of, 22–23
Plant-growth-promoting rhizobacteria, 330
Planting date, influence on disease development, 316
Plasmalemma, 76
Plasmidiophora brassicae, 84
Plasmodia, 14, 76
Plasmodiophora, brassicae, 14, 91, 93–94
experiment on susceptibility of crucifers to, 99–102
life history of, 94
Plasmodiophorids, 92
Plasmodiophoromycota, 14, 84. *See also* parasitic slime molds
Plasmogamy, 80
Plasmopara, 84, 174, 178
viticola, 84
Pleiotropic symbiosis, 239
Plum pox virus, 24, 244
control of, 30
Plums
black knot disease, 316
Taphrina diseases of, 112
Podosphaera, 124
Poinsettia, canker caused by *Curtobacterium flaccumfaciens,* 43
Pollen, as a virus vector, 27
Polycyclic diseases, 9, 283
Polyetic pathogens, 283
Polygenic resistance, 297, 309
Polymerase chain reaction, 217
for disease diagnosis, 346
generation of molecular markers by, 211
viral identification by, 29
Polymorphisms, 210
Polymxya graminis, 92
Polymyxa, transmission of viruses by, 27
Polynucleotide chains, 210
Poplar, wetwood caused by *Clostridium,* 43
Population improvement, 304
Poria placenta, 170
Potato
bud and leaf nematodes and, 66
cholera, 4
colonization of lenticels by *Pectobacterium carotovora,* 48
cyst nematodes, 63
early death of caused by *Pratylenchus penetrans* and *Verticillium dahliae,* 64
experiment for determining physical defenses against soft rotting bacteria, 267–268
late blight, 4, 76, 84, 239, 281
crop losses due to, 290
powdery scab of, 92
ring rot caused by *Clavibacter sepedonicus,* 43
roots, colonization of by *Pseudomonas fluorescens,* 45
scab caused by *Streptomyces scabies and acidiscabies,* 43, 48, 50
soft rot of caused by *Claustridium,* 43
wart disease, 93
wilt of caused by *Ralstonia solanacearum,* 42
yellow dwarf virus, transmission of by leafhoppers, 26

Potting media, 314
Potyviruses, nonpersistent transmission of, 25
Powdery mildews, 8
case study, 265
description of, 117–118
host relationships, 118
host responses to, 119
identification of, 120
infection process of, 118–119
inhibition of by free water on foliage, 313
taxonomy of, 119
Pratylenchus, 63–65
Preformed resistance, 50
Prepenetration, 261–262
Primary inoculum, 8–9
Programmed cell death, 264
Prokaryotes, 8. *See also* bacteria
characteristics of phytopathogenic, 41
damage of plants by, 46–48
experiments regarding, 53–60
nature of in the phyllosphere, 49
nutritional classes of endophytic, 48–49
rapid identification of phytopathogenic, 44
Propagation techniques, 313
Proteases, bacterial. *See also* proteinases
plant pathogenesis by, 47
Protein isolation and fungal growth, experiment for, 251–254
Proteinases. *See also* proteases
bacterial, plant pathogenesis by, 47
Proteins, 45
Protista, 84
Protomyces, 111–112
Protomycetaceae, 111–112
Protomycopsis, 111–112
Protozoa, 5, 8
flagellate, 198–199
Prunus taphrina diseases of, 112
Pseudomonas, 42
cytokinin production by, 245
fluorescens
colonization of potato roots by, 45
Pf-5, 329, 332–333
Q2-87, 328
indole-3-acetic acid production by, 245
rubrisubalbicans, 42 (*See also Herbaspirillum*)
savastanoi, auxin production by, 272
soft rot pathogenesis by, 46
syringae, 46, 273
angular leafspot of cucumbers caused by, 49
colonization of bean leaves by, 45
endophytic residence phase of, 48
olive and oleander knot caused by, 48
production of cutinases and suberin esterases by, 47
syringotoxin produced by, 47
Pseudoperonospora, 174, 178
Pseudothecia, 86, 130
Pseudozyma flocculosa, use for control of *Podosphaera,* 125
Psilocybe cubensis, 167
Puccinia graminis, 155–157
Puccinia path, 158, 308
Puffballs, 13–14, 165
Punctodera, 63
Pure line cultivars, 305
Purine, 210
Pycnidium, 11, 80, 135
Pyrenomycetes, 131

Pyrenophora tritici-repentis, 273
Pyricularia, grisea, 81
Pyrimidine, 210
Pyrularia pubera, 193
Pythiaceae, 174
Pythium, 14, 84, 174, 314
    aphanidermatum, 175
    blight, 174
    debaryanum, 175
    disruption of absorption and translocation of water and nutrients due
        to, 271–272
    experiment on root rot caused by, 183–184
    experiment to determine effect of fertilization on root rot of
        geraniums, 317–318
    experimental production of sporangio and oospores by, 186–187
    gametes of, 14
    in flood conditions, 202
    planting date influence on development of, 316
    root disease, 174
    ultimum, 175
        inhibition of by *Pseudomonas fluorescens* Pf-5, 329

Q

Qualitative disease resistance, 298
Quantitative disease resistance, 298, 307
Quarantine, as a method to control viruses, 30

R

R-genes, 211–212
Race surveys, 307
Race-specific resistance, 299
Radish, *Plasmodiophora brassicae* as a pathogen of, 94
Ralstonia, 42
    solanacearum
        degradation of pectin by, 46
        disruption of absorption and translocation due to, 271
        extracellular polysaccharide production by, 245
        Granville wilt caused by, 45, 46, 48
        increase in incidence of over time, 286
Randomly amplified polymorphic DNA, 211
Rape, *Plasmodiophora brassicae* as a pathogen of, 94
Raphanus, as host for *Plasmodiophora brassicae,* 94
Reaction mixture, 218
Reddish-purple leaves, due to nutrient deficiencies, 203
Removal of noncrop host plants, 316
Reniform nematodes, 61, 66
Repeating spores, 155
Reproduction, 8
Resistance, 50–51
    breeding for, 303–306
    engineered, 51
    genes, deployment of, 308–309
    heritability of, 305
    incorporating, 304
    types of, 295–300
Resistant cultivars, 9
Resistant hybrids, breeding of, 5
Respiration, disruption of, 271
Restriction fragment length polymorphic markers, 211
Rhabdoviridiae, transmission of by aphids, 26
Rhizobacter, 42
Rhizobitoxin, 47

Rhizobium, nitrogen-fixing nodules on legume roots caused by, 48
Rhizoctonia, 166, 170–171
    experimental determination of anastomosis grouping for isolates of,
        146–148
    experimental determination of number of nuclei in hyphal cells, 146
    experimental identfication of using hyphal characteristics, 145–146
    solani, 81, 138
        experiment on biocontrol of damping off of cotton seedlings,
            335–336
        experiment on parsitism, 333–335
        experiment to suppress mycelial growth of, 332–333
        hyphae of, 12
Rhizomonas, 42
Rhizomorphs, 77
Rhizoplane, 49
Rhizopus
    niger
        sporangia of, 12
        zygotes of, 13
    stolonifer, 80, 85
Rhizosphere, 49
Rhodococcus, 43
    fascians, 48
Rhododendron, wilt, 84
Rhododendron, parasitic algae infection of, 199
Riboflavin-induced systemic resistance, 51
Ribose, 209
Rice, *Xanthomonas* pathogenicity of, 42
Rigid rod viruses, 24
Ring nematodes, 66
    disruption of absorption and translocation of water and nutrients
        due to, 271
Ringspots, as a symptom of viral infection, 24
RNA, 209
Rockwool, 314
Roguing, 315
Rome, wheat crops in early, 4
Root knot, 6
Root necrosis, due to flooding, 202
Root proliferations, caused by *Agrobacterium,* 41
Root rot, 84
    annosum, 169
    experiment on oomycete-caused, 183–184
Root-knot nematodes, 62–63
Rorippa, as host for *Plasmodiophora brassicae,* 94
Rose
    autoecious rusts of, 158
    mosaic virus, 24
    powdery mildew disease of, 124
        inhibition of by free water on foliage, 313
Rosette, due to nutrient deficiencies, 203
Rot, definition of, 9
Rotylenchulus reniformis, 66
Row crops
    lesion nematodes and, 63
    sting nematodes and, 65
Rust, effects of on photosynthetic competence, 270
Rusts, 8, 13, 155–161
    coffee, 4–5, 76
    white pine blister, 76
Rutabaga, *Plasmodiophora brassicae* as a pathogen of, 94
Rye
    ergotism caused by *Claviceps purpurea,* 81
    ergotism from consumption of, 4
    seed gall nematodes and, 66
Ryegrass staggers, 87

S

Salem Witch Trials, 4
Salicaceae, 167–169
Salicylic acid, systemic acquired resistance signaled by, 51
Sanitation, 315
Sap extracts, preparation of, 37
Saprophytes, 49, 127, 165
    facultative, 8
    passive mobility and energy dependency of, 237
Sclerospora, 178
Sclerotia, 12, 77, 82
    of *Claviceps purpurea,* 4
Sclerotinia minor, 285
Sclerotium rolfsii, 138, 283
Scorches, caused by *Xylella fastidiosa,* 42, 49
Scouting, in integrated pest management, 337–338
Seagrass, die-off of due to *Labyrinthula,* 97
Seed, as a virus vector, 27
Seed catalogs, resistance and tolerance information in, 310
Seed certification programs, 30
Seed gall nematode, 66
Seedling infection, 153
Selection pressure, 236, 306
Self-pollination, effect on disease development, 284
Semipersistent transmission of viruses, 25
Septoria glycines, 138
Septoria leaf spot, use of TOM-CAST for control of, 291
Serology, of viruses, 28
Serratia, 43
Setae, 11
    of *Colletotrichum,* 135–136
Shoestring fungus-honey mushrooms, 168
Short-season cultivar, as disease control measure, 289
Shortened internodes, due to drought stress, 202
Signal recognition/transduction, 262
Signal words, on pesticide labels, 319
Signs, 9
Simple sequence-repeat, 211
Sinapis, as host for *Plasmodiophora brassicae,* 94
Single-gene resistance, 285, 297
Sisymbrium, as host for *Plasmodiophora brassicae,* 94
Slime molds, 76, 84, 91–92
Smuts, 13
    diseases, 151–153
    flag, 153
    floral infection by, 153
    loose, 153
        control of by fungicide seed treatments, 341
    mycelium of, 153
    seedling infection by, 153
    wheat, stinking, 82
    white, 112
Sobemovirus
    transmission of by beetles, 26
    transmission of by thrips, 26
Soft rot, 6
    caused by *Clostridium,* 43
    caused by *Erwinia amylovora,* 43
    caused by *Klebsiella,* 43
    caused by *Pectobacterium,* 43, 48
    due to *Pythium,* 84
    experiment to determine physical defenses of potato against, 267–268
    formation of from pectate lyases and proteases, 46
Soil
    disease suppressive, 327
    effect on disease development of, 284
    electrical conductivity, for determining total soluble salt, 352
    pH
        determination for disease diagnosis, 352
        effect on diseases, 317
        effect on nutrient deficiency and toxicity, 203
    salinity of, 203
Soil preparation methods, 313
Soil saprophytes, 49
Soil water mold, 14, 84
Soilless potting media, 314
Solanaceous crops, Granville wilt of caused by *Ralstonia solanacearum,* 42, 48
Solar radiation, effect on disease development of, 284
Solarization, 314–315
Southern corn blight, 5, 133, 273, 282
Southern magnolia, parasitic algae infection of, 199
Southern procedure, 210
Soybean
    anthracnose pathogen of, 78
    binding of *Bradyrhizobium* by, 46
    brown spot of, 138
    charcoal rot of, 285
    cyst nematode, 63, 302
    disease of caused by *Bradyrhizobium,* 41–42
    infections of by *Eremothecium coryli,* 111
    lance nematodes and, 66
    root rot, 283
    rot of caused by *Bacillus,* 43
    southern blight of, 315
Spanish moss, 199
Spatial modeling, of epidemics, 290
Specific induced resistance, 51, 299. *See also* hypersensitive reaction
Specific virulence, 301
Sphacelia segetum, 81
Sphaerotheca
    fuliginea, 124
    macularis, 283
Spilocaea pomi, 81
Spiroplasma, 44
    citri, 46
Spongospora
    subterranean, 92
    transmission of viruses by, 27
Sporangiophores, 178
Sporangiospores, 11, 80
Sporangium, 11, 80
    oomycotic, 173
Spores, 11
Sporocarps, 77
Sporodochium, 11, 135–136
Sporothrix rugulosa, use for control of *Podosphaera,* 125
Sprayer calibration, experiment to determine effect of nozzle size on water output, 324–325
Squawroot, 193
St. Anthony's Fire. *See* ergotism
Stand establishment, fungicide seed treatments for, 341
Stem and bulb nematodes, 66
Stem rust fungus, 239
Sterility, as a symptom of viral infection, 24
Sterol-inhibiting fungicides, 342
Stewart's wilt, 291
    caused by *Pantoea stewartii,* 43, 48
Sting nematodes, 65
Stinkhorns, 165

Stinking smut, 153
Strains, of viruses, 19
Stramenopila, 14, 84, 173. *See also* Chromista
Strangle weed, 193. *See also* dodder
Straw, as mulch, 314
Strawberry
    bud and leaf nematodes and, 66
    stem and bulb nematodes and, 66
Streaks, as a symptom of viral infection, 23
Streptomyces, 43
    as an antibiotic used for biological control, 329
    scabies, production of cutinases and suberin esterases by, 47
Streptomycin, use of to control fire blight, 51
Striga, as a limiting factor in crop production, 5
Striga lutea, 193
Stripes, as a symptom of viral infection, 23
Strobilurins, 320
Strobilurus esculentus, 167
Stunting
    as a symptom of viral infection, 24
    due to flagellate protozoa infection, 199
    due to nutrient deficiencies, 203
    due to plant-parasitic nematodes, 61
Styrofoam, 314
Suberin esterases, bacterial, plant pathogenesis by, 47
Sugar-beet, nematode, 63
Sugarcane, smut diseases of, 151–153
Suillus, 167
Sulfur dioxide, plant injury due to, 205–206
Sun scald, 201
Sunflower, rust, 158
Susceptability, 296
Sweet potato
    Albugo ipomoeae-panduranae as a pathogen of, 181
    stem and bulb nematodes and, 66
Symptoms, 9
    caused by fungi, 78–80
Synchytrium
    endobioticum, 93
    macrosporum, 91, 94–95
        experiment of plant infection by, 102–106
        life history, 95–96
Synnema, 11, 80, 135–137
Syringomycein E, 273
Syringopeptin, 273
Syringotoxin, 47, 273
Systemic acquired resistance, 51
    chemical-induced, 323
Systemic fungicides, 321
Systemic resistance, 264
Systemic acquired resistance, 264

T

T-toxin, 273
Tabtoxinine, 47
Tagetitoxin, 47
Take-all patch diseases, 201
Take-all root disease, 327
Talaromyces flavus Tf1, use in biocontrol, 329
Tan spot, 273
Taphridium, 111–112
Taphrinaceae, 112–114
Taphrini, diseases caused by, 114–116
Teleomorph, 81

Telemorph stage of fungi, 12
Temperature extremes, 8
    effect on disease development of, 284
Texas cytoplasmic male sterility gene, 5
Thanatephorus, 165
    cucumeris, 81, 171
Thielaviopsis, 314
    disruption of absorption and translocation of water and nutrients due to, 271–272
Thigmotropism, 239
Thiram, 320
Thlaspi, as host for *Plasmodiophora brassicae,* 94
Thrips, as virus vectors, 26
Tickseed, 112
Tilaceae, 167–169
Tillandsia recurvata, 199
Tilletia, 153
Tilletia caries, 82
Tilletiopsis, use for control of *Podosphaera,* 125
Tissue incubation, for disease diagnosis, 352
Tobacco
    blue mold of, 180
    coculture experiment, 187–190
    diseases of due to broomrapes, 199
    experiment to show effect of phytotoxin on seed germination, 373–374
    experimental induction of pathogenesis-related proteins in, 3 75–376
    Granville wilt of caused by *Ralstonia solanacearum,* 45, 48
    in vitro production of wildfire-resistant plants, 367
    mosaic virus, 6, 315
        case study, 265
        experiment for transmission of, 33–37
        transmission of in laboratory studies, 22–23
        transmission of in seed coats, 27
    rattle virus, 66
    ringspot virus, 65
        experiment for mechanical transmission of, 35
        Ochterlony double diffusion experiment for detection of, 37–39
    rot of caused by *Bacillus,* 43
    wilt of caused by *Ralstonia solanacearum,* 42
Tobravirus, transmission of by nematodes, 27
Tobraviruses, 66
Tolerance frequency, 310
TOM-CAST, 291–292
Tomato
    blight of, 84
    canker caused by *Clavibacter michiganensis,* 43, 48
    early blight of, 135
    ringspot virus, 65
    rot of seedlings caused by *Bacillus,* 43
    spot, treated with Pto locus, 51
    spotted wilt virus, 24, 283
    wilt, decrease of water flow in the xylem due to, 271
    wilt of caused by *Ralstonia solanacearum,* 42
Tospovirus, transmission of by thrips, 26
Toxins
    bacterial, plant pathogenesis by, 47
    chemical attacks on plants by, 245
Trametes versicolor, 170
Transcription, 273–274
Translation, 273–274
Tree bark, composted, 314
Trichoderma

as a fungal antagonist, 329
atroviride, 212
Trichodorus, 65
    transmission of viruses by, 27
True tolerance, 309–310
Trypanosomatidae, 198–199
Tulipomania, 22
Tumorigenesis, 48
Tumors, as a symptom of viral infection, 24
Turf
    fungal holomorphs in, 81
    ring nematodes and, 66
    sting nematodes and, 65
    stubby-root nematodes and, 65–66
Turfgrass endophytes, 87
Turnip, *Plasmodiophora brassicae* as a pathogen of, 94
Tyloses, 272
Type IV secretion of proteins and nucleic acids by pili, 45

U

Uredinales, 151
Urediniomycetes, 166
Urocystis agropyri, 153
Urophlyctis, 93
USDA Cereal Disease Laboratory, 307
Ustilaginales, 151
Ustilaginomycetes, 166
Ustilago, 153

V

Vectors, 22
    control of, 29–30
    insects as, 25–26
Vegetables
    lesion nematodes and, 63
    pathogenicity of mushroom-forming genera on, 166
    sting nematodes and, 65
    stubby-root nematodes and, 65–66
Vegetative propagation, as a virus vector, 27
Vein banding, as a symptom of viral infection, 24
Vein clearing, as a symptom of viral infection, 24
Venturia, 132
    inaequalis, 81, 239, 283
Vermiculite, 314
Vertical resistance, 299
Verticillium
    effect of fertilization rates on, 317
    increase in incidence of over time, 286
Verticillium lecanii, use for biocontrol of powdery mildew, 125
Vertifolia effect, 305–306
Violets, smut diseases of, 151–153
Viroids, 5, 8, 15
Virulence, 300
    of pathogens, 7
Virulence formula, 301
Viruses, 5, 8, 15. *See also* specific viruses
    architecture of, 24
    control of, 29–30
    definition of, 22
    detection and identification of, 27–29
    families and genera of, 20–21
    identification of based on nucleic acid genome, 28–29

    identification of by physical properties, 28
    mechanical transmission of, 22–23, 33
    molecular properties of, 28–29
    naming of, 19
    passive mobility and energy independency of, 243–244
    symptomatology of, 23–24
    transmission of, 24–27
Viscin, 195
Viscum, 197
Volatile compounds, use of as antagonists for biological control, 329
Volkartia, 111–112

W

Walnut, *Brennaria* pathogenicity of, 43
Warm temperature frost damage, 48
    caused by *Pseudomonas,* 42
Water
    disruption of absorption and translocation of, 271–272
    experiment on disruption of translocation of, 276–277
    stress, 202
Water management, 313–314
Water molds, 93. *See also* chytrids
Water-dispersable granules, 320
Water-soluble pouches, 320
Watermelon, *Acidovorax avenae* pathogenicity of, 41
Watson-Crick complementary rules of base pairing, 210
Wettable powders, 320
Wetwood, caused by pectolytic *Clostridium,* 43
Wheat
    disease management, 339–344
    foliar fungicides for, 342–344
    leaf rust, 157–158
        experiment, 160–161
    powdery mildew diseases of, 122
    rusts, 76
    seed gall nematodes and, 66
    smut diseases of, 151–153
    stem rust, 155–157, 284
    stinking smut of, 82
    streak mosaic virus, 23
        transmission of by mites, 27
        yield loss from, 24
    take-all root disease of, 327
    white stripe of caused by *Bacillus,* 43
White pine blister rust, 76
White rots, 170
White rusts, 8, 174, 180–181
Whiteflies, as virus vectors, 26
Wild species, 303
Willow, cricket-bat, *Brennaria* pathogenicity of, 43
Wilt, 48
    bacterial, case study, 265
    caused by *Clavibacter michiganense,* 45
    caused by extracellular polysaccharides, 45
    caused by fungi, 78
    caused by plant-parasitic nematodes, 61
    caused by *Ralstonia solanacearum,* 42
    definition of, 9
    due to abiotic stresses, 6
    due to drought stress, 202
    due to flagellate protozoa, 199
    due to flooding, 202
    Fusarium, 136, 307
    Granville, 45

cellulases of, 46
increase of over time, 286
tomato, 271
Witches' broom, 48, 112, 195
Witchweed, 193, 198
as a limiting factor in crop production, 5
Wound tumor virus, 24

X

Xanthan gum, 45
Xanthomonas, 42
campestris, 283
blackrot of crucifers caused by, 48
extracellular polysaccharides in, 45
soft rot pathogenesis by, 46
Xerompholina, 166, 168
Xiphinema, 65
transmission of viruses by, 27
Xylella, 8, 42
fastidiosa, 49
Xylophilus, 42

Y

Yeasts, 76
ascomycetous, 111
fission, 111
Yellows
caused by *Phytoplasma*, 44, 49
caused by plant-parasitic nematodes, 61
Yield loss, as a symptom of viral infection, 24

Z

Ziram, 320
Zoogloea, 50
Zoospores, 11
oomycotic, 173
viruses carried by, 27
Zoosporic fungi, 91
Zygomycota, 12, 80, 85–86
Zygospores, 12, 81
Zygotes, formation of, 12–13